ON BIOCULTURAL DIVERSITY

ON BIOCULTURAL DIVERSITY

linking language, knowledge, and the environment

edited by luisa maffi

SMITHSONIAN INSTITUTION PRESS

Washington and London

COPY EDITOR: Jean Eckenfels
PRODUCTION EDITOR: Duke Johns
DESIGNER: Amber Frid-Jimenez

LIBRARY OF CONGRESS CATALOGING-IN-PUBLICATION DATA

On biocultural diversity : linking language, knowledge,
  and the environment / edited by Luisa Maffi.
      p.   cm.
    Includes bibliographical references and index.
    ISBN 1-56098-905-X (alk. paper) — ISBN 1-56098-930-0
    (paper : alk. paper)
      1. Ethnoscience. 2. Language attrition. 3. Biological
    diversity conservation. 4. Pluralism (Social sciences).
    5. Nature—Effect of human beings on. 6. Globalization.
    I. Maffi, Luisa.
    GN476.O5   2001
    306.44—dc21   00-061262

BRITISH LIBRARY CATALOGUING-IN-PUBLICATION DATA AVAILABLE

Manufactured in the United States of America
08 07 06 05 04 03 02 01   5 4 3 2 1

To the world's indigenous and traditional peoples,
who hold the key to the inextricable link
between language, knowledge, and the environment,
and to Darrell and Michael,
two of their greatest champions.

CONTENTS

ILLUSTRATIONS

MAPS

TABLES

FOREWORD

When I was working in Indonesia back in the late 1970s, some of my most interesting fieldwork was done on the island of Siberut. At 400,000 ha, it was about the same size as Bali, but it had just 18,000 people. More interesting, these 18,000 spoke 15 distinct languages, with each linguistic group tending to occupy an entire watershed. Perhaps even more interesting from the point of view of biodiversity, Siberut was occupied by four species of primates that were found nowhere else except on a few adjacent islands; one of these was an endemic genus (the pig-tailed langur *Simas concolor*), and another was the only species of gibbon (*Hylobales klossii*) found on a small island. The people of Siberut have been hunting the primates and other species on the island for thousands of years, yet the populations of these primates have continued to prosper. With their great cultural diversity, which parallels their linguistic diversity, these people had discovered ways of living in balance with their available resources.

This story is almost too good to be true, and indeed, modern developments on Siberut have tended to replace the 15 indigenous languages with Bahasa Indonesia, the lingua franca of the country. The traditional ways of life are being replaced by logging and plantation agriculture, cultural approaches to conservation are being lost, and the survival of the primates is increasingly precarious.

This is far from the only example of how language, culture, and biodiversity come together, and where the loss of one leads to the loss of the others. Many other Indonesian islands demonstrate very similar processes. On Sumba, an island in the eastern part of the country, fortified ancestral villages in precolonial times tended to have their own ceremonial languages, often used as a cultural means of letting off steam by expressing generalized anger and to build social cohesion among the populace and, incidentally, to keep them under the control of the village chief, who continued to provide benefits in the form of effective management of resources. Following colonization and incorporation of the islands of the Timor Sea into the Indonesian state, the ceremonial languages of Sumba are no longer used to express cultural values but rather they are employed to convey the contrary attitude of humility, as ritual is transformed into popular entertainment and politics begins to take a back seat to more socially acceptable song and dance routines.

All of this relates very much to biodiversity, because the precolonial Sumba society was based on management of resources that involved traditional feasting through which ritual-speech names could be earned through demonstrating the capacity to earn a surplus; rather like the potlatch of the Pacific Northwest Coast of North America, the traditional feasting was a way of dealing with unpredictable and irregular surpluses produced by the seas and forest. As part of ceremonial languages, ritual-speech names are still found in Sumba, but today they are applied to more modern consumer goods such as motor vehicles, race horses, and businesses—yet another indication of how language adapts to changing conditions of declining biodiversity.

Another example worth contemplating is Chinese kinship terms. Chinese culture historically has been based very strongly on family relationships, and a rich vocabulary of kinship terms developed over time, enabling the speaker to clearly relate others to his or her lineage in terms of blood and age relationships. With the new one-child policy of the modern Chinese government, such kinship terms will very soon be relevant only as a literary form. This quite justifiable policy has been implemented to ensure that the human population does not outstrip the productivity of the Chinese countryside, and therefore is closely linked to biodiversity. The interesting point here is that this ecologically based policy change also has profound implications for the social integration of Chinese society, perhaps requiring the state to step in where families once served the purpose of adapting to changing conditions.

As a final example, I recall from the two years I spent in eastern Nepal that each valley tended to be occupied by a quite distinct cultural and linguistic group and that historically very high levels of animosity existed between adjacent valleys, as people were living very close to the carrying capacity of the land. Often competing for limited resources, the different cultures used different means to build sustainable relationships with their available biological resources, generating new cultural innovations that were tested against the reality of constantly changing conditions. Their primary trading relationships were with the lowlands, not with each other. Today, they all speak Nepali, and while most are still aware of their cultural attachments, these are weakening as roads help to bring greater trading opportunities to people who formerly had to walk days, even weeks, to reach a market. These trading opportunities are reducing the diversity of approaches to resource management, making people more dependent on systems over which they have no control and putting more pressure on native biodiversity (as well as local languages).

These various brief examples all demonstrate that biological and cultural diversity are related in complex and important ways. *Language, Knowledge, and the Environment* is a much more thorough and coherent exploration of this important and underappreciated topic. An extremely useful compilation of papers put together by Luisa Maffi to explore thoroughly the issue of how biological and cultural diversity relate, this book shows that languages are even more threatened by modern society than biodiversity is, and that our species will be the poorer for losing the many ways that people have developed to express themselves and share information, especially about their relationship with the living environment. With numerous examples from all parts of the world, this book is a clear demonstration of how important language is in helping people relate in productive ways with the rest of nature. I hope and expect that the many messages contained within this book will contribute in a significant way to global efforts to recognize the importance of cultural diversity at a time of rapid globalization of knowledge. For globalized knowledge is highly unlikely to generate the subtle new answers that are required to enable people to live in a productive balance with their locally available resources and adapt to the changing conditions that are certain to come. Building on the strong foundation of this book, we must use every possible means to promote cultural diversity and the greatest possible wealth of human languages, and harness such efforts to conserving the biological diversity upon which human welfare ultimately depends.

Jeffrey McNeely
Chief Scientist
IUCN—The World Conservation Union
Gland, Switzerland

ACKNOWLEDGMENTS

Several institutions generously supported the work that led to the making of this book. The international working conference Endangered Languages, Endangered Knowledge, Endangered Environments, held in Berkeley, California, 25–27 October 1996, on which this book is based in part, was funded by the Wenner-Gren Foundation (conference grant no. CONF-197), the Office for Research and the College of Letters and Sciences of the University of California at Berkeley (UCB), and the UNESCO, WWF, and Kew Gardens' People and Plants Initiative. The conference was sponsored by the nongovernmental organization Terralingua: Partnerships for Linguistic and Biological Diversity and by UCB's Department of Integrative Biology and University and Jepson Herbaria. Preparation of the book was accomplished during the editor's tenure of a National Institutes of Health Individual National Research Service Award (fellowship no. MH11573-02) at Northwestern University. All of this support is gratefully acknowledged.

The people who influenced my thinking about and guided my exploration of the interdisciplinary field at the root of both the Berkeley conference and this book are legion, and I will make no attempt to thank them all here. They include, among others, the contributors to this volume and other conference participants, as well as many of the authors cited in the references to the book's introduction. One person I do wish to single out for special recognition: my former UCB graduate advisor and mentor Brent Berlin. It was through his masterful training and intellectual guidance that my mind opened up to the interconnectedness of language, knowledge, and the environment. Without that training and guidance, neither the conference nor the book would have seen the light. I am also forever thankful to my family all, in space and time, for giving me the strength to pursue my enduring quest for learning.

My deep appreciation goes to the contributors to this book for their perseverance in sticking with me through the lengthy process of getting the manuscript ready for submission and then edited after review. They were all remarkably forthcoming in response to my repeated requests for revised texts, improved graphics, missing bibliographic items, and so forth, all through busy professional lives, fieldwork spells, health and family crises, and whatever else life brought. In spite of being scattered at the four corners of the world (email helping a great deal, of course!), they managed to function as an extraordinary team. In more ways than one, this book is testimony to their commitment to promoting an innovative interdisciplinary endeavor.

I owe a great debt of gratitude to Bill Merrill, of the Smithsonian Institution's National Museum of Natural History, for examining a preliminary version of the manuscript and encouraging me to submit the book to Smithsonian Institution Press, as well as for his friendly advice and enthusiastic support of the ideas embodied in the book. I am likewise indebted to Scott Mahler, the anthropology acquisitions editor at Smithsonian Institution Press, for believing in a book of this topic and scope in spite of its challenging size, and for being the kind of professional reader who will seek to uncover the circulatory system of a book. Final presentation of the manuscript also benefited from the much appreciated comments of two anonymous reviewers, and from the careful and expert editing of Jean Eckenfels. The Smithsonian Institution Press's production and marketing teams were also a pleasure to work with.

It grieves me terribly to report that in December 1997, shortly after delivering a draft of his chapter, Mike Warren met his sudden and untimely death at his African home in Ara, Nigeria. The news was received with shock by the contributors to this book, most of whom had seen Mike for the last time at the Berkeley conference in 1996 and admired his energetic enthusiasm for making the value of indigenous knowledge understood and appreciated throughout the world, and for promoting it as a fundamental basis for sustainable development. He is deeply missed, but his legacy lives on through the work of the many Indigenous Knowledge resource centers he helped found worldwide, whose activities are described in his chapter, and through the development projects he had initiated in Nigeria, which his wife Mary Salawuh Oyelade Warren has undertaken to carry on.

As of this writing, another extraordinary life devoted to the cause of indigenous peoples, some of which was poured into this book, hangs in the balance. Darrell Posey lies in a hospital with what is believed to be terminal illness. He may yet surprise us, for as his close collaborator and friend Graham Dutfield has put it, Darrell did not succeed in life by being a conformist. We all dearly beg to be surprised. Whatever happens, however, the present circumstances have only made everyone who knows Darrell more keenly aware of what has been so aptly expressed by Graham: "Darrell has achieved more in his short life than most of us ever will, however long we live. And in his way he has done his bit to make the world a better place. Many of his ideas that seemed so radical at first are increasingly accepted and respected, and it wouldn't be an exaggeration to say he has changed the world as much as one person can."

Of both men one could say what William James wrote in tribute to his friend Frederic Myers (a quote sent to me by Dave Harmon): "When a man's pursuit gradually makes his face shine and grow handsome, you may be sure it is a worthy one." This book is, in part, dedicated to Mike and Darrell.

Last but not least, the indigenous colleagues and friends with whom I have worked in East Africa and Mexico, and the many others whose words I have had the opportunity to read or listen to, taught me, in their wisdom, the lessons that would give living substance to what, at the outset, were only intuitively perceived links in my mind. Above all, this book is dedicated to them and to their brothers and sisters all over the world.

I

INTRODUCTION

On the Interdependence of Biological and Cultural Diversity

Luisa Maffi

> We are among the rarest of the rare not because of our numbers, but because of the unlikeliness
> of our being here at all, the pace of our evolution, our powerful grip on the whole planet, and
> the precariousness of our future. We are evolutionary whiz kids who are better able to transform
> the world than to understand it. Other animals cannot evolve fast enough to cope with us.
> It is possible that we may also become extinct, and if we do, we will not be the only species
> that sabotaged itself, merely the only one that could have prevented it.
>
> Diane Ackerman, *The Rarest of the Rare* (1997:xviii–xix)

PROLOGUE

On a sunny but crisp winter morning in the Highlands of Chiapas, Mexico, I was
standing outside the field clinic in one of the hamlets comprising the Tzeltal Maya
municipio of Tenejapa. It was the early 1990s, and I was there to do my doctoral
research on Tzeltal ethnosymptomatology—the Tzeltal language of signs and
symptoms of illness (Maffi 1994). As a part of this research, I was also interested
in issues of culture change: how traditional medical knowledge was being affected by Tenejapans' ever increasing contact with national Mexican society
through the rapid expansion of communications with the outside nonindigenous
world, Western media, formal schooling, and access to biomedical care. I had
come to this hamlet on the designated day for the periodic visit of the Mexican

Health Services *pasante* (medical trainee) to the field clinic, in order to witness his interactions with Tenejapan patients. I hoped to build my understanding of the local dynamics of medical systems in contact by engaging some of the patients in casual conversation about their health-seeking attitudes and behaviors.

Long before the clinic opened its doors, Tenejapan men, women, and children had been lining up in wait—an ideal circumstance for striking up conversations. After identifying myself and explaining the purpose of my study, with the aid of my Tzeltal collaborator, the hamlet's *promotor de salud* (health promoter), I started chatting with some of the people in line. Once it was clear to them that I was a student and not related to the Mexican Health Services or other national or state agencies, they did not mind discussing the health reasons that had brought them there. I expected to hear complaints about some of the more serious or uncommon illnesses known to the Tzeltal, many of them imported by colonizers from the Old World, for which the Tzeltal medical tradition had not developed effective treatments. But, to my surprise, the overwhelming majority of complaints referred to some of the most common ailments recorded among the Highland Maya—diarrheas, coughs, colds, skin problems—for which an abundant, efficacious traditional pharmacopoeia (mostly botanical in nature) was readily available (Berlin et al. 1990; Berlin and Berlin 1996).

Perhaps, then—my next assumption was—these people had already tried to treat their ailments the traditional way without success and were now submitting these stubborn syndromes to biomedical attention for treatment with more potent synthetic drugs. I asked the question, but again I was wrong. Prior to coming to the clinic, I was informed, my interlocutors had either self-medicated with drugs purchased at a pharmacy or done nothing at all. I looked around: true, most of the people in line were younger men and women, but by Tzeltal standards they were adults, fully developed and functional members of society, already with family and other customary adult responsibilities; they lived and worked in the village, spoke Tzeltal fluently (if peppered with Spanish words). It could not be that they had *not yet* acquired the traditional medical knowledge. Could it be that they had not acquired it *at all?*

I turned to a young man who was carrying in his arms his two-year-old daughter suffering from diarrhea; he had already struck me by telling me he had started out at dawn from his isolated household to get to the clinic—hours of walking, and now hours of waiting, with his sick child in his arms, hours of delay in getting treatment, a delay that might well prove fatal to her. With mounting anguish I asked him whether he knew of any plants or other local remedies for diarrhea, even if he had not tried to administer them to his daughter. He searched his mind, apparently in vain, then looked to another, slightly older man nearby, and started an animated discussion in Tzeltal with him. It became clear that between the two of them they were trying to dredge up and piece together scattered fragments of latent ethnomedical knowledge—knowledge perhaps only imperfectly learned, never concretely used, and now almost forgotten. I heard them question each other: "What's its name, the grasshopper thing?" The "grasshopper thing": *yakan k'ulub wamal* 'grasshopper leg herb' (*Verbena litoralis*), one of the commonest diarrhea remedies in the Highlands. They could hardly remember its name, let alone master its use.

The *pasante* had finally arrived, and people started filing in. I watched the young man walk off with his daughter in his arms, sent my *promotor* collaborator after him to try and ensure the man would get the best possible for his daughter out of what I already knew was almost invariably unsympathetic, superficial, culturally (and often even medically) inappropriate biomedical care, and stood there, with a sinking feeling. No doubt, that young man had to have the omnipresent *yakan k'ulub wamal* growing right in his back yard, but perhaps he could not recognize it, or if he did, he clearly did not know how to use it. Or maybe he had actually pulled it out as a weed, as my own collaborator had told me he had unwittingly done with medicinal plants his late father, a traditional healer, once kept in his house garden—only to later become aware of their virtues, paradoxically, through his work with ethnobiologists and other Tzeltal traditional healers.

I do not relate this episode to indulge in the postmodern anthropological penchant for colorful vignettes and reflections on fieldwork, but because it has everything to do with the genesis of this book. It was when for the first time I became aware—and acutely so—of the relationship between language, knowledge, and the environment, of the breakdown of these ties under the pressure of "modernization," and of the far-reaching implications of this breakdown for indigenous and other local peoples, and for humans at large. The memory of that day continued to haunt me long afterward, as I embarked on the course of reflection, research, and action that led, among other things, to this book.

THREATS TO THE WORLD'S DIVERSITY

Over the past decade, scholars and advocates from a variety of fields have been increasingly pointing to the detrimental effects of current global socioeconomic and environmental processes on the very objects of their concern: biological species, the world's ecosystems, and human cultural and linguistic groups and their traditional knowledge. Their respective work indicates that these various manifestations of life on Earth are facing comparable threats of radical diversity loss.

Biodiversity loss is by now a well-known phenomenon. From the late 1980s on (Wilson 1988), biosystematists and conservation biologists have successfully brought it to the level of public consciousness (see, e.g., McNeely et al. 1989; IUCN 1991; Groombridge 1992; WRI, IUCN, and UNEP 1992; Systematics Agenda 2000

1994; Ehrenfeld 1995; Heywood 1995; Reaka-Kudla, Wilson, and Wilson 1997; Mishler this volume), and we now hear about it almost daily in the media. Although estimates of the number of species on Earth vary greatly, experts consider it a strong possibility that 20 percent of the world's existing species will be lost over the next thirty years (Wilson 1992), and some estimates are even higher (Stork 1997). Calculating that perhaps only 10 percent of existing species have been discovered and named (Mishler this volume), scholars point out that scores of species will go extinct before they are scientifically identified and studied.[1] Biologists stress that, while species extinctions have occurred throughout the history of life on Earth, the present extinction crisis is the first to be overwhelmingly due to the direct or indirect impact of humans on the environment—to human disruption or destruction of ecosystems and the habitats of plants and animals—and that it is proceeding at an alarmingly increasing rate. Many also warn that we do not have the key to what holds the web of life together; as a consequence, we do not know at what point ecosystems under stress may stop performing the vital services on which humans, like all other living species, depend—and we may be experimenting at our peril.

Less widely known, although attracting increasing attention, is the diversity loss that is affecting the world's languages and cultures (see, e.g., Burger 1987; Dorian 1989; Robins and Uhlenbeck 1991; Hale et al. 1992; Durning 1993; Goehring 1993; Miller 1993; Bodley 1994; Hinton 1994; Hill 1995; Bobaljik, Pensalfini, and Storto 1996; Dixon 1997; Grenoble and Whaley 1998; Lewis 1998; Maffi 1998; Maffi, Skutnabb-Kangas, and Andrianarivo 1999; *National Geographic Magazine* 1999).[2] There are an estimated 6000 or more oral languages spoken today.[3] Most of these languages, however, pertain to small communities of speakers—the indigenous and minority groups of the world.[4] Harmon (1995) calculates that about half of the world's languages are spoken by communities of 10,000 speakers or less; half of these languages, in turn, are spoken by communities of 1000 or fewer speakers. Overall, linguistic communities with up to 10,000 speakers total about 8 million people, less than 0.2 percent of an estimated world population of 5.3 billion.[5] On the other hand, of the remaining half of the world's languages, a small group of less than 300 (such as Chinese, English, Spanish, Arabic, Hindi, and so forth) are spoken by communities of 1 million speakers or more, accounting for a total of over 5 billion speakers, or close to 95 percent of the world's population (Harmon 1995). The top ten of these alone actually comprise almost half of this global population. Taken together, these figures show that, while more than nine out of ten people in the world are native speakers of one or other of only about 300 languages, most of the world's linguistic diversity is carried by small communities of indigenous and minority people.

These are the linguistic communities that have been and continue to be under threat, due to the ever growing assimilation pressures that promote incorporation into "mainstream" society and the collective abandonment of the native languages in favor of majority languages (the phenomenon known as "language shift"). Harmon (1995) suggests that virtually all languages with fewer than 1000 speakers are threatened in this sense, although even more widely spoken languages are fully susceptible to the same pressures. Many of these smaller languages are already at risk of disappearing because of a drastic reduction in the number of their speakers; younger generations decreasingly or no longer learn their language of heritage. Many more have reached a stage of near extinction ("moribundity"), and only a few elderly speakers are left. "Nearly extinct" languages are estimated to make up between 6 percent and 11 percent of the currently spoken languages (Harmon 1995). In some projections, as many as 90 percent of the world's languages may disappear or become moribund during the course of the twenty-first century (Krauss 1992). These figures portray a threat to linguistic diversity that may be far greater in magnitude than the threat facing biodiversity.

Languages, like biological species, have undergone extinction before. Informed guesses suggest that the peak of linguistic diversity on Earth may have occurred at the beginning of the Neolithic (10,000 years b.p.), at which time more than twice the current number of languages may have been spoken (Robb 1993; also see Hill 1997, this volume). Population movements and political and economic expansion have long contributed to reducing linguistic diversity everywhere in the world, even well before the era of European colonization and empire building. As with biodiversity, however, what is unprecedented is an extinction crisis of the present scale and pace. It has been estimated that there may already be 15 percent fewer languages now than 500 years ago, when European colonization began (Bernard 1992). Losses have been especially marked in the Americas and Australia, and the trend is now accelerating throughout the world, with Australia and the Americas (especially the United States) still in the lead.

Historically, the main waves of colonial and imperial expansion (of both European and other major civilizations) have often come not only to the detriment of local peoples' sovereignty and control over their ancestral territories and resources, but also to the detriment of their ancestral languages and cultural traditions. Whenever assimilation into the dominant culture has been the goal, as it has mostly been, this assimilation has been effected crucially by way of *linguistic* assimilation: through the imposition of the dominant language in schooling, the media, government affairs, and most other public contexts; through the denigration of the local languages and the cultures they embody as "defective," "primitive," unfit for the "modern world"; as well as through the severe restriction of their contexts of use and even explicit governmental prohibition of their use, resulting in punishment, at times corporal, for violations. Perhaps a majority of these languages are unwritten and undocumented, so that, as for biological species, their disappearance often represents a total and irretrievable loss to their former speakers, and to humanity as a whole.

Along with the languages, much of the cultural knowledge and wisdom, the ways of life and worldviews of their speakers are under threat. While not all knowledge may be linguistically encoded, language does represent the main tool for humans to elaborate, maintain, develop, and transmit knowledge (see Fishman 1982, 1996; Bernard 1992; Diamond 1993; Woodbury 1993; Harmon 1996a; Maffi 1998, in press; also Hill, Pawley this volume). When external forces begin to undermine traditional societies, pushing them into the "mainstream"—whether this process is propelled by dispossessing local peoples of their sovereignty over land and resources, trampling their cultural traditions, or promoting linguistic assimilation (generally, all three occur at once and are mutually reinforcing)—local peoples generally end up losing control over, and ultimately contact with, their traditional natural and cultural environments. Global socioeconomic change disrupts traditional ways of life, promoting poverty, population growth, overexploitation of the environment by outside forces and by local groups themselves, as well as tension and conflicts over local peoples' land and resource rights. Under such conditions of rapid and drastic change, traditional knowledge, beliefs, and wisdom, and the languages in which they are encoded, tend to lose their functions for local peoples and begin to erode. External pressures also foster change in perceptions and attitudes on the part of local peoples, often leading to the disvaluing and abandonment of traditional knowledge and behaviors and of the languages that are the repositories and means of transmission of such knowledge. Furthermore, local knowledge does not "translate" easily into the majority languages to which minority language speakers switch. Generally, the replacing language does not represent an equivalent vehicle for linguistic expression and cultural maintenance (see Woodbury 1993; Hunn this volume), and along with the dominant language usually comes a dominant cultural framework that begins to take over and displace the traditional one. Because in most cases indigenous knowledge is only carried by oral tradition, when rapid shift toward "modernization" and dominant languages occurs and oral tradition in the native languages is not kept up, local knowledge ends up being lost (see Maffi this volume).

At an especially high risk of disappearing, due to its place-specific and subsistence-related nature, is traditional ecological knowledge, that is, local peoples' classification, knowledge, and use of the natural world, their ecological concepts, and their resource management institutions and practices, whose in-depth nature and value for sustainability social scientists have documented over several decades (e.g., Berlin, Breedlove and Raven 1974; Hunn 1977, 1990; Majnep and Bulmer 1977, 1990; Brokensha, Warren, and Werner 1980; Grenand 1980; Williams and Hunn 1982; Hames and Vickers 1983; Alcorn 1984; Brown 1984; Nabhan 1989; Posey and Balée 1989; Taylor 1990; Atran 1990, 1993; Ostrom 1990; Berlin 1992; Blackburn and Anderson 1993; Williams and Baines 1993; Balée 1994; Norgaard 1994; Warren, Slikkerveer, and Brokensha 1995; Anderson 1996; Berlin and Berlin 1996; Berkes

1999; Blount and Gragson 1999; Medin and Atran 1999; also Warren this volume). As local peoples are removed from their lands, or subsist in severely degraded ecosystems, and are absorbed into a market economy in which normally there is little room for traditional subsistence practices and resource use, local ecological knowledge and beliefs and the accumulated wisdom about human-environment relationships begin to lose their relevance to their lives.

This phenomenon has been called the "extinction of experience": the radical loss of the direct contact and hands-on interaction with the surrounding environment that traditionally comes through subsistence and other daily life activities (Nabhan and St. Antoine 1993; see also Hill, Zent, Lizarralde, Florey, Majnep, Molina this volume). Both the intergenerational transmission and the contexts of use of knowledge in the native languages begin to erode. These processes affect not only the younger generations, but older people as well (Zent, Hill this volume), and over time a decline in "cultural support" for nature knowledge may be observed (Wolff and Medin this volume). Under these concerted pressures, local human subsistence activities that often did not affect biodiversity negatively, and in some cases may even have fostered it, begin to be replaced by other, environmentally unsound practices.

The consequences of such global and local processes of cultural and ecological change are everywhere to be seen, in the progressive depletion of the world's forests and ranges, deterioration of water and air quality, mounting violations of indigenous and other local peoples' land, resource, cultural, and linguistic rights, and humans' increasing inability to live sustainably and harmoniously on Earth (see Norgaard 1994, this volume).

LINKS BETWEEN LINGUISTIC, CULTURAL, AND BIOLOGICAL DIVERSITY

Interestingly, when in the late 1980s linguists and anthropologists were beginning to voice concern about the state of the world's languages and of traditional knowledge, they sometimes drew parallels with the loss of biodiversity, as a way of suggesting comparable damage to humanity's heritage (see, e.g., Hale 1992; Krauss 1992). But in these initial pronouncements, linguists made no significant attempt to go beyond such parallels and ask whether there might be more than a metaphorical relationship between linguistic and biological diversity and the loss thereof (but see Krauss 1996). And by and large they were not engaging in joint discussion of the threats to diversity with either anthropologists or biologists.

For their part, anthropologists and other scholars and advocates did point to connections between biological and cultural diversity, but in most cases did not devote attention to linguistic diversity (e.g., Dasmann 1991; Gray 1991; Oldfield and Alcorn 1991; Shiva et al. 1991; Chapin 1992; Durning 1992; Nietschmann 1992; Castilleja et al. 1993; Colchester 1994; Toledo 1994, 1995; McNeely and Keeton 1995;

Poole 1995; Wester and Yongvanit 1995; Wilcox and Duin 1995; Alcorn 1996; Carruthers 1996; Schoonemaker, von Hagen, and Wolf 1997). In particular, many ethnobiologists showed keen awareness of the simultaneous extinction threats facing tropical and other biodiversity-rich ecosystems, on the one hand, and indigenous peoples on the other; they stressed indigenous peoples' stewardship over the world's biological resources and affirmed the existence of an "inextricable link" between cultural and biological diversity on Earth.[6] Yet, language was not mentioned as a part of this "inextricable link."

In turn, conservation biologists and ecologists were becoming familiar with the concept of biocultural diversity, especially since the role of indigenous and other local peoples in the conservation of biodiversity found its way into international instruments such as the Convention on Biological Diversity (CBD) after the 1992 UN Conference on Environment and Development, the Rio Summit (see Glowka, Burhenne-Guilmin, and Synge 1994). In this context too, however, the relevance of linguistic diversity was not in focus.

It is only more recently that this issue has been explicitly raised and the idea has begun to emerge that, along with cultural diversity and as a central part of the latter, linguistic diversity should also be seen as inextricably linked to biodiversity.[7] A 1995 study (Harmon 1996a) showed some striking global correlations between linguistic and biological diversity. Starting from the observation that the majority of the world's smaller languages (which, as we have seen, account for most of the world's linguistic diversity) can be labeled as "endemic"—in that they are spoken exclusively within a given country's borders—and comparing a list of countries by number of endemic languages with the World Conservation Union (IUCN, formerly the International Union for the Conservation of Nature) list of biological "megadiversity" countries (McNeely et al. 1989), this study found that 10 out of the top 12 megadiversity countries (or 83 percent) also figure among the top 25 countries for endemic languages. A global cross-mapping of endemic languages and endemic higher vertebrate species carried out by Harmon (1996a) brought out the remarkable overlap between linguistic and biological diversity throughout the world (see map 1.1). Similar results emerged from a global comparison of endemic languages and flowering plant species.

Harmon (1996a) pointed out that several large-scale biogeographical factors might account for these correlations, in that they might comparably affect both biological and linguistic diversity and especially endemism (such as extensive land masses with a variety of terrains, climates, and ecosystems; island territories, especially with internal geophysical barriers; tropical climates, fostering higher numbers and densities of species). All these factors might increase linguistic diversity by increasing mutual isolation among human populations and thus favoring linguistic diversification (see Smith this volume; see Nettle 1996, Mühlhäusler this volume for different views on the role of biogeographical vs. socioeconomic

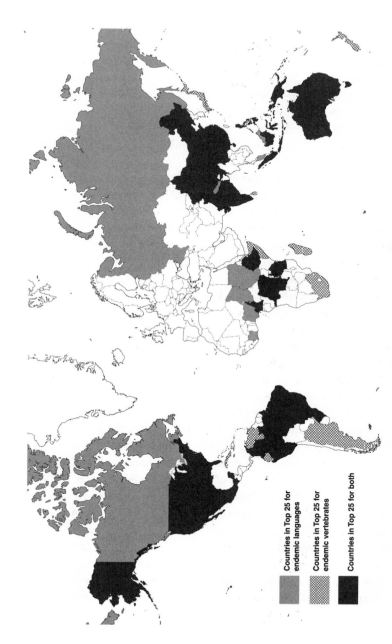

Map 1.1. Overlap of biological and cultural diversity: Endemism in higher vertebrate species and languages. Figures for Ethiopia include Eritrea. Higher vertebrates include mammals, birds, reptiles, and amphibians; reptiles not included for the United States, China, and Papua New Guinea because the numbers were not reported in the source table. Original art by David Harmon.
Source: Harmon (1996a), based on data from Groombridge (1992:139–141) for species and Grimes 1992 (passim) for languages.

Countries in Top 25 for endemic languages

Countries in Top 25 for endemic vertebrates

Countries in Top 25 for both

factors). In addition, Harmon proposed a small-scale ecological phenomenon as also possibly accounting for biodiversity-linguistic diversity correlations: a process of coevolution of small human groups with their local ecosystems, in which over time humans interacted closely with the environment, modifying it as they adapted to it and developing specialized knowledge of it (cf. Thompson 1999). In order to convey this vital knowledge and ways of acting upon it for individual and group survival, Harmon suggested, humans would have also developed specialized ways of talking about it. Thus the local languages through which this knowledge was encoded and transmitted would in turn have became molded by and specifically adapted to their socioecological environments (see Mühlhäusler 1995, 1996; Balée this volume).

At the same time, scholars have also observed that linguistic and cultural distinctiveness have often developed even in circumstances of nonisolation, for example, among human groups who occupy the same or contiguous ecological niches and who belong to the same broadly defined cultural area, or whose languages are considered to be historically related or to have undergone extensive mutual contact (Mühlhäusler 1996; see also Milton this volume). Such circumstances—high concentrations of linguistically distinct communities coexisting side by side in the same areas—appear to have occurred frequently throughout human history (Hill 1997) and still do today in many parts of the world (the most notable case being that of New Guinea, which has more than 800 different languages and ethnically distinct groups). Complex networks of multilingualism in several local languages and pidgins or lingua francas have been a commonplace way of dealing with cross-language communication in these situations. Mühlhäusler (1996) has identified in this extensive multilingualism another key factor in the maintenance of linguistic diversity historically, countering the increasing pressures of linguistic assimilation. Based on his research in the Pacific region, Mühlhäusler (1996) has proposed that the functional relationships that develop in space and time among linguistic communities that communicate across language barriers should be understood in terms of "linguistic ecologies" (see Haugen 1972) and that within an ecological theory of language the diversity of languages as such, its functions and the mechanisms that sustain it in the history of humanity, should be the focus of investigation (see Mühlhäusler this volume).

Mühlhäusler has also pointed out that the study of traditional linguistic ecologies reveals that they encompass not only the linguistic and social environment, but the physical environment as well, within a worldview in which physical reality and the description of that reality are not seen as separate phenomena, but instead as interrelated parts of a whole. As he puts it: "Life in a particular human environment is dependent on people's ability to talk about it" (Mühlhäusler 1995: 155). Such relationships, Mühlhäusler (this volume) notes, require extended periods of intensive human-environment interactions (of the coevolutionary kind postulated by Harmon 1996a) to become established.

While it is now commonly acknowledged that humans have always modified, to a greater or lesser extent, the environments in which they live (contra the belief in "pristine wildernesses"), a controversy exists on the precise circumstances, currently and historically, of both positive and negative impacts of human activities on biodiversity and ecosystem health, in small-scale societies as well as with the rise of complex civilizations (e.g., Williams and Hunn 1982; Hames and Vickers 1983; Diamond 1987, 1991; Harris and Hillman 1989; Posey and Balée 1989; Ponting 1991; Bahn and Flenley 1992; Blackburn and Anderson 1993; Williams and Baines 1993; Flannery 1995; Kirch and Hunt 1996; Kirch 1997; Smith this volume). Doubts are also sometimes expressed about whether indigenous peoples actually possess conscious conservation practices—the so-called myth of the ecologically noble savage (Redford 1991; also see Bulmer 1982; Diamond 1986; Johnson 1989; Hames 1991; Denevan 1992; Ellen 1994; Nations this volume). But the overall consensus is that small local groups with a history of continued and unchallenged occupation of given territories will over time tend to develop and maintain detailed and accurate knowledge about their ecological niches, as well as about sustainable ways of extracting and managing natural resources (see Hunn, Nabhan, Atran, Padoch and Pinedo Vasquez, Carlson, Brush this volume). What has been said of Australian Aboriginal tribes might be said in hundreds of other cases of local peoples around the world: "Coincidences of tribal boundaries to local ecology are not uncommon and imply that a given group of people may achieve stability by becoming the most efficient users of a given area and understanding its potentialities" (Tindale 1974:133; quoted in Mühlhäusler this volume). From this perspective, linguistic and cultural diversity and the loss thereof, with its frequent corollary of loss of traditional ecological (and other) knowledge and practices, can be seen as an integral part of the overall ecological processes affecting biodiversity on Earth.

In affirming the interdependence of biological and cultural diversity, Nietschmann (1992:3) proposed a "Rule of Indigenous Environments": "Where there are indigenous peoples with a homeland there are still biologically-rich environments." The research discussed above suggests that this rule should be understood as including also viable indigenous languages and cultural traditions among the conditions for the persistence of biologically rich environments. In turn, indigenous and other local peoples, struggling for survival and self-determination with secure land bases and means of subsistence, increasingly see their languages and cultural traditions as essential elements in this struggle. Language, cultural traditions, and land are considered by most of them as equally constitutive of their identity as distinct peoples and of their right to live as such (see Majnep, Molina, Manriquez, Posey, Skutnabb-Kangas, Maffi, Toledo, Blount, Brush this volume). In this light, it is also possible to suggest that landscapes are anthropogenic not only in the sense that they are physically modified by human intervention, but also because they are symbolically brought into the sphere of human communication by language: by the words, expressions, stories, legends, songs, and verbal interactions that encode and convey human relationships with the environment and inscribe the history of those relationships onto the land (Maffi 1998; see also Wilkins 1993; Pawley, Wollock this volume).

A MOMENT OF REFLECTION

As these various strands of thinking and research were beginning to converge, a number of relevant activities took place in 1996. One was the creation of Terralingua: Partnerships for Linguistic and Biological Diversity, an international non-governmental organization specifically devoted to research, information, education, and advocacy on issues of linguistic diversity and its relationships with cultural and biological diversity (see Appendix 2). Time was also ripe for a moment of reflection on this emerging integrated perspective on the various manifestations of the diversity of life, the threats they are facing, the foreseeable consequences of massive disruption of such long-standing interactions, and the possible courses of action to ensure the perpetuation and continued development of all forms of diversity on Earth. Two scholarly meetings, of which Terralingua was among the sponsors, set out to address this need: the symposium Losing Species, Languages, and Stories: Linking Cultural and Environmental Change in the Binational Southwest (Tucson, Arizona, 1–3 April 1996), organized by Gary Nabhan (see Nabhan 1996), and the international working conference Endangered Languages, Endangered Knowledge, Endangered Environments (Berkeley, California, 25–27 October 1996), organized by the editor of this book (see Maffi 1996, 1997).[8] The latter conference was the first international meeting of representatives of a wide array of disciplines and concerns dealing with the threats faced by the linguistic, cultural, and biological diversity of the planet—linguists, anthropologists, ethnobiologists, cognitive psychologists, biologists, ecologists, natural resource conservationists and managers, economists, and indigenous experts and advocates. All participants had done relevant work in some of the areas of highest linguistic, cultural, and biological diversity in the Americas, Oceania, Southeast Asia, Africa, and Europe, and several of them were natives of such areas. Among the issues addressed at the Berkeley conference were the following:

1. Are biological, cultural, and linguistic diversity related, and if so, what is the nature of these relationships?

2. What is the evidence of sustainable human-environment relationships among indigenous and other local groups around the world, and under what cir-

cumstances, currently and historically? What kinds of human activities, and in which institutional settings, result in the maintenance or disruption of such relationships?

3. What role does the conservation or loss of traditional ecological knowledge have in the maintenance or disruption of sustainable human-environment relationships?

4. What role does language have in the acquisition, accumulation, maintenance, and transmission of this knowledge, and how does language loss affect these processes?

5. What socioeconomic factors underlie processes of language and culture shift and changes in environmental knowledge and behavior?

6. What are the cognitive underpinnings of attrition caused by contact between different linguistic and cultural models, as well as different knowledge systems?

7. How do these phenomena of change affect individual and societal choices and decision making (with special reference to activities affecting the environment)?

8. How can our understanding of these issues best inform systematic studies of ethnoecological knowledge change and loss as well as action aimed at biocultural diversity conservation and promotion?

9. How can this understanding best be made available to local communities as a tool for informed decision making?

10. How can it best be used to educate the general public on the global threats to linguistic and cultural diversity and their relationships to ecosystem endangerment?

The present volume was inspired significantly by the Berkeley conference and the intensive discussion that took place among participants, as well as by the developments it sparked.[9] Since the conference, the connections between biodiversity and cultural, especially linguistic, diversity have been receiving increasing attention in conservation circles, including the IUCN, the World Wide Fund for Nature (WWF), and the United Nations Environment Programme (UNEP) (see, e.g., Posey 1999; *Nation* 1999; WWF International and Terralingua 2000; Maffi and Carlson forthcoming). The topic is ripe for the kind of integrated, interdisciplinary approach this book begins to set forth.

ORGANIZATION AND CONTENT OF THIS VOLUME

The chapters in this book are organized in four parts: a series of chapters that present some of the main issues and perspectives in each field involved and point to emerging links, a number of case studies, a set of chapters focusing on action strategies, and two final chapters that call for change in Western epistemological models and ways of thinking in order to achieve the goals of biocultural diversity conservation and sustainability on Earth. A range of theoretical frameworks and approaches is represented, but each author attempts to engage other perspectives in an unusual effort to build bridges across disciplines and other institutional and intellectual boundaries.

Part One. Language, Knowledge, and the Environment: Interdisciplinary Framework

In the opening chapter, Harmon sets the scene by discussing the concept of diversity in both nature and culture, basing his argument on William James's psychological and philosophical analysis of diversity. He notes that James saw diversity as the basic condition of life, one to which humans, like all other species, are born and which forms the very substance (or rather function) of human consciousness. James pointed to the way in which human consciousness is compelled to build sameness out of the existing diversity perceived by our senses. Harmon agrees with James that diversity is the means through which the human "consciousness function" operates, suggesting that, if consciousness is what makes us human, it then follows that diversity makes us human. Harmon explores the way this process operates in various domains of human experience, from the conceptualization of biological species to that of languages and cultures. The author argues that this suggests an affinity between biological and cultural diversity, an affinity stemming from the long-range evolutionary processes that have shaped the diversity of life. This "deep" diversity, Harmon contends, can never be replaced by the superficial diversity promoted by market-based ideologies as an artificial substitute for the genuine diversity, both natural and cultural, now being wiped out by globalization processes. He adds that, if we acknowledge that humans, like all other species, are the products of evolutionary processes of which diversity is the quintessential expression, then we have the basis of a moral imperative for preserving diversity in both culture and nature that transcends the interests of humankind to affirm the dignity of all life on Earth.

The next two chapters introduce the reader to some of the basics of the notions of biodiversity and linguistic diversity, respectively. Mishler discusses the phylogenetic species concept, which rests on the historical relationships between taxa (instead of on blood relations and similarities among individuals, as in the still widely used biological species concept; see Harmon this volume). The author presents the main processes ("divergence" and "reticulation") and entities ("replicators" and "interactors") involved in biological evolution, leading to life being organized as a hierarchy of nested lineages. Mishler argues for the central relevance of the phylogenetic approach in addition to the ecological approach that is gaining currency in conservation biology (and is favored by several contributors to this volume). He points out that biological lineages, at one and the same time, are the fruit and bear the legacy of the diversification of life on Earth. Therefore, they carry invaluable information about these diversification processes that can be totally lost when lineages go extinct. The author also observes that languages and cultures form lineages as well and examines analogies and discrepancies between linguistic and cultural evolution, on the one hand, and biological evolution, on the other. He concludes that, while the affinities suggest commonalities in the way the different forms of diversity evolve, the discrepancies appear to result in an acceleration of cultural and linguistic evolution over biological evolution, and ponders the implications of this phenomenon.

Corbett's purpose in his chapter is to characterize linguistic diversity from a theoretical standpoint, again to provide elements for comparison with the other manifestations of diversity. He points out that there are different aspects of the "linguistic enterprise" for which diversity matters, one being historical linguistic reconstruction (reconstruction of linguistic lineages, in Mishler's sense), in order to understand genetic relationships among languages and the ways in which languages change over time. His own approach, instead, is that of the typologist, for whom the goal is to define the notion of "possible human language," that is, the range of typological features that can be said to characterize language as a specific human faculty. Dealing with this issue requires having access to as much of the world's linguistic diversity as possible, as shown by the fact that over and over again earlier generalizations are overthrown by the identification of novel typological features in little-known languages, many of which are currently endangered or whose unusual typological features are especially subject to erosion. Corbett notes that many of the threatened languages belong to small language families, so that their disappearance represents a radical loss of linguistic information, both genetically and typologically. He illustrates these points by presenting examples of typological features whose full range has only come to be recognized by analyzing a wide array of genetically diverse languages.

The chapters by Smith and Hunn both attempt to define the notion of culture as a basis for discussing the nature of cultural diversity, then delve into the links between biological, linguistic, and cultural diversity. The evolutionary dimensions of human-environment relationships and the issue of the possible coevolution of cultural, linguistic, and biological diversity are the specific focus of Smith's chapter. The author first sets forth his definition of culture, as "socially transmitted information"—an ideational definition by which culture is recognized as a system of inheritance (see Mishler this volume). At the same time, Smith argues that cultural diversity should not be strictly understood in terms of variation in culturally heritable information because of the essential unboundedness of cultural units or lineages and the extent of diffusion among them (reticulation, in Mishler's terms). On the other hand, he is in essential agreement with Harmon (this volume) on the usefulness of the concept of culture(s), in spite of the lack of sharp boundaries (just as is the case with biological species). In this connection, he favors an adaptationist over a phylogenetic perspective on cultural change, which he views as "a form of coevolution between cultural information and the social and natural environment." Smith then puts to the test the global correlations that have been found between the various manifestations of diversity (Harmon 1996a) by examining the case of North American indigenous languages, culture areas, and ecoregions. He finds significant correlations on some measures but not on others. This fact points to the need for better, more fine-grained ways to define and identify linguistic and cultural diversity, as well as to the need to take into account historical data on population movements and contacts (including displacement or extinction of ethnolinguistic groups) and on processes of environmental change. Assuming general correlations, however, Smith examines possible factors accounting for them, in terms of various ways in which cultural diversity might directly enhance biodiversity and vice versa. He also looks for factors that might account for an observable correlation between low-diversity cultural systems and low biodiversity.

Hunn's chapter is also concerned with the general issue of the links between biological, linguistic, and cultural diversity, as well as the definition, measurement, and comparison of these "diversities." While species richness (number of different species) is commonly used as a measure of biodiversity, Hunn points out that all species are not equal, in that endemic species (ones limited in geographical range) contribute more in terms of genetic information than widely distributed ones. Thus, biodiversity must ultimately be understood in terms of this genetic information. Hunn sees languages and cultures as systems of information, as Smith does, and asks questions about the proper units of language and culture to be compared to biological units at various levels. He finds this somewhat more difficult for cultures than for languages. In spite of phenomena of hybridization among languages (again, reticulation in Mishler's terms), he considers languages to be more discrete as basic units, on a par with species, than "cultures." The author wonders whether, at least in the present, one can actually talk of a diversity of relatively autonomous "cultural species," or rather of a "single massive hybrid swarm." He speculates on this issue by contrasting circumstances in a prestate "tribal" epoch to those in the epoch of the state. As Hunn points out, however, even today there are local communities that maintain their cultural knowledge— and particularly traditional ecological knowledge—intact: ones in which cultural knowledge is intimately bound to specific places and the biogeography of such places, which he calls "endemic cultures" (see Nabhan 1996, where the notion of "ethnobiological endemism" is proposed). Hunn illustrates this point with the case of two Mixtepec Zapotec–speaking communities in Oaxaca, Mexico, that have retained vast and detailed knowledge of the local environment, encoded uniquely in their language, over a long period of time, despite five hundred years of Spanish linguistic and cultural influence on the region. The author discusses the global relevance of this diversity of local cultural information and the conditions favoring its retention.

Such close relationships between humans and their ecosystems, Mühlhäusler reminds us in his chapter, take a long time to establish and are tremendously difficult to recover once lost. Mühlhäusler discusses this issue in the context of ecolinguistics, an approach that views language as ecologically embedded, that is, not as a self-contained system but as an integral part of a larger ecosystem (including both the natural and the social environments, as well as the context represented by the interaction of the various languages used by linguistic communities in situations of contact). Rather than seeing the diversity of languages as the mechanical outcome of language splits in circumstances of mutual isolation between human populations, ecolinguistics focuses on the functions of language diversity—functions that go beyond transmitting information to include such metacommunicative goals as marking and sustaining group identity. And rather than looking at languages as separate entities related phylogenetically in a way comparable to biological species, ecolinguistics looks at language networks and the speech repertoires used in intergroup communication. Finally, the ecolinguistic perspective also focuses on the "adaptation of ways of speaking to specific environmental conditions" and how this process develops over time. This question, the author proposes, is best addressed by studying cases of populations moving or being moved to a new environment and thus having to learn from scratch about its flora and fauna, the relationships among species (including between humans themselves and other species), and how to talk about them. Mülhäusler examines two cases of recent settlements of small populations on small islands, Pitcairn Island and Norfolk Island in the Pacific Ocean, both of which show evidence of overexploitation of resources as well as of an impoverished lexicon for local species and ecological processes. He then compares them with other known cases (such as the Marquesas Islands) with a longer history of human colonization, showing enormous environmental destruction at the beginning, until over time "an approximation between the contours of language and knowledge and the contours of the environment" was achieved.

The following two chapters focus on the "approximation between the contours of language and knowledge and the contours of the environment" in situations in which local peoples have a long history of residence in their environment. Both chapters concentrate particularly on ethnoecological knowledge (local people's understanding of their relationships with the environment as well as of the interactions between plants and animals in the environment) and stress the importance of such knowledge for sustainable resource use and for biodiversity conservation. Nabhan points out that one reason why claims that indigenous peoples are "the first ecologists" have met with skepticism is that little systematic work has actually been done on indigenous peoples' ethnoecological knowledge and its effects on biodiversity, while few studies of biogeography take humans into account as agents of coevolution. The author stresses that many of the ecological relations recognized by indigenous peoples are often barely if at all known to Western science, particularly because they tend to be place-specific, and that many of these interactions involve threatened or rare species, often endemic ones. Yet lack of appreciation for such indigenous knowledge has severely limited Western-trained scientists' ability to test its validity and examine whether and how it could be successfully incorporated into biodiversity conservation strategies. Nabhan attributes this state of affairs to a prevailing focus of Western bioscience on species rather than interactions, which he finds largely reproduced in the original focus of ethnobiology. And, like Mühlhäusler, he identifies the same problem also in linguistics, a problem that has made it difficult to understand how knowledge of such interactions is encoded and retained or eroded in a language (see Hill this volume). The author illustrates this point with examples drawn from his research on the Seri of the Sonoran Desert of the U.S.-Mexican border. These examples demonstrate a rich and complex understanding of (and sometimes influence over) ecological phenomena relating to, among other things, seed dispersal and animal diet. In all cases, this knowledge is codified in the Seri language, not just in terminology, but in stories, songs, and other forms of oral tradition. Nabhan shows the vulnerability of this knowledge to external processes that are bringing about the "extinction of experience" and advocates the revalidation of such knowledge and the pursuit of integrated biocultural conservation efforts.

In turn, Atran elaborates on the need to study traditional ecological knowledge systematically and comparatively and how this knowledge correlates with reasoning about and action vis-à-vis the environment (as in the extraction and use of natural resources). The work he reports on concentrates on three local populations occupying and making use of the same environment, the lowland tropical forest of the Petén region of Guatemala, and cross-culturally compares their respective "mental models" of the environment (conceptions of plant-animal interactions and human-environment interactions) and resource use practices. These three groups, the Itzaj Maya, Ladinos (Spanish-speaking mestizos), and Q'eqchi' Maya, have a different history of occupation of the land: the Itzaj have lived in the region since pre-Columbian times, while Ladinos and Q'eqchi' are recent immigrants. All three base their subsistence on *milpa* (cornfield) cultivation and use of products from the forest, but their respective agroforestry practices present striking differences in their relative sustainability: the Itzaj are at the sustainable end and the Q'eqchi' at the nonsustainable end (see also Nations this volume). The Ladinos are in-between, showing some practices in common with the Itzaj. The three groups' respective folk-ecological models correlate in significant ways with these differences in resource management practices. Atran examines these findings in light of the different social networks within and between these groups and how these networks favor or disfavor the circulation of knowledge and particularly the formation of what he calls an "emergent knowledge structure"—a fluid theorylike belief system that arises as an individual's cultural upbringing "primes one to pay attention to certain observable relationships at certain levels of complexity," allowing both for independent discovery of relevant relationships and for learning about them from others. It is worth pointing out that the notion of emergent knowledge structure is relevant to a better understanding of the nature of ethnoecological knowledge: skepticism about its existence and validity (as lamented in Nabhan this volume) may well also derive from the erroneous expectation that it should be an explicit "culturally stipulated, conventionally learned or rigorously formulated" theory. Atran's research shows that it may not be so, while at the same time it begins to capture this emergent knowledge structure through experimental and statistical methods. Yet, it appears that this knowledge structure is at risk, as younger Itzaj are no longer learning the language and how to "walk alone" (i.e., behave appropriately) in the forest.

But what actually happens when local languages and traditional knowledge begin to erode? The next three chapters tackle various aspects of this issue. Hill examines the processes of language decay (as distinct from regular language change) that lead to language death. These processes occur as speakers of local languages, and in many cases also of major regional languages, shift to "world languages" such as English or Spanish, and fewer and fewer people speak the local languages in a progressively reduced variety of contexts. Language decay manifests itself in a radical loss of both functions of language use and structural features of the language. Hill points out that, while in ordinary language change structural reduction in one domain is counterbalanced by development in another, in language decay such counterbalancing does not occur, as people instead develop their competence in the replacing language, and so the system undergoes actual impoverishment. The author notes that the interruption of intergenerational transmission has been mentioned most frequently as the main cause of language decay, but phenomena of language decay are observed also among fluent speakers. She argues that in such cases the determining factor is rather the loss of contexts in life in which given functions or structures of the language are appropriate. Hill illustrates these processes with examples of erosion in two Uto-Aztecan languages, Nahuatl (Mexico) and Tohono O'odham (United States). Her examples cover aspects of attrition across functional domains as well as decay in structural features at various levels, and she analyzes both the observable changes and the circumstances that lead to them. In particular, she discusses the loss of folk biological nomenclature in Tohono O'odham found not only among younger speakers (as observed by Nabhan and St. Antoine 1993 among O'odham children) but also among some older speakers, which appears to cut off speakers from re-call of environmental knowledge and even from environment-related aspects of life histories and from affective reactions.

Ethnobiological knowledge loss is the specific focus of Zent's chapter. Zent re-marks that so far the loss of traditional ecological knowledge (TEK) has mostly been the object of theorizing, as well as of advocacy and salvage operations. In-stead, he concentrates on field methods that make it possible to document and measure knowledge loss systematically and to identify relevant factors, as the nec-essary basis for informed efforts toward biocultural diversity conservation. The author notes that one main factor that has so far hampered the systematic study of traditional knowledge loss is the dearth of historical documentation of such knowledge in traditional cultures that would afford a diachronic perspective on knowledge change. On the other hand, he points out that linguistic and cognitive studies have shown how synchronic variability of cultural knowledge can be con-sidered an indicator of diachronic change. Therefore, Zent sets out to infer his-torical processes of TEK change among the Piaroa of Venezuela by looking at current patterns of intracultural TEK variability and by investigating the correla-tions between these patterns and other social variables related to events whose temporal dimension is known or that are themselves indicative of cultural and en-vironmental changes. Zent measures ethnobotanical knowledge variation among the Piaroa by means of four methods: the ethnobotanical plot survey, the con-trolled interview technique, consensus analysis, and linear regression analysis. His study reveals a dramatic decline in ability to provide the correct Piaroa names for plants among Piaroa individuals under 30 years of age. That these results are not simply the effect of a normal "learning-with-age" curve is shown by correla-tion with two other variables used as representative of acculturation, that is, bilin-gual ability in Spanish and degree of formal education. In addition, Zent finds a significant correlation between the ability to provide the Piaroa name for a given plant species and the correct identification of the cultural uses of that species—a finding that parallels Hill's on the O'odham, confirming the function of ethnobi-ological labels in anchoring knowledge about natural kinds.

The lack of diachronic data on ethnobiological knowledge change lamented by Zent can be remedied in the case of languages for which historical documenta-tion does exist. Wolff and Medin investigate the evolution and devolution of folk biological terminology in the English language in a longitudinal study of British English from the fourteenth to the twentieth century, using historical dictionary quotations. The authors set out to test the hypothesis that, with modernization, knowledge of natural kinds has devolved and that this may be due not only to de-creasing exposure to the natural world, but also to diminished "cultural support," that is, the extent to which a society actively promotes a given domain of knowl-edge. Focusing on the life form "tree," Wolff and Medin use as a measure of cultural support the extent to which trees are mentioned in written sources over the history of British English. They also investigate at what level of taxonomic specificity trees are mentioned (life-form, generic, and specific levels). They find that overall mention of tree terms over time increases until the nineteenth century, then drops dramatically in the twentieth; mention of trees at all three levels of specificity follows a similar pattern. The turning point appears to be the Industrial Revolution, when England's population became increasingly urban and cut off from the natural environment. Yet the decline in the twentieth century is less marked for folk specifics than the other two ranks. Wolff and Medin speculate that, rather than to the presence of a subset of tree specialists, this lesser decrease in specific terms may be due to the lingering of labels in the language after the knowledge about their referents has eroded, so that these terms may represent "loose categories," decoupled from their actual referents in the world. Loss of the labels may come next—while at the same time, whatever traces of the knowledge may still be indirectly carried in the language may offer hope for its future reacquisition.

The last two chapters of Part One raise the question: If language plays such a key role in the relationship between knowledge and the environment, why is it, then, that the language sciences have been so reluctant to acknowledge this link and have done such a poor job of recording linguistically encoded cultural knowledge? Pawley addresses this question in the context of a comparison between two models of language: the "grammar-based model," currently prevalent in linguistics, and the "subject matters model," which draws inspiration from humanistic views of language. Grammar-based models conceive of languages as autonomous systems, independent of any specific beliefs about and knowledge of the world; their notion of what belongs in a lexical description of a language is that of a very small subset of all the possible words and expression of that language— in essence, only the unanalyzable elements of the lexicon. Therefore, linguistic descriptions based on this model do not take into account cultural knowledge; as a consequence, they have had little to contribute to documenting and conserving that knowledge. Instead, a subject matters model focuses on ways of talking about given subject matters, such as traditional ecological knowledge. Pawley echoes other authors in this book in pointing out that much of this knowledge (where "knowledge," he argues, should be understood as including "perceptions" and "beliefs") is not just about entities per se, such as natural kinds, but about natural processes and relations among entities, such as the relationships among plants and animal species or between humans and the ecosystem. He stresses that this knowledge is not simply carried by nomenclature; instead, it pervades the whole linguistic system, all aspects of grammar and language use. The author illustrates these points with examples from the Kalam language of Papua New Guinea, including excerpts of Kalam discourse about ecology. He shows that this discourse is replete with highly conventionalized expressions that conveniently package enormous amounts of cultural knowledge. He argues that the effort involved in unpacking this knowledge in linguistic description should give pause to those who are skeptical that the notion of "linguistic diversity" carries any deep meaning and are wont to claim that "anything that can be said in one language can be said in any other language."

Wollock addresses similar issues from a philosophical perspective. He suggests that, if Western linguistic science has largely been silent about linguistic diversity and has ignored or even denied any connection between language and the real world, it is because it is born of the nominalist philosophical tradition that has taken gradual hold in the history of Western thought. Nominalism treats all universal concepts (including "nature" and "community") as arbitrary social constructs with no connection to the real world. Within this tradition, language itself is seen as an arbitrary system of signs that bears little or no relation to the extralinguistic world. Such a conception of language, Wollock argues, is inherently incapable of dealing with the relationship between language and the environment or with the ways in which language may orient the mind in certain directions—including ones that may be either beneficial to or destructive of the environment. According to Wollock, nominalist philosophy, in fact, lies behind most of the discourse of "colonizing cultures" about both language and the environment and behind the increasing tendency for this discourse to treat diversity as an epiphenomenon at best, a nuisance and a threat at worst. On the other hand, the author also contends that the response does not lie in the recent intellectual trend known as postmodernism, which in reaction to the centralizing and homogenizing tendencies of modernism denies the existence of any overarching system of meaning and only admits of diversity, decrying unity as an illusion. Wollock observes that all great metaphysical traditions recognize endless diversity as the reality of the planet, and indeed the universe, while perceiving a fundamental unity in it—the unity of the Logos, whose likeness can only be approximated through the maximum diversity. The author argues that only a shift from viewing language as grammar to viewing it as action within the social and natural world can make it possible to talk adequately about the relationship of linguistic diversity to biodiversity, of how languages as repositories of cultural memory and guides to action can influence the landscape and its biodiversity. In understanding and celebrating unity in diversity, he concludes, lies our best hope for a sustainable future.

Part Two. Biocultural Diversity Persistence and Loss: Case Studies

The chapters included in Part Two present case studies of indigenous peoples and other local groups (such as migrant settlers) from Africa, Asia, Oceania, and the Americas, focusing on the interactions between linguistic, cultural and biological diversity. Issues of language shift and knowledge loss are examined in light of the role of various factors of culture change and acculturation. Cases of cultural and linguistic diversity persistence are also illustrated, along with the factors and circumstances supporting these. Overall, these studies exemplify the high degree of resolution that is required to identify cultural variation relevant to the study of biocultural diversity correlations, that is, variation reflecting specific local adaptations. Each of these studies also raises issues of general theoretical significance.

The state of linguistic diversity in Native South America and its correlations with biodiversity are assessed by Lizarralde, with the help of maps charting the current locations of South American Indian ethnolinguistic groups, the existing biodiversity reserves and other high biodiversity areas, and their overlap—most native groups living in the remaining areas of high biodiversity of South America, including national parks and biosphere reserves. Lizarralde also provides figures about the dramatic reduction of Native South American population and ethnolinguistic groups since pre-Columbian times. This reduction has resulted from the socioeconomic and cultural pressures brought about by European colonization and from the wholesale loss of indigenous peoples' territories. While this process has recently begun to reverse in some countries, the gains remain at risk. Lizarralde then zooms in on the case of the Barí of Venezuela, whose extensive ecological knowledge he has documented, and identifies and measures a number of factors that are affecting ethnobotanical knowledge, including length of residency in a given territory, change in subsistence practices, bilingual ability, and degree of formal education. Overall, he finds a radical knowledge loss from the older to the younger, more acculturated generation (see Zent this volume). Habitat loss in turn affects these processes by drastically reducing the very ability to have contact with and to learn about the environment (the "extinction of experience"). Lizarralde underscores the crucial need to involve indigenous peoples in biodiversity conservation efforts as a way of also ensuring the persistence of their cultural traditions.

Milton's chapter, on the other hand, points to persisting cultural and ecological diversity among forest-based Amazonian groups in Brazil. The author notes that recent studies have refuted previous representations of the Amazon Basin as a relatively homogeneous and stable environment and of Amazonian Indians as correspondingly homogeneous in their ecological adaptations. Her comparative research on the ecological practices of four small indigenous groups of the Brazilian Amazon (Parakaná, Araweté, Mayoruna, and Matis) who have a relatively short history of contact with the outside world reveals remarkable differences in cultural and ecological knowledge and practices even between neighboring groups pertaining to the same linguistic stock. Milton finds a diversity of staple crops and preys, different uses of forest products for food and medicine, variations in basic technologies, and so forth, as apparent, for example, in the fact that one indigenous society often makes no use of a commonly available plant or animal species that figures prominently in the cultural practices of another. This suggests to her, on the one hand, that the biological richness of the neotropical forest is reflected in the diversity of local adaptations; on the other, that some of the variation (e.g., in the use of foods) may be reinforced by cultural factors, such as marking a group's identity and social boundaries. But Milton also points to evidence that some of the currently observable differences in ecological practices (e.g., in the use of forest plants for medicine) may be recent developments due to outside influences. Therefore, she cautions against automatically assuming that they represent "traditional" (in the sense of "long-standing") reality. Reality appears to be fluid and dynamic—with signs, on the one hand, of breakdown of intergenerational transmission of knowledge about the forest in the group with the shortest history of contact and, on the other, signs of revival of traditional culture in a more acculturated one. The author examines the implications of the observed great diversity of practices in terms of the magnitude of any loss of traditional knowledge in these societies. She also discusses the prospects for more recently contacted groups in Brazil and elsewhere in light of the current, more favorable climate for indigenous peoples.

Balée's study of the Sirionó of Bolivia is concerned with linguistic and cultural variation in a diachronic rather than a synchronic perspective: the perspective of historical ecology, which considers languages and landscapes as mutually constitutive entities formed in the course of historical human-environment relationships analyzable only in terms of their mutual interactions. The author investigates similarities and differences observed in the Sirionó plant nomenclature and its meanings over time, as compared with other languages of the same family (Tupí-Guaraní). He argues that similarities show Sirionó to be a full member of the Tupí-Guaraní family (which was once questioned), while the divergences can only be accounted for in terms of the Sirionó's specific historical interactions with the environment. The Sirionó had originally been characterized as hunter-gatherers. Balée instead provides evidence that the Sirionó's traditional technology is better described as "trekking," an environmental adaptation involving cultivation of some domesticated plants in short-term swiddens along with high mobility. His study illustrates the kinds of cultural historical inferences that can be drawn from the interaction of linguistic and ecological data, as in the use of Sirionó's ethnobiological vocabulary to guide the reconstruction of the history of their movements across the Amazon and of their contacts with other populations. On these bases, Balée suggests hypotheses about prehistoric land management, the emergence of trekking, the persistence of ecologically variable species in Amazonia, the reflexes of these factors on ethnobiological vocabulary, and the relative robustness or vulnerability of different domains of native knowledge about the environment.

Causes and consequences of language endangerment and knowledge loss in Africa are discussed by Batibo, with a special focus on the case of Botswana. Africa is a center of high linguistic diversity, each language also being a store of cultural and ecological knowledge. Yet, Batibo observes, many languages are threatened by the pressures of modernization and the dominance not only of ex-colonial languages such as English and French, but also, and especially, of indigenous lingua francas. The author finds that these languages are imbued with high social prestige deriving from their being seen as tools for access to modern living and much needed information (such as on matters of subsistence and health). Minority language speakers strive to acquire them in the context of what Batibo calls "marked bilingualism," that is, bilingualism in two languages of unequal prestige, ultimately leading to shift to the more prestigious language. These problems are compounded by the tendency of many African states to see minority languages as a threat to national unity. All these phenomena are observable in Botswana, where minority language speakers (including Khoesan speakers, the original inhabitants of Botswana, long adapted to life in the Kalahari desert) strive to learn, and often shift to Setswana, the majority lingua franca, while, ironically, Setswana speakers strive to learn English. Batibo describes the patterns of language shift and death in Botswana and at the same time points to some recent positive developments through a change in attitude on the part of the government toward minority languages, including the use of these languages at the formative level in education. In light of these developments, he discusses the prospects for Botswana's minority languages to expand their range of use and raise their status, reaching a more balanced bilingual situation ("unmarked bilingualism") that may offer minority language speakers the way out of their critical dilemma: either abandon their native languages (and the knowledge and practices that go with them) in order to gain access to the wider society *or* conserve their languages but remain marginalized from national affairs.

The relationship between language obsolescence and the endangerment of traditional sociocultural and ecological knowledge is examined by Florey among the Alune people of the island of Seram, in the Maluku region of eastern Indonesia. Florey describes the sociopolitical, cultural, linguistic, and ecological changes that are bringing about language shift toward the majority language, a regional variety of Malay, and threatening the maintenance of indigenous knowledge of the environment. The author shows how these processes are affecting Alune communities differentially depending on the more or less remote location of villages, by comparing two Alune villages: one that has been protected, until very recently, from such changes by its relative remoteness in a mountain location and another, created by a breakaway group at a site nearer to the coast, that has been subject to more intense pressure and a faster pace of change. Florey finds that shifts in patterns of subsistence economy are both reducing reliance on forest products and degrading the environment, making many of these resources scarce and the related knowledge obsolete. Relocation to ecologically different coastal areas is having an equally detrimental effect on traditional ecological knowledge. Formal education is taking children and youth away from subsistence activities and daily contact with the local environment. Religious conversion to Christianity has resulted in suppression of pre-Christian spiritual practices, including healing, so that ethnomedical knowledge is being lost too. And the use of Malay, the language of Christianity, is being actively promoted as the sign that the Alune have exited the pre-Christian "dark era." On the other hand, Florey points to the Indonesian government's recent opening up of opportunities for the inclusion of language and cultural studies in the school curriculum and to a renewed interest among the Alune in their language and cultural traditions as factors that may eventually lead to a reversal or slowing down of language shift and traditional knowledge loss.

Ultimately, it may be argued that the maintenance and promotion of traditional knowledge and languages, and with them the conservation or restoration of local ecosystems, rest centrally on the initiative of indigenous and local peoples themselves and on continued use and application of their languages and knowledge systems. The next two chapters present insiders' perspectives on such issues, coming from a member of an indigenous group from Oceania and a Native North American. Majnep describes the state of traditional ecological knowledge among the Kalam people of Papua New Guinea. He gives a telling account of his encounter and long-lasting collaboration with the late anthropologist Ralph Bulmer, with whom he co-authored several books on Kalam knowledge and use of local plants and animals. Majnep describes aspects of this knowledge in detail, with special regard to Kalam views of ecological zones, the plants and animals found in them, plant-animal interactions, animal behavior, and hunting techniques. He stresses that hunting is both a subsistence activity and a recreational one—one that men and women also resort to in order to get away from daily worries and to enjoy being in the forest. (For anyone espousing strictly utilitarian views on the foundations of ethnoecological knowledge, Majnep's image of Kalam hunters sitting on a clump of epiphytes high up in a tree, scanning the landscape and delighting in the beauty of the distant woodlands so much as to find it "hard to bring oneself to climb down and return home," is unforgettable evidence that fulfillment of one's subsistence needs is in no way incompatible with aesthetic appreciation of the surrounding environment.) Most older people, the author reports, still display extensive knowledge of the natural world and of traditional resource use, although this is changing rapidly. The younger generations go away to school and lose contact with and interest in the local environment and the related knowledge. Moreover, population is on the rise and people settle closer to the main road, concentrating on planting crops and raising livestock to pay for the cost of formal education, which is bringing about environmental degradation along with the loss of traditional ecological knowledge and practices. Majnep offers his views on action that might be taken to conserve such knowledge among the Kalam and other peoples of Papua New Guinea, by bringing informal and formal education closer together, and by building on what children can informally learn about the natural world.

In his chapter, Molina introduces us to the spiritual truth that has traditionally guided the Yoeme (Yaqui) people of the arid lands on the U.S.-Mexican border, in the states of Arizona and Sonora. This truth teaches the Yoeme to respect all forms of life and to live in the right way on Earth. It is expressed in traditional stories, sayings, and ceremonial songs that talk and teach about the plants and animals of the desert. Molina provides examples of this oral tradition with two Yoeme deer songs about the Huya Ania, the Wilderness World. This rich tradition, the author notes, is still being maintained, but cultural and ecological changes are jointly threatening it. Molina describes processes through which some of the plant and animal species described in the stories and songs and used in the home and in ritual are disappearing or are being replaced by others, as the desert environment undergoes rapid degradation because of increasing settlement of the desert by non-Yoeme, unsound harvesting practices, and conversion of land to different uses. This, he argues, is making it more difficult for the Yoeme to interpret the traditional truth and to maintain the appropriate behavior in relation to the environment. In addition, many younger Yoeme are not learning their native language anymore. For this and other reasons the intergenerational transmission of knowledge is at risk, although several youngsters interviewed by Molina expressed a desire to learn the language as an essential component of being Yoeme, communicating with the elders, and continuing Yoeme traditions. The author proposes various measures that could be taken by Yoeme communities to bridge this growing gap between the older and younger generations, again by integrating formal and informal ways of learning and teaching.

In the final chapter of Part Two, Padoch and Pinedo-Vasquez remind us of one too often neglected fact: it is not only indigenous peoples who may be found to possess complex resource management technologies; other local people, such as migrants, may, too, if they have a long enough history of residence in a given area and of dependence on it for their subsistence (see also Mühlhäusler, Atran this volume). This issue is illustrated by the authors through the case of the Portuguese-speaking *ribeirinhos* (or *caboclos*) of Brazil and Spanish-speaking *ribereños* of the Peruvian Amazon, that is, the rural folk of Amazonia who are neither tribally organized indigenous Amazonians nor recent immigrants. Some are descendants of indigenous peoples, others of both indigenous Amazonians and immigrant Europeans, Africans, or Asians. Many are settled in the floodplains along the banks of the Amazon and its tributaries, which are fertile but changeable, because of the annual floods, and thus require specially adapted production systems. Padoch and Pinedo present several of these technologies—including forms of agriculture, agroforestry, and forestry—underscoring their diversity, complexity, and dynamism. The authors point out that, if such local people's management practices have been overlooked or ignored, it is probably because they are subtle in nature and elude existing classifications of types of resource management. Padoch and Pinedo argue that, in combining indigenous Amazonian with other resource management traditions, *ribeirinho/ribereño* practices maintain and develop some of the indigenous knowledge that is rapidly being lost elsewhere in Amazonia as an increasing pace of integration of indigenous groups threatens the continuity of their traditions and their ability to adapt to change.

Part Three. Perpetuating the World's Biocultural Diversity: Agenda for Action

Pronouncements about the importance of diversity often conclude emphatically on some universalistic note. Yet it is time to go beyond these general (and generic) statements, true as they may be. There is little doubt that we need diversity—cultural, linguistic, biological—for the benefit of humanity. But far too often, as indigenous and local peoples are the first to know, the hailed "benefit of humanity" has actually meant the benefit (and especially the *economic* benefit) of a very small, privileged subset of said humanity, not including that vast majority of humans among whom most of this diversity resides.

Advocates of the joint perpetuation of cultural, linguistic, and biological diversity must be constantly aware of these risks and recognize that today research, applied work, and advocacy must go hand in hand: that theoretical and applied issues are two sides of the same coin, as are scientific and ethical issues, and that there must be a genuine commitment to dealing with the two sides together. This is not to say that basic research is no longer needed. It is to say, however, that it can no longer proceed in a vacuum and that scientists need to educate themselves and others about the nature and implications of what they do and about the local and global contexts of their work. This also implies strengthening the links among the various fields of research and action involved in diversity conservation. Experts from many different specialties need to overcome disciplinary and other institutional and intellectual barriers that exist among themselves as well as in their relations to indigenous and minority peoples. They also need to learn to work together in interdisciplinary and intercultural teams to understand, and devise solutions for, what is perhaps the single most complex issue facing humanity at the dawn of the third millennium. Furthermore, it should become increasingly common for individual specialists to be competent in two or more of the fields involved. All of these developments will require a profound rethinking of approaches to teaching and training at all levels of education, to promote intercommunication among the biological sciences, the social sciences, and the humanities. It will also be essential to broadcast these views to the general public, in order to gain support for research and action in this domain.

In addition, such a shift in perspective requires acknowledging that no effective solutions for biocultural diversity conservation—and for the safety and well-being of hundreds of millions of people on Earth—can ultimately be found without taking into account the rights of the world's indigenous and other traditional peoples: their right to continue to exist as distinct peoples with their own land bases, languages, and cultural traditions, and their right to control their resources and to choose their own forms of development—in sum, their right to self-determination. The concerted efforts of thousands of grassroots and other nongovernmental organizations, as well as concerned individuals, are beginning to make a difference, and recognition of the "inextricable link" between linguistic and biological diversity is beginning to emerge internationally. International bodies, such as the United Nations Educational, Scientific, and Cultural Organization (UNESCO), the UN Centre for Human Rights, the United Nations Environment Programme (UNEP), the World Intellectual Property Organization (WIPO), the Secretariat of the Convention on Biological Diversity (CBD), and others, are turning their attention to issues of indigenous languages and knowledge within the framework of biocultural diversity conservation and the protection of the rights of indigenous peoples.

The chapters in Part Three are devoted to discussion of some of these issues: the protection of indigenous and other traditional peoples' land and resource rights, linguistic human rights, and heritage rights; large-scale efforts to document and support indigenous languages and traditional knowledge; both local and international initiatives aimed at biocultural diversity conservation and development; and the dynamics of use of local knowledge and resources by local peoples themselves as well as by outside parties.

Posey points to the recognition of the link between biological and cultural diversity in the 1992 Convention on Biological Diversity in terms of the role that indigenous peoples and local communities "embodying traditional lifestyles" can play, as holders of traditional ecological knowledge, in the conservation of biodiversity. This role, Posey notes, has long been advocated by indigenous peoples themselves. They argue that they have often sustainably managed natural resources in some of the world's most biodiversity-rich and fragile environments for hundreds, even thousands of years, and continue to do so through a "tradition of invention and innovation"; therefore, biodiversity conservation strategies need to involve them and require respect for and protection of their rights. Posey reviews the evidence for sustainability in indigenous and traditional livelihood systems and the holistic social, cultural, and spiritual values that underlie such systems, embodied in indigenous knowledge and in the concept of indigenous "stewardship" of nature. He stresses that the extent of indigenous management of the environment has often been misunderstood because indigenous technologies subtly mimic natural biodiversity. This has led to labeling as "wildernesses" (with a connotation of *terra nullius*) areas that are in fact "cultural landscapes." The author discusses the key features of the rights (most of them now enshrined in the 1994 UN Draft Declaration on the Rights of Indigenous Peoples) for which indigenous peoples are striving, subject to the principle of self-determination: from territorial and land rights to other forms of community control and empowerment, to cultural, religious, and linguistic rights. In the latter connection, he points out that, because traditional knowledge is crucially dependent on language for cultural transmission, the importance of promoting linguistic diversity along with cultural diversity must be raised in the international fora in which the role of indigenous peoples in biodiversity conservation is being debated and codified.

The chapter by Skutnabb-Kangas deals specifically with linguistic human rights, stressing that such rights are as crucial to self-determination as land rights. The author points out that linguistic rights are especially needed in education, if indigenous and minority language maintenance and development is to be achieved in a world in which more and more children get access to formal education and do more and more of their learning in schools. She argues that discrimination is increasingly being practiced on the basis of language (rather than "race," for example), and indicates that linguistic genocide (as defined by the United Nations in 1948) is still amply practiced in relation to indigenous and minority languages around the world, either actively (by suppressing languages) or passively (by not supporting them). This phenomenon, Skutnabb-Kangas contends, is based on states' ideology of "monolingual reductionism" (requiring the acquisition of majority languages at the expense of, instead of in addition to, minority ones) and on the explicit or implicit claim that unassimilated indigenous or minority groups represent a threat to the state's sovereignty. The author sees these ideological positions reflected in states' educational policies—and even in the main international human rights instruments, which give much poorer treatment to language (and especially language rights in education) than to any other basic human attribute, such as gender, "race," or religion. Skutnabb-Kangas also analyzes the pros and cons of the 1996 Draft Universal Declaration of Linguistic Rights (a nongovernmental initiative), the first attempt at a specific universal formulation of linguistic rights, and reviews other relevant instruments (including the Draft UN Declaration on the Rights of Indigenous Peoples). She then sets forth what she considers to be the necessary individual rights that a universal convention of linguistic human rights should guarantee.

Maffi attempts to bring the protection of indigenous knowledge and languages together within the same frame of reference, in light of international processes that suggest an emerging convergence of different bodies of international law (human rights, labor rights, environmental protection, cultural heritage protection, and intellectual property rights). She suggests that cross-fertilization and hybridization among these different legal frameworks might favor the elaboration of an integrated, holistic sui generis system for the protection of indigenous peoples' heritage, such as indigenous peoples themselves have been advocating. Based on the recognition of the central role of language in encoding, transmitting, and sustaining cultural knowledge, Maffi makes a case for the inclusion of language as a part of indigenous peoples' heritage (along with cultural traditions and natural resources). She argues that the protection of indigenous languages and oral traditions should be considered intimately linked to the protection of indigenous knowledge, including traditional ecological knowledge, which has recently been the specific focus of international instruments. The author critically examines the hypothesis, currently under discussion at the international level, that such integrated protection might come in part through the evolution of the existing system of intellectual property rights (IPR) toward meeting indigenous peoples' needs. Maffi also points out that languages and knowledge should be recognized and protected as dynamic processes and stresses the fundamental oral character of these processes for indigenous peoples, particularly in the informal learning of cultural traditions, including human-environment interactions. On these bases, she advocates the recognition, along with other fundamental indigenous peoples' rights, of a "right to orality," as a key instrument for cultural integrity and continuity.

Another major kind of effort that is being carried out, informed by awareness both of the links among linguistic, cultural, and biological diversity, and of the threats this diversity is undergoing, is the intensive documentation of indigenous languages and cultural knowledge, particularly traditional ecological knowledge. The next two chapters present examples of such efforts. Moore discusses a large-scale project being set up in Brazil for the systematic audio and video documentation of all 180 indigenous languages spoken in that country, which are for the most part poorly if at all described and the viability of whose predominantly small and encroached-upon speech communities is seriously threatened. The author describes the conceptual, methodological, and technical details of the project, as well as the content of the documentation, in order to make this information available for similar projects elsewhere. Moore points out that, in addition to phonological and grammatical data, this project will pay special attention to culturally relevant vocabulary, including ethnobiological vocabulary, and that information will be collected on a vast array of aspects of each speech community's social structure and organization, kinship, cosmology, ecology, and subsistence, as well as the history and prehistory of each group and other elements useful for the reconstruction of linguistic relationships and culture history. Moore stresses that these materials are meant not only for linguists, but for the speech communities themselves, by providing a durable record of their languages and cultural knowledge, which can raise the prestige of their languages and cultures and to which they can resort in their maintenance or revitalization efforts. In addition, nonstandardized audio and video taping will be done at the request of speech communities, on topics of their choice, and the tapes will be made available to the communities, many of which have cassette recorders and VCRs, although usually they lack the equipment and expertise for videotaping.

An international effort to document indigenous knowledge (IK) and its application in sustainable development, in the form of a worldwide and rapidly expanding network of IK resource centers, is described by Warren. The author points to the international processes and instruments that have led to the recognition of the link between cultural and biological diversity and of IK systems as "cultural capital" and as central to biodiversity conservation. He surveys the many aspects of IK that have been documented, with special recognition for their diversity and variability across and within communities, as well as for their nature as systems of experimentation and innovation in the management of natural resources and even in the fostering and creation of biodiversity. Warren also describes the dynamics of such systems as a continuous cycle in which knowledge forms the basis for individual and community decision making, leading to experimentation and innovation and hence to new knowledge. These features of IK systems, he notes, are being increasingly recognized, validated, and applied in community development, training, and policy making through the work of the IK resource centers network. These centers provide national and international links between individuals and institutions concerned with IK. They are involved in the documentation of IK and its use in sustainable development through research, workshops, publication of books and magazines, educational programs and teaching modules, and popularization via the media. Warren describes the extensive IK resources made available through the centers and via the internet and the ways in which this network acts as a major repository of information about IK, as well as for the protection of IK and its promotion in both national and international development policies.

But what are the realities of biocultural diversity conservation work on the ground? What are the challenges, the reasons for failure, the conditions for success? The following two chapters tackle these issues from two distinct points of view: the efforts of conservation organizations and those of local communities themselves. Both chapters focus on Mexico and, in part, its southern neighbors—one of the most diverse regions of the globe, both biologically and culturally, which is experiencing the whole range of social, economic, and political tensions that characterize today's global world and which has recently come to the world's attention as the theater of a historic indigenous and rural mobilization movement. Nations's chapter analyzes the relationship between indigenous peoples and conservation in the Maya tropical forest of southern Mexico and northern Guatemala. Nations warns that widespread convictions that "where there are indigenous people, there are tropical forests" (and vice versa) and that "the only forests to survive in the future will be those controlled by indigenous communities" must be qualified. According to the author, the validity of these statements depends first and foremost on whether or not the lands on which the forests stand are indigenous peoples' homelands (see Nietschmann 1992) and lands of which those peoples have secure tenure. In the Maya lowlands, Nations observes, the dramatic rate of deforestation is coming in part—along with the establishment of large-scale logging and ranching concerns—from land-hungry immigrant Maya pushed out of the highlands by loss of their own lands to nonindigenous coffee plantations and by explosive population growth (see Atran this volume). The author points out that these developments are threatening the integrity of the environments and cultures of the lowlands' original Maya occupants and their ability to live sustainably on their lands; they are also often pitting indigenous people against other indigenous people. In this connection, Nations discusses the added complexities and contradictions brought about by rising ethnic awareness with the spread of the Pan-Mayan movement. He then suggests actions that conservationists can take to address these concerns realistically in order to achieve biodiversity conservation and sustainable development: from assisting indigenous peoples in obtaining secure tenure of their homelands, to helping sustain conservation traditions and adapt agricultural systems, to favoring population control, to fostering formal education with native languages and cultural traditions included in it.

Toledo further probes into these issues, suggesting that those rural areas in Mexico and elsewhere in which globalization is having the effect of destroying or severely degrading biological and cultural diversity are those in which individuals and local communities are unable to resist what he sees as the three main mechanisms of globalization: dependency, specialization, and centralization of power. He then goes on to review the experiences of those hundreds of rural indigenous communities in Mexico that have been successful in resisting these trends and to identify the sources of this success. Diversity, self-sufficiency, grassroots democracy, equity, and decentralization of power are seen by the author as the key mechanisms in this context, leading to maintenance of both biological and cultural diversity. Toledo stresses and illustrates the intimate link between these two faces of diversity in Mexico, a country whose biogeographical and ecological heterogeneity has created a rich mosaic of both environments and cultures, and a wealth of linguistically encoded ecological knowledge and values. He then describes the nonspecialized, diversity-based (and diversity-enhancing), multiuse production strategies employed by indigenous peasants. These sustainable forms of resource use, the author argues, can be maintained and developed if, through endogenous mechanisms, a community can establish or reestablish control over the processes affecting it. This implies first of all control over its territory, both legally and ecologically, as well as various forms of cultural, social, economic, and political control. According to Toledo, these are the elements that lead to local empowerment and self-reliance through community consciousness, hence to sustainable community development. All these processes, he contends, in turn suggest the path to a genuine alternative modernity.

The final three chapters in Part Three deal with the relationships between the local and the global from the point of view of specific natural resources developed, used, or managed by indigenous and local peoples, such as traditional medicinals and crop genetic resources, or scenic and biologically rich landscapes. Each case raises issues of conservation of biological and cultural diversity, as well as of resource rights and the just sharing of benefits from the use of local resources, when these are accessed by outside parties such as international corporations interested in large-scale drug and food production or ecotourism ventures. Traditional botanical medicine, Carlson notes, is receiving ever growing attention in the Western world as a possible source of new drugs, and indeed the effectiveness both of the traditional pharmacopoeias and of drug discovery strategies based on ethnomedical knowledge are being increasingly demonstrated. But, he argues, the maintenance and continued development of local medical knowledge systems and the languages, cultures, and environments that support them should be viewed first and foremost in terms of the provision of primary health care for rural peoples. The author reports World Health Organization estimates according to which 80 percent of the world's population use plants as their primary form of medicine. He observes that, while biomedical treatment is not readily available to or affordable for rural communities, especially in the tropics, traditional botanical medicine offers these communities an accessible and low-cost or free, as well as effective and more culturally appropriate, form of medical care. Medicinal plant materials are usually available locally, and the knowledge of how to use these plants as medicine in the treatment of specific diseases is present within the community and culture. Expensive and more invasive biomedical care can be left to those cases that traditional medical systems are not well equipped to treat. Carlson also points out that traditional medical knowledge has proven to be an invaluable resource in the health care of refugee and displaced populations. Because much of this knowledge is passed on orally from one generation to the next, he argues, it is essential to conserve the local cultures and languages along with the biological systems and species so that the knowledge of how to use these plant species is maintained. This requires working with local communities and host conservation organizations to support projects that will conserve the biocultural diversity in these regions and helping the host country's scientists and communities maintain and optimally use safe, effective, and inexpensive traditional botanical medicines.

The growth and impact of ecotourism in areas of the world that are characterized by both high biodiversity and high cultural and linguistic diversity are the topic of Blount's chapter. Tourism, he observes, is one of the world's largest industries, and in developing countries it can generate up to 50 percent of the GNP. Especially in such countries (many of them tropical), ecotourism—which Blount defines as travel to natural areas with the purpose of understanding the culture and natural history of the place, not altering the local ecosystem, and generating revenues that can aid the conservation of natural resources for the benefit of local people—has become a major focus of the tourism industry. This has led to the "greening" of ecotourism, that is, its extensive commercial exploitation, generating a host of economic, environmental, and sociocultural problems. Blount reviews some of these problems, such as who really benefits from the significant revenues generated by ecotourism; how the economic structure of a country is modified by the large influx of tourists wealthier than the majority of the population; how to deal with the environmental degradation caused by such influx (waste disposal, soil erosion, disturbance of flora and fauna, etc.); and how it impacts local people—by and large indigenous—both in socioeconomic terms and in terms of their sense of cultural identity and cultural integrity. In the latter connection, the author discusses findings that he considers to be mixed at best. He then examines solutions that have been proposed and attempted and stresses that a key factor is empowerment of local people to be fully and equally involved in decision-making processes concerning ecotourism plans that will affect their livelihoods.

Finally, Brush analyzes the conflicting agendas of indigenous peoples, outside private and public users of biological resources, and states over access, control, compensation, and conservation of crop genetic resources and the related knowledge developed and maintained by small-scale farmers practicing traditional agriculture. He notes that one solution that is currently being promoted, as a way of curbing the uncontrolled outside exploitation of traditional knowledge and resources (commonly labeled "biopiracy") and of achieving equitable compensation of the holders of such knowledge and resources, consists in bioprospecting agreements (contracts regulating the search for commercially viable genetic sequences and natural compounds). But Brush argues that such agreements are based on an idea of ownership and commodification of knowledge and crop genetic resources that involves the individualistic appropriation of cultural property pertaining to the public domain. The author contends that this implies a misconception of the nature and manner of creation and management of genetic resources by indigenous farmers, which occurs in essentially open cultural and biological systems, and that such an approach will backfire by tearing apart the cultural fabric that supports the maintenance and continued creation of crop genetic resources. Instead, if the goals of biological and cultural diversity conservation and equity for indigenous people providing genetic resources are to be achieved, Brush proposes that the conservation and compensation debate be shifted toward "biocooperation." He observes that "in the villages and farms where crop genetic resources thrive 'biocooperation' is the common ethic" and that this ethic already reaches beyond the local level in cooperative research between scientists and farmers seeking to improve crop varieties for the public good. Brush advocates extending biocooperation between the stewards of genetic resources and external users in ways that will keep knowledge and resources in the public domain while giving just recognition to their traditional holders.

Part Four. A Vision for the Future, and a Plea

But can we achieve genuine "biocooperation" on a global scale? Or global *biocultural* cooperation, for that matter? In light of the current globalization trends that are having such a dramatic impact on the world's biological, cultural, and linguistic diversity, is this utopia? Many fear that without a radical reconsideration of prevailing economic and social models efforts to conserve and foster this diversity for the benefit of both local communities and humanity at large may be doomed to failure. Norgaard's chapter suggests that the answers to these questions will depend to a large extent on how this latest phase in the Western story of "progress"—based on the assumption of "one right way of knowing" and on an overarching materialism, and leading toward global homogenization via economic globalization—will play out. Norgaard notes that this course has been driven by the use of fossil fuel and nuclear energy in the case of industry and by the replacement of ecological processes with chemical inputs in the case of agriculture. Both trends are now in serious environmental trouble. Furthermore, he points out, the ideas themselves of "progress," "development," and "growth" are being questioned, while the notion of sustainability is taking center stage. In addition, the processes of cultural homogenization are being resisted throughout the world through alternative processes of "reculturalization," with people choosing to maintain and even enhance their cultural identities in the midst of economic globalization. Grassroots efforts are being aided by the parallel process of "NGOization," and nongovernmental organizations are responding to new needs and taking up roles that existing institutions cannot fill. As the monopoly of Western science on knowing is being questioned, the author observes, scholars from different disciplines, as well as scholars and advocates, are coming together to learn from one another and consider "possibilities after progress." Norgaard sees these emerging trends as providing the impulse for countering the loss of cultural, linguistic, and biological diversity and envisions a new economic model, based on a coevolutionary social and ecological framework, as the context in which humanity at the beginning of the millennium could strive to achieve sustainability on Earth.

Reaching this goal requires that concerned people join forces, both locally and globally. Indigenous and local peoples, who see their very survival as distinct peoples at stake, have been at the forefront of resistance against cultural homogenization and forced incorporation into the global economy. Much of what is involved is finding or regaining their own native voices—which had been silenced by other, dominant voices—and making them heard in the arenas in which the future is being shaped. In her chapter, Manriquez, a Tongva-Ajachmem Native Californian, indicates that in many cases this implies first of all reclaiming their own languages and cultural traditions by means of community initiatives aimed at bringing indigenous world views back into play through language. Like many other indigenous groups the world over, Native Californians are engaging in integrated biocultural diversity conservation efforts. The linguistic and cultural revival activities in which they are involved go hand in hand with advocacy for environmental restoration of their lands and the renewed use of native plants for traditional handicrafts and other purposes. Manriquez also points out that, while indigenous peoples are committed to these efforts with or without external help, they do recognize the potential and actual benefits of collaborating with academics and other outside professionals. But she stresses that they often encounter a major obstacle in the apparent inability or reluctance of many such professionals, even those with the best intentions, to take the time just to listen to what native peoples have to say—to what their views, needs, and aspirations actually are. This final chapter voices a plea for Western ears to open up to the kind of intent listening to indigenous and other local peoples that alone can bring about genuine mutual understanding and true collaboration in facing the common threats to the world's linguistic, cultural, and biological diversity.

CODA

Both anthropology and biology have made use of evolutionary theory, if controversially, to propose accounts for the emergence of human culture (Wilson 1975, 1978; Cavalli-Sforza and Feldman 1981; Lumsden and Wilson 1981; Boyd and Richerson 1985; Durham 1990, 1991a, b) as well as to seek correlations of genetic trees of human populations with genetic trees of human languages, assuming language diversification as the consequence of human groups drifting apart in space and time (Cavalli-Sforza et al. 1988; Cavalli-Sforza 1991; Cavalli-Sforza, Menozzi, and Piazza 1994). Widespread use has also been made of biological metaphors in linguistics (Hoenigswald and Wiener 1987).

When human diversity itself becomes the focus of study rather than being an assumed condition, however, novel questions arise, for example, about prehistoric human population movements and interactions in given bioregions at great historic depths (Nichols 1992) or about the functions of diversity in the relationships of human groups among themselves and with the environment (Mühlhäusler 1996). It also becomes possible to ask, as the present book does, whether cultural and linguistic diversity and diversification may share substantive (not merely metaphorical) characteristics with biological diversity and diversification—characteristics that ultimately are those of all life on Earth. At issue here is the adaptive nature of variation in humans, as well as in other species, and the role of language and culture as providers of diversity in humans. Human culture is a powerful adaptation tool, and language at one and the same time enables and conveys much cultural behavior and both expresses and evokes certain abits of thought. In this light, as Bernard (1992:82) puts it, "Linguistic diversity . . . is at least the correlate of (though not the cause of) diversity of adaptational ideas." Models of variation and selective retention, as applied to cultural evolution and language change, come to the fore (see Campbell 1960, 1965, 1974; also Barnett 1953; Stross 1975; Handwerker 1989; Barkow, Cosmides, and Tooby 1992).[10]

It is true that diversity characterizes languages (and cultures) not just with respect to one another, but also internally, with patterns of variation by geographical location, age grade, gender, social status, and a host of other variables. This internal variation combines with the variation resulting from historical contact among human populations in propelling language and culture change and all manner of innovation. But as diverse languages and cultural traditions are increasingly overwhelmed by more dominant ones, and progressive homogenization ensues, one of the two main motors of change and innovation—the observation of cross-linguistic and cross-cultural difference—breaks down, or is seriously damaged. The end result is a global loss of diversity.

So far, questions about the consequences of loss of linguistic and cultural diversity have been raised mostly in terms of ethics and social justice and of maintaining the human heritage from the past (see Thieberger 1990). Rightly so. At the same time, when we consider the interrelationships between linguistic, cultural, and biological diversity, we may also begin to ask these questions as questions about the future, about the continued viability of humans on Earth: "any reduction of language diversity diminishes the adaptational strength of our species because it lowers the pool of knowledge from which we can draw" (Bernard 1992:82; see also Fishman 1982; Diamond 1993). From this perspective, issues of linguistic and cultural diversity conservation may be formulated in the same terms as for biodiversity conservation: as a matter of "keeping options alive" (Reid and Miller 1993) and of preventing "monocultures of the mind" (Shiva 1993). Mühlhäusler (1995) argues that convergence toward majority cultural models increases the likelihood that more and more people will encounter the same "cultural blind spots"—undetected instances in which the prevailing cultural model fails to provide adequate solutions to societal problems. Instead, "it is by pooling the resources of many understandings that more reliable knowledge can arise," and "access to these perspectives is best gained through a diversity of languages" (ibid.:160). Or, simply stated, "Ecology shows that a variety of forms is a prerequisite for biological survival. Monocultures are vulnerable and easily destroyed. Plurality in human ecology functions in the same way" (Pattanayak 1988:380).

Biological evolution is a process of learning—the cell learns, the genes encode the learning of the species. Cultural evolution is also a process of learning and of memory encoding, largely occurring through language. Now economic globalization processes are being touted as the ultimate, inevitable step in human evolution. If so, one should expect them to enhance human memory correspondingly. But quite to the contrary, they are crucially based on the effacing, the annihilation of memory: biological memory, by wiping out species and environments; cultural memory, by wiping out, either physically or through assimilation, whole distinct human groups, with their diverse stores of knowledge, beliefs, and practices and the languages in which the latter are encoded and by which they are transmitted; and even individual memory, as everything we know is at constant risk of being washed out by the rising tide of homogenization by which the forces of economic globalization are fostering shorter and shorter memory spans and more and more mindless living.

Shall we, then, continue to "sabotage ourselves," as Diane Ackerman points out in the epigraph to this chapter, or rather move to do something to prevent this outcome, since it is in our power to do so? There is already enough memory loss in the world. One of the most essential tasks we have as individuals and groups standing up against this tide is one of memory: to keep remembering who we are, where we come from, and where we want to go; not to let ourselves forget the wealth of diverse local and communal ways of living and knowing and communicating that humans throughout the world still have—or did have within the confines of our living memories; and to let that remembrance, enriched by what we have learned inbetween, guide our path toward the future.

The chapters in this book go a long way in telling us why diversity matters. Together—in a way that no individual scholar could accomplish—the authors begin to assemble the pieces of the giant puzzle that is the diversity of life in its various manifestations and complex interrelationships. Overall, the book offers both a common foundation and practical avenues for dealing with some of the most urgent problems facing science and policy at the beginning of the twenty-first century: helping preserve biodiversity and maintain and develop the human wealth represented by indigenous and minority languages and cultures and the knowledge they embody—for the benefit of local linguistic and cultural communities, humanity at large, and the world's ecosystems.

There is a lesson to draw. The adage goes that diversity is the spice of life. Perhaps we are beginning to learn that it is both spice and food—the food of survival.

NOTES

1. Estimates of the number of distinct species extant on the planet vary from 2 million to as many as 50 million, based on different calculations of how much of the world's biodiversity still remains to be discovered. A recent consensus document (Heywood 1995) provides a working figure of over 13 million species. Estimated rates of species extinction also vary widely, with consensus (Heywood 1995) currently standing at about 1000 to 10,000 times the average expected background extinction rates (rates of species extinction prior to humans' introduction of agriculture and rapid population growth). Some biologists predict that between one-third and two-thirds of all plant and animal species might be lost during the second half of the twenty-first century, a loss equaling those that occurred during the previous five mass extinction events registered during the history of the Earth (Raven 1999).

2. This introduction draws in part on Maffi (1998) and Maffi's sections in Maffi, Skutnabb-Kangas, and Andrianarivo (1999).

3. *Ethnologue,* the most comprehensive catalogue of the world's languages (Grimes 1996) reports 6703 languages (mostly oral), of which 32 percent are found in Asia, 30 percent in Africa, 19 percent in the Pacific, 15 percent in the Americas, 3 percent in Europe. This figure does not include most sign languages, which may be almost as numerous as oral languages (Supalla 1993). In recent work, the diversity of human languages has been used as the best available indicator of human cultural diversity (e.g., Clay 1993; Durning 1993). Harmon (1996a) argues that, while no individual aspect of human life can be taken as a diagnostic indicator of cultural distinctiveness, languages are carriers of many cultural differences, indeed "the building blocks of cultural diversity, arguably the fundamental 'raw material' of human thought and creativity" (Harmon 1996a:95). Furthermore, Harmon (1998:4) suggests that using language as a proxy "allow[s] a comprehensible division of the world's peoples into constituent groups." A proxy is, admittedly, an imperfect tool. Linguistic diversity as a proxy for cultural diversity works better on the global scale than it may work in any specific local or regional instances. In many cases, distinctiveness of languages does not correspond to distinctiveness of cultures or sameness of language to sameness of culture. What matters in this context, however, is the possibility of identifying general trends, rather than the ability to satisfactorily account for every single case. It should also be noted that the definition of what a "language" is (as distinct from a "dialect" or a "family of related languages") is in itself far from universally agreed upon. Different criteria lead to different results, so calculations vary among sources. Again, provided the criteria are made clear, one needs to acknowledge the degree of imprecision and approximation that all concepts and definitions used in scientific work carry with them and accept that conclusions drawn on such bases are always provisional and represent one's best educated guesses. (The same considerations apply for species; cf. Harmon 1996a, b.) *Ethnologue* tends to favor social over linguistic boundaries when distinguishing languages from dialects. In a way, this may be closer to how language communities tend to think of their tongues rather than to how linguists do: linguists use structural criteria, while for a linguistic community social, cultural, and pragmatic elements are in focus—what they speak is "their lan-

guage," no matter how linguists would classify it. On that criterion alone, of course, the number of different languages would be enormously larger.

4. The term "indigenous" is used here as a shorthand for "indigenous and tribal," according to the definition in Article 1 of the International Labour Organization's Convention 169 on Indigenous and Tribal Peoples in Independent Countries (ILO 169), which states that the Convention applies to:

> (a) Tribal peoples in independent countries whose social, cultural and economic conditions distinguish them from other sections of the national community, and whose status is regulated wholly or partially by their own customs or traditions or by special laws or regulations;
> (b) Peoples in independent countries who are regarded as indigenous on account of their descent from the populations which inhabited the country, or a geographical region to which the country belongs, at the time of conquest or colonisation or the establishment of present State boundaries and who, irrespective of their legal status, retain some or all of their own social, economic, cultural and political institutions.

Article 1 of ILO 169 also states: "Self-identification as indigenous or tribal shall be regarded as a fundamental criterion for determining the groups to which the provisions of this Convention apply." These criteria are followed in various other international instruments and by many indigenous and tribal peoples themselves. (See also Toledo 2000 for other useful criteria.) The expression "local peoples," also used in this chapter, refers to "local communities embodying traditional lifestyles" (as per Article 8j of the Convention on Biological Diversity).

5. This is a 1995 world population estimate, as used in Harmon 1995. Gray (1999) calculates that there are at least 300 million people at present who are indigenous (see Posey this volume). This does not imply a disparity with Harmon's figures, but rather that several indigenous groups have larger populations than the cut-off figure of 10,000 chosen by Harmon as representative of a small linguistic community.

6. As expressed in the Declaration of Belém of the International Society of Ethnobiology (ISE), elaborated in 1988 at the First International Congress of Ethnobiology in Belém, Brazil. Available on the ISE web site at http://guallart.dac.uga.edu/ISE.

7. As now enshrined in the 1998 International Society of Ethnobiology's Code of Ethics (approved at the Sixth International Congress of Ethnobiology in Whakatane, Aotearoa/New Zealand, November 1998), which states: "Culture and language are intrinsically connected to land and territory, and cultural and linguistic diversity are inextricably linked to biological diversity." See Appendix 1 for the full text.

8. Additional information about this conference can be found on the web site of the University of California at Berkeley's Herbaria, hosts of the conference, at http://ucjeps.berkeley.edu/Endangered_Lang_Conf/Endangered_Lang.html. Information about Terralingua is available at http://www.terralingua.org.

9. Most contributors to this volume were participants in the Berkeley conference. Some others were invited to write specifically for the book, to cover relevant topics that had not been touched upon at the conference. All authors received guidelines for the production of their chapters, based on what was learned at the conference, and their drafts were jointly discussed and revised with the editor. Conference participants who did not contribute to the book included Alejandro de Avila, Ignacio Chapela,

Leanne Hinton, Dominique Irvine, Willett Kempton, Gary Martin, Johanna Nichols, and Mark Poffenberger.

10. It is worth noting with Cosmides, Tooby, and Barkow (1992:5) that in an evolutionary perspective "cultural variability is not a challenge to claims of universality [of human nature], but rather data that can give one insight into the structure of the . . . mechanisms that helped generate it." Trivial as it may sound, the point is that recognizing that human cultures, and languages, universally share certain features—be it of structure, content, or function—is by no means incompatible with recognizing that they also differ in significant ways, and vice versa. This point seems to be largely overlooked in the heat of the relativist vs. universalist debates in anthropology and linguistics, whether because it sounds *too* trivial or because it gets obfuscated. The issue is an empirical one: it is not a matter of scoring the highest points for one view over the other; it is a matter of empirically understanding *how* cultures and languages are similar and *how* they are different, and *why*. To draw a significant parallel (see Harmon 1996a, b, this volume), when biologists talk about species, recognizing their diversity in aspects of structure, function, or behavior in no way prevents them from recognizing the ways in which they also share fundamental aspects of structure, function, and behavior, and vice versa. Biologists are aware that, while the differences may reflect different evolutionary paths, adaptations to different ecological niches, and the like, the similarities more often than not reflect common evolutionary history ("more often than not" because natural history also presents plenty of cases of analogy, or independent emergence of similar traits, that are not the fruit of common evolutionary history; therefore one cannot simply assume that similarity has only one interpretation; see Mishler this volume). It seems that by and large the same might be said of cultures and languages. The human capacity for culture and language emerged as one of the outcomes of human evolution, and although language and culture have taken on lives of their own (Mishler this volume), they have not become radically divorced from their evolutionary roots as tools for human adaptation. A "language and culture in environment" perspective (including both the natural and the social environments, as well as the environments represented by co-existing and interacting languages and cultures themselves), along with a notion of language and culture as adaptive tools developed by humans, may provide us with enough instruments to understand, and account for, local differences as well as global similarities and to overcome age-old Whorfian relativism vs. universalism debates (see Hunn this volume).

REFERENCES

Ackerman, D. 1997 [1995]. *The Rarest of the Rare.* New York: Vintage Books.

Alcorn, J.B. 1984. *Huastec Mayan Ethnobotany.* Austin: University of Texas Press.

Alcorn, J.B. 1996. Is biodiversity conserved by indigenous peoples? In *Ethnobiology in Human Welfare,* ed. S.K. Jain. Pp. 234–238. New Delhi: Deep Publications.

Anderson, E.N. 1996. *Ecologies of the Heart: Emotion, Belief, and the Environment.* New York and Oxford: Oxford University Press.

Atran, S. 1990. *Cognitive Foundations of Natural History.* Cambridge: Cambridge University Press.

Atran, S. 1993. Itza Maya tropical agro-forestry. *Current Anthropology* 34:633–700.

Bahn, P., and J.R. Flenley. 1992. *Easter Island, Earth Island.* London: Thames & Hudson.

Balée, W. 1994. *Footprints of the Forest: Ka'apor Ethnobotany—The Historical Ecology of Plant Utilization by an Amazonian People.* New York: Columbia University Press.

Barkow, J.H., L. Cosmides, and J. Tooby, eds. 1992. *The Adapted Mind: Evolutionary Psychology and the Generation of Culture.* New York: Oxford University Press.

Barnett, H. 1953. *Innovation: The Basis of Cultural Change.* New York: McGraw-Hill.

Berkes, F. 1999. *Sacred Ecology: Traditional Ecological Knowledge and Resource Management Systems.* Philadelphia: Taylor & Francis.

Berlin, B. 1992. *Ethnobiological Classification: Principles of Categorization of Plants and Animals in Traditional Societies.* Princeton, N.J.: Princeton University Press.

Berlin, B., et al. 1990. *La Herbolaria Médica Tzeltal-Tzotzil en los Altos de Chiapas.* Series Nuestros Pueblos. Tuxtla Gutiérrez, Chiapas: Gobierno del Estado de Chiapas-CEFIDIC, DIF-Chiapas, and Instituto Chiapaneco de Cultura.

Berlin, B., D.E. Breedlove, and P.H. Raven. 1974. *Principles of Tzeltal Plant Classification: An Introduction to the Botanical Ethnography of a Mayan-Speaking Community in Highland Chiapas.* New York: Academic Press.

Berlin, E.A., and B. Berlin. 1996. *Medical Ethnobiology of the Highland Maya of Chiapas, Mexico: The Gastrointestinal Diseases.* Princeton, N.J.: Princeton University Press.

Bernard, R. 1992. Preserving language diversity. *Human Organization* 51(1):82–89.

Blackburn, T.C., and K. Anderson, eds. 1993. *Before the Wilderness: Environmental Management by Native Californians.* Menlo Park, Calif.: Ballena Press.

Blount, B., and T. Gragson, eds. 1999. *Ethnoecology: Knowledge, Resources, and Rights.* Athens: University of Georgia Press.

Bobaljik, J.D., R. Pensalfini, and L. Storto, eds. 1996. *Papers on Language Endangerment and the Maintenance of Linguistic Diversity.* MIT Working Papers in Linguistics vol. 28. Cambridge, Mass.: MIT, Department of Linguistics.

Bodley, J.H. 1994. *Cultural Anthropology: Tribes, States, and the Global System.* Mountain View, Calif.: Mayfield Publishing Co.

Boyd, R., and P.J. Richerson. 1985. *Culture and the Evolutionary Process.* Chicago: University of Chicago Press.

Brokensha, D., D.M. Warren, and O. Werner, eds. 1980. *Indigenous Knowledge Systems and Development.* Washington, D.C.: University Press of America.

Brown, C. 1984. *Language and Living Things: Uniformities in Folk Classification and Naming.* New Brunswick, N.J.: Rutgers University Press.

Bulmer, R.N.H. 1982. Traditional conservation practices in Papua New Guinea. In *Traditional Conservation in Papua New Guinea: Implications for Today,* ed. L. Morauta, J. Pernetta, and W. Heaney. Pp. 59–77. Bokoro, Papua New Guinea: Institute of Applied Social and Economic Research.

Burger, J. 1987. *Report from the Frontier: The State of the World's Indigenous Peoples.* Atlantic Heights, N.J.: Zed Books.

Campbell, D.T. 1960. Blind variation and selective retention in creative thought as in other knowledge processes. *Psychological Review* 67(6):380–400.

Campbell, D.T. 1965. Variation and selective retention in socio-cultural evolution. In *Social Change in Developing Areas,* ed. H.R. Barringer, G.I. Blanksten, and R.W. Mack. Pp. 19–49. Cambridge, Mass.: Schenkman Publishing Co.

Campbell, D.T. 1974. Evolutionary epistemology. In *The Philosophy of Karl Popper,* ed. P.A. Schilpp. Pp. 413–463. Evanston, Ill.: Open Court Publishing.

Carruthers, D.V. 1996. Indigenous ecology and the politics of linkage in Mexican social movements. *Third World Quarterly* 17(5):1007–1028.

Castilleja, G., et al. 1993. *The Social Challenge of Biodiversity Conservation.* Working Paper no. 1, Global Environment Facility. Washington, D.C.: World Bank.

Cavalli-Sforza, L.L. 1991. Genes, peoples, and languages. *Scientific American* 265(5):104–110.

Cavalli-Sforza, L.L., and M. Feldman. 1981. *Cultural Transmission and Evolution: A Quantitative Approach.* Princeton, N.J.: Princeton University Press.

Cavalli-Sforza, L.L., P. Menozzi, and A. Piazza. 1994. *The History and Geography of Human Genes.* Princeton, N.J.: Princeton University Press.

Cavalli-Sforza, L.L., A. Piazza, P. Menozzi, and J. Mountain. 1988. Reconstruction of human evolution: Bringing together genetic, archeological, and linguistic data. *Proceedings of the National Academy of Sciences* 85(16):6002–6006.

Chapin, M. 1992. The co-existence of indigenous peoples and environments in Central America. *Research and Exploration* 8(2), inset map.

Clay, J.W. 1993. Looking back to go forward: Predicting and preventing human rights violations. In *State of the Peoples: A Global Human Rights Report on Societies in Danger,* ed. M.S. Miller. Pp. 64–71. Boston: Beacon Press.

Colchester, M. 1994. *Salvaging Nature: Indigenous Peoples, Protected Areas, and Biodiversity Conservation.* Geneva: United Nations Research Institute for Social Development.

Cosmides, L., J. Tooby, and J.H. Barkow. 1992. Introduction: Evolutionary psychology and conceptual integration. In *The Adapted Mind: Evolutionary Psychology and the Generation of Culture,* ed. J.H. Barkow, L. Cosmides, and J. Tooby. Pp. 3–15. New York: Oxford University Press.

Dasmann, R.F. 1991. The importance of cultural and biological diversity. In *Biodiversity: Culture, Conservation, and Ecodevelopment,* ed. M.L. Oldfield and J.B. Alcorn. Pp. 7–15. Boulder, Colo.: Westview Press.

Denevan, W.M. 1992. The pristine myth: The landscape of the Americas in 1492. *Annals of the Association of American Geographers* 82(3):369–385.

Diamond, J.M. 1986. The environmentalist myth. *Nature* 344:19–20.

Diamond, J.M. 1987. The worst mistake in the history of the human race. *Discover* May 1987:64–66.

Diamond, J.M. 1991. *The Rise and Fall of the Third Chimpanzee.* New York: Harper & Collins.

Diamond, J.M. 1993. Speaking with a Single Tongue. *Discover* February 1993:78–85.

Dixon, R.M.W. 1997. *The Rise and Fall of Languages.* Cambridge, UK: Cambridge University Press.

Dorian, N., ed. 1989. *Investigating Obsolescence: Studies in Language Contraction and Death.* New York: Cambridge University Press.

Durham, W.H. 1990. Advances in evolutionary culture theory. *Annual Review of Anthropology* 19:187–210.

Durham, W.H. 1991a. *Coevolution: Genes, Culture, and Human Diversity.* Stanford, Calif.: Stanford University Press.

Durham, W.H. 1991b. Applications of culture evolutionary theory. *Annual Review of Anthropology* 21:331–355.

Durning, A.T. 1992. *Guardians of the Land: Indigenous Peoples and the Health of the Earth.* Worldwatch Paper no. 112. Washington, D.C.: Worldwatch Institute.

Durning, A.T. 1993. Supporting indigenous peoples. In *State of the World 1993: A Worldwatch Institute Report on Progress toward a Sustainable Society.* Pp. 80–100. New York: Norton & Co.

Ehrenfeld, D., ed. 1995. *To Preserve Biodiversity: An Overview.* Readings from *Conservation Biology.* Cambridge, Mass.: Society for Conservation Biology and Blackwell Science.

Ellen, R. 1994. Rhetoric, practice, and incentive in the face of changing times: A study in Nuaulu attitudes to conservation and deforestation. In *Environmentalism: The View from Anthropology,* ed. K. Milton. Pp. 127–143. London and New York: Routledge.

Fishman, J.A. 1982. Whorfianism of the third kind: Ethnolinguistic diversity as a worldwide societal asset. *Language in Society* 11:1–14.

Fishman, J.A. 1996. What do you lose when you lose your language? In *Stabilizing Indigenous Languages,* ed. G. Cantoni. Pp. 80–91. Flagstaff, Ariz.: Center for Excellence in Education, Northern Arizona University.

Flannery, T. 1995. *The Future Eaters: An Ecological History of the Australasian Lands and People.* New York: George Braziller.

Glowka, L., F. Burhenne-Guilmin, and H. Synge. 1994. In collaboration with J.A. McNeely and L. Gundling. *A Guide to the Convention on Biological Diversity.* IUCN Environmental Policy and Law Paper no. 30. Gland, Switzerland: IUCN.

Goehring, B. 1993. *Indigenous Peoples of the World: An Introduction to Their Past, Present, and Future.* Saskatoon, Saskatchewan: Purich Publishing.

Gray, A. 1991. *Between the Spice of Life and the Melting Pot: Biodiversity Conservation and Its Impact on Indigenous Peoples.* IWGIA Document no. 70. Copenhagen: IWGIA.

Gray, A. 1999. Indigenous peoples, their environments and territories: Introduction. In *Cultural and Spiritual Values of Biodiversity,* ed. D.A. Posey. Pp. 61–66. London and Nairobi: Intermediate Technology Publications and UNEP.

Grenand, P. 1980. *Introduction à l'Étude de l'Univers Wayãpi: Ethnoécologie des Indiens de Haut-Oyapock (Guyane Française).* Langues et Civilisations à Tradition Orale no. 40. Paris: SELAF.

Grenoble, L.A., and L.J. Whaley, eds. 1998. *Endangered Languages.* Cambridge, UK: Cambridge University Press.

Grimes, B., ed. 1992. *Ethnologue: Languages of the World.* 12th ed. Dallas: Summer Institute of Linguistics.

Grimes, B., ed. 1996. *Ethnologue: Languages of the World.* 13th ed. Dallas: Summer Institute of Linguistics. (Available on the World Wide Web at: http://www.sil.org/ethnologue/.)

Groombridge, B., ed. 1992. *Global Biodiversity: Status of the Earth's Living Resources.* Compiled by the World Conservation Monitoring Centre. London: Chapman & Hall.

Hale, K. 1992. On endangered languages and the safeguarding of diversity. *Language* 68:1–3.

Hale, K., et al. 1992. Endangered languages. *Language* 68(1):1–42.

Hames, R.B. 1991. Wildlife conservation in tribal societies. In *Biodiversity: Culture, Conservation, and Ecodevelopment,* ed. M.L. Oldfield and J.B. Alcorn. Pp. 172–199. Boulder, Colo.: Westview Press.

Hames, R.B., and W.T. Vickers, eds. 1983. *Adaptive Responses of Native Amazonians.* New York: Academic Press.

Handwerker, W.P. 1989. The origins and evolution of culture. *American Anthropologist* 91(2): 313–326.

Harmon, D. 1995. The status of the world's languages as reported in *Ethnologue. Southwest Journal of Linguistics* 14:1–33.

Harmon, D. 1996a. Losing species, losing languages: Connections between biological and linguistic diversity. *Southwest Journal of Linguistics* 15:89–108.

Harmon, D. 1996b. The converging extinction crisis: Defining terms and understanding trends in the loss of biological and cultural diversity. Paper presented at the colloquium Losing Species, Languages, and Stories: Linking Cultural and Environmental Change in the Binational Southwest, Arizona-Sonora Desert Museum, Tucson, Arizona, 1–3 April 1996.

Harmon, D. 1998. Sameness and silence: Language extinctions and the dawning of a biocultural approach to diversity. *Global Biodiversity* 8(3):2–10.

Harris, D.R., and G.C. Hillman, eds. 1989. *Foraging and Farming: The Evolution of Plant Exploitation.* London: Unwin Hyman.

Haugen, E. 1972. *The Ecology of Language.* Selected and introduced by A.S. Dil. Stanford, Calif.: Stanford University Press.

Heywood, V.H., ed. 1995. *Global Biodiversity Assessment.* Cambridge and New York: Cambridge University Press and UNEP.

Hill, J.H. 1995. The loss of structural differentiation in obsolescent languages. Paper presented at the symposium Endangered Languages, AAAS Annual Meeting, Atlanta, Georgia, 18 February 1995.

Hill, J.H. 1997. The meaning of linguistic diversity: Knowable or unknowable? *Anthropology Newsletter* 38(1):9–10.

Hinton, L. 1994. *Flutes of Fire: Essays on California Indian Languages.* Berkeley, Calif.: Heyday Books.

Hoenigswald, H.M., and L.F. Wiener, eds. 1987. *Biological Metaphor and Cladistic Classification: An Interdisciplinary Perspective.* Philadelphia, Pa.: University of Pennsylvania Press.

Hunn, E.S. 1977. *Tzeltal Folk Zoology: The Classification of Discontinuities in Nature.* New York: Academic Press.

Hunn, E.S. 1990. *Nchi'-Wana, The Big River: Mid-Columbia Indians and Their Land.* Seattle: University of Washington Press.

IUCN. 1991. *Caring for the Earth: A Strategy for Sustainable Living.* Cambridge, UK: IUCN Publication Services.

Johnson, A. 1989. How the Machiguenga manage resources: Conservation or exploitation of nature? In *Resource Management in Amazonia: Indigenous and Folk Strategies,* ed. D.A.

Posey and W. Balée. Pp. 213–222. Advances in Economic Botany vol. 7. The Bronx: New York Botanical Garden Press.

Kirch, P.V. 1997. Microcosmic histories: Island perspectives on "global" change. *American Anthropologist* 99(1):30–42.

Kirch, P.V., and T.L. Hunt, eds. 1996. *Historical Ecology in the Pacific Islands: Prehistoric Environmental and Landscape Change*. New Haven, Conn.: Yale University Press.

Krauss, M. 1992. The world's languages in crisis. *Language* 68(1):4–10.

Krauss, M. 1996. Linguistics and biology: Threatened linguistic and biological diversity compared. In *CLS 32, Papers from the Parasession on Theory and Data in Linguistics*. Pp. 69–75. Chicago: Chicago Linguistic Society.

Lewis, P. 1998. Too late to say "extinct" in Ubykh, Eyak, and Ona. *The New York Times* 15 August 1998: A13–A15.

Lumsden, C., and E.O. Wilson. 1981. *Genes, Mind, and Culture*. Cambridge, Mass.: Harvard University Press.

Maffi, L. 1994. A Linguistic Analysis of Tzeltal Maya Ethnosymptomatology. Ph.D. diss., Ann Arbor: University Dissertation Services.

Maffi, L. 1996. Position paper for the working conference Endangered Languages, Endangered Knowledge, Endangered Environments, Berkeley, Calif., 25–27 October 1996. Available electronically at http://www.terralingua.org/paper001.html.

Maffi, L. 1997. Language, knowledge and the environment: Threats to the world's biocultural diversity. *Anthropology Newsletter* 38(2):11.

Maffi, L. 1998. Language: A resource for nature. *Nature and Resources: The UNESCO Journal on the Environment and Natural Resources Research* 34(4):12–21.

Maffi, L. 2001. Linking language and the environment: A co-evolutionary perspective. In *New Directions in Anthropology and Environment: Intersections*, ed. C.L. Crumley. Pp. 24–48. Walnut Creek, Calif.: AltaMira Press.

Maffi, L., and T. Carlson, eds. Forthcoming. *Ethnobotany and Conservation of Biocultural Diversity*. Submitted to Advances in Economic Botany Series. The Bronx: New York Botanical Garden Press.

Maffi, L., G. Oviedo, and P.L. Larsen. 2000. *Indigenous and Traditional Peoples of the World and Ecoregion Conservation: An Integrated Approach to Conserving the World's Biological and Cultural Diversity*. A WWF Report. Gland, Switzerland: WWF International.

Maffi, L., T. Skutnabb-Kangas, and J. Andrianarivo. 1999. Linguistic diversity. In *Cultural and Spiritual Values of Biodiversity*, ed. D.A. Posey. Pp. 21–57. London and Nairobi: Intermediate Technology Publications and UNEP.

Majnep, I.S., and R.N.H. Bulmer. 1977. *Birds of My Kalam Country*. Auckland: Auckland and Oxford University Presses.

Majnep, I.S., and R.N.H. Bulmer. 1990. *Kalam Hunting Traditions*. University of Auckland: Department of Anthropology.

McNeely, J.A., and W.S. Keeton. 1995. The interaction between biological and cultural diversity. In *Cultural Landscapes of Universal Value: Components of a Global Strategy*, ed. B. von Droste, H. Plachter, and M. Rossler. Pp. 25–37. Jena and New York: Fischer Verlag and UNESCO.

McNeely, J.A., et al. 1989. *Conserving the World's Biological Diversity*. Gland, Switzerland and Washington, D.C.: IUCN, WRI, CI, WWF-US, World Bank.

Medin, D.L., and S. Atran, eds. 1999. *Folkbiology*. Cambridge, Mass., and London: MIT Press.

Miller, M.S., ed. 1993. *State of the Peoples: A Global Human Rights Report on Societies in Danger*. Boston: Beacon Press.

Mühlhäusler, P. 1995. The interdependence of linguistic and biological diversity. In *The Politics of Multiculturalism in the Asia/Pacific*, ed. D. Myers. Pp. 154–161. Darwin, Australia: Northern Territory University Press.

Mühlhäusler, P. 1996. *Linguistic Ecology: Language Change and Linguistic Imperialism in the Pacific Rim*. London: Routledge.

Nabhan, G.P. 1989. *Enduring Seeds: Native American Agriculture and Wild Plant Conservation*. San Francisco: North Point Press.

Nabhan, G.P. 1996. Discussion paper for the colloquium Losing Species, Languages, and Stories: Linking Cultural and Environmental Change in the Binational Southwest, Arizona-Sonora Desert Museum, Tucson, Arizona, 1–3 April 1996.

Nabhan, G.P., and S. St. Antoine. 1993. The loss of floral and faunal story: The extinction of experience. In *The Biophilia Hypothesis*, ed. S.R. Kellert and E.O. Wilson. Pp. 229–250. Washington, D.C.: Island Press.

Nation. 1999. Environmental destruction a threat to languages: UN Environment Programme, 7 September 1999. Distributed via Africa News Online by Africa News Service.

National Geographic Magazine. 1999. Special issue on Global Culture, 196(2):2–89, and Millennium supplement map on Languages of the World.

Nettle, D. 1996. Language diversity in West Africa: An ecological approach. *Journal of Anthropological Archaeology* 15:403–438.

Nichols, J. 1992. *Linguistic Diversity in Space and Time*. Chicago: University of Chicago Press.

Nietschmann, B.Q. 1992. *The Interdependence of Biological and Cultural Diversity*. Occasional Paper no. 21, Center for World Indigenous Studies, December 1992.

Norgaard, R.B. 1994. *Development Betrayed: The End of Progress and a Coevolutionary Revisioning of the Future*. London and New York: Routledge.

Oldfield, M.L., and J.B. Alcorn, eds. 1991. *Biodiversity: Culture, Conservation, and Ecodevelopment*. Boulder, Colo.: Westview Press.

Ostrom, E. 1990. *Governing the Commons*. Cambridge: Cambridge University Press.

Pattanayak, D.P. 1988. Monolingual myopia and the petals of the Indian lotus: Do many languages divide or unite a nation? In *Minority Education: From Shame to Struggle*, ed. T. Skutnabb-Kangas and J. Cummins. Pp. 379–389. Clevedon, UK: Multilingual Matters.

Ponting, C. 1991. *A Green History of the World*. London: Sinclair-Stevenson.

Poole, P. 1995. *Indigenous Peoples, Mapping, and Biodiversity Conservation*. BSP People and Forests Program Discussion Paper Series. Washington, D.C.: Biodiversity Support Program.

Posey, D.A., ed. 1999. *Cultural and Spiritual Values of Biodiversity*. London and Nairobi: Intermediate Technology Publications and UNEP.

Posey, D.A., and W. Balée, eds. 1989. *Resource Management in Amazonia: Indigenous and Folk Strategies*. Advances in Economic Botany vol. 7. The Bronx: New York Botanical Garden Press.

Raven, P.H. 1999. Plants in peril: What should we do? Address presented at the Millennium Symposium, Sixteenth International Botanical Congress, St. Louis, Missouri, 1–7 August 1999.

Reaka-Kudla, M.L., D.E. Wilson, and E.O. Wilson, eds. 1997. *Biodiversity II: Understanding and Protecting Our Biological Resources.* Washington, D.C.: Joseph Henry Press.

Redford, K.H. 1991. The ecologically noble savage. *Cultural Survival Quarterly* 13(1):46–48.

Reid, W.V., and K.R. Miller. 1993. *Keeping Options Alive: The Scientific Basis for Conserving Biodiversity.* Washington, D.C.: World Resources Institute.

Robb, J. 1993. A social prehistory of European languages. *Antiquity* 67:747–760.

Robins, R.H., and E.M. Uhlenbeck, eds. 1991. *Endangered Languages.* Oxford: Berg.

Schoonemaker, P.K., B. von Hagen, and E.C. Wolf, eds. 1997. *The Rainforests of Home: Profile of a North American Bioregion.* Covelo, Calif.: Island Press.

Shiva, V. 1993. *Monocultures of the Mind: Perspectives on Biodiversity and Biotechnology.* London and Atlantic Heights, N.J.: Zed Books.

Shiva, V., et al., eds. 1991. *Biodiversity: Social and Ecological Perspectives.* London and Atlantic Heights, N.J.: Zed Books.

Stork, N.E. 1997. Measuring global biodiversity and its decline. In *Biodiversity II: Understanding and Protecting Our Biological Resources,* ed. M.L. Reaka-Kudla, D.E. Wilson, and E.O. Wilson. Pp. 41–68. Washington, D.C.: Joseph Henry Press.

Stross, B. 1975. Variation and natural selection as factors in linguistic and cultural change. In *Linguistics and Anthropology: In Honor of C.F. Voegelin,* ed. M.D. Kinkade, et al. Pp. 607–632. Lisse, The Netherlands: Peter de Ridder.

Supalla, T. 1993. *Report on the Status of Sign Language.* Helsinki: World Federation of the Deaf.

Systematics Agenda 2000. 1994. *Systematics Agenda 2000: Charting the Biosphere.* Technical Report. New York: Systematics Agenda 2000.

Taylor, P.M. 1990. *The Folk Biology of the Tobelo People: A Study in Folk Classification.* Smithsonian Contributions to Anthropology no. 34. Washington, D.C.: Smithsonian Institution Press.

Thieberger, N. 1990. Language maintenance: Why bother? *Multilingua* 9(4):333–358.

Thompson, J.N. 1999. The evolution of species interactions. *Science* 284:2116–2118.

Tindale, N.B. 1974. *Aboriginal Tribes of Australia.* Berkeley: University of California Press.

Toledo, V.M. 1994. Biodiversity and cultural diversity in Mexico. *Different Drummer* 1(3): 16–19.

Toledo, V.M. 1995. *Mexico: Diversity of Cultures.* México, D.F.: CEMEX and Sierra Madre.

Toledo, V.M. 2000. Biodiversity and indigenous peoples. In *Encyclopedia of Biodiversity,* 3:451–463. San Diego: Academic Press.

Warren, D.M., L.J. Slikkerveer, and D. Brokensha, eds. 1995. *The Cultural Dimension of Development: Indigenous Knowledge Systems.* London: Intermediate Technology Publications.

Wester, L., and S. Yongvanit. 1995. Biological diversity and community lore in northeastern Thailand. *Journal of Ethnobiology* 15(1):71–87.

Wilcox, B.A., and K.N. Duin. 1995. Indigenous cultural and biological diversity: Overlapping values of Latin American ecoregions. *Cultural Survival Quarterly* 18(4):49–53.

Wilkins, D. 1993. Linguistic evidence in support of a holistic approach to traditional ecological knowledge. In *Traditional Ecological Knowledge: Wisdom for Sustainable Develop-*

ment, ed. N.M. Williams and G. Baines. Pp. 71–93. Canberra: Centre for Resource and Environmental Studies, Australian National University.

Williams, N.M., and G. Baines. 1993. *Traditional Ecological Knowledge: Wisdom for Sustainable Development.* Canberra: Centre for Resource and Environmental Studies, National Australian University.

Williams, N.M., and E.S. Hunn, eds. 1982. *Resource Managers: North American and Australian Hunter-Gatherers.* AAAS Selected symposium no. 67. Boulder, Colo.: Westview Press.

Wilson, E.O. 1975. *Sociobiology: The New Synthesis.* Cambridge, Mass.: Harvard University Press.

Wilson, E.O. 1978. *On Human Nature.* Cambridge, Mass.: Harvard University Press.

Wilson, E.O., ed. 1988. *Biodiversity.* Washington, D.C.: National Academy Press.

Wilson, E.O. 1989. Threats to biodiversity. *Scientific American* September 1989:108–116.

Wilson, E.O. 1992. *The Diversity of Life.* Cambridge, Mass.: Harvard University Press.

Woodbury, A.C. 1993. A defense of the proposition, "When a language dies, a culture dies." In *SALSA I: Proceedings of the First Annual Symposium about Language and Society,* ed. R. Queen and R. Barrett. Pp. 101–129. Texas Linguistic Forum 33.

WRI, IUCN, and UNEP. 1992. *Global Biodiversity Strategy: Guidelines for Action to Save, Study, and Use Earth's Biotic Wealth Sustainably and Equitably.* Washington, D.C.: World Resources Institute.

Part 1

LANGUAGE, KNOWLEDGE, AND THE ENVIRONMENT

Interdisciplinary Framework

2

ON THE MEANING AND MORAL IMPERATIVE OF DIVERSITY [1]

David Harmon

It is often said that any coherence that once may have existed in the Western scholarly tradition is long gone. Gone too, so the reasoning holds, is any hope for a broad, intelligible view of what is going on in this particular region of thought. The few remaining would-be generalists must skulk through a fragmented, fractious intellectual landscape, picking their way on cat's feet through minefields laid down by increasingly specialized and insular disciplines, moving gingerly so as not to detonate the latest fashionable theory. Erosion, of a kind, is responsible for the dominant feature of this landscape: a chasm between science and the humanities, now grown so wide and deep that it is often given up as unbridgeable. [1] Actually, "given up" is a mild way of putting it. There are plenty of people who positively relish the distance, thankful of any opportunity to dismiss the other side, on guard always against any attempts at bridge building.

This imagined scene is a caricature, but, like all caricatures, the exaggerated details serve to highlight a core truth. Anyone who takes an interdisciplinary view of biological and cultural diversity must deal with the chasm. As a matter of procedure, then, it would seem pretty risky to postulate any kind of continuity between the two great realms of difference we live with. But, apart from intellectual rifts, why should anyone care one way or the other? Because by seeking a holistic understanding of diversity, we gain a more accurate picture of how each of us, as individuals, shares in the collective life of humankind. If there is such a thing as "our common humanity," we would do well to examine what it is made of. It is a question of the first importance, for "we cannot be unaware, in the twentieth century, of the extremes—called 'apartheid' and 'Wnal solution', among other things—to which giving up the ideal of the unity of the human race can lead" (Todorov 1993:88). The paradox is that we can only grasp what is universal by first recognizing what is different. To do the reverse puts us in the blind alley of ethnocentrism, "the unwarranted establishing of the specific values of one's own society as universal values" (Todorov 1993:1). (It is not difficult to see that a similar fallacy has helped create the schism between science and the humanities.) Every person, from the tribal member in New Guinea to the most case-hardened New Yorker, has personally to arrange the natural and cultural facets of existence. Surely, our differing responses to nature and culture—to the variety they embody—must be accounted for in the search for commonalities.

Nonetheless, it has to be admitted that an intellectual program focused on the relationship between biological and cultural diversity will always strike some as impossibly diffuse and therefore pointless. Alfred North Whitehead once gave his fellow philosophers some good advice that we can employ in response to that criticism. When evaluating the thought of any given period, he counseled, look for what is left unsaid: "There will be some fundamental assumptions which adherents of all the variant systems within the epoch unconsciously presuppose. Such assumptions appear so obvious that people do not know what they are assuming because no other way of putting things has ever occurred to them" (Whitehead 1925:71). Diversity—the fact that, conceptually, perceptually, ontologically, things are different—is the ultimate unconscious presupposition. We do not often think about the meaning of diversity because it is the only "way of putting things" human beings have. To understand this better, we can turn to Whitehead's intellectual forebear, William James.

WHAT JAMES KNEW

With characteristic vigor, James's work in psychology and philosophy established that real-world diversity and the structure of the mind exist in a state of reciprocity. In contrast to absolutist and idealist philosophies, which try to subsume individual existences into one seamless, metaphysically perfect totality, James stood for the messy, the vague, the disjunct, the unfinished—for the grit and tumble of what life is really like. His work validates our daily experience of variety in the world. "It is surely a merit in a philosophy," he wrote, "to make the very life we lead seem real and earnest. Pluralism, in exorcising the absolute, exorcises the great de-realizer of the only life we are at home in, and thus redeems the nature of reality from essential foreignness. Every end, reason, motive, object of desire, ground of sorrow or joy that we feel is in the world of finite multifariousness, for only in that world does anything really happen, only there do events come to pass" (James 1909:49–50). Diversity, James believed, is not some projection of human consciousness, but rather the very means through which consciousness operates. Of the many traits said to be definitive of humanity, none is more basic than our innate capacity for discerning and classifying difference—and thus resolving portions of it into sameness. James returned to this point again and again, embellishing it, recasting it, burnishing it like a master sculptor.

We are, first of all, born to diversity: "Consciousness, from our natal day, is of a teeming multiplicity of objects and relations, and what we call simple sensations are results of discriminative attention, pushed often to a high degree" (James 1890, 1:224). From then on until the day we die the mind is forever choosing to attend to one thing or another, but "few of us are aware how incessantly" this faculty is at work because it is the very substance, and therefore not often also the object, of consciousness.[2] "Accentuation and Emphasis are present in every perception we have. We find it quite impossible to disperse our attention impartially over a number of impressions." In fact, the only way we can deal with great tracts of the variety thrust at us by the world is by ignoring most of the phenomenal terrain (James 1890, 1:284). Luckily, the course of evolution happens to have furnished us with five senses yoked to our consciousness, senses that give us the consummate ability to divest the world of its primordial confusions. This is fundamental:

> To begin at the bottom, what are our very senses themselves but organs of selection? Out of the infinite chaos of movements, of which physics teaches us that the outer world consists, each sense-organ picks out those which fall within certain limits of velocity. To these it responds, but ignores the rest as completely as if they did not exist. . . . Out of what itself is an undistinguishable, swarming *continuum*, devoid of distinction or emphasis, our senses make for us, by attending to this motion and ignoring that, a world full of contrasts, of sharp accents, of abrupt changes, of picturesque light and shade. (James 1890, 1:284–285)

Endowed with human nature, we are compelled to chisel identity out of objectively existing diversity. We have no choice; that's the way our consciousness is. James called this "sense of sameness" the "very keel and backbone of our thinking," "the most important of all the features of our mental structure" (James 1890, 1:459, 460).[3] It is what enables us to confirm that a reality beyond our mind truly exists.

> The judgment that *my* thought has the same object as *his* thought is what makes the psychologist call my thought cognitive of an outer reality. The judgment that my own past thought and my own present thought are of the same object is what makes *me* take the object out of either and project it by a sort of triangulation into an independent position, from which it may *appear* to both. *Sameness* in a multiplicity of objective appearances is thus the basis of our belief in realities outside of thought. (James 1890, 1:272)

We can here only skim the surface of this, one of the core ideas in James's thought. Actually, it is a constellation of ideas, aspects of which he would expand upon in such philosophical works as *Pragmatism* (1907), *A Pluralistic Universe* (1909), his valedictory *Some Problems of Philosophy* (1911), and even in his monumental analytic study, *The Varieties of Religious Experience* (1902). So impressed was James by the questions surrounding the meaning of diversity, of the "one and the many," that he declared them "the most central of all philosophic problems" (James 1907:129). On his authority, I believe, we can take the meaning of diversity to be an issue worthy of pursuit.

If James's pluralism is on the mark, then it is literally quite natural for humans to be intrigued by diversity. This is patently so in terms of culture, and the perennial interest field studies hold for both amateur and professional biologists balances the ledger. It should never be forgotten that Darwin's years of lonely toil were sustained by this fascination. Closer to our own time, a foundation work of evolutionary biology's "modern synthesis" of genetics and the Darwinian theory of natural selection (cf. Curtis 1983:892; Wilson 1994:112), Theodosius Dobzhansky's *Genetics and the Origin of Species,* begins with the observation that "for centuries the diversity of living things has been a major interest of mankind" (1941:3). Quite unexpectedly, this landmark of twentieth-century science provides a model of thinking about diversity, both biological and cultural, in an integrated way.

THE "SPECIES PROBLEM" IN NATURE

The germ of Dobzhansky's approach in this book is his recognition that nature is not painted in a continuous wash of variety. Rather, "a more intimate acquaintance with the living world discloses a fact almost as striking as the diversity itself," namely, "the discontinuity of the variation among organisms":

> If we assemble as many individuals living at a given time as we can, we notice at once that the observed variation does not form any kind of continuous distribution. Instead, a multitude of separate, discrete, distributions are found. In other words, the living world is not a single array of individuals in which any two variants are connected by unbroken series of intergrades, but an array of more or less distinctly separate arrays, intermediates between which are absent or at least rare. Each array is a cluster of individuals, usually possessing some common characteristics and gravitating to a definite modal point in their variations. Small clusters are grouped together into larger secondary ones, these into still larger ones, and so on in an hierarchical order. (Dobzhansky 1941:3–4)

The entire Western scientific taxonomy is built on this last fact, and "evidently the hierarchical nature of the observed discontinuity lends itself admirably to this purpose":

For the sake of convenience the discrete clusters are designated races, species, genera, families, and so forth. The classification thus arrived at is to some extent an artificial one, because it remains for the investigator to choose, within limits, which cluster is to be designated a genus, family, or order. But the classification is nevertheless a natural one in so far as it reflects the objectively ascertainable discontinuity of variation, and in so far as the dividing lines between species, genera, and other categories are made to correspond to the gaps between the discrete clusters of living forms. Therefore the biological classification is simultaneously a man-made system of pigeonholes devised for the pragmatic purpose of recording observations in a convenient manner and an acknowledgment of the fact of organic discontinuity. (Dobzhansky 1941:4)

As an example of how this works in practice, Dobzhansky invites us to compare a house cat with a lion. He begins by observing that, as different as individual house cats (and lions) are from each other, no house cat is *so* different that it could be mistaken for a lion, or vice versa. Why? Because "in common as well as in scientific parlance the words 'cat' and 'lion' frequently refer neither to individual animals nor to all the existing individuals of these species, but to certain modal points toward which these species gravitate." While these "modal points are statistical abstractions having no existence apart from the mind of the observer," the species *Felis domestica* and *F. leo* are, nevertheless, decidedly real, existing independently of "any abstract modal points which we may contrive. No matter how great may be the difficulties in finding the modal 'cats' and 'lions', the discreteness of species as naturally existing units is not thereby impaired."[4] There may be troubles in defining what a species is, but they are not due to any artificiality of the species themselves (Dobzhansky 1941:4–5).

There is much more going on here than the mere observation that species can be characterized by the bell-shaped "normal curve" most commonly found in the measurement of individual differences (Anastasi 1958:26–27).[5] Dobzhansky is here expressing a correspondence between our perceptions of what is distinctive—the "abstract modal points"—and what actually *is* distinctive—the "naturally existing units." In so doing he hit upon a fundamental model that explains both how diversity is actually structured and how it is perceived. For any entity to be truly discrete, it must embody some indissoluble set of core characteristics that it does not share exactly with other entities, no matter how closely related they may otherwise be. From an ontological standpoint, this set of core characteristics is definitive. In other words, the set of characteristics—not taken separately and individually, but as a unique group—is what actually makes the entity distinctive, regardless of whether or not human beings perceive it so. Any two or more entities that share the defining set are, in actuality, "the same." Dobzhansky's concern happened to be a group entity—the species—and therefore for an individual organism to partake of that species, it must exhibit the species' core defining set of characteristics, namely, those that isolate it reproductively from otherwise similar individuals.

Now, it is well known that the criterion of separate gene pools upon which this classic biological species concept is based is contradicted at various places throughout nature. The validity of there being true species among bacteria and some other simple organisms is highly questionable, for example (Curtis 1983:372). More generally, Templeton (1989) and Cracraft (1989), among many others, argue forcefully that the biological species concept, while intuitively attractive, does not apply to great numbers of organisms. They make two basic objections: first, that the concept simply ignores the large class of organisms that reproduce asexually or self-mate, and second, that there are so many exceptions to the concept, even among sexually reproducing organisms, that it loses much of its validity. These exceptions involve evidence that there is considerably more gene flow between species than should take place if they were truly reproductively isolated. Even so, the biological species concept (as opposed to such alternatives as the phylogenetic species concept, in which differentiation is signified by historical relationships between whole taxa rather than current blood relationships of individuals) has not been overthrown; in fact, it is "probably the most widely accepted view of the species held by biologists today" (Groombridge 1992:14). The inability of biologists to define species precisely, while at the same time accepting them as "the fundamental unit" (Wilson 1992, chap. 4, esp. 37–38), is something of an institutionalized self-reproach within the profession (see Groombridge 1992:13; Harmon 1996).

The endurance of the biological species concept in the face of these seemingly fatal contradictions can be attributed to Dobzhansky's principal insight: that there is a basically accurate correspondence between what we perceive to be separate groups of fundamentally similar organisms and the actual existence in nature of such separate groups. The exceptions to the rule do not disprove it; they merely illustrate the higher-order difficulty intrinsic to any analysis of diversity versus identity. The difficulty is how to separate discrete from continuous variation at the margins, in the gray area away from the modal points—the points which, perceptually as well as actually, anchor the "sense of sameness" that James wrote of.

THE "SPECIES PROBLEM" IN CULTURE

This model applies equally well to religions, languages, kinship systems, artistic genres, and the other basic elements of human culture as it does to biological species. Thus we find versions of the "species problem" cropping up all over the place. What, for example, defines the Roman Catholic religion? Is it the set of doctrines handed down by the Vatican? If so, what do we make of the fact that some of these are regularly contravened by large numbers of people who consider themselves and, what is even more important, are still considered by the Church hierarchy to be Catholic? When does deviation from doctrine grade into heresy, and from there into total dissolution? These puzzles notwithstanding, no one would deny that Catholicism exists in actuality as a religion. There is evidently a "modal point" of Catholicism, a defining set of characteristics around which the meaning "Catholic" coheres. We perceive it to be a "naturally existing unit" of the world's faiths, and, for all practical purposes, it is.

Working from a premise similar to Dobzhansky's, the scholar of religion John Hick has offered a "family resemblance" analogy to explain the plurality of religious traditions alive today. He likens religious diversity to "a complex continuum of resemblances and differences analogous to those found within a family": while no two members of a family are exactly alike, "nevertheless there are characteristics distributed sporadically and in varying degrees which together distinguish this from a different family" (1989:4). In other words, faiths are individuated because the variation among them is discontinuous, yet they all are recognizably "religions" in that their practices and beliefs are imbued with a sense of deep, permanent, ultimate importance transcending mundane life. Hick calls this the "starting point" for charting the range of religious phenomena, and it performs the same function for comparative religion as the criterion of reproductive isolation does for taxonomy. The unique elaborations of transcendent importance manifested in different religions become their "modal points," as described in the example of Catholicism above (Hick 1989:242–245).[6] As Dobzhansky found within the species concept, Hick sees in religious belief a fundamental correspondence between perceived experience and reality. Such correspondence allows us freely, naturally, even innately to apprehend that there are differences among religions even though we may not be able to pinpoint precisely what they are. And, as it does for species, in the end the concept of religion emerges undefined—and unscathed.

Languages, too, are intuitively considered to be naturally existing units even though linguists have difficulty defining what a language is and how to distinguish among closely related languages and dialects. Leonard Bloomfield long ago pointed out that "we speak of French and Italian, of Swedish and Norwegian, of Polish and Bohemian as separate languages, because these communities are politically separate and use different standard languages, but the differences of local speech-forms at the border are in all these cases relatively slight and no greater than the differences which we find within each of these speech-communities" (1933:54). This is as good a summary of the language-or-dialect problem as one is apt to find. The "oneness" of many languages is geopolitical, not linguistic. For instance, how can there be a single language "Italian" when the Ligurian spoken in the north may well sound like a scramble to a Sicilian (Grimes 1992:465–468)? The answer is obvious: at this point in history there happens to exist a bounded entity named "Italy," and it serves many purposes for there to be a single, officially recognized language to match. Away from the centers of power and the linguistic standards they promote, we soon encounter Bloomfield's anomalies. There are, as the linguist Peter Mühlhäusler has noted, dozens of "controversial tongues" (e.g., Aromanian, Frisian, Gagauz, Lallans, Sater, Tsakonian, and Zyrian, to run through the Roman alphabet) in the rest of Europe alone (1996:258). Language or dialect? The dispute—and it is often highly charged—extends to speech forms around the world. Like species, like religions, here again the problem is how to distinguish what is happening at the margins.

Matters of ethnicity are just as perplexing. The neighboring Nuer and Dinka peoples (Naath and Jieng, respectively, in their self-appellations), pastoralists of East Africa, are widely recognized as distinct groups, but the closeness of their languages (divided, though, into greatly differing dialects), the absence of any fixed boundaries between their territories, and their propensity to intermarry all make it very hard to pin down any set of distinguishing marks (see discussion in Bodley 1994:95–96). Nonetheless, the perception of difference persists among outside observers, and, decisively, the Dinka and Nuer see themselves as being distinct. Evidently each group recognizes within itself enough shared, identifying characteristics—enough to establish "ethnic modal points"—to keep the two groups separate, no matter how jumbled the boundary between them is.

Finally, there is *culture* itself, a term just as problematic and resistant to easy definition as is *species*. Once again, there is a conceptual similarity between the two. Dobzhansky observed that variation in the natural world is discontinuous, and built from there. The social anthropologist Fredrik Barth, in his study of the social organization of culture difference, lays the same foundation: "Practically all anthropological reasoning rests on the premise that cultural variation is discontinuous: that there are aggregates of people who essentially share a common culture, and interconnected differences that distinguish each such discrete culture from all others" (1969:9). Later, Barth again echoes Dobzhansky when he observes that "the ethnic label subsumes a number of simultaneous characteristics which no doubt cluster statistically, but which are not absolutely interdependent and connected. Thus there will be variations between members, some showing many and some showing few characteristics. Particularly where people change their identity, this creates ambiguity since ethnic membership is at once a question of source of origin as well as of current identity." But this fluidity does not vitiate the basic model:

> What is then left of the boundary maintenance and the categorical dichotomy, when the actual distinctions are blurred in this way? Rather than despair at the failure of typological schematism, one can legitimately note that people *do* employ ethnic labels. . . . What is surprising is not the existence of some actors that fall between these categories, and of some regions of the world where whole [peoples] do not tend to sort themselves out in this way, but the fact that variations tend to cluster at all. We can then be concerned not to perfect a typology, but to discover the processes that bring about such clustering. (Barth 1969:29)

So it is not necessary to go so far as to subscribe to Herder's monadic conception of the *Volk,* each endowed with its own "group mind," to believe nonetheless that cultures are objectively existing units, albeit complicated ones. This notion is expressed in some of the most prominent English-language definitions, from Tylor, who spoke of culture as a "complex whole," to Sapir's "assemblage" and Geertz's "pattern of meaning" (cited in Fleischacker 1994:127–128). Again, it is evident that some defining set of characteristics is at the center of these complex unities.

DIVERSITY IN PERIL

These examples begin to show that in both nature and culture variation is discontinuous, with diversity described by modal points of identity,[7] and suggest that there is an inherent affinity (if not necessarily a continuity) between biological and cultural diversity. Whether or not one accepts this conclusion on theoretical grounds alone, however, it is hard to ignore the similarities between the practical forces driving biological extinctions and cultural homogenization (Wurm 1991:2–3; Harmon 1996). And it is in this practical realm that a genuine sense of crisis has taken hold, quite independent of theoretical considerations. The feeling of crisis is driven by the conviction that we soon will reach a momentous threshold, a point of no return beyond which a critical amount of biological and cultural diversity will have been lost, never to be regenerated on any time scale significant to the development of humankind.

Without such a long-range, encompassing view, touting a diversity crisis becomes merely polemical. So we must consider where the world stands today in relation to the course of life over the ages. Though proportionally most species that have ever existed are now extinct (to the nearest order of magnitude, so runs the joke, *all* species are extinct), the momentum of evolution to the present day has been toward a fecundity of life forms. Taxonomy is still a viable career choice because we have millions of species, not thousands or hundreds. Likewise, the variety of human expression and organization still extant is, by any standard, little short of astonishing. There are, depending on how one counts them, something on the order of 6000 oral languages still spoken as mother tongues, over 4000 "distinct cultures known to anthropology" (as cited in Durham 1990:194), perhaps a similar number of religious denominations, an untold number of artistic forms, kinship systems, and the like—at least tens of thousands of cultural differentiae in all. The biological and cultural diversity now existing is the preeminent fact of life, a first-order condition of Earth's being. Nothing less than this is now at stake, according to proponents of diversity. It is the essential nature of diversity, combined with a looming threshold of irreversible diminishment, that justifies the label *crisis.*

We see this in the quick rise to prominence of the term *biodiversity,* a neologism dating from the 1980s.[8] Etymologically, it is nothing more than a contraction of *biological diversity,* a term that had itself been preceded by such variants as "natural," "biotic," and "organic" diversity. But *biodiversity* is no mere synonym of these: it carries an explicit, unprecedented sense of urgency, of impending catastrophe. The evidence for the extinction crisis was accumulated through years of spade-work by scientists studying individual species and natural communities, and was synthesized (beginning, roughly speaking, in the late 1970s), by Paul Ehrlich, Thomas Lovejoy, Norman Myers, Peter Raven, and Edward O. Wilson, to name just a few of the best known. It is now known that the world has experienced five previous "extinction spasms" over the course of millions of years (Raup and Sepkoski 1982:1502; Wilson 1992:188–192). But the rebounds from these epochal events all took place in a biosphere not dominated by human activity, and never before have we faced the prospect of massive extinctions at the hands of humans. Thus the term *biodiversity* corresponds to a new reality.

Moreover, the prospect is unfolding against a new backdrop: a rapidly emerging global culture fueled by the telecommunications revolution, the near-collapse of Communism and the unbridled diffusion of market-based ideologies, the unprecedented reach and political influence of transnational corporations, the primacy accorded Western scientific and technical knowledge, the availability of worldwide travel, and on and on. Globalization affects biodiversity in many ways: by easing the spread of exotic species, making the logging of remote rain forests profitable, undercutting proven indigenous land-management systems. Its effects on culture are also well known: global broadcasting of Western (especially North American) popular music drives out or hybridizes traditional folk music; English displaces indigenous languages; processed foods brought in on container ships replace traditional, locally produced fare in daily diets; proselytizing missionary work imposes monotheism on native spiritual beliefs. In some instances the change is innovative, to be sure. But because of its unprecedented scale and pace much of it is destructive to less politically powerful forms of cultural expression and organization.

Cultures need a base of variety from which to work if they are to generate new differences and thereby avoid stagnation (see the discussion of the evolutionary implications of linguistic diversity in Maffi, Skutnabb-Kangas, and Andrianarivo 1999). The diversity must be genuine, that is, it cannot rest on narrow pecuniary considerations. As the political scientist Benjamin Barber points out, today's global marketing strategy "depends on a systematic rejection of any genuine consumer autonomy or any costly program variety—deftly coupled, however, with the appearance of infinite variety."

> Selling depends on fixed tastes (tastes fixed by sellers) and focused desires (desires focused by merchandisers). Cola companies . . . can no more afford to encourage

the drinking of tea in Indonesia than Fox Television can encourage people to spend evenings at the library reading books they borrow rather than buy; and Paramount, even though it owns Simon & Schuster, cannot really afford to have people read books at all unless they are reading novelizations of Paramount movies. By the same logic, for all its plastic cathedrals, Disneyland cannot afford to encourage teenagers to spend weekends in a synagogue or church or mosque praying for the strength to lead a less materialistic, theme-park-avoiding, film-free life. Variety means at best someone else's product or someone else's profit, but cannot be permitted to become no product at all and thus no profit for anyone. (Barber 1995:116)

We are headed for a future where, in Barber's memorable phrase, "velocity is becoming an identity all its own" (1995:194), a future where genuine cultural diversity is truncated. Overlaid upon this, however, and confusing the issue, will be the sham diversity and illusory freedom of choice promoted by the burgeoning global consumer culture. Genuine cultural diversity is a counterbalance to all this, necessary "for the constant rehumanization of humanity in the face of materialism" (Fishman 1982:6).

So the diversity crisis is genuine. Yet, based on current knowledge, it cannot be said to issue from the likelihood of a descent into what might be called "terminal scarcity." No one knows how many species exist, but a figure of 5–15 million is considered reasonable, and most recent estimates of near-term extinction prospects put the impending global species loss at 1–10 percent per decade (Stork 1997:62–65). If we extend any combination of these per-decade projections over the next hundred years, we are still left with millions of species intact. Now let us turn to what I consider to be the most accessible indicator of global cultural diversity: the status of the world's languages. Here, a similar conclusion awaits those who speculate on the possible extent of mother-tongue extinctions over the next century. Some estimates run very high, from 20–25 percent to, in worst-case scenarios, 90 percent (Krauss 1992:7).[9] Yet no matter how one counts languages (as opposed to dialects), even a 90 percent reduction would still leave several hundred intact (Harmon 1995). Using these numbers, no one can say that we face the prospects of an absolutely depauperate planet or a true global monoculture.

But it would be a grave mistake to take comfort from this or use it as an excuse for inaction. We have no idea how many species we can lose before essential ecosystem services (purification of water, mitigation of floods and droughts, pollination of plants, and many others discussed in Daily 1997) and overall environmental quality are seriously compromised. As Stephen Kellert rightly observes, "removing 10 percent or even 1 percent of the planet's species" would be akin to "randomly destroying pieces of an extremely complex mechanism while blindly hoping not to damage some vital element or process" (1996:31). Nor do we know how much cultural distinction can be lost before we begin slipping toward a de facto totalitarianism. Whitehead considered the "Gospel of Uniformity" almost as dangerous a threat to social progress as the "might makes right" doctrine of brute force:

> The differences between the nations and races of mankind are required to preserve the conditions under which higher development is possible. . . . A diversification among human communities is essential for the provision of the incentive and material for the Odyssey of the human spirit. Other nations of different habits are not enemies: they are godsends. Men require of their neighbors something sufficiently akin to be understood, something sufficiently different to provoke attention, and something great enough to command admiration. (Whitehead 1925:297, 298)

BIOCULTURAL DIVERSITY AND THE FLOW OF LIFE

We have so far explored part of the theoretical groundwork underlying diversity and considered some of the practical implications of a biocultural extinction crisis. At this point a fundamental question arises: Does the existence of diversity itself, and the fact that it is facing a crisis, carry any ethical implications? Is there, in short, a moral imperative to preserve diversity?

Many people would say the answer is simple: humans have an innate need for diversity and therefore we must preserve it. In light of James's conception of consciousness, this well-intentioned reasoning does not go nearly far enough. Diversity is not just one need on a par with dozens of others. If it is the means through which our consciousness function operates, and if consciousness is what makes us human, then diversity makes us human. When we act in ways that reduce diversity, whether in the nonhuman world or in our own cultures, we corrode our essential humanity. This quality of humanity is achieved through the continual mental activity of evaluating differences and resolving them (albeit often provisionally) into individualities. This largely unconscious process of comparison, which we all carry out countless times each day, is, upon analysis, almost inexpressibly sophisticated. Whatever it means to "be human," this process cannot be omitted.[10] To the extent that its richness is impaired—as would happen if biological and cultural diversity were sharply curtailed—that much are we prevented from enacting the full potentiality of being human. James knew that diversity is the field against which we identify things and events as being "the same" and that, without our archetypal notion of sameness, deprived of the "psychological sense of identity" to deal with objective diversity, "the most important of all the features of our mental structure," the human mind would be fundamentally changed (James 1890, 1:460). When the field of diversity shrinks, the potential for distilling sameness from difference is impoverished commensurably. If, at some point, the impoverishment becomes acute enough, then, as I interpret it, our species will have passed a threshold. We will have become something other than human. By this view, then, diversity ought to be preserved because it provides the grounds for the continuance of *Homo sapiens*.

But even this far-reaching rationale does not fully establish the value of continuing the biologically diverse systems that support all life on Earth. For the contemporary environmental philosopher Holmes Rolston III, any ethical treatment of species (including ours) has as its prerequisite the acceptance that they "be objectively there as living processes in the evolutionary ecosystem," for "if species do not exist except embedded in a theory in the minds of classifiers, it is hard to see how there can be duties to save them. No one proposes duties to genera, families, orders, or phyla; everyone concedes that these do not exist in nature" (1985:721). Rolston takes the varied criteria for species (descent, reproductive isolation, morphology, gene pool) as evidence of their objective existence.

> At this point, we can anticipate how there can be duties to species. What humans ought to respect are dynamic life forms preserved in historical lines, vital informational processes that persist genetically over millions of years, overleaping short-lived individuals. It is not *form* (species) as mere morphology, but the *formative* (speciating) process that humans ought to preserve, although the process cannot be preserved without its products. Neither should humans want to protect the labels they use, but the living process in the environment. (Rolston 1985:722)

Though species in general, unlike *Homo sapiens* in particular, are not moral agents, Rolston defends their biological identity, declaring that "the dignity resides in the dynamic form; the individual inherits this, instantiates it, and passes it on" (1985:722).

This is not to say that the evolutionary process itself is teleological or invested with morality. Evolution by natural selection predates *Homo sapiens* by millions of years and will likely continue beyond the time our species has disappeared, barring a planetary cataclysm. The process churns on, distilling new forms of life, heedless of human wishes, unguided by purpose on or beyond the Earth. An explanation of its significance will fit on a bumper sticker: Speciation Happens. The production of new species is a warrant that nature really exists, that "The Big Outside," in the clever phrase of the environmental activists Dave Foreman and Howie Wolke, really is at once singular, large, and "out there" apart from subjective human perceptions.[11]

Nonetheless, as Rolston demonstrates, we can derive morality from the fact of evolution. I extend his reasoning to biological and cultural diversity as a whole. The moral imperative for preserving diversity is not just the continuance of humankind but the safekeeping of the biocultural evolutionary process that produced us and every other species and brought us all to where we are—together.[12]

Now we can finally glimpse the crisis at its deepest level: it is founded on the realization that ours is the first generation with the capacity to reverse the age-old momentum toward fecundity in both nature and culture. "Extinction shuts down the generative process," warns Rolston. "The wrong that humans are doing, or allowing to happen through carelessness, is stopping the historical flow in which the vitality of life is laid" (1985:723). That flow and vitality are surely both biological and cultural at the same time.

The human species did not evolve in a world of drab monotony. Our brains, the consciousness function they produce, and the cultural variety expressing that function have evolved over millions of years within a lavish, enveloping environment of biological riches. And, conversely, the natural environment has been widely transformed by differentiated human action. Our evolutionary history tells us that cultural diversity is intimately related to the biological diversity of the nonhuman world. Current events tell us they face the same threats. The only effective way to meet them is with a cohesive, biocultural response. Through it we would find, at last, that unity does not require uniformity. Once this insight is truly grasped, we will finally be ready to start making real progress toward that elusive ideal of a common humanity.

NOTES

1. C.P. Snow's celebrated lecture on the "two cultures" did much to bring the split before the public, but, as he said (1959:19), the rift was evident, in England at least, by the 1930s, and its roots certainly reach far deeper. Actually, Snow's main concern in the lecture was not to contrast science with the humanities in general but to compare the attitude of scientists with that of the literary set in particular. The image of two opposing cultures was a conscious simplification on Snow's part (1959:9–10).

2. In reference to James's thought, one can only speak metaphorically of the "substance" of consciousness, because he considered consciousness to be a function, not an entity. See "Does Consciousness Exist?" in James 1912.

3. Seconded by the eminent contemporary philosopher W.V. Quine in his essay "Natural Kinds" (1969:116): "There is nothing more basic to thought and language than our sense of similarity; our sorting of things into kinds."

4. Later on, Dobzhansky characterizes the modal points in more technical terms, building on ideas put forth by his fellow geneticist Sewall Wright. To summarize: If we consider the total number of potential gene combinations for a particular organism (which is the "'field' within which evolutionary changes can be enacted"), it is apparent that the adaptive values for some will be greater than for others. Probably a vast majority of the combinations are "discordant and unfit for survival." Closely similar gene combinations will tend to be similar in adaptive value. "If, then, the field of the possible gene combinations is graded with respect to adaptive value, we may find numerous 'adaptive peaks' separated by 'valleys.' The 'peaks' are the groups of related gene combinations that make their carriers fit for survival in a given environment; the 'valleys' are the more or less unfavorable combinations. Each living species or

race may be thought of as occupying one of the available peaks in the field of gene combinations" (Dobzhansky 1941:337). The adaptive peaks, then, are associated with the species' modal points.

5. The work of the nineteenth-century statistician Adolphe Quetelet provided the foundation for subsequent advances in differential psychology as well as for the pivotal discovery of the analysis of variance by Dobzhansky's fellow pioneer of the "modern synthesis," the geneticist R.A. Fisher (Anastasi 1958:6–8, 28; see also Eiseley 1958:227).

6. The family resemblance analogy derives from Wittgenstein 1953, sec. 67; the original is *Familienähnlichkeiten*, which, as Baker and Hacker note, was a term used by Nietzsche (1980:326 n. 7). The concept has since been widely used and adapted, both by professional philosophers within the huge body of Wittgenstein exegesis and in related fields, as evidenced by Hick's usage. With the family resemblance analogy as a platform, Hick builds an elaborate case for the "veridical" nature of religious experience, namely, that faith as it is thought of and experienced verifies the existence of an objectively existing 'Real', which, depending on the tradition, takes a personal, theistic form (Islam's Allah, Christianity's Holy Trinity, the God of Israel, etc.) or is conceived of as an impersonal, nontheistic Absolute, such as Nirvana or the Brahman. Impressive though his argument is, I have a number of serious reservations about how Hick develops it, stemming from my personal views on the utility of religion as well as because of his undue emphasis on "post-axial" religious forms (i.e., the large "world religions") that promise personal salvation and liberation, leading him to give short shrift to indigenous ("pre-axial") spiritual beliefs that tend to emphasize harmony, Earth stewardship, and the like. But the important point here is that he has plausibly characterized religious diversity.

7. We have been discussing this in the context of James's philosophy, Dobzhansky's model of organic diversity, and Hick's "pluralistic hypothesis" of religion, but similar insights are found further afield. There is Whitehead's dictum on quantum mechanics: "For when we penetrate to these final entities, this startling discontinuity of spatial existence discloses itself. . . . Accordingly, in asking where the primordial element is, we must settle on its average position at the centre of each period" (1925:53, 54). And Tzvetan Todorov, discussing the *First and Second Discourses,* says that "Rousseau repeatedly advocates the paradoxical enterprise of discovering [universal] properties by way of difference" (1993:11).

8. The landmark event for the term *biodiversity* came in September 1986 with the National Forum on BioDiversity (so spelled), a high-profile conference that drew not only top scientists but reporters from leading news outlets (see Tangley 1986). Heywood (1995:5) reports that *biodiversity* was coined by Walter G. Rosen at the first planning meeting for the conference. Papers from the forum were published in Wilson 1988. It is no coincidence that the field of conservation biology, a self-described "crisis discipline" focused on the study and preservation of biodiversity, emerged during this same period.

9. The worst-case numbers may seem hysterically high, but at least one archaeologist has warned that as many as 90 percent of the world's archaeological sites may be ir-

revocably damaged or destroyed within the next 50–60 years (Harrington 1991, cited in Wylie 1996:162–163). The recognition of an impending crisis has been current (at least among professional archaeologists in America) since the mid-1970s (Wylie 1996:162).

10. Whether or not one *values* the process is another story. According to Hick, Zen Buddhism teaches that the human propensity for "continually distinguishing, comparing and evaluating" obscures the true nature of reality by falsely placing the individual consciousness at the center of existence. The process thus becomes a "distorting screen" through which the world is viewed, and only by "ending or suspending this self-centred discriminative activity" can we finally experience the world as it truly is (Hick 1989:288–289). In contrast, I think the continual discriminative process is something to be understood and affirmed, not overcome and negated, but in any event I construe the Zen viewpoint as confirming the *humanness* of the process itself.

11. Cf. the discussions of deconstruction in Soulé and Lease 1995, passim, and of subjectivism in Whitehead 1925:130.

12. The ethical undercurrent throughout this discussion is that we should value what has evolved over millennia and without conscious human direction, rather than what will likely emerge over decades with a utilitarian end in view. That is why no proponent of biodiversity is assuaged by the prospect of biotechnology being able to create new life forms or clone existing ones, nor of the growing capacity of zoos, aquariums, and botanical gardens to preserve genetic material or individual specimens ex situ for possible cloning or reintroduction into the wild. Similarly, no number of programmatically invented languages (Esperanto, Volapük, Klingon, etc.) will suffice for lost "natural" languages.

REFERENCES

Anastasi, A. 1958. *Differential Psychology: Individual and Group Differences in Behavior.* 3d ed. New York: Macmillan.

Baker, G.P., and P.M.S. Hacker 1980. *Wittgenstein: Understanding and Meaning.* (An analytical commentary on the *Philosophical Investigations,* vol. 1.) Chicago: University of Chicago Press.

Barber, B.R. 1995. *Jihad vs. McWorld.* New York: Times Books.

Barth, F. 1969. Introduction. In *Ethnic Groups and Boundaries: The Social Organization of Culture Difference,* ed. F. Barth. Pp. 9–38. Boston: Little, Brown & Co.

Bloomfield, L. 1933. *Language.* New York: Henry Holt.

Bodley, J.H. 1994. *Cultural Anthropology: Tribes, States, and the Global System.* Mountain View, Calif.: Mayfield.

Cracraft, J. 1989. Speciation and its ontology: The empirical consequences of alternative species concepts for understanding patterns and processes of differentiation. In *Speciation and Its Consequences,* ed. D. Otte and J.A. Endler. Pp. 28–59. Sunderland, Mass.: Sinauer.

Curtis, H. 1983. *Biology.* 4th ed. New York: Worth.

Daily, G.C., ed. 1997. *Nature's Services: Societal Dependence on Natural Ecosystems.* Washington, D.C., and Covelo, Calif.: Island Press.

Dobzhansky, T. 1941. *Genetics and the Origin of Species.* 2d ed. Columbia Biological Series no. 11. New York: Columbia University Press.

Durham, W.H. 1990. Advances in evolutionary culture theory. *Annual Review of Anthropology* 19:187–210.

Eiseley, L. 1958. *Darwin's Century: Evolution and the Men Who Discovered It.* Garden City, N.Y.: Doubleday.

Fishman, J.A. 1982. Whorfianism of the third kind: Ethnolinguistic diversity as a worldwide societal asset. *Language in Society* 11:1–14.

Fleischacker, S. 1994. *The Ethics of Culture.* Ithaca, N.Y., and London: Cornell University Press.

Grimes, B.F., ed. 1992. *Ethnologue: Languages of the World.* 12th ed. Dallas: Summer Institute of Linguistics.

Groombridge, B., ed. 1992. *Global Biodiversity: Status of the Earth's Living Resources.* Compiled by the World Conservation Monitoring Centre. London: Chapman & Hall.

Harmon, D. 1995. The status of the world's languages as reported in Ethnologue. *Southwest Journal of Linguistics* 14:1–28.

Harmon, D. 1996. Losing species, losing languages: Connections between biological and linguistic diversity. *Southwest Journal of Linguistics* 15:89–108.

Harrington, S.P.M. 1991. The looting of Arkansas. *Archaeology* 44(3):22–31.

Heywood, V.H., ed. 1995. *Global Biodiversity Assessment.* Cambridge, U.K.: Cambridge University Press.

Hick, J. 1989. *An Interpretation of Religion: Human Responses to the Transcendent.* Basingstoke, Hampshire: Macmillan.

James, W. 1890. *The Principles of Psychology.* 2 vols. New York: Henry Holt & Co. Facsimile reprint, 1950. New York: Dover.

James, W. 1907. *Pragmatism: A New Name for Some Old Ways of Thinking.* New York: Longmans, Green & Co.

James, W. 1909. *A Pluralistic Universe: Hibbert Lectures at Manchester College on the Present Situation in Philosophy.* New York: Longmans, Green & Co. Facsimile reprint, 1996. Lincoln and London: University of Nebraska Press.

James, W. 1912. *Essays in Radical Empiricism.* New York: Longmans, Green & Co. Facsimile reprint, 1996. Lincoln and London: University of Nebraska Press.

Kellert, S. 1996. *The Value of Life: Biological Diversity and Human Society.* Washington, D.C.: Island Press.

Krauss, M. 1992. The world's languages in crisis. *Language* 68(1):4–10.

Maffi, L., T. Skutnabb-Kangas, and J. Andrianarivo 1999. Linguistic diversity. In *Cultural and Spiritual Values of Biodiversity,* ed. D.A. Posey. Pp. 21–57. London and Nairobi: Intermediate Technology Publications and UNEP.

Mühlhäusler, P. 1996. Review of *Linguistic Human Rights: Overcoming Linguistic Discrimination,* ed. T. Skutnabb-Kangas and R. Phillipson. *Lingua* 99:257–260.

Quine, W.V. 1969. *Ontological Relativity and Other Essays.* New York: Columbia University Press.

Raup, D.M., and J.J. Sepkoski, Jr. 1982. Mass extinctions in the marine fossil record. *Science* 215:1501–1503.

Rolston, H., III. 1985. Duties to endangered species. *BioScience* 35(11):718–726.

Snow, C.P. 1959. *The Two Cultures and the Scientific Revolution: The Rede Lecture, 1959*. New York: Cambridge University Press.

Soulé, M.E., and G. Lease, eds. 1995. *Reinventing Nature? Responses to Postmodern Deconstruction*. Washington, D.C.: Island Press.

Stork, N.E. 1997. Measuring global biodiversity and its decline. In *Biodiversity II: Understanding and Protecting Our Biological Resources*, ed. M.L. Reaka-Kudla, D.E. Wilson, and E.O. Wilson. Pp. 41–68. Washington, D.C.: Joseph Henry Press.

Tangley, L. 1986. Biological diversity goes public. *BioScience* 36(11):708–715.

Templeton, A.R. 1989. The meaning of species and speciation: A genetic perspective. In *Speciation and Its Consequences*, ed. D. Otte and J.A. Endler. Pp. 3–27. Sunderland, Mass.: Sinauer.

Todorov, T. 1993. *On Human Diversity: Nationalism, Racism, and Exoticism in French Thought*. Catherine Porter, trans. Cambridge, Mass.: Harvard University Press.

Whitehead, A.N. 1925. *Science and the Modern World: Lowell Lectures, 1925*. New York: Macmillan.

Wilson, E.O. 1992. *The Diversity of Life*. Cambridge, Mass.: Belknap Press of Harvard University Press.

Wilson, E.O. 1994. *Naturalist*. Washington, D.C.: Island Press.

Wilson, E.O., ed. 1988. *Biodiversity*. Washington, D.C.: National Academy Press.

Wittgenstein, L. 1953. *Philosophical Investigations*. G.E.M. Anscombe, trans. New York: Macmillan.

Wurm, S. 1991. Language death and disappearance: Causes and consequences. In *Endangered Languages*, ed. R.H. Robins and E.Uhlenbeck. Pp. 1–18. Oxford and New York: Berg.

Wylie, A. 1996. Ethical dilemmas in archaeological practice: Looting, repatriation, stewardship, and the (trans)formation of disciplinary identity. *Perspectives on Science* 4(2):154–194.

3

BIODIVERSITY AND THE LOSS OF LINEAGES

Brent D. Mishler

Pressures of development are causing an alarming increase in the rate of extinction in biodiversity. This is widely recognized as a problem by the general public, but its full impact has not yet been realized. Add to this problem the interrelated (but currently less lamented) problem of extinction in cultural and linguistic diversity, and we clearly have a crisis on our hands. This is a crisis both of fact (lineages are disappearing at a high rate) and of perception (neither the magnitude of diversity nor its peril is widely enough known).

In this chapter, I will concentrate primarily on biodiversity, its definition, origin, and perils, to provide a framework for comparisons that come elsewhere in this book. However, I will touch briefly on similar scientific and ethical issues in cultural and linguistic diversity. There are four main questions I will address: (1) What are the basic units of biodiversity? (2) What is the value of biodiversity? (3) What should be conserved in the face of political and economic constraints? (4) What are the valid parallels between biological diversity and cultural-linguistic diversity?

THE BASIC UNITS OF BIODIVERSITY

Given the reasonable estimate that systematists have only discovered and named perhaps 10 percent of the species on earth, and the fact that only a tiny fraction of those have been studied in any detail, there is much scientific work to be done in a short time. Many species will go extinct before we even know them; a double tragedy.

We all have a sense of something slipping away, but what exactly? What exactly is being lost? We need to solve this problem before we can possibly begin to set conservation priorities. Fortunately, biologists have been making considerable progress in their understanding of the nature of biological classification. Newly developed theories and methods for gathering data and analyzing phylogenetic relationships (i.e., the genealogy of species) position us on the threshold of a deep understanding of the history of the biological world, which will in turn allow a better-focused approach to biological conservation.

Systematics

The debate over biological classification has a long history (Hull 1988; Stevens 1994). It is generally agreed that there should be one consistent, general-purpose reference system, for which the Linnaean hierarchy should be reserved. But many issues have been at stake in deciding what should be used for the general-purpose system, foremost of which is the nature of taxa. Are they just convenient groupings of organisms with similar features, or are they lineages, marked by homologies?

It is clear in general that the most effective and natural classification systems are those that "capture" entities resulting from processes generating the things being classified. Evolution is the single most powerful and general process underlying biological diversity. The major outcome of the evolutionary process is the production of an ever-branching phylogenetic tree, through descent with modification along the branches. This results in life being organized as a hierarchy of nested lineages. For these reasons, a general, if not completely universal, consensus has been reached that phylogeny is the best criterion for the general-purpose classification (Hennig 1966; Nelson 1973; Wiley 1981; Farris 1983; Sober 1988). This is justified both theoretically (the tree of life is the single universal outcome of the evolutionary process) and practically (phylogenetic relationship is the best criterion for summarizing known data about attributes of organisms and predicting unknown attributes).

The prevailing paradigm in biological classification has become known as Hennigian phylogenetic systematics, based on what has been called the *Hennig Principle* (Hennig 1965, 1966). Hennig noted that in a system evolving via descent with modification and splitting of lineages, characters that changed state along a particular lineage can serve to indicate the prior existence of that lineage, even after further splitting occurs. *Homology* is defined as a feature shared by two organisms because of descent from a common ancestor that had that feature (Roth 1991). A *transformation* is a heritable change in a homology along a lineage from a prior state (termed a *plesiomorphy*) to a posterior state (termed an *apomorphy*). Homologous similarities among organisms come in two basic kinds: *synapomorphies,* due to immediate shared ancestry (i.e., a common ancestor at a specific phylogenetic level), and *symplesiomorphies,* due to more distant ancestry. Only the former are useful for reconstructing the relative order of branching events in phylogeny.

The phylogenetic tree is reconstructed using the preponderance of the postulated synapomorphies (called a parsimony analysis). A feature that appeared to be homologous before the analysis, but that is incongruent with the majority of the characters following the analysis is called a *homoplasy.* Such characters can be due to convergence, reversal, or simple random matching of characteristics in unrelated lineages (a problem that is exacerbated when some branches in the analysis are much longer than others and when characters have only a few character states).

Divergence refers to the splitting of one lineage into two lineages. *Reticulation* is the blending of two lineages into one lineage. The parsimony analysis assumes that divergence is occurring; when it is not occurring, however, the existence of reticulation can be postulated by the occurrence of certain patterns of homoplasy (McDade 1992). Thus reticulation, which is not uncommon in many plant groups, can still be detected by means of phylogenetic analysis.

Figure 3.1 illustrates the main terms and concepts discussed above. The main branching diagram shows the phylogenetic relationships (relative recency of common ancestry) of six terminal taxa at the tips. The distribution of three states of one homologous character is shown above the tips. The two transformations in that character on the tree are shown by cross-bars. Note that the apomorphy relationship is relational, rather than absolute: the gray dot is a synapomorphy for five of the terminal taxa, relative to the white dot which is plesiomorphic for the group. But when a further transformation occurs in this character, the gray dot condition is plesiomorphic with respect to the new black dot condition, which is a synapomorphy for two terminal taxa. Two kinds of phylogenetic groupings are shown: a monophyletic group (a synchronic time slice capturing all and only the descendants of a common ancestor) and a lineage (a diachronic series of ancestor-descendant relationships). Two kinds of phylogenetic events are shown: divergence (splitting of one lineage into two), and reticulation (union of two lineages to form one).

A corollary of the Hennig Principle is that classification should reflect reconstructed branching order; only *monophyletic groups* should be formally named. A strictly monophyletic group is one that contains all and only the descendants of a common ancestor. Monophyletic taxa are thus "natural" in the sense of being the result of the evolutionary process. A *paraphyletic* group is one that excludes some of the descendants of the common ancestor—these groups are to be avoided in the Hennigian system because of their unnatural status. Paraphyletic groups were used traditionally to recognize "grades," i.e., groups of organisms that share a similar level of evolution (such as "reptiles" or "green algae"), while Hennigians name only "clades," or monophyletic groups (such as mammals or flowering plants).

A discussion of the meaning of phylogenetic systematics for the fundamental

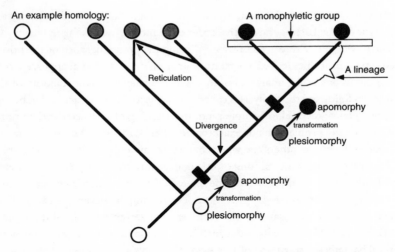

Figure 3.1. Phylogenetic tree illustrating terms and concepts of phylogenetic systematics.

species rank is beyond the scope of this chapter. It will have to suffice here to say that the same arguments for the importance of discovering and naming natural monophyletic groups applies at the species level, although problems such as reticulation tend to occur more commonly at this level. Many Hennigians follow one version or another of the phylogenetic species concept (Cracraft 1983; Mishler and Brandon 1987; Nixon and Wheeler 1990; Mishler and Theriot 2000a, b, c)

Evolution

The briefest explanation of evolution by natural selection is warranted in this discussion as well, for comparative purposes with languages and cultures. The most important distinction for our purposes, because of its generality, is that between *replicators* and *interactors* (Hull 1980). These two kinds of entities together make up the process of evolution by natural selection.

A *replicator* is any entity that passes its structure on with high fidelity. The most obvious biological example is a whole DNA molecule, but there are many other entities, at various hierarchical levels (known as the *genealogical hierarchy;* Brandon 1990), that can meet this definition. These include a short stretch of DNA (in the case of recombination in sexual reproduction), or a whole genome or whole organism (in the case of asexual reproduction), or even a population of organisms (in cases of group selection). It is important to emphasize that not all replication occurs at the DNA level; various more holistic levels in this genealogical hierarchy can be replicators, depending on the actual process of reproduction that is occurring. A succession of replicators forms a *lineage* (which can now be defined precisely as a sequence of ancestor-descendant replicators). This concept of replicator gives the connection from evolutionary process to the phylogenetic systematic patterns referred to above (i.e., the splitting of lineages gives rise to monophyletic groups, and transformations in characteristics along a lineage give rise to apomorphies).

An *interactor* is an entity that interacts with other entities so that replication is differential. The most familiar interactor is the individual organism competing with other individuals. But, again, there is a whole nested hierarchy (known as the *ecological hierarchy*) of interactors ranging from parts of organisms to large groups of them. Examples could range from groups of cells competing for greater replication within one organism (as in growth of a tumor), to individual organisms competing for a limited number of nesting sites, to whole populations (or even species) competing to determine which one will occupy a certain habitat. These functional groupings provide the impetus for adaptive change in traits.

But a *complete* understanding of the process of adaptation also requires attention to the level at which replication is occurring (not necessarily the same level at which interaction is occurring), the phylogenetic context, and the environment in which the interaction is occurring (Brandon 1990). This is because evolution by natural selection has three basic stages: (1) heritable variation (i.e., transformation) in a trait causing (2) differential reproductive success of one replicator lineage over others (3) because of competition among interactors within a common environment. An important point is that the initial variation in a trait is random with respect to what might be selected for. *Adaptation* is the process of increasing fit of an apomorphic trait to some particular problem posed by the environment, because of natural selection on the plesiomorphic variation that happened to exist.

Note that not all evolutionary change in a feature occurs by adaptation and natural selection; in fact, some think that most change does not (e.g., Gould and Lewontin 1979). Random drift is considered important in many situations (Kimura 1983). In addition, various constraints can cause a character to either change or remain the same because of its developmental or physical correlation with other characters (Oster and Alberch 1982; Smith et al. 1985; Raff and Raff 1987). The bottom line is that, while all adaptations are apomorphies at some level, not all apomorphies are adaptations.

THE VALUE OF BIODIVERSITY

There are many ways to put values on biodiversity, depending on the point of view one takes. These range from the practical to the sublime.

One viewpoint is *economic:* Regardless of one's politics or ethical value system, it is abundantly clear that natural lineages are a potential source for a myriad of products of direct economic benefit (medicines, food, shelter, esthetics, etc.). Loss of biological diversity is a disaster from an economic standpoint. How many useful organisms for food, medicine, or technology will go extinct? There are many recent examples of discoveries of natural products from biodiversity prospecting that are of major economic importance (e.g., taxol, the drug used to control ovarian and breast cancer, was discovered in the bark of *Taxus brevifolia,* the Pacific Yew, and a search based on an understanding of phylogenetic relationships led quickly to the European Yew [*Taxus baccata*], which was a better source since the leaves could be used; see Systematics Agenda 2000 1994 for many other examples).

A second viewpoint is *ecological:* a diversity of interactors is needed for proper functioning of ecosystems; there are many roles to play. Diverse ecosystems tend to be much more stable than simple ones, because of built-in redundancies in their functional roles. For example, drastic reduction in numbers of one decomposing species (say one particular fungus) may have little effect on a system if there are many other decomposers but would have a catastrophic effect if that fungus happened to be the only decomposer.

A third viewpoint is *evolutionary:* a diversity of replicators is needed as the raw material for natural selection (future evolutionary potential). Evolution proceeds faster if more different variants are present initially. Important examples include disease and pest resistance.

A fourth viewpoint is *intellectual:* we have a basic need to understand the world, how it came to be, and where we fit in it. How did the diversity of species, including *Homo sapiens,* come to be the way it is? Differences among lineages represent the legacy of biological diversification on earth and are our only means for learning about the history of life. The more data points the better, for accurate phylogeny reconstruction (since the presence of overly long branches in a phylogenetic analysis causes problems with homoplasy being mistaken for homology).

A fifth viewpoint is *ethical:* we have no right to despoil a world of living things that has taken nearly 4 billion years to develop. Each lineage existing today should be viewed as a thread in an heirloom fabric that we have been given as a legacy and that we have a responsibility to pass on to future generations. While ideally we would want to pass this fabric on intact and unchanged, if circumstances required losing a few threads, they should be chosen carefully so as to leave the major warp threads intact. Thus, the ethical ideal does not preclude prioritizing.

WHAT TO CONSERVE IN THE FACE OF POLITICAL AND ECONOMIC CONSTRAINTS?

All "species" are not equal in a phylogenetic sense (Mishler and Donoghue 1982). In an ideal world all lineages could be preserved. But in the real world of limited resources (time, money, and public goodwill), we need to set priorities (Mishler

1995). Setting priorities for conservation biology is not easy. Which taxonomic, ecological, or evolutionary units deserve the highest priority? There is a tension in biology, brought on by current political and financial limitations, among three poles of opinion: the traditional endangered species approach, the current push for conservation via community protection, and the recent calls for setting conservation priorities using phylogeny.

The first of these concentrates on taxonomic groups that have been given the rank of species and develops arguments for conservation priority on a case-by-case basis. It can be effective but is inevitably biased toward more charismatic organisms and ones whose biology is well-known (because of the burden of proof needed in legal proceedings to establish endangerment). In addition to this taxon bias, this process is slow and cumbersome and does not take into account the modern advances in phylogenetic systematics described above.

Because of the inherent slowness of the endangered species approach, a number of ecologists have taken the strategy that by preserving unique or unusual communities, one is saving many taxonomic groups at once, which will probably be unusual as well. This approach is also beneficial in that ecosystem functional properties can be taken fully into account. This approach, however, is based more on interactors than replicators, and the latter tend to be treated as "black boxes" when only communities or ecosystems are taken into account.

Thus it is also important to pay attention to the conservation of replicators, which as indicated above are the basic fabric of current biodiversity and the building blocks for future evolutionary progress. To pursue this approach, it is important first to recognize that a phylogenetic classification of organisms is necessary. With phylogenetically natural taxa, one can rationally talk about issues such as evolution, biogeography, and extinction. With unnatural taxa (i.e., artificial assemblages of unrelated populations) such issues are meaningless, and conservation efforts are hampered at best (and misguided at worst). Without knowing the relationships of populations and species, there is no practical way to conserve them.

While the ecological (interactor-based) approach has a lot to recommend it, and should therefore be pursued as well, I would argue that conservation priorities can best be set by a consideration of the phylogenetic relationships among species (a replicator-based approach). When we are forced to make hard choices, we can prioritize our efforts by concentrating on taxa that represent the "warp" of the fabric of biodiversity. This is because all attributes of organisms (genetic similarities, ecological roles, morphological specializations) tend strongly to be associated with phylogeny. From the standpoint of preserving the maximum phylogenetic diversity (and its associated attributes), saving a "long-branch" species—for example, one such as the Coast Redwood, with much change along the terminal branch, due either to previous extinction or to rapid evolution along that branch—should carry a higher priority than saving a "short-branch" species, for

example, a dandelion clone differing in only a few minor features from near relatives. Thus, phylogenetic considerations should play a much more important role
in conservation biology than they have to date. Indices based on phylogeny are
just being developed to help us preserve the maximal genetic, morphological,
chemical, and ecological diversity (Vane-Wright, Humphries, and Williams 1991;
Faith 1992).

POSSIBLE PARALLELS BETWEEN BIOLOGICAL DIVERSITY AND CULTURAL-LINGUISTIC DIVERSITY

Since languages and cultures also form lineages, it is fruitful to examine whether
these principles, derived from studies of biological lineages, can be applied to
them. It is very tempting to see analogies between biological diversity and cultural-
linguistic diversity (e.g., Dawkins 1976; Boyd and Richerson 1985; Hull 1988; Mishler 1991). They can and are described in similar terms, show similar patterns of
treelike historical relationships, and possibly evolve by similar mechanisms. Like
all analogies, however, one must look carefully to be sure there is a good fit.

There are indeed some close analogies, it appears. Replication is a very general
process in linguistic and cultural evolution. As in biology, it occurs at various hierarchical levels, ranging from transmission of individual elements ("memes" to
use the terminology of Dawkins 1976) of language or culture from parent to off-
spring within a single family, to wholesale education or conversion of large populations. These lines of replicators form lineages, and the changes in lineages (a novel
behavior or new word, for example) are like transformations in biological evolution between plesiomorphic and apomorphic characteristics. Divergence occurs
in cultural-linguistic lineages, as when a party of colonists departs, or when a
mountain range separates halves of an originally uniform culture. The combination of transformation and divergence leads to the presence of homologous features shared among different lineages, that can be used for phylogenetic reconstruction (see review by Cavalli-Sforza, Menozzi, and Piazza 1993).

Interaction also occurs in linguistic and cultural evolution, and again it occurs
at a series of nested hierarchical levels. Different words or behaviors vie for use
within an individual's mind. Family groups vie for status. Tribes and nations compete for dominance. Interaction at each of these levels, if it is coupled with differential replication, can lead to a process analogous to evolution by natural selection. This in turn can lead to a process analogous to adaptation, if particular
"memes" become more common because their possession enhances replication.

There are also some possible nonanalogies between biological diversity and
cultural-linguistic diversity. One is the unusually high degree of reticulation seen
in cultures and languages, where there are frequent, direct horizontal movements
of homologies through merging of replicator lineages. This is in contrast to the

vertical transmission via splitting events that is the rule in biological evolution (with the important exceptions discussed above). Probably the most severe non-analogy are the directly adaptive transformations seen in cultural evolution. In other words, unlike biological evolution where the initial variation is "blind" with respect to what might be beneficial to changing environments, in cultural evolution conscious participants can select variations that appear likely to succeed.

The net effect of both of these nonanalogies would appear to be an acceleration of cultural-linguistic evolution relative to biological evolution. Competing social groups (even though they tend to be conservative and hold on to their own characteristics) can pick and choose "memes" from each other and try them out, in a way that is unknown in biological evolution. Thus cultural-linguistic evolution should be able to outstrip biological evolution in most situations, which should give pause to those who advocate strong biological constraints on human behavior (e.g., Wilson 1978).

These likely analogies, as well as the nonanalogies, are relevant to the issue of conservation in cultural anthropology and linguistics. It is essential to conserve both interactors and replicators in the cultural-linguistic realm, and for the same spectrum of value-based reasons as described above. There are the same pragmatic economic, ecological, and evolutionary benefits, as well as the more esoteric intellectual and ethical benefits. In closing, however, I want especially to call attention to the importance of replicator-based conservation of lineages, because this is often neglected. A purely functional ("ecological") basis for conservation can be problematic if it neglects the historical particularities of replicator lineages. We must do our best to preserve as much of the heirloom fabric of cultural and linguistic diversity as possible, where each thread represents a lineage.

REFERENCES

Boyd, R., and P.J. Richerson. 1985. *Culture and the Evolutionary Process.* Chicago: University of Chicago Press.

Brandon, R.N. 1990. *Adaptation and Environment.* Princeton, N.J.: Princeton University Press.

Cavalli-Sforza, L.L., P. Menozzi, and A. Piazza. 1993. *The History and Geography of Human Genes.* Princeton, N.J.: Princeton University Press.

Cracraft, J. 1983. Species concepts and speciation analysis. *Current Ornithology* 1:159–187.

Dawkins, R. 1976. *The Selfish Gene.* Oxford: Oxford University Press.

Faith, D.P. 1992. Conservation evaluation and phylogenetic diversity. *Biological Conservation* 61:1–10.

Farris, J.S. 1983. The logical basis of phylogenetic analysis. In *Advances in Cladistics,* vol. 2, ed. N. Platnick and V. Funk. Pp. 7–36. New York: Columbia University Press.

Gould, S.J., and R.C. Lewontin. 1979. The spandrels of San Marco and the Panglossian paradigm: A critique of the adaptationist programme. *Proceedings of the Royal Society of London* Ser. B 205:581–598.

Hennig, W. 1965. Phylogenetic systematics. *Annual Review of Entomology* 10:97–116.

Hennig, W. 1966. *Phylogenetic Systematics.* Urbana, Ill.: University of Illinois Press.

Hull, D.L. 1980. Individuality and selection. *Annual Review of Ecology and Systematics* 11:311–332.

Hull, D.L. 1988. *Science as a Process: An Evolutionary Account of the Social and Conceptual Development of Science.* Chicago: University of Chicago Press.

Kimura, M. 1983. *The Neutral Theory of Molecular Evolution.* Cambridge: Cambridge University Press.

McDade, L.A. 1992. Hybrids and phylogenetic systematics II. The impact of hybrids on cladistic analysis. *Evolution* 46:1329–1346.

Mishler, B.D. 1991. Phylogenetic analogies in the conceptual development of science. *PSA* 1990(2):225–235.

Mishler, B.D. 1995. Plant systematics and conservation: Science and society. *Madroño* 42:103–113.

Mishler, B.D., and R.N. Brandon. 1987. Individuality, pluralism, and the phylogenetic species concept. *Biology and Philosophy* 2:397–414.

Mishler, B.D., and M.J. Donoghue. 1982. Species concepts: A case for pluralism. *Systematic Zoology.* 31:491–503.

Mishler, B.D., and E. Theriot. 2000a. The phylogenetic species concept sensu Mishler and Theriot: Monophyly, apomorphy, and phylogenetic species concepts. In *Species Concepts and Phylogenetic Theory: A Debate,* ed. Q.D. Wheeler and R. Meier. Pp. 44–54. New York: Columbia University Press.

Mishler, B.D., and E. Theriot. 2000b. A critique from the Mishler and Theriot phylogenetic species concept perspective: Monophyly, apomorphy, and phylogenetic species concepts. In *Species Concepts and Phylogenetic Theory: A Debate,* ed. Q.D. Wheeler and R. Meier. Pp. 119–132. New York: Columbia University Press.

Mishler, B.D., and E. Theriot. 2000c. A defense of the phylogenetic species concept sensu Mishler and Theriot: Monophyly, apomorphy, and phylogenetic species concepts. In *Species Concepts and Phylogenetic Theory: A Debate,* ed. Q.D. Wheeler and R. Meier. Pp. 179–184. New York: Columbia University Press.

Nelson, G. 1973. Classification as an expression of phylogenetic relationships. *Systematic Zoology* 22:344–359.

Nixon, K.C., and Q.D. Wheeler. 1990. An amplification of the phylogenetic species concept. *Cladistics* 6:211–223.

Oster, G., and P. Alberch. 1982. Evolution and bifurcation of developmental programs. *Evolution* 36:444–459.

Raff, R.A., and E.C. Raff, eds. 1987. *Development as an Evolutionary Process.* New York: Adan R. Liss.

Roth, V.L. 1991. Homology and hierarchies: Problems solved and unresolved. *Journal of Evolutionary Biology* 4:167–194.

Smith, J.M., et al. 1985. Developmental constraints and evolution. *Quarterly Review of Biology* 60:266–287.

Sober, E. 1988. *Reconstructing the Past.* Cambridge, Mass.: MIT Press.

Stevens, P.F. 1994. *The Development of Biological Systematics.* New York: Columbia University Press.

Systematics Agenda 2000. 1994. *Systematics Agenda 2000: Charting the Biosphere.* Technical Report. New York: Systematics Agenda 2000.

Vane-Wright, R.I., C.J. Humphries, and P.H. Williams. 1991. What to protect? Systematics and the agony of choice. *Biological Conservation* 55:235–254.

Wiley, E.O. 1981. *Phylogenetics: The Theory and Practice of Phylogenetic Systematics.* New York: John Wiley and Sons.

Wilson, E.O. 1978. *On Human Nature.* Cambridge: Harvard University Press.

4

WHY LINGUISTS NEED LANGUAGES

Greville G. Corbett

To an outsider, it must seem self-evident that linguists would have a key role in any investigation of the interdependence of linguistic, cultural, and biological diversity. And yet there are many professional linguists who are not concerned about diversity and its imminent reduction. This is surprising, given the situation evoked by this chilling quote:

> Obviously we must do some serious rethinking of our priorities, lest linguistics go down in history as the only science that presided obliviously over the disappearance of 90% of the very field to which it is dedicated. (Krauss 1992a:10)

This chapter makes the basic but essential point that linguistic diversity is central for linguistics. There are strong cultural and moral reasons for concern about language endangerment; and there are pertinent questions about how linguists can best use their expertise for the benefit of communities whose languages are under threat (see, for instance, Wilkins 1992; Bach 1995; for an overview of the issues of endangerment see Hale et al. 1992). Here we tackle the easier question of why linguists need languages (and more than most of them realize), in a way intended to "continue the campaign" within linguistics, while at the same time giving a window for those more concerned with cultural or biological diversity on what linguists do. Before giving some examples of the potential loss for linguists, I will outline the linguistic enterprise for the benefit of the general reader and then consider the problem of data.[1]

THE LINGUISTIC ENTERPRISE

One way to characterize the goal of linguistics is to say that it seeks is to define the notion of "possible human language." That is certainly a challenging goal; if ever a linguist has the illusion of coming close to it (in the vaguest intuitive terms), a talk to a field linguist is usually sufficient reminder that languages are even more amazing than previously thought. The goal is also a useful one: if we could define "possible human language," this would be of value for second-language pedagogy and for speech therapy, and it would be of evident interest to psychologists, neurologists, and others.

Let me give just a few of the suggestions made for constraining the notion. For instance, Chomsky has claimed that "all known formal operations in the grammar of English, or of any other language, are structure-dependent" (Chomsky 1972a:28). This means that it is structure that matters, not numerical position in a linear sequence, which is surprising when we remember that language comes to us as a string of sound. To see what is intended, imagine we have a Martian visitor. It has good processing skills and has learned elements of English, but it cannot yet ask "yes-no" questions. Could we give instructions? If we have:

(1) Mary can speak Russian.

we might then instruct our Martian to form a yes-no question by putting the second word first (this assumes that the Martian already knows what a word is):

(2) Can Mary speak Russian?

This seems to work, as in the following example too:

(3) John is at home reading Dostoevsky.
(4) Is John at home reading Dostoevsky?

Unfortunately, our Martian will have problems with such a rule:

(5) Your friend can speak Russian.
(6) *Friend your can speak Russian?

(An asterisk indicates an unacceptable sentence.) The rule based on position in linear sequence will not do. We could try a more complex solution: the first verb is put at the front of the sentence (this requires that we explain what verbs are, or list them). Then we would obtain (7) instead of (6):

(7) Can your friend speak Russian?

With this rule our Martian could handle all the examples discussed so far, but not this one:

(8) Your friend, who is studying French, can also speak Russian.

Our rule would produce this question:

(9) *Is your friend, who studying French, can also speak Russian?

Such a question would give away our Martian as not being a native speaker of English. It might be argued that no one in their right mind would produce a question like (9); indeed not—this illustrates well just how much we "know" about language, but at a subconscious level. Even if we could not externalize the information, we know that the auxiliary in initial position in the question must be that of the main clause:

(10) Can your friend, who is studying French, also speak Russian?

This area of English syntax turns out to be far from simple. (For a helpful discussion see Smith and Wilson 1979:85–89, and for a formal account see Gazdar et al. 1985:60–65.)

It may be asked why, since no one would produce (9), anyone would discuss it. We would like to characterize and understand language, and at a more modest level to find out why (9) is impossible. People who successfully digest their breakfast are not thereby biologists, and being able to decipher these signs on the page does not make the reader an expert on vision. Simply not producing examples like (9) as speakers does not mean that we understand what is involved. After all, most humans speak a language; indeed in many parts of the world it is normal to speak several. Linguists would like to know how it is that we manage that, for what it will tell us about language, and what it will tell us about us.

Let us look more quickly at some other claims about what constitutes a possible human language. One is that syntax is "phonology free" (Pullum and Zwicky 1988). That is to say that sentence structure is not affected by the sound shape of words. For instance, there could not be a language in which questions were formed in one way if the main verb started with a vowel and another if it started with a consonant. (Of course not, one might object, but then "of course" objects fall downwards.)

Another claim is that gender is always predictable for the vast majority of nouns (Corbett 1991:68). Note that unlike the previous two claims this one is not applicable to all languages. Not all languages have gender systems. But all that do, it is claimed, obey the constraint. To discover it means looking at languages of a particular type; many languages would provide no evidence for or against the claim.

A further typological claim, that is, one relating to language types, is the following: "No language has a trial number unless it has a dual. No language has a dual unless it has a plural" (Greenberg 1966:94). Duals are forms for two individuals, and trials are for three. Here we need to think of four language types (see table 4.1). Ngan'gityemerri is a Daly family language, found 300 miles SW of Darwin, Australia (Reid 1990); it has around 100 speakers. It has a dual and a trial.

Table 4.1 The Interrelation of Dual
and Trial Numbers

Has Trial	Has Dual	
	YES	No
YES	Ngan'gityemerri	NOT FOUND
No	Upper Sorbian	English

Upper Sorbian, a Slavonic language spoken around Bautzen in Lusatia, has at most 60,000 speakers; all except the youngest also speak German—a reminder that Europe too has its endangered languages (on which see Salminen 1996). It has a dual but no trial. Then there is English, with neither dual nor trial. The fourth type, with a trial but not a dual, is the one which it is claimed does not and cannot exist.

Of course, this sort of claim arises only when languages with a trial number are discovered (and that means looking outside Europe and North America) and validating the claim requires us to look at a large number of languages. And though this particular correlation is not surprising, there are many other subtler ones, some explicable from "outside," some suggestive of what goes on "inside" speakers.

DIFFICULTY WITH THE DATA

The first problem is the number of languages. Of course there are difficulties of definition and of lack of data in some areas, but many accept the figure of around 6000 for the present languages of the world (Krauss 1992a:5). (For details see Grimes, Pittman, and Grimes 1992, who have a higher figure; another source for statistical data is Kloss and McConnell 1974–1984.) It would appear that linguists have ample data, at least in principle. But if two languages are genetically related (if they are descended from a common parent), a phenomenon observed in both may be, in a sense, the same phenomenon observed twice. According to the figures in Ruhlen (1987), who reports classifications for somewhat under 5000 languages, a few families account for substantial proportions of the total number of languages, as seen in table 4.2 (note that we disregard Ruhlen's more speculative higher groupings; for discussion see, for instance, Ringe 1996). If we stick with the "splitters," those who are most careful before accepting genetic connections,[2] we are still left with many fewer points of comparison than at first seemed to be the case. Moreover, the trends of language loss seem to be working almost deliberately against us. The "safe" languages are not evenly distributed: of the 250 "safe" languages, as discussed by Krauss (1992b:2), over half belong to Indo-European or Niger-Kordofanian. This means that whole families, and whole linguistic phe-

Table 4.2 The Two Largest Language Families

Family	Number of Languages	Percentage of Total (4794)
Niger-Kordofanian	1,064	22.2
Austronesian	959	20.0

nomena, are likely to die out. For instance, take Khoesan, the family in southern Africa famous for its click consonants (once considered undescribable, these can be dental, alveo-palatal, palatal, and lateral, and can be produced as voiced, voiceless, or nasal). According to Vossen (1987), there were 150 Khoesan languages and dialects some 140 years ago, of which only one-third survive today. Vossen expected that many more might die out before the end of the twentieth century.

We should also consider the New Guinea area, since "the c. 750 non-Austronesian languages spoken in New Guinea and neighboring islands provide the richest and most complex language situation in the world" (Dixon 1991:229). If we look just at Papua New Guinea, we find 50–60 distinct families and the distribution of speakers shown in table 4.3. This table is taken from a set of surveys of the world's languages (Robins and Uhlenbeck 1991), which give an account of the differing situations in different parts of the world, based on a large number of sources.

We have seen that genetic relations are a problem for certain types of linguistic research. A second problem is areal connections. Languages that are unrelated or distantly related may still share phenomena because of contact. Thus there are features shared between the various languages of the Balkans, which the individual languages do not share with their closer relations. In such cases the languages are said to form a *Sprachbund* (see Comrie 1989:204–208). There are several such areas in the world, and again they have the effect of reducing the number of distinct data points. (From another point of view, of course, genetic relationships and areal phenomena are themselves of enormous interest, as Nichols's [1992] work shows.)

The final type of problem I will mention here is that sometimes we wish to investigate interrelationships between features that are unusual. Very soon we run out of languages. For instance, I pointed out earlier that trial number is not found without the dual. In attempting a typology of number systems (Corbett 2000), I have been investigating all the possible systems of number values. We find the following:

SINGULAR-PLURAL. Numerous examples, such as English and French.

SINGULAR-DUAL-PLURAL. Many examples, like Upper Sorbian or Central Alaskan Yup'ik.

SINGULAR-PAUCAL-PLURAL. Bayso. The paucal is for a small number, here above one.

SINGULAR-DUAL-PAUCAL-PLURAL. A fairly common system: an example is Ungarinjin
(Rumsey 1982). Here the paucal is for a small number above two.

SINGULAR-DUAL-TRIAL-PLURAL. Relatively few certain examples;[3] one is
Ngan'gityemerri (Reid 1990).

In addition to these, there are languages with even larger systems, comprising five
number values; there is clear evidence for this in Sursurunga (Hutchisson 1986), an
Austronesian language with 2700 speakers in southern New Ireland (Papua New
Guinea), the closely related Tangga (Malcolm Ross, personal communication 1994),
and Marshallese (a Micronesian language with some 20,000 speakers on the Mar-
shall Islands; Bender 1969:8–9). It is difficult to establish exactly what the values
are (for instance, the form usually labelled "quadral" in Sursurunga is used for
groups of four and above). Given this small number of languages, the claims we
make about the possible systems having five number values must be tentative.

In view of these problems, linguists have to think carefully about their samples.
There is a useful literature on the subject.[4] But at the simplest level, just as it
would help the typology of number if there were more languages comparable to
Sursurunga, so also much other work on language would be better or easier if
there were more languages available.

THE LOSS FOR LINGUISTS

Consider first one of the many gems attributed to Anonymous:

"Life is very strange" said Jeremy. "Compared with what?" replied the spider.
(From N. Moss, *Men Who Play God,* quoted in Mackay 1977:2.)

Part of the success of modern linguistics has been to show that the remarkable diversity we see coexists with shared constraints. In other words, it shows why certain linguistic phenomena are strange. This perspective may be lost if we get too
close, particularly if we always look at the same type of language. As Chomsky

Table 4.3 The Languages of Papua New Guinea

Number of Speakers	Austronesian Family	Non-Austronesian Families	Total
More than 100,000	—	3	3
10,000 to 100,000	7	30	37
1,000 to 10,000	60	240	300
200 to 1,000	40	250	290
Fewer than 200	20	110	130
Total	127	633	760

Source: Dixon 1991:245.

puts it: "Phenomena can be so familiar that we really do not see them at all" (Chomsky 1972b:24). I will describe some strange things I have worked on. Other linguists would have different but equally strange phenomena to report.

Gender

Gender is something that fascinates nonlinguists and linguists alike (Corbett 1991). And indeed, there are numerous evidently intriguing leads: Diyari, a language of South Australia, has one gender for nouns with female referents (women, girls, doe kangaroos), and the other is for all remaining nouns (Austin 1981:60). This relatively unusual system evokes for some people questions about social organization. The Rikvani dialect of Andi (a Daghestanian language) has a gender for nouns denoting insects (Khaidakov 1980:57–66). And a language from which disproportionately much has been learned about gender systems is Dyirbal, a language of North Queensland, Australia, with four genders (Dixon 1982:178–183). These are primarily for male humans and nonhuman animates, female humans, nonflesh food, and, finally, others. There are many apparent exceptions. For example, the moon is in the first, masculine gender and the sun is in the second, feminine gender. The reason is that in Dyirbal mythology, as indeed in much of Australia, the moon is the husband of the sun; in Dyirbal the role in mythology determines gender. Of these three languages, Diyari is now probably extinct, Dyirbal is approaching extinction, Andi as a whole (all dialects) has perhaps 9000 speakers.

Particularly in the nineteenth century, but continuing into the twentieth, a good deal of energy was expended on the origin of gender systems. The language considered was usually Indo-European, where the relevant data lie so far beyond reach that the topic is largely a matter of speculation. But the Daly languages of northwest Australia appear to be in the process of developing a new gender system and present a much better way of seeing how gender systems are born (Reid 1990, 1997). Unfortunately, the languages are likely to die out before the gender system is fully developed.

This case also illustrates the importance of diversity for another part of the linguistic enterprise, namely, historical linguistics. One of the ways of trying to understand the complex systems of natural language is to observe the possible ways in which they can change. The loss of a language also represents the loss of a set of changes in progress, each of which might have added something to our understanding of possible linguistic structures.

Number

A substantial proportion of the "safe" languages have an obligatory singular-plural opposition. (Hale 1997 makes an eloquent case for the importance of diversity, also referring to grammatical number.) Thus in English we are usually forced to choose between singular and plural whenever we use a noun. There are, however, languages for which number is less dominant, languages in which the meaning of the noun can be expressed without reference to number. We shall call such uses "general," by which we mean that they are outside the number system. In working on number systems, one of the most significant languages I have come across is the Cushitic language Bayso, which has a remarkable system (Corbett and Hayward 1987). Nouns have a special form, precisely to mark general number. Thus *lúban* 'lion' denotes a particular type of animal, but the use of this form does not commit the speaker to a number of lions—there could be one or more than that. Other forms are available, as we noted above, for indicating reference specifically to one lion, a small number of lions (the paucal) and more than that (the plural), when required. At the last count this language had a few hundred speakers, possibly as few as two hundred, and Dick Hayward (personal communication 1996) was pessimistic about its chances of survival.

Number is not used only in the straightforward way; consider this particularly unusual use in the Finno-Ugric language Mansi, which has something under four thousand speakers on the left bank of the River Ob' in western Siberia. In this language, according to the account in Rombandeeva (1973:42), noun phrases referring to two separate single but closely related items are found in the dual:[5]

(11) ēkwa-γ   ōjka-γ   ōl-ē̄γ
     woman-DUAL   man-DUAL   live-DUAL

   'a wife and husband live'

Each noun is in the dual, though there is one woman and one man; in this sentence the form of the verb makes clear that there are only two and not four people involved. The next example is comparable. (The apparent difference in the dual forms results from the form of the stem: the dual marker is -γ for stems in -*a* and -*e*, and -*a*γ otherwise [Rombandeeva 1973:58].)

(12) taw   mis-aγ-e   luw-aγ-e   iŋ   ōn's'-ij-aγ-e
     he   cow-DUAL-3.SG.POSS   horse-DUAL-3.SG.POSS   still   keeps-PRES-DUAL.OBJ-3.SG

   'he still keeps his cow and horse'

Here the morphology on the nouns suggests that four animals are involved, whereas in fact there are only two, as the dual object agreement on the verb indicates. (For further examples, from Mansi as well as from the related Khanty [Ostyak], see Ravila 1941:14.) It is not too difficult to find languages with duals. But for the construction illustrated above, these are the only two currently spoken languages for which I have data.

Color Terms

Color terms have proved a fertile area of research, which has generated a substantial literature. Much of that work was stimulated by Berlin and Kay (1969), whose research led them to propose an implicational hierarchy consisting of the following positions, constraining the possible inventories of *basic color terms* (from Berlin and Kay 1969:5):

white		green						purple / pink
< red <			< blue <		brown	<		
black		yellow						orange / gray

white   green   purple
                                          pink
    <   red   <   <   blue   <   brown   <
black   yellow   orange
                                          gray

The constraint is that the presence of any given color term in a language implies the existence of all those to the left in the hierarchy (thus a language with a basic term for "yellow" will have basic terms for "white," "black," and "red"). The hierarchy is also taken to make diachronic predictions, in the sense that languages are expected to acquire color terms in the order sanctioned by the hierarchy. The hierarchy has been considerably modified in later work, but the main ideas held up, and the modifications are not of direct relevance here.

My colleague Ian Davies and I have worked on a variety of languages, including the Daghestanian language Tsakhur. This is one of the healthier Daghestanian languages with slightly under 30,000 speakers. But for investigating color terms we may have been just in time. For focal colors, consultants had little difficulty. In more problematic cases, however, some consultants considered first what they would call the color in Russian and then attempted to translate it into Tsakhur. Thus this area of the lexicon is under threat.

Davies and I had seen the urgency of documenting color vocabulary in work on the Bantu language Setswana: adult speakers have an interesting color system with probably six basic terms: *bontsho* 'black', *bosweu* 'white', *bohibidu* 'red', *botala* 'grue' (green plus blue), *bosetlha* 'yellow', and *borokwa* 'brown', and, especially for the men, a large number of other terms, particularly for describing cattle (Davies et al. 1992). There are over 3 million speakers, and the language appears safe, at least in the short term. But the situation is not so clear for our purposes. When we came to investigate children, we found they offered more terms of English origin than traditional Setswana terms (Davies et al. 1994). We are therefore unlikely to be able to observe the further development from the six-term system because this part of the lexicon is being invaded by English.

This leads us to a particularly interesting aspect of the Berlin and Kay work: color terms seem to be an example of evolutionary development. Many other areas of interest to typologists (such as gender) are cyclical in nature: they rise, develop, decay, and are lost. They can rise again in the same or other languages. But color inventories seem only to grow; this means, unfortunately, that we cannot expect surviving languages to "cycle round" the phenomenon, as they do with gender, determiners, and so on. When current languages have died out, or have lost the original color terms system, the chance to observe will have gone for good. The other clear example of this type of development is numeral systems, as shown by Hurford (1987). Again the small systems (i.e., those with few numerals) are found in endangered languages, and there too there is the threat of the importation of numeral systems from other languages. With numeral systems as with color systems, the interesting system may be endangered, even if the language as a whole is not.

We should also recall Rosch's work on the Dani (Papua New Guinea), whose color system is claimed to have just two basic color terms (Rosch Heider 1972). Rosch's results can be found in most of the introductory textbooks on human cognition; there are problems with them, but they have never been replicated. Dani and potentially similar languages are endangered, and unless the replication is attempted very soon, the intrusion of Western color terms will make this impossible.[6]

It may seem that I have tackled a narrow band of concerns here. I have chosen topics intended to be of interest without too much explanation. With a little more discussion, however, almost any other grammatical category—case, person, tense, aspect, and so on—would serve just as well.

CONCLUSION

In order to do linguistics properly, we need every last language that there is. And this is a message for linguists as much as for others. While linguists have a strong professional interest in the issue, there is a much wider concern at stake. If we want to understand ourselves, language gives away a good deal:

> In the study of language, we embrace the very definition of what it means to be human and to be a developing member of the species. (John L. Locke 1993:4)

NOTES

1. The support of the International Association for the Promotion of Cooperation with Scientists from the New Independent States of the Former Soviet Union under grant 932378 "Typology of Grammatical Categories," for work on endangered languages in the Caucasus, and of Britain's Economic and Social Research Council (ESRC) under grant R000222419, for work on number, is gratefully acknowledged.
2. See McMahon and McMahon (1995) for a helpful recent contribution, of special interest here in that they discuss explicitly the analogy with biology.

3. Some of the examples cited in the literature actually have a paucal rather than a trial.
4. Bell 1978 gives a very helpful discussion of the problems of sampling, still of considerable value. Tomlin worked on a relatively large sample, which was better justified than in some comparable work (1986:17–36). Methodology was revisited by Rijkoff et al. 1993 and in Bakker 1994:84–91. A way of tackling areal bias is discussed in Dryer 1989. For a good discussion of the different methods of sampling for different purposes, see Nichols 1992:25–44.
5. I am grateful to Hannu Tommola and Ulla-Maija Kulonen for help with these examples.
6. A project led by Jules Davidoff (involving Debi Roberson, Ian Davies and myself, funded by the ESRC) is attempting the replication with other languages with small color inventories in Papua New Guinea.

REFERENCES

Austin, P. 1981. *A Grammar of Diyari, South Australia.* Cambridge: Cambridge University Press.

Bach, E. 1995. Endangered languages and the linguist. In *Linguistics in the Morning Calm 3: Selected Papers from SICOL-1992,* ed. I.-H. Lee. Pp. 31–43. Seoul: Hanshin Publishing.

Bakker, D. 1994. *Formal and Functional Aspects of Functional Grammar and Language Typology.* Studies in Language and Language Use 5. Amsterdam: IFOTT.

Bell, A. 1978. Language samples. In *Universals of Human Language II: Method and Theory,* ed. J.H. Greenberg, C.A. Ferguson, and E.A. Moravcsik. Pp. 123–156. Stanford: Stanford University Press.

Bender, B.W. 1969. *Spoken Marshallese: An Intensive Language Course with Grammatical Notes and Glossary.* Honolulu: University Press of Hawaii.

Berlin, B., and P. Kay. 1969. *Basic Color Terms: Their Universality and Evolution.* Berkeley and Los Angeles: University of California Press. Reprinted 1991, with additional bibliography of color categorization research 1970–1990 by Luisa Maffi.

Chomsky, N. 1972a. *Problems of Knowledge and Freedom: The Russell Lectures.* New York: Vintage Books. Originally published 1971 in *Cambridge Review.*

Chomsky, N. 1972b. *Language and Mind.* Enlarged edition. New York: Harcourt Brace Jovanovich. First edition 1968.

Comrie, B. 1989. *Language Universals and Linguistic Typology: Syntax and Morphology.* 2d ed. Oxford: Blackwell. First edition 1981.

Corbett, G.G. 1991. *Gender.* Cambridge: Cambridge University Press.

Corbett, G.G. 2000. *Number.* Cambridge: Cambridge University Press.

Corbett, G.G., and R.J. Hayward. 1987. Gender and number in Bayso. *Lingua* 73:1–28.

Davies, I.R.L., et al. 1992. Color terms in Setswana: A linguistic and perceptual approach. *Linguistics* 30:1065–1103.

Davies, I.R.L., et al. 1994. A developmental study of the acquisition of colour terms in Setswana. *Journal of Child Language* 21:693–712.

Dixon, R.M.W. 1982. *Where Have All the Adjectives Gone? And Other Essays in Semantics and Syntax.* Berlin: Mouton.

Dixon, R.M.W. 1991. The Endangered languages of Australia, Indonesia, and Oceania. In *Endangered Languages,* ed. R.H. Robins and E.M. Uhlenbeck. Pp. 229–255. Oxford: Berg.

Dryer, M.S. 1989. Large linguistic areas and language sampling. *Studies in Language* 13: 257–292.

Gazdar, G., et al. 1985. *Generalized Phrase Structure Grammar.* Oxford: Blackwell.

Greenberg, J.H. 1966. Some universals of grammar with particular reference to the order of meaningful elements. In *Universals of Language,* ed. J.H. Greenberg. Paperback edition. Pp. 73–113. Cambridge, Mass.: MIT Press. First published 1963.

Grimes, B.F., R.S. Pittman, and J.E. Grimes. 1992. *Ethnologue: Languages of the World.* 12th ed. Dallas: Summer Institute of Linguistics.

Hale, K. 1997. Some observations on the contributions of local languages to linguistic science. *Lingua* 100:71–89.

Hale, K., et al. 1992. Endangered languages. *Language* 68:1–42.

Hurford, J.R. 1987. *Language and Number: The Emergence of a Cognitive System.* Oxford: Basil Blackwell.

Hutchisson, D. 1986. Sursurunga pronouns and the special uses of quadral number. In *Pronominal Systems* (Continuum 5), ed. U. Wiesemann. Pp. 217–255. Tübingen: Narr.

Khaidakov, S.M. [Xajdakov, S.M.] 1980. *Principy imennoj klassifikacii v dagestanskix jazykax.* Moscow: Nauka.

Kloss, H., and G.D. McConnell. 1974–1984. *Linguistic Composition of the Nations of the World.* 5 vols. Québec: Les Presses de L'Université Laval (International Centre for Research on Bilingualism).

Krauss, M. 1992a. The world's languages in crisis. *Language* 68:4–10.

Krauss, M. 1992b. The language extinction catastrophe just ahead: Should linguists care? Paper read at the Fifteenth International Congress of Linguists, Québec, Université Laval, September 1992.

Locke, J.L. 1993. *The Child's Path to Spoken Language.* Cambridge, Mass.: Harvard University Press.

Mackay, A.L. 1977. *The Harvest of a Quiet Eye: A Selection of Scientific Quotations.* Bristol and London: Institute of Physics.

McMahon, A.M.S., and R. McMahon. 1995. Linguistics, genetics, and archaeology: Internal and external evidence in the Amerind controversy. *Transactions of the Philological Society* 93:125–225.

Nichols, J. 1992. *Linguistic Diversity in Space and Time.* Chicago: University of Chicago Press.

Pullum, G.K., and A. Zwicky. 1988. The syntax-phonology interface. In *Linguistics: The Cambridge Survey. I: Linguistic Theory: Foundations,* ed. F.J. Newmeyer. Pp. 255–280. Cambridge: Cambridge University Press.

Ravila, P. 1941. Über die Verwendung der Numeruszeichen in den uralischen Sprachen. *Finnisch-ugrische Forschungen (Zeitschrift für finnisch-ugrische Sprach- und Volkskunde)* 27:1–136.

Reid, N.J. 1990. Ngan'gityemerri: A Language of the Daly River Region, Northern Territory of Australia. Ph.D. diss., Australian National University, Canberra.

Reid, N.J. 1997. Class and classifier in Ngan'gityemerri. In *Nominal Classification in Aboriginal Australia,* ed. M. Harvey and N.J. Reid. Pp. 165–228. Studies in Language Companion Series 37. Amsterdam: John Benjamins.

Rijkoff, J., et al. 1993. A method of language sampling. *Studies in Language* 17:169–203.

Ringe, D.A. 1996. The mathematics of "Amerind." *Diachronica* 13:135–154.

Robins, R.H., and E.M. Uhlenbeck, eds. 1991. *Endangered Languages*. Oxford: Berg.

Rombandeeva, E.I. 1973. *Mansijskij (vogul'skij) jazyk*. Moskva: Nauka.

Rosch Heider, E. 1972. Universals in color naming and memory. *Journal of Experimental Psychology* 93:10–20.

Ruhlen, M. 1987. *A Guide to the World's Languages. I: Classification*. Stanford: Stanford University Press.

Rumsey, A. 1982. *An Intra-sentence Grammar of Ungarinjin, North-Western Australia*. Pacific Linguistics, Series B, no. 86. Canberra: Department of Linguistics, Research School of Pacific Studies, Australian National University.

Salminen, T. 1996. *UNESCO Red Book on Endangered Languages: Europe*. Available at http://www.helsinki.fi/tasalmin/endangered.html.

Smith, N., and D. Wilson. 1979. *Modern Linguistics: The Results of Chomsky's Revolution*. Harmondsworth: Penguin.

Tomlin, R.S. 1986. *Basic Word Order: Functional Principles*. London: Croom Helm.

Vossen, R. 1987. Am Anfang steht der Schnalz: Afrikanische Buschmann-Sprachen untersucht. *Forschung* 3:16–17.

Wilkins, D. 1992. Linguistic research under Aboriginal control: A personal account of fieldwork in Central Australia. *Australian Journal of Linguistics* 12:171–200.

5

ON THE COEVOLUTION OF CULTURAL, LINGUISTIC, AND BIOLOGICAL DIVERSITY

Eric A. Smith

The present volume is concerned with the connections between linguistic, cultural, and biological diversity. My chapter focuses in particular on whether and how these three types of diversity might coevolve or influence each other. I begin by considering some fundamental issues. First, what *is* cultural diversity? To answer this, we obviously must start with a definition of culture, as well as a consideration of the nature of cultural systems and their boundaries (e.g., are human populations characterized by distinct "cultures" akin to species?). Second, what generates cultural (including linguistic) diversity? I briefly distinguish two perspectives on this issue: phylogenetic and adaptationist.

I then offer an empirical examination of the correlation between cultural, linguistic and biological diversity using a continental-scale case study. Specifically, I examine the correlation between measures of biodiversity and measures of both linguistic and cultural diversity for indigenous culture areas in America north of Mexico. My results, while preliminary, indicate that Native North American linguistic and cultural diversity are independently correlated with at least one measure of biodiversity (but not some others). The correlation between linguistic and cultural diversity in Native North America is more problematic, since many linguistically similar peoples have culturally diverged, while many groups with divergent linguistic heritage developed cultural similarity through cultural diffusion and ecological adaptation. Thus, analysis of the Native North American data raises numerous questions about the relationships between biological and cultural and linguistic diversity that will require much further theoretical and empirical work to resolve.

Using the Native American case study results as a springboard, I examine three general explanations for why biodiversity and cultural and linguistic diversity might be correlated: (a) small-scale, culturally diverse societies conserve or enhance biological diversity; (b) biological diversity directly enhances cultural diversity; (c) large-scale, centralized cultural systems require or generate low cultural, linguistic, *and* biological diversity. It is important to note that these are not mutually exclusive explanations. Thus, if (a) and (b) were both correct, the result would be a co-evolutionary process of mutual reinforcement between cultural and biological diversity. And if either (a) or (b) were correct, this might produce a pattern of correlation in many parts of the world that has been increasingly obliterated by the expansion of states and empires, as proposed in (c).

ON THE MEANING OF CULTURAL DIVERSITY

To define what cultural diversity is and where it comes from, we obviously must decide what we mean by the term "culture." Despite the centrality of this concept for their discipline, anthropologists have perennially disagreed about its meaning. Many scholars interested in cultural evolution choose to define culture as *socially transmitted information,* where "information" refers to beliefs, values, knowledge, and the like (Boyd and Richerson 1985; Durham 1991). This definition has several features. It distinguishes culture from behavioral phenotypes, including artifacts, while recognizing that these can be products of cultural information. It recognizes that behavioral phenotypes can be jointly shaped by genes, culture, and nonsocial environment. By highlighting the criterion of social transmission, it both emphasizes that culture is a system of inheritance and distinguishes culture from genetic inheritance. This last point implies that culture, like genetic information, is subject to evolutionary change (through drift, natural selection, and possibly other means). While many people disagree with the idea that culture is ideational (in the sense just described) and hence distinct from behavior, I believe this is the best way to reconcile the concept with the genotype-phenotype distinction. While culture is transmitted via behavior (speech, visual symbols, gesture, etc.), since any behavior is phenotypic, it cannot be purely cultural because it is necessarily shaped by gene-environment interaction too.

What Is Cultural Diversity?

By analogy with measures of biodiversity, we might suppose that cultural diversity can be defined in terms of the variation in culturally heritable information and its distribution across cultural lineages (cf. Mishler this volume). Indeed, there has been a recent upsurge in phylogenetic analyses of cultural variation (e.g.,

Mace and Pagel 1994) and its correlation with linguistic and genetic phylogeny (e.g., Cavalli-Sforza et al. 1988, 1989; Cavalli-Sforza, Minch, and Mountain 1992). But there are some serious drawbacks with this phylogenetic approach to culture and cultural diversity. First, humanity is a single species, and there are few barriers to the flow of cultural information (diffusion) between cultural units or lineages relative to those limiting gene flow between species. While a phylogeny requires a set of discrete taxa, it is by no means clear that cultural variation (including cultural innovations) is bounded into discrete cultures. Of course, the same could be said about biological species, but in general the taxonomy of biospecies is much less problematic than is the case for cultural analogues.

On the other hand, I do not think we need to have sharp, clear boundaries between entities in order to find it useful to distinguish them; if that were the case, we could never differentiate day from night, or summer from winter. As long as we take care not to reify these "constructed" entities or view them as strictly bounded and impermeable, a notion of discrete cultures might be a reasonable approximation for many times and places (though increasingly less so in the "global-transnational-postcolonial" era). Although I am not a linguist, and while I am aware of pidgins and creoles and various forms of linguistic borrowing, my sense is that boundaries between languages are more easily distinguished than those between other types of cultural variation. In any case, sophisticated measures of cultural diversity will need to take diffusion into account and consider cultural microdiversity or trait distribution between individuals as well as at larger geographical scales.

Where Does Cultural Diversity Come From?

Two broad approaches to this issue can be found in the literature. The *phylogenetic* perspective sees it as a branching process of speciation and extinction, akin to biogenetic phylogenies. This perspective gives priority to isolation and to chance historical factors (cultural "drift," so to speak) in generating cultural diversity. As I argued above, this may work reasonably well for languages but is of limited utility for nonlinguistic cultural diversity, as a result of extensive cultural diffusion. Alternatively, many have taken an *adaptationist* perspective, viewing cultural change as a form of coevolution between cultural information and the social and natural environment. In this view, cultural diversification occurs through various processes of cultural adaptation, including niche partitioning (à la Barth 1956) and in some cases direct competition between cultural entities.

THE NATIVE NORTH AMERICAN CASE

Since analysis of the relationships between biodiversity and cultural or linguistic diversity is in its infancy, there is clearly a need for empirical assays to determine which of the various theoretical possibilities have in fact been commonly realized.

Table 5.1   Linguistic Diversity in Native North America

Culture Area	Language Phyla or Families	Languages
Arctic	Eskimo-Aleut	9
Basin	Uto-Aztecan	6
California	Algonkian, Athapaskan, Hokan, Penutian, Uto-Aztecan, Yukian	74
Northeast	Algonkian, Iroquoian	> 38
Northwest	Chimukan, Na-Dene, Penutian, Salishan, Wakashan	44
Plains	Algonkian, Aztec-Tanoan, Caddoan, Siouan	18
Plateau	Sahaptian, Salishan	15
Southeast	Caddoan, Iroquoian, Muskogean	> 32
Southwest	Aztec-Tanoan, Yuman, Athapaskan	27
Subarctic	Algonkian, Athapaskan	12
Total	8 phyla, 59 families and isolates	> 275

Sources: Goddard 1978; Hale and Harris 1979; Hunn 1990; Kehoe 1992; Kendall 1983; Lounsbury 1978; Miller 1986; Sherzer 1991; Shipley 1978; Thompson and Kinkade 1990; Young 1983.

As a small contribution toward this end, I have chosen to examine the ethnographic region I know best, Native North America.

There is a long tradition of considering the relation between environmental factors and Native American cultural variation, beginning with Wissler's (1924, 1926) attempt to define culture areas on the basis of subsistence, and continuing with the much more detailed work of Kroeber (1939) and Jorgensen (1980) on the correlation between environmental zones (defined by climate and vegetation) and culture areas (defined on the basis of cultural features as recorded in the ethnographic literature).

In the analyses presented below, I utilize this literature to examine the degree to which biodiversity correlates with linguistic and cultural diversity in Native North America. To do this, we must first decide how to measure the three different kinds of diversity. Native American cultural diversity is conventionally categorized according to subcontinental regions known as "culture areas."[1] In recent years, the most common culture-area scheme has consisted of ten areas (e.g., Kehoe 1992), as listed in table 5.1. While these areas differ in size, population density, and degree of cultural heterogeneity, they are at least roughly comparable units, and prior to European invasion each contained dozens of ethnolinguistic units and hundreds of local groups.[2] Thus, the culture area scheme provides a handy if imperfect framework for analyzing aboriginal cultural and linguistic diversity on the North American continent.

To quantify linguistic diversity, I chose a crude but feasible measure, the minimum estimated number of distinct languages spoken in each culture area at time of contact, although table 5.1 also provides some information on diversity of higher-level taxonomic units (language families and phyla).

Table 5.2   Cultural Diversity in Western Native
North America

Culture Area	Ethnolinguistic Groups	Similarity Index[a]
Basin	24	68%
California	55	51%
Northwest	35	55%
Plains	36	n.a.
Plateau	21	62%
Southwest	37	43%

Sources: Kehoe 1992 (Plains); Jorgensen 1980 (all others).

[a]Between-group within-area coefficient of similarity for 292 variables containing 1,577 attributes (Jorgensen 1980); the higher the value, the lower the areal
diversity.

For nonlinguistic cultural diversity, I used two measures. The simplest, following
the example of Wilcox and Duin (1995), is number of ethnolinguistic groups. The
second assay of cultural diversity I employed is a statistical measure of culture
trait diversity, which quantifies the proportion of traits shared between the ethnolinguistic groups in a given culture area (table 5.2). Unfortunately, comparable data
on these two measures are only available for a subset of Native American culture
areas (Jorgensen 1980).[3] Of course, the validity of the entire procedure depends on
how good a basis there is for selecting and measuring a set of culture traits. Many anthropologists feel there is no good basis for doing so; while I recognize the problems,
I feel that some quantitative measure of cultural variation is better than none at all.

Biodiversity can be measured in a variety of different ways; for present purposes I settled on a crude but feasible measure, species richness. Given the available data, I chose to measure species richness for selected taxonomic categories,
namely trees, native vascular plants, and homeotherms (birds and mammals). I
used maps plotting these data for North America developed by the World Wildlife
Fund (Ricketts et al. 1997) and superimposed them on a map of Native North
American culture areas of the same scale.[4] I then visually estimated the average
species richness for each culture area. The resulting estimates (table 5.3) are not
very precise, but since my analysis requires only ordinal rankings, I believe the
data and methods are sufficient for present purposes.

Linguistic Diversity

What, then, do we find when we examine the correlations between these various
measures of diversity? First, linguistic diversity shows a moderate degree of positive correlation with tree species diversity (fig. 5.1). Indeed, the four culture areas

Table 5.3 Biodiversity Measures for Native North American Culture Areas

Culture Area	Biodiversity (Mean Species Richness)			
	Trees	Vascular	Mammals	Birds
Arctic	10	551	30	97
Basin	30	2292	77	216
California	55	2035	60	213
Northeast	80	1584	61	206
Northwest Coast	32	1233	59	193
Plains	56	1874	74	213
Plateau	26	1428	63	198
Southeast	150	2401	61	215
Southwest	45	2220	88	223
Subarctic	20	893	37	159

Source: Calculated by author from map data in Ricketts et al. 1997.

with the lowest tree-species diversity also have the lowest linguistic diversity, while the three highest tree-diversity areas include two of the three areas with highest linguistic diversity (tables 5.1 and 5.3). The major outlier for this correlation is the Southeast culture area, which, because of its moist and temperate to subtropical climate, has very high tree diversity (averaging 150 species) but only intermediate linguistic diversity. The precontact linguistic diversity of the Southeast is difficult to establish, however, since early and massive depopulation undoubtedly extinguished a number of Southeastern languages before they were ever recorded (hence the notation in table 5.1 of "> 32" languages). If the actual number of languages spoken in the Southeast in 1492 were as much as 50 percent above the recorded value, this would give it the second-highest linguistic diversity and increase the correlation with tree diversity somewhat.[5] One other notable divergence between arboreal and linguistic diversity is the low tree diversity rank (fifth) for Northwest Coast, an area with the second-highest linguistic diversity; this may reflect the fact that Northwest Coast Indian subsistence is focused on marine, not terrestrial, resources, making tree diversity a poor indicator of relevant biodiversity.

When we turn to native vascular plants, we find that species richness shows only a very low and statistically insignificant correlation with Native North American linguistic diversity (fig. 5.2). The correlation is greatly weakened by the two outliers, Great Basin (upper-right point) and Northwest Coast (upper-left); if we remove these two regions from the sample, the correlation becomes moderately strong (r_s = 0.738) and statistically significant (p < .02, one-tailed, n = 8).

In the case of faunal (bird and mammal) diversity, no correlation at all with lin-

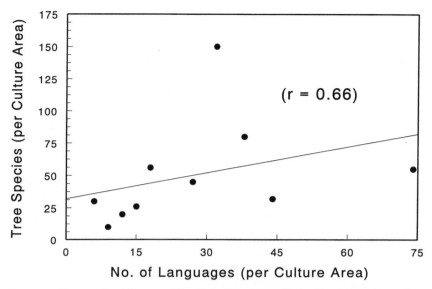

Figure 5.1. Tree species richness and linguistic diversity per Native North American culture area (correlation is Spearman's r_s coefficient for rank order, $p < .05$; plotted line is simple linear regression).

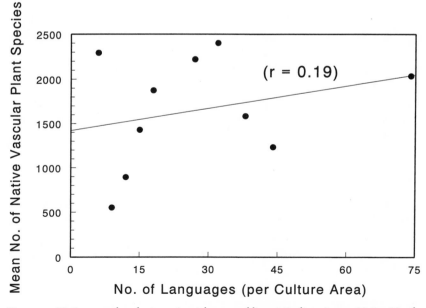

Figure 5.2. Native vascular plant species richness and linguistic diversity per Native North American culture area (correlation is Spearman's r_s coefficient for rank order, and is not significant; plotted line is simple linear regression).

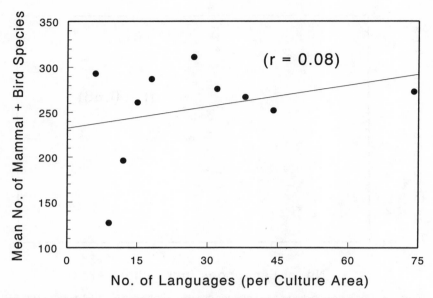

Figure 5.3. Faunal species richness and linguistic diversity per Native North American culture area (correlation is Spearman's r_s coefficient for rank order, and is not significant; plotted line is simple linear regression).

guistic diversity is evident (fig. 5.3).[6] For example, the Northwest Coast, despite its very high linguistic diversity, is at the low end of faunal diversity; again, this lack of correlation may reflect the relative unimportance of terrestrial species in Northwest Coast aboriginal economies. The Southwest and Great Basin areas have the two highest values for both avian and mammalian species richness, yet are intermediate to very low in linguistic diversity. Clearly, faunal diversity, at least as measured by avian and mammalian species, has little or no bearing on linguistic diversity in Native North America.

Cultural Diversity

Analysis of the relationships between biodiversity and nonlinguistic measures of cultural diversity reveals the same patterns found with linguistic measures. Tree species diversity is highly correlated with the number of ethnolinguistic groups found in each culture area (fig. 5.4) and with a quantitative measure of the degree of culture-trait similarity found within each of these areas (fig. 5.5). These correlations retain statistical significance, despite the fact that problems with data availability reduce the sample to a handful of culture areas (see n. 3).

Faunal diversity, by contrast, is not significantly correlated with either measure of cultural diversity (table 5.4), although both correlations are positive. If we sep-

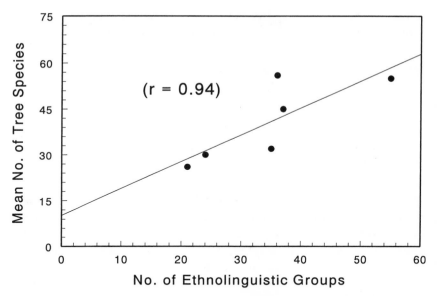

Figure 5.4. Tree species richness and number of ethnolinguistic groups per Native North American culture area (correlation is Spearman's r_s coefficient for rank order, $p < .05$; plotted line is simple linear regression).

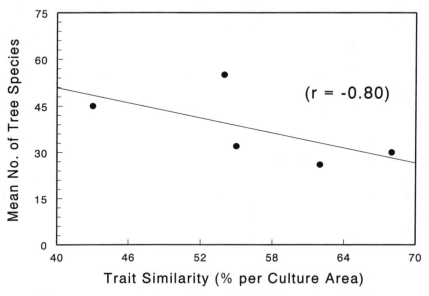

Figure 5.5. Tree species richness and coefficient of culture trait similarity per Native North American culture area (correlation is Spearman's r_s coefficient for rank order, $p < .01$; plotted line is simple linear regression).

Table 5.4 Summary of Correlations between Cultural, Linguistic, and Biological Diversity for Native North American Culture Areas

Species	Languages	Ethnolinguistic Groups	Cultural Traits
Tree species	0.661*	0.943*	0.800†
Native vascular plant species	0.188	0.257	−0.100
Faunal species	0.079	0.314	0.300

Sources: Data from tables 5.1–5.3. Spearman rank order correlations; statistical significance indicated by † ($p < .05$) or * ($p < .01$) (one-tailed, n−2 degrees of freedom).

arate faunal species into birds and mammals, mammalian species richness does correlate significantly with cultural trait diversity ($r_s = 0.9$, $p < .01$, one-tailed), but not with number of ethnolinguistic groups. Native vascular plant diversity shows no correlation at all with the Jorgensen measure of cultural diversity or with number of ethnolinguistic groups (table 5.4).

Although the various measures of linguistic and cultural diversity employed here have weaknesses, the fact that tree diversity correlates well with each of them is interesting. The fact is that the correlation between linguistic and cultural diversity in Native North America is quite problematic. This is because many linguistically related groups have culturally diverged, while many groups with divergent linguistic heritage developed cultural similarity through cultural diffusion and ecological adaptation. For example, speakers of Athapaskan languages are spread across four different culture areas, and in each area they exhibit marked cultural adjustment to the particular social and ecological setting, despite the fact that some of the languages involved seem to have diverged only about a millennium ago. The opposite phenomenon of cultural convergence despite divergent linguistic origins is also widely exemplified in Native North America; the California and Plains culture areas provide particularly striking examples. In sum, linguistic diversity, at least as measured phylogenetically, cannot be assumed to covary with cultural diversity. Hence, it is quite remarkable that at least one measure of biodiversity is independently correlated with both cultural and linguistic diversity in Native North America. Other than that, however, the findings do not seem to support the idea that biodiversity is robustly correlated with either cultural or linguistic diversity.

WHY MIGHT CULTURAL AND BIOLOGICAL DIVERSITY CO-EVOLVE?

According to Maffi (1996:3), "evidence is emerging of remarkable overlaps between areas of greatest biological and greatest linguistic/cultural diversity around the world. These striking correlations require close examination and must be accounted for." Indeed, a number of authors (e.g., Chapin 1992; Toledo 1994; McNeely and Keeton 1995; Poole 1995; Wilcox and Duin 1995; Harmon 1996; Lizarralde this volume) have suggested that cultural (including linguistic) and biological diversity tend to be correlated, and a larger number have argued that small-scale or indigenous social systems, which tend to occur in areas of high cultural and linguistic diversity, are more likely to conserve or coexist with biodiversity, although this latter argument is controversial (e.g., compare Alcorn 1996 with Hames 1991).

While examination of the Native North American case gives mixed results, let us suppose that biodiversity does tend to correlate with cultural and/or linguistic diversity. How might we explain such a correlation? Three general hypotheses to answer this question are considered here: *(H1)* small-scale societies conserve or enhance biological diversity; *(H2)* biological diversity directly enhances cultural diversity; *(H3)* large-scale social systems reduce both cultural and biological diversity. These hypotheses are not mutually exclusive—indeed, a coevolutionary model in which cultural and biological diversity mutually reinforce each other combines *(H1)* and *(H2)*—but for analytical purposes it is fruitful to consider them separately.

(H1) Small-scale societies enhance or conserve both cultural and biological diversity: This position is argued by a number of ethnobiologists and anthropologists who view small-scale societies as reservoirs of cultural and linguistic diversity and as preservers of biodiversity (see review in Bodley 1996). The general argument here is that small-scale societies protect local biological resources because they rely on these for their subsistence and for their very survival and that a regional aggregate of these societies, because of their small scale and localized nature, will tend to contain high levels of linguistic and cultural diversity as well. Some have criticized this argument for advancing a naïve and romantic view of indigenous peoples as "ecologically noble savages" (Redford 1991; Alvard 1993). But an empirically rigorous and theoretically sophisticated case for *(H1)* can be made.

There are several mechanisms that have been ethnographically described by which small-scale societies directly enhance biodiversity, including:

(a) subsistence-related enhancement of biodiversity (e.g., plant breeding, active transplanting, soil enrichment, maintenance of habitat patchiness or successional disturbance through clearing and burning);

(b) ceremonial and secular proscriptions against resource depletion or utilization (e.g., sacred groves, controls on the timing or location of resource harvesting); and

(c) epiphenomenal conservation (Hunn 1982) as a byproduct of subsistence-driven mobility, low population density, or habitat avoidance (e.g., buffer zones between territories of hostile neighboring groups).

The overall prevalence and effectiveness of these mechanisms among small-scale societies in general is not well known at present, although a substantial number of cases have been described (e.g., Johannes 1978; Berkes et al. 1989; Nabhan 1989; various authors in Posey and Balée 1989; Bye 1993; Sponsel and Loya 1993; Balée 1994; Brush 1995; Russell-Smith et al. 1997). On the other hand, an increasing number of instances where prestate societies (particularly those colonizing new habitats or engaging in subsistence innovation) have depleted biodiversity have also been documented (Brightman 1987; Kohler 1992; Redford and Stearman 1993; Alvard 1994, 1998; Broughton 1994; Steadman 1995; various authors in Kirch and Hunt 1997). Determining the overall balance between these two opposing trends, and the conditions which favor one outcome or the other, will require much more research. My own preliminary reading of the historical and ethnographic evidence is that even cases of biodiversity enhancement or preservation involve actions or institutions aimed at enhancing something other than biodiversity, such as density or productivity of resources useful for human consumption (e.g., Johnson 1989; Low 1996; Ruttan 1998). But this is an issue on which various anthropologists and other scholars (e.g., conservation biologists, environmental historians) are currently rather strongly polarized, and a more systematic assessment of the evidence is needed than I can provide here before any firm conclusions can be drawn.

(H2) Biodiversity directly enhances cultural diversity: Whatever the validity of *(H1),* it seems unlikely to explain such intriguing phenomena as the concentration of much linguistic diversity in areas of high biodiversity such as the moist tropical forest of Central America, Amazonia, Central Africa, Southeast Asia, and Australia, and Melanesia (table 5.5), given that most of the biodiversity of these forests has natural rather than anthropogenic causes.

(H2) provides an alternative or complementary explanation for such patterns. At least three possible mechanisms might underlie this hypothesis:

(a) high biodiversity, by providing an increased number of niches, may encourage greater cultural diversification through niche partitioning;

(b) high biodiversity may occur in areas less subject to environmental fluctuations, resulting in more stable resource populations which permit smaller, more localized human societies to be relatively self-sufficient (and hence, given reduced cultural diffusion and perhaps increased cultural "drift," more divergent through time);

(c) high biodiversity may covary with high biological productivity, which in turn allows the coexistence of a variety of production systems and associated sociocultural patterns.

We do not have robust tests of any of these possible mechanisms, but some brief and tentative assessments can be made. With regard to (a), high biodiversity may not result in niche divergence unless it is the "right kind" of biodiversity. Indeed, I suspect that niche diversification among sympatric populations (groups sharing the same habitat) may actually be more common among intensive agro-

Table 5.5 Endemic Languages in Tropical Forest Regions

Region	Number of Countries	Percent of Languages	World Total
Central Africa	8	881	13.0
Southeast Asia	8	1046	15.5
Amazonia	5	362	5.4
Mesoamerica	8	285	4.2
PNG/Melanesia	4	1056	16.2
(Europe, for comparison)	(38)	(75)	(1.1)
Totals (tropics only)	33	3630	55.6

Source: Tabulations by the author from data in Grimes 1992.

pastoralists, in areas of moderate to low linguistic diversity and substantially degraded biodiversity, than in the "hot spots" of biological and linguistic diversity
such as tropical forests. My logic here is that complex socioeconomic systems of
production and exchange should expand the number of ways in which various
groups can utilize the environment to make a living, and hence the opportunities
for sympatric niche diversification (e.g., Barth 1956). It also seems probable that
seasonality plays a large role in sympatric niche diversification, as when the same
area is used successively by agriculturalists, pastoralists, fishermen, and so on
(Monique Borgerhoff Mulder, personal communication 1998).

Mechanism (b) is essentially a phylogenetic approach, since it assumes that cultural diversity accumulates as a function of time and (relative) isolation or self-
sufficiency. Since biodiversity in terms of endemic species may largely be a function
of (genetic) isolation of species and (ecological) fragmentation of communities,
both of which might build up more readily in stable (e.g., tropical low-seasonality)
environments, it could be that correlations between cultural and biological diversity often reflect coevolutionary isolation of both kinds of "information." Harmon (1996) has shown that linguistic and biological endemism (the latter for vertebrate species) are quite coincident around the globe, measured at the country
level. But the detailed reasons why this coincidence may exist have not been examined; mechanism (b) is one candidate for explaining this.

As for mechanism (c), I am skeptical of it for the simple reason that high biodiversity does not necessarily mean higher net primary productivity (NPP) overall, let alone higher availability of resources consumable by humans.[7] I am not
aware of any systematic attempts to correlate human population density and biodiversity, but we can get some sense of the problems with (c) by examining the
data for Latin American ecoregions compiled by Wilcox and Duin (1995). They
provide data on NPP and two measures of biodiversity ("forest tree genetic resources" and "centers of plant diversity") for 218 ecoregions grouped into 11 habi-

Table 5.6 Summary of Correlations between Cultural and Biological
Diversity for Latin American Ecoregions

	Number of Indigenous Populations	Net Primary Productivity
Tree genetic resources	0.961*	0.070
Centers of plant diversity	0.849*	−0.100
Net primary productivity	0.588†	—

Source: Calculated by the author from data in Wilcox and Duin 1995. Spearman rank order correlations, with statistical significance indicated by † ($p < .05$) or * ($p < .001$) (one-tailed, $n − 2 = 9$ degrees of freedom).

tat types (after Dinerstein et al. 1995). These data indicate that NPP is completely uncorrelated with biodiversity across habitats and is only marginally correlated with the number of indigenous populations per habitat, Wilcox and Duin's measure of cultural diversity (table 5.6). NPP is very high in some habitats (e.g., mangrove swamps, tropical moist forests) where human population density is low, because the biological productivity is locked up in forms (e.g., woody plant tissue) that are not directly consumable by humans or even by their herbivorous prey. To put it colloquially, neither people nor their herbivorous prey eat trees.

Indeed, the converse of (c) may often hold: many human populations may *reduce* biodiversity in order to channel a larger proportion of NPP into edible products, the most extreme expression of this being the monocropping and landscape modification practiced in some forms of intensive agriculture. For example, most of the Javanese landscape has been transformed by human action over the last several centuries from tropical forest to highly managed irrigated and terraced wet-rice agroecosystems (Geertz 1963; Lansing 1991). But that argument takes us into the domain of *(H3)*.

(H3) Large-scale social systems reduce both cultural and biological diversity: The basic idea here is that large-scale (stratified, centralized, nation-state, high population density) social systems have political-economic dynamics that consume an increasing proportion of the biological productivity in areas they occupy, and in turn such systems expand at the expense of small-scale (decentralized, subsistence-based, low-density) systems. This historical process is also an ecological process with significant implications for biodiversity (Norgaard 1988). If cultural and linguistic diversity is maximized in regions inhabited by small-scale social and economic systems *(H1),* and withers when incorporated into large-scale systems (via direct conquest or economic domination), then cultural and biological diversity will become increasingly correlated at the very same time that both are suffering unprecedented declines. That seems a fair (if depressing) description of much of recent planetary history (e.g., Diamond 1997; Ponting 1991). Indeed, this process is continuing or even accelerating over huge regions such as lowland South Amer-

ica (e.g., Painter and Durham 1995) and the circumpolar north (e.g., Smith and McCarter 1997).

Several specific mechanisms might underlie the dynamic envisioned under *(H3)*:

(a) certain kinds of high biodiversity environments (e.g., some moist tropical forests) are not conducive to intensive agriculture, and hence inhibit the spread of states;

(b) areas favorable for intensive agriculture fostered the development of large-scale political economies, losing much cultural diversity and at least some biodiversity in the process;

(c) state systems both require and foster high population densities.

The assumption behind (a) is that state systems require intensive agriculture for their support, and that tropical forest ecosystems will not generally permit this form of production. Thus, state economies either convert tropical forests into other kinds of ecosystems, or they fail to expand into areas where such conversion is not politically or ecologically feasible. In most cases of which I am aware, the political, cultural, and linguistic expansion of centralized state systems has indeed been incompatible with preservation of high-diversity forest ecosystems.

Thus, in South and Southeast Asia and the Andean region, the boundaries of state expansion have reached their limits in areas where clearing the forest for intensive agriculture was not effective; beyond this boundary smaller-scale societies continued to exist in semiautonomy, alternately resisting state control and seizing opportunities to benefit from trade with states and empires. A similar pattern of competitive exclusion (Durham 1979) can be seen in the colonization of Central and South America by European states. This may be the major reason why the Latin American data show such a strong correlation between biodiversity and indigenous populations (Chapin 1992; Wilcox and Duin 1995; Lizarralde this volume). With the current rise of multinational corporate capitalism, its high mobility of capital facilitated by institutions such as the World Bank and free trade agreements, it has become economically possible (if ecologically unsustainable) to extract timber, minerals, and short-term agropastoral products from these tropical forest refugia. This, plus in-migration of impoverished and displaced peasants, is fueling a process that is drastically reducing both cultural and biological diversity (e.g., Painter and Durham 1995).

Eurasian and recent North American history suggests that even temperate biodiversity is reduced by the expansion of nation-state political economies. The loss of biodiversity, particularly that associated with forests and forest margins, is increasingly documented in studies of environmental history of ancient China, the Mediterranean empires, and elsewhere (e.g., Ponting 1991, and references therein; Hughes 1996). Indeed, the only major exception to the incompatibility between states-empires and forest biodiversity of which I am aware is the case of the Classic Maya. But even in this case there is ongoing debate about the extent to which

Classic Maya states might have produced drastic ecological change in their low-land forest habitats, which in turn precipitated their collapse (Wiseman 1985; various authors in Fedick 1996; Paine and Freter 1996).

As a corollary of (a), (b) proposes that state systems were most likely to develop in areas ecologically favorable to intensive agriculture, given a certain level of technoeconomic development. Such development is then expected to reduce cultural and biological diversity, along the lines sketched above.

Mechanism (c) proposes that state systems both require and foster high population densities. That is, the conditions allowing states to arise or persist may include densities sufficient to facilitate the kinds of economic organizations (including surplus production) upon which political centralization depends. In turn, by increasing the intensity and scope of regional trade, states may often encourage population increase. These denser populations in turn have greater average impact on plant and animal populations, making depletion or even extinction of some species more likely, thus reducing biodiversity. By incorporating the mosaic of more locally adapted societies into the state political economy through economic incentive, political coercion, and conversion of their environmental base, high-density populations with relatively low (per capita and per unit area) cultural diversity come to be associated with areas of anthropogenically low biodiversity.

CONCLUSIONS

I can only draw very tentative conclusions from the above analysis of Native North American data, and my brief survey of more widespread patterns. Linguistic diversity seems to be driven by both ecological factors (environmental stability, low mobility, localized resources) and sociopolitical ones (decentralized political and economic organization, ethnic boundary maintenance, local endogamy), though these two categories do not separate neatly. Supporting these somewhat speculative generalizations, Nichols has shown for a global sample of languages that linguistic diversity (as measured by phylogenetic divergence, not simply number of languages) is highest in coastal, tropical and subtropical, and some montane regions, and in areas lacking "large-scale economies and/or societies such as empires whose languages spread with their political/economic systems" (1992:234). In essence, high linguistic (and perhaps nonlinguistic cultural) diversity should be favored by any factors that lead to *localization of speech communities,* whether this localization is due to environmental factors favoring isolation or self-sufficiency or sociopolitical dynamics favoring boundary definition in order to control resource access, marriage, and group membership (Luisa Maffi, personal communication 1998).

Although I have not been able to explore such factors in depth, most of the areas of high linguistic diversity in Native North America are places where resources important to aboriginal economies were relatively dense and localized—just the conditions under which we expect low mobility and territorial defense (restricted access to communally owned resources) to be favored (Dyson-Hudson and Smith 1978). Thus, the salmon streams, acorn groves, shellfish beds, and cornfields of the Northwest Coast, California, the Eastern Woodlands, and the agricultural parts of the Southwest are precisely where we would expect dense and localized populations to evolve considerable linguistic, and perhaps cultural, diversity.

It is doubtful that states or empires developed aboriginally north of Mesoamerica, but parts of the eastern woodlands (Southeast and Northeast culture areas) did develop relatively centralized systems of trade, tribute, and military alliance that joined local groups over larger regions than elsewhere in Native North America. Perhaps these systems played enough of a role in eroding prior linguistic (and cultural?) diversity to produce some discordance between the high floral diversity of the eastern woodlands and its only moderate linguistic diversity; in any case, these areas certainly suffered considerable language loss with the onset of European colonization.

The best ways to define and measure cultural diversity, the reasons for variation in such diversity, and why it may or may not covary with biodiversity, are all much harder to assess. The results from North America as well as those from some other areas (see tables 5.4 and 5.6) suggest that tree species richness does serve as an indicator of ecological situations favoring cultural diversification (as measured by number of distinct groups or even cultural trait diversity), while diversity of other life forms (vascular plants, mammals, and birds) does not. But why this should be the case, given that most populations do not structure their subsistence around trees, is puzzling. Clearly, more detailed ecological and ethnological analysis is needed to clarify this matter.

The broader discussion of how biological and cultural diversity might influence each other or even coevolve also identified some unresolved issues. Is the oft remarked upon correspondence between small-scale societies and areas of high biodiversity due to active systems of resource conservation and commons management, or does it more commonly reflect the unintended consequences of low population density, lack of export markets, and preindustrial technology? Do small-scale societies actively foster biodiversity, or merely coexist with it until displaced by expanding centralized states based on intensive agriculture and resource extraction? Opinions on these matters are abundant, but hard data relatively scarce and in need of careful analysis. Whatever the final answers, it seems apparent that the links between biological and cultural (including linguistic) diversity are tangled and indirect, involving social and political factors as much as environmental ones. Clearly there is ample room for further theoretical and empirical work on these intellectually fascinating and socially significant topics.

ACKNOWLEDGMENTS

For helpful comments on versions of this paper, I am grateful to Dee Boersma, Jason Clay, Pete Coppolillo, Jane Hill, Leanne Hinton, Gene Hunn, Luisa Maffi, Monique Borgerhoff Mulder, and Johanna Nichols. Thanks also to Eric Dinerstein, David Olson, and Gordon Orians for providing access to the WWF North American biodiversity maps.

NOTES

1. While the number of culture areas defined and the criteria used for defining them vary from one scholar to another, the general concept is that a culture area is a geographically bounded region within which societies or local groups have more in common than they do with units outside the culture area. This is obviously quite imprecise, though some (e.g., Jorgensen 1980) have used quantitative measures of trait similarity to define areal boundaries.

2. The variation in size of Native North American culture areas is roughly counterbalanced by variation in population density, with the smallest areas (California, Northwest Coast) having some of the highest aboriginal population densities, while the largest areas (Subarctic, Arctic) are near or at the bottom in population density. By "ethnolinguistic unit" I mean a set of people sharing a single language (or a small set of closely related ones) and many other cultural features; these are often referred to as "tribes" (e.g., Pomo, Kwakiutl). In most cases these ethnolinguistic units were originally composed of a number of local, politically autonomous groups (e.g., "Pomo" refers to a set of people north of San Francisco Bay who once spoke seven closely related languages and were divided into approximately thirty local groups).

3. Jorgensen (1980) provides a comprehensive listing of 172 ethnolinguistic groups for Western North America, comprising five of our ten culture areas, and quantitatively analyzes cultural similarity between various sets of these groups. To measure similarity, Jorgensen employs Driver's coefficient of similarity, defined as $a/(\sqrt{a+b})(\sqrt{a+c})$, where a = the number of variable attributes (e.g., matrilineal kinship) that occur in both units (e.g., two ethnolinguistic groups) being compared, b = those found in the first unit but not the second, and c = those found in the second unit but not the first (Jorgensen 1980:311). Values of this coefficient, which varies from 0 to 1.0 (0–100%), are then averaged for the entire set of groups in a culture area to obtain the mean degree of cultural similarity. For the tally of ethnolinguistic groups, I added the Plains culture area to Jorgensen's set, using data in Kehoe (1992:297ff). Given sufficient time, one could no doubt use the published literature to do this for the Arctic and Subarctic areas as well; the ethnographic record for the Northeast and Southeast is probably too fragmentary to construct comparably reliable estimates.

4. The species richness maps for these taxonomic categories were generated by the Conservation Science Division of the World Wildlife Fund–U.S. (Ricketts et al. 1997). I am grateful to my colleague Gordon Orians (Zoology, University of Washington)

and to Eric Dinerstein (WWF U.S.) for making these available to me prior to their publication.

5. Specifically, if the number of languages indigenous to the Southeast were > 44 (the number for the Northwest Coast), then the Spearman rank-order correlation used here would increase from 0.661 ($p < .05$) to 0.733 ($p < 0.02$). I do not have sufficient knowledge of the Southeast culture area to judge whether this is a realistic possibility, although after reading some of the ethnohistoric and archaeological literature on aboriginal population density and postcontact decline (e.g., Ramenofsky 1987), I have the impression that it is.

6. This lack of correlation holds for bird diversity and mammal diversity separately ($r_s = 0.136$ and -0.103, respectively, $p > 0.5$ in both cases), as well as with the combined measure of avian-mammalian diversity (i.e., the simple sum of species richness for the two taxa) shown in figure 5.3.

7. Net primary productivity (NPP) is defined as the amount of energy (in calories/m^2/yr) or biomass (dry organic matter, in grams/m^2/yr) fixed or produced by photosynthesizing plants, minus the energy or biomass utilized by these plants in respiration (self-maintenance). NPP varies widely according to sunlight, moisture, temperature, and nutrient availability, ranging from > 3000 g/m^2/yr in young tropical forests and rice paddies to < 200 g/m^2/yr in dry or cold deserts or the open ocean.

REFERENCES

Alcorn, J.B. 1996. Is biodiversity conserved by indigenous peoples? In *Ethnobiology in Human Welfare*, ed. S.K. Jain. Pp. 234–238. New Delhi: Deep Publications.

Alvard, M.S. 1993. Testing the "ecologically noble savage" hypothesis: Interspecific prey choice by Piro hunters of Amazonian Peru. *Human Ecology* 4:355–387.

Alvard, M.S. 1994. Conservation by native peoples: Prey choice in a depleted habitat. *Human Nature* 5:127–154.

Alvard, M.S. 1998. Evolutionary ecology and resource conservation. *Evolutionary Anthropology* 7:62–74.

Balée, W. 1994. *Footprints of the Forest: Ka'apor Ethnobotany: The Historical Ecology of Plant Utilization by an Amazonian People.* New York: Columbia University Press.

Barth, F. 1956. Ecologic relationships of ethnic groups in Swat, North Pakistan. *American Anthropologist* 58:1079–1089.

Berkes, F., D. Feeny, B.J. McCay, and J.M. Acheson. 1989. The benefits of the commons. *Nature* 340:91–93.

Bodley, J.H. 1996. *Anthropology and Contemporary Human Problems.* Mountain View, Calif.: Mayfield Publishing Co.

Boyd, R., and P.J. Richerson. 1985. *Culture and the Evolutionary Process.* Chicago: University of Chicago Press.

Brightman, R.A. 1987. Conservation and resource depletion: The case of the boreal forest Algonquians. In *The Question of the Commons*, ed. B.J. McCay and J.M. Acheson. Pp. 121–141. Tucson: University of Arizona Press.

Broughton, J.M. 1994. Late Holocene resource intensification in Sacramento Valley, California: The vertebrate evidence. *Journal of Archaeological Science* 21:501–514.

Brush, S.B. 1995. In situ conservation of landraces in centers of crop diversity. *Crop Science* 35:346–354.

Bye, R. 1993. The role of humans in the diversification of plants in Mexico. In *Biological Diversity of Mexico: Origins and Distribution,* ed. T.P. Ramamoorthy et al. Pp. 707–731. New York and Oxford: Oxford University Press.

Cavalli-Sforza, L.L., A. Piazza, P. Menozzi, and J. Mountain. 1988. Reconstruction of human evolution: Bringing together genetic, archaeological, and linguistic data. *Proceedings of the National Academy of Sciences* 85:6002–6006.

Cavalli-Sforza, L.L., A. Piazza, P. Menozzi, and J. Mountain. 1989. Genetic and linguistic evolution. *Science* 244:1128–1129.

Cavalli-Sforza, L.L., E. Minch, and J. Mountain. 1992. Coevolution of genes and language revisited. *Proceedings of the National Academy of Sciences* 89:5620–5624.

Chapin, M. 1992. The co-existence of indigenous peoples and environments in Central America. *Research and Exploration* 82 [inset map].

Diamond, J.M. 1997. *Guns, Germs, and Steel: The Fates of Human Societies.* New York: W.W. Norton.

Dinerstein, E., D.M. Olson, D.J. Graham, A.L. Webster, S.A. Primm, M.P. Bookbinder, and G. Ledec. 1995. *A Conservation Assessment of the Terrestrial Ecoregions of Latin America and the Caribbean.* Washington, D.C.: World Bank.

Durham, W.H. 1979. *Scarcity and Survival in Central America: Ecological Causes of the Soccer War.* Stanford, Calif.: Stanford University Press.

Durham, W.H. 1991. *Coevolution: Genes, Culture, and Human Diversity.* Stanford, Calif.: Stanford University Press.

Dyson-Hudson, R., and E.A. Smith. 1978. Human territoriality: An ecological reassessment. *American Anthropologist* 80:21–41.

Fedick, S.L., ed. 1996. *The Managed Mosaic: Ancient Maya Agriculture and Resource Use.* Salt Lake City: University of Utah Press.

Geertz, C. 1963. *Agricultural Involution.* Berkeley: University of California Press.

Goddard, I. 1978. Eastern Algonquian languages. In *Handbook of North American Indians,* vol. 15: *Northeast,* ed. B. Trigger. Pp. 70–77. Washington, D.C.: Smithsonian Institution.

Grimes, B.F., ed. 1992. *Ethnologue: Languages of the World.* 12th edition. Dallas: Summer Institute of Linguistics.

Hale, K., and D. Harris. 1979. Historical linguistics and archeology. In *Handbook of North American Indians,* vol. 9: *Southwest Pueblo,* ed. A. Ortiz. Pp. 170–177. Washington, D.C.: Smithsonian Institution.

Hames, R.B. 1991. Wildlife conservation in tribal societies. In *Biodiversity: Culture, Conservation, and Ecodevelopment,* ed. M.L. Oldfield and J.B. Alcorn. Pp. 172–199. Boulder, Colo.: Westview Press.

Harmon, D. 1996. Losing species, losing languages: Connections between biological and linguistic diversity. *Southwest Journal of Linguistics* 15:89–108.

Hughes, J.D. 1996. *Pan's Travails: Environmental Problems of the Ancient Greeks and Romans.* Baltimore: Johns Hopkins University Press.

Hunn, E.S. 1982. Mobility as a factor limiting resource use in the Columbia Plateau of North America. In *Resource Managers: North America and Australian Hunter-Gatherers*, ed. N.M. Williams and E.S. Hunn. Pp. 17–43. Boulder, Colo.: Westview Press.

Hunn, E.S. 1990. *Nch'i-Wana "The Big River": Mid-Columbia Indians and Their Land*. Seattle: University of Washington Press.

Johannes, R.E. 1978. Traditional marine conservation methods in Oceania and their demise. *Annual Review of Ecology and Systematics* 9:349–364.

Johnson, A. 1989. How the Machiguenga manage resources: Conservation or exploitation of nature? In *Resource Management in Amazonia: Indigenous and Folk Strategies*, Advances in Economic Botany vol. 7, ed. D.A. Posey and W. Balée. Pp. 213–222. The Bronx, N.Y.: New York Botanical Garden Press.

Jorgensen, J.G. 1980. *Western Indians: Comparative Environments, Languages, and Cultures of 172 Western American Indian Tribes*. San Francisco: W.H. Freeman and Co.

Kehoe, A.B. 1992. *North American Indians: A Comprehensive Account*. 2d ed. Englewood Cliffs, N.J.: Prentice-Hall.

Kendall, M.B. 1983. Yuman languages. In *Handbook of North American Indians*, vol. 10: *Southwest Non-Pueblo*, ed. A. Ortiz. Pp. 4–12. Washington, D.C.: Smithsonian Institution.

Kirch, P.V., and T.L. Hunt, eds. 1997. *Historical Ecology in the Pacific Islands: Prehistoric Environmental and Landscape Change*. New Haven: Yale University Press.

Kohler, T. 1992. Prehistoric human impact on the environment in the upland North-American Southwest. *Population and Environment* 13:255–268.

Kroeber, A.L. 1939. *Cultural and Natural Areas of Native North America*. University of California Publications in American Archaeology and Ethnology no. 38. Reprinted 1963 by University of California Press, Berkeley.

Lansing, J.S. 1991. *Priests and Programmers: Technologies of Power in the Engineered Landscape of Bali*. Princeton, N.J.: Princeton University Press.

Lounsbury, F.G. 1978. Iroquoian languages. In *Handbook of North American Indians*, vol. 15: *Northeast*, ed. B. Trigger. Pp. 334–343. Washington, D.C.: Smithsonian Institution.

Low, B.S. 1996. Behavioral ecology of conservation in traditional societies. *Human Nature* 74:353–379.

Mace, R., and M. Pagel. 1994. The comparative method in anthropology. *Current Anthropology* 35:549–564.

McNeely, J.A., and W.S. Keeton. 1995. The interaction between biological and cultural diversity. In *Cultural Landscapes of Universal Value: Components of a Global Strategy*, ed. B. von Droste, H. Plachter, and M. Rossler. Pp. 25–37. Jena and New York: Fischer Verlag and UNESCO.

Maffi, L. 1996. Position paper for the interdisciplinary working conference Endangered Languages, Endangered Knowledge, Endangered Environments, Berkeley, California, October 25–27, 1996. Institute of Cognitive Studies, University of California, Berkeley.

Miller, W.R. 1986. Numic languages. In *Handbook of North American Indians*, vol. 11: *Great Basin*, ed. W.L. d'Azevedo. Pp. 98–106. Washington, D.C.: Smithsonian Institution.

Nabhan, G.P. 1989. *Enduring Seeds: Native American Agriculture and Wild Plant Conservation*. San Francisco: North Point Press.

Nichols, J. 1992. *Linguistic Diversity in Space and Time*. Chicago: University of Chicago Press.

Norgaard, R.B. 1988. The rise of the global exchange economy and the loss of biological diversity. In *Biodiversity*, ed. E.O. Wilson. Pp. 206–216. Washington, D.C.: National Academy Press.

Paine, R.R., and A.C. Freter. 1996. Environmental degradation and the Classic Maya collapse at Copan, Honduras a.d. 600–1250: Evidence from studies of household survival. *Ancient Mesoamerica* 71:37–47.

Painter, M., and W.H. Durham, eds. 1995. *The Social Causes of Environmental Destruction in Latin America*. Ann Arbor: University of Michigan Press.

Ponting, C. 1991. *A Green History of the World: The Environment and the Collapse of Great Civilizations*. New York: Penguin Books.

Poole, P. 1995. *Indigenous Peoples, Mapping, and Biodiversity Conservation*. BSP People and Forests Program Discussion Paper Series. Washington, D.C.: Biodiversity Support Program.

Posey, D.A., and W. Balée, eds. 1989. *Resource Management in Amazonia: Indigenous and Folk Strategies*. Advances in Economic Botany 7. The Bronx, N.Y.: New York Botanical Garden Press.

Ramenofsky, A.F. 1987. *Vectors of Death: The Archaeology of European Contact*. Albuquerque: University of New Mexico Press.

Redford, K.H. 1991. The ecologically noble savage. *Orion* 9:24–29.

Redford, K.H., and A.M. Stearman. 1993. Forest-dwelling native Amazonians and the conservation of biodiversity: Interests in common or in collision? *Conservation Biology* 7:248–255.

Ricketts, T., et al. 1997. *A Conservation Assessment of Terrestrial Ecoregions of North America*. Draft report. Washington, D.C.: WWF U.S. and WWF Canada.

Russell-Smith, J., et al. 1997. Aboriginal resource utilization and fire management practice in western Arnhem Land, monsoonal northern Australia: Notes for prehistory, lessons for the future. *Human Ecology* 252:159–195.

Ruttan, L. 1998. Closing the commons: Cooperation for gain or restraint? *Human Ecology* 26:43–66.

Sherzer, J. 1991. Genetic classification of the languages of the Americas. In *America in 1492*, ed. A.M. Josephy, Jr. Pp. 445–448. New York: Vintage Books.

Shipley, W.F. 1978. Native languages of California. In *Handbook of North American Indians*, vol. 8: *California*, ed. R.F. Heizer. Pp. 80–90. Washington, D.C.: Smithsonian Institution.

Smith, E.A., and J. McCarter, eds. 1997. *Contested Arctic: Indigenous Peoples, Nation States, and Circumpolar Environments*. Seattle: University of Washington Press.

Sponsel, L.E., and P. Loya. 1993. "Rivers of hunger"? Indigenous resource management in the oligotrophic ecosystems of the Rio Negro, Amazonas, Venezuela. In *Tropical Forests, People, and Food: Biocultural Interactions and Applications*, ed. C.M. Hladik et al. Pp. 435–446. Paris: UNESCO and Parthenon.

Steadman, D.W. 1995. Prehistoric extinctions of Pacific Island birds: Biodiversity meets zooarchaeology. *Science* 267:1123–1131.

Thompson, L.C., and M.D. Kinkade. 1990. Languages. In *Handbook of North American Indians*, vol. 7: *Northwest Coast*, ed. W. Suttles. Pp. 30–51. Washington, D.C.: Smithsonian Institution.

Toledo, V.M. 1994. Biodiversity and cultural diversity in Mexico. *Different Drummer* 13:96–99.

Wilcox, B.A., and K.N. Duin. 1995. Indigenous cultural and biological diversity: Overlapping values of Latin American ecoregions. *Cultural Survival Quarterly* 184:49–53.

Wiseman, F.M. 1985. Agriculture and vegetation dynamics of the Maya collapse in central Peten, Guatemala. *Papers of the Peabody Museum of Archaeology and Ethnology* 77:63–71.

Wissler, C. 1924. The relations of nature to man as illustrated by the North American Indian. *Ecology* 5:311–318.

Wissler, C. 1926. *The Relation of Nature to Man in Aboriginal America.* Oxford: Oxford University Press.

Young, R.F. 1983. Apachean languages. In *Handbook of North American Indians,* vol. 10: *Southwest Non-Pueblo,* ed. A. Ortiz. Pp. 393–400. Washington D.C.: Smithsonian Institution.

6

PROSPECTS FOR THE PERSISTENCE OF "ENDEMIC" CULTURAL SYSTEMS OF TRADITIONAL ENVIRONMENTAL KNOWLEDGE

A Zapotec Example

Eugene S. Hunn

The guiding premise of this volume is that there is an essential link between bio-diversity, on the one hand, and linguistic and cultural diversity, on the other. This linkage cannot be assumed but must be demonstrated. This requires that we clearly understand what is meant by "diversity" in each case and that we develop measures of diversity applicable to each that will allow meaningful comparisons. Then we may investigate how biological diversity is related to linguistic and cultural diversity and the causal relationships responsible for those connections.

WHAT DO WE MEAN BY "DIVERSITY"?

Diversity is from one perspective simply a measure of "density of difference," which may be realized by counting how many kinds of something exist within a bounded area of a certain size. Let us take the example of my current research site, the Zapotec-speaking *municipio* of San Juan Mixtepec in the Sierra de Miahuatlán of Oaxaca, Mexico. Oaxaca is said to be the Mexican state with the greatest bio-diversity because it sustains an estimated 9000+ species of plants and more than 700 species of birds in 95,364 km². Likewise, Mexico is ranked among the half-dozen most biodiverse nations on earth with some 22,000 vascular plant species and over 1000 species of birds in 1,972,544 km² (Ramamoorthy et al. 1993; Rzedowski 1993; Howell and Webb 1995). Oaxaca also ranks high in linguistic diversity. More than 100 mutually unintelligible languages are spoken within the state (Rendon 1993).

A measure of diversity is useful for comparative purposes if it employs units of difference that are clearly defined, for example, biological species, and geographic units of comparable size, since the relationship between density and area is non-linear.[1] Other units of taxonomic difference may be employed to measure bio-diversity: for example, genera or families—moving up the taxonomic hierarchy—or lineages, populations, or cultivars—moving downwards.[2]

Moreover, not all species are equal in their contribution to biodiversity. In particular, *endemic* species, that is, species strictly limited in their geographic range, are believed to contribute more to biodiversity than widely distributed species.[3] The extinction of an endemic species represents a greater loss of genetic information than the extirpation of a population of the same size of a widely distributed species. A species, regardless of its total population and geographic range, represents a quantity of genetic variability an order of magnitude greater than the genetic variability within a species. A species is a pool of genetic information sufficient to define a unique "way of life," a unique survival strategy.

To appreciate this point it is useful to imagine two ideal extremes. Consider a world in which all species are of universal distribution. Every region would have an equally diverse biota, regardless of size and location, and each species would contribute equally to global and regional biodiversity. There would be no endemic species. Of course ours is not such a world. At the opposite extreme, consider a world in which the floral and faunal inventories of each region are entirely distinct. The biodiversity of such a world would equal the biodiversities of each region. In this world also each species would contribute equally to global biodiversity since every species would be endemic.

Instead, our world contains species of varying distributional capacities, from those found on all continents, such as bracken fern (*Pteridium aquilinum*) and the osprey (*Pandion haliaetus*), to those restricted to particular oceanic islands a few hundred meters wide, such as the Laysan Rail (*Porzana palmeri*).[4] The most widely distributed species are generalists. Many of these are "weedy" species adapted to colonizing disturbed sites. Species with more limited ranges are specialized for the limited possibilities of particular conditions and particular places and are thus particularly vulnerable to extinction. At the extreme, such species are endemic to particular countries. They occupy "islands" of appropriate habitat, which may be actual islands or habitat "islands" such as mountain tops or isolated semideserts. The fascinating irregularity of the earth's surface is ultimately responsible for assuring that the evolutionary process will generate seemingly endless novelty.

COMPARING DIVERSE DIVERSITIES

Can we generalize the concept of biodiversity to encompass linguistic and cultural phenomena? I believe so, but only with appropriate qualifications. Biologi-

cal, linguistic, and cultural systems consist of information that governs the behavior of living organisms. Furthermore, these systems of information have
evolved by means of descent with modification, a basic Darwinian principle. The
evolutionary processes in each case have through long spans of time generated
patterns of diversity across the space of our planet. In the biological case, there is
a broad consensus that the diversity of living things is of the essence, a priceless
treasure. The question before us is: Should our appreciation of biodiversity extend to linguistic and cultural diversity? If so, can we measure linguistic and cultural diversity by the same means we use to measure biological diversity? Will the
measures we have devised to conserve biological diversity be as useful in defense
of linguistic and cultural diversity? Finally, should we extend the special concern
we hold for endemic species to "endemic" languages and cultures?

If the assessment of biodiversity is complex and problematical, the assessment
of linguistic and especially of cultural diversity is considerably more so (table 6.1).
First we must resolve the question of which units of language or culture are
properly analogous to established units of biological diversity, that is, typically a
local population of a biological species. A local population of a species is genetically isolated from the populations of other species with which it interacts within
the local habitat. A habitat supports a *community* of interacting species populations, each of which represents a unique adaptive complex that has evolved over
millions of years. The essence of each species is embodied in the genetic information encoded in the DNA of its component individuals. In the final analysis,
biodiversity is this genetic *information*. Thus E.O. Wilson speaks of species extinction as the burning of a library (1989).

Languages resemble species in a number of important respects for our present
argument. They are classified on the basis of overall similarity into "families,"
"stocks," and "phyla" according to reconstructed evolutionary histories of descent with modification from ancestral forms. The time-depth of separation of a
linguistic stock is an index of its distinctiveness. Though languages may "hybrid-
ize" more readily than species, in particular by borrowing vocabulary, one prominent theorist, at least, has argued that however extensive the borrowings from one
language to another, each retains its essential ancestral integrity (Sapir 1921). Be
that as it may, hybridization is for both languages and species an exception that
proves the rule. Genomes and grammars both are highly complex systems that
powerfully resist radical restructuring through exchanges of information. De-
fenders of linguistic diversity have raised the alarm in recent years, projecting
massive extinctions in the immediate future among the 6000 extant human languages (Krauss 1992). Is the loss of a language not also the "burning of a library"?

But what of cultural diversity? Granted that a culture may be construed as a
specieslike system of information, a blueprint guiding the behavior of the individuals who "bear" the culture in question (Goodenough 1957), or as a system of

Table 6.1   Systems of Diversity

Units/Measures	Biodiversity	Linguistic Diversity	Cultural Diversity
Taxonomic units	families	families	culture areas?
basic units:	*species*	*languages*	*cultures?*
	subspecies	dialects	subcultures?
elementary particle:	gene	word	idea
qua system:	genome: genetic information coded in DNA generates organisms and directs behavior	grammar: generates sentences	cultural system: symbolic information coded in language directs behavior
Ecological units	communities of many interacting species populations; food webs; niche packing	*prestate:* dialect mosaics; *modern:* languages of empire displace tribal languages	*prestate:* mosaic of community-based cultural traditions; *modern:* global culture challenges autonomy of local traditions
Measures of diversity	taxa/area + indices of endemism	languages/area + indices of time depth	cultural endemism, autonomy of local traditions

"memes" (Dawkins 1976)—more properly, I believe, of "ideas" about how to live a human life. Granted also that a multitude of diverse human cultural adaptations have evolved on the planet over the past few hundred thousand years through an evolutionary process of descent with modification, often hand in hand with the languages by means of which these cultural traditions were elaborated and transmitted. These adaptations evolved in response to that same creative irregularity of the earth's surface that gave rise to biological diversity, by virtue of physical (or, occasionally, social) isolation that at least temporarily prevented the free exchange of ideas between the inheritors of historically distinct traditions.

Can we consider these adaptations as "cultural species," and does the human species today constitute a multitude of cultural species more or less autonomous as cultural systems, or is it a single massively complex system? Our answer is critical for the questions before us, the value of cultural diversity and the problem of its conservation. In fact, it is rare to find a localized multicultural community even remotely comparable in its functional complexity to a biological community such as a pond or forest. Where will we find the cultural analogue of a tropical rainforest where a single hectare may encompass hundreds of species of flowering plants and thousands of species of insects all intensely interacting within a stable multispecies community? Rarely do we find hints of such cultural "niche packing."

Barth's classic account of the Pathan region is one suggestive example. Here three distinct ways of life have coexisted for some time by virtue of each exploiting the subsistence resources of the region in complementary ways (Barth 1956). The rarity of such cases, however, highlights the fundamental organizational differences between cultural and biological communities.

Despite the continued presence of both physical and language barriers, humans' mobility and their capacity to learn multiple languages subvert those barriers and open the floodgates to intercultural communication and influence. Often an expanding culture absorbs or destroys its less "successful" neighbors. Yet it is just as clear that cultural diversity persists. Perhaps cultural diversity is more closely analogous to the genetic diversity *within* a species than that preserved in interspecific relationships. It is a diversity of *adaptive strategies* drawn from a single exceedingly diverse "meme pool" rather than a diversity of discrete adaptations. One might argue that the culture of today's world represents a single massive hybrid swarm rather than a complex community of interacting cultural species.

A HISTORICAL ASIDE

Some historical perspective is useful here. Let us divide human history into two epochs, a prestate or "tribal" epoch during which human communities were small enough so that everyone knew everyone else and in which each person was kin to most, all sharing a common language and dialect to the margins of the band or tribe. Each community lived within its traditional territory largely on its own terms. The second epoch saw the subjugation of one community by another resulting in societies organized by class, encompassing ethnic and occupational diversity, with each person, family, neighborhood, guild, and status dependent on the state for their mutual coordination. This trend continues, having now reached a peak of intense interaction in which nations are but pieces in the global mosaic.

In the first epoch, tribal communities approximated autonomous cultural units, occupying isolated territories, interacting with neighbors only at their edges. Collectively they resembled a jigsaw puzzle of allopatric sibling species, as, for example, the species of the Yellow-bellied Sapsucker (*Sphyrapicus varius*) superspecies. We might then have legitimately spoken of cultural species. In the epoch of the state, however, the isolation of allopatric cultural species was overcome by force. As a result, the cultures of the world today function more as a single organism. Cultural evolution has produced not a complex community of diverse cultural species but a single massively complex multicellular individual.

This point of historic transition also represents the apogee of linguistic diversity, estimated at 10,000 distinct languages, each spoken by a few thousand individuals (of a total prestate human population of some 10,000,000). Linguistic diversity has been in decline since, but at a rate that has drastically accelerated in the last hundred years when the reach of national governments and of the world market has penetrated the remaining bastions of tribal autonomy. The 6000 languages that still exist today are sharply divided between languages of empire, such as English, Mandarin, Arabic, and Spanish, each spoken by several hundreds of millions of humans, and "tribal languages" now in a great many cases restricted to a remnant few speakers, often elderly, whose grip on their language is weakening daily. The vast majority of the world's languages are literally drowning in a sea of English. While in some notorious cases tribal languages have been aggressively and systematically targeted for extinction, in most cases they are dying because they have lost their survival value for the present generation, beset by the enticements and exigencies of global capitalism. As autonomous tribal cultures have gone, tribal languages now follow. This has been a typical consequence of the cultural evolutionary process in its later stages. But is it an inevitable consequence thereof?

A RATIONALE FOR PRESERVING CULTURAL DIVERSITY

Given that cultural diversity is not the necessary final outcome of the cultural evolutionary process, should we nevertheless seek to preserve it? Biologists have devised various rhetorical strategies in defense of biodiversity, such as, the loss of as yet undiscovered medicinal marvels; the loss of potential new crops better adapted than the handful we now depend on for the great bulk of our nutritional needs to conditions we may face down the road; the loss of the biological grist we need for our theoretical mills. Less anthropocentrically, to lose biodiversity is to lose a measure of evolutionary flexibility to the detriment of the grand evolutionary project that is life on earth (Wilson 1989).

The arguments for conserving linguistic diversity have been somewhat less widely persuasive, I fear. Utilitarian arguments are hard to come by in defense of linguistic diversity, though concerned linguists decry the loss of this diversity as preventing a more complete theoretical understanding of the phenomenon of human language. More often, we resort to arguments in terms of the *right* of people to retain their natal language, relying on moral suasion to convince the 600,000,000 English speakers that they should grant the same right of expression to the few hundreds or thousands who learned Pomo, Lushootseed, or Navajo at their mother's knee.

Let us propose a somewhat different line of defense, however: that linguistic diversity is of value to us all because of the cultural information it conveys. Specifically, in my case, the argument is that Zapotec is important, and more specifically, Mixtepec Zapotec—one of more than 40 mutually unintelligible dialects of

Zapotec (Reeck 1991)—is important because of an intrinsic connection between the diversity of languages and the richness of our understanding of the natural world, and that that richness of understanding is a resource of value to us all now and in the future.

In one aspect this argument is a simple extension of the utilitarian argument for preserving biodiversity. This argument has been widely popularized by Mark Plotkin (1993), whose heroic exploits in search of the ethnomedical secrets of Amazonian shamans elevated his face to the cover of *Time* magazine. But there is a serious weakness in this approach, to wit, that once Plotkin has done his ethnobotanical work and the secrets extracted, the shaman's language is no longer of much interest. Can we devise stronger arguments for the preservation of linguistic diversity that do not depend on the faith that exploitable secrets about the natural world might be discovered by conversing with speakers of rare languages?

Here we encounter an ironic twist. To argue that linguistic diversity is worth preserving not only because it is a human right but also because we need it would seem to require that the world's languages differ fundamentally from one another, so that a Navajo or Zapotec understanding of the world is fundamentally distinct from—and thus not replaceable by—a Spanish or an English understanding of that world. In short, our defense would appear to rest on a Whorfian foundation, which in recent decades has begun to crack before the big guns of Chomskyan theory (Pinker 1994:59–64). Ethnobiological theory has likewise seriously undermined the relativist position. I refer in particular to Berlin's universalist arguments (1992) and the mounting psychological evidence that ethnobiological categories are grounded less in any particular language than in a specieswide capacity for apprehending the real structure of the living world.

Consider the specifics of my ethnobiological case study in San Juan Mixtepec, Oaxaca. Should I conclude that it really makes no difference if the biological diversity of the Sierra Sur of Oaxaca is inventoried by indigenous speakers of Mixtepec Zapotec or by Spanish or English-speaking professional biologists?

I must grant that I believe that the evidence indicates that Zapotec, Spanish, and English-speaking observers are capable of *seeing* the same biological reality in this local landscape. I have been putting this to the test, in fact, during the past year, the first of a two-year ethnobiological research project in two Mixtepec Zapotec villages. I have been working with several bilingual native Zapotec speakers, male and female, of San Juan Mixtepec to create an inventory of the plants and animals of their community and its lands. This inventory must be recorded in four languages: Mixtepec Zapotec, Spanish, English, and scientific Latin. The some 1300 named categories we have so far recorded in the local Zapotec language are readily *translated* into their corresponding scientific taxonomic denotata. It has been more difficult to find corresponding Spanish and English terms.

Of what essential value, we must ask, is the Zapotec intermediary? Is my scientific understanding of the biodiversity of the Mixtepec Zapotec terrain significantly enhanced by the participation of Mixtepec Zapotec speakers in the project? Could not a Western trained biologist simply camp out there for a year or so and then produce an inventory of the biodiversity of this region more comprehensive and less subject to error than will be produced by this collaboration? The answer is clearly, "no." The local Zapotec input contributes in unique and significant ways to the grand human project of scientifically characterizing the living world (of which our project is but one small part). I will provide some examples below.

First, their many eyes take in far more than my two. This is the basis for the rapid environmental assessments described by Gary Martin in his ethnobotanical methods handbook (1995). They point out distinctions I might well miss, such as the existence of at least half a dozen varieties of marigolds, known throughout Mexico by the Aztec loan word *cempasúchil*. This is the *flor de muerto*, which adorns the graves of their ancestors amidst a grand profusion of flowers of many colors, shapes, and proveniences on Todos Santos (1 November). The Mixtepec Zapotec classification of marigolds is somewhat more refined than the botanical in that infraspecific distinctions are recognized in at least one case, two cultivars of *Tagetes erecta* L., which they call *guièe-cöb-mzhïg* and *guièe-cöb-yâg*. Zapotec people today are, it seems, obsessed with flowers. So were the Aztecs (Heyden 1983).

They also distinguish all local species of "century plants" or magueys (*Agave*) and sort them by characteristic habitats and primary uses, whether for fiber, pulque, mezcal, medicine, or as habitat for animals. The local Zapotec classification of oaks is highly elaborated. Oaks dominate the ecotone between the deciduous woodland below the village and the pine forest above, where their villages are by preference located. Oaks are highly valued generally for firewood, but that scarcely justifies recognizing at least five folk-generic level categories within this single genus (see table 6.2).

Given that no comprehensive flora exists as yet for Oaxaca nor any recently revised keys to the oaks of Mexico, I believe the Mixtepec Zapotec terminology is at present the most adequate taxonomic treatment available. Our oak example shows that the Mixtepec Zapotec oak terminology is considerably more refined—and corresponds more closely with the biological facts on the ground here—than does the local Spanish vernacular, in which all are called *encinos,* with some inconsistent attempts to distinguish *encino negro, encino blanco,* and *encino corriente.*[5] This deficit is apparent also in the sixteenth-century Spanish terminology. Despite nearly 500 years of residence in Oaxaca, Spanish glosses of indigenous lexical distinctions remain much as they were in the first century of Spanish colonial occupation. Compare the examples from Fray Juan de Córdova's *Vocabulario en Lengua Çapoteca,* published in 1578. Though I have not been able to determine precisely

Table **6.2** Mixtepec Zapotec Oak Classification

Mixtepec Zapotec	Mitla Zapotec	Córdova[a]	Latin
yàg-xìid[-làs]	yag-baxhuii ?	yàga-pìto ?	Quercus conspersa Née
yàg-xìid-diè	yag-zijn̲	yàga-níça, yàga-níta ?	Q. obtusata Humb. and Bonp.
yàg-xìid-sêd*			Myrica cerifera L.
yàg-pxù[-nrùdz]	yag-bixujy	yága-pixóhui	Q. glaucoides Mart. and Gal.
yàg-pxù-diè			Quercus laeta Liebm.?
yàg-pxù-làs			Quercus sp.
yàg-lbiis			Q. acutifolia Née
			Q. laurina Humb. and Bonp.
yàg-zhòg	yag.quechaa	yága-yóo ?	Quercus urbani Trel.
yàg-zhòg-yëets			Quercus rugosa Née?
yàg-rèdz	yag.zajtz ?	yàga-záchi	Quercus magnoliifolia Née
yàg-rèdz-bëy*	yag.fres [<Sp. fresno]	yàga-quillàa	Fraxinus uhdei (Wenzig) Linglesheim

* These non-oak taxa might better be analyzed as folk generic taxa, coordinated rather than subordinated to their respective folk generics, as in English "chestnut" versus "horse chestnut."

[a]Fray Juan de Córdova's Vocabulario en Lengua Çapoteca (1578).

how the oak terms for each variety of Zapotec correspond to each other and to the scientific nomenclature, it is striking that the basic structure of five contrasting folk-generic categories persists, despite the passage of four hundred years since Cordova's work and the fact that the three systems reflect three different ecological contexts. (Cordova's vocabulary reflects Valley and Isthmus contributions.)

It seems clear that if Zapotec, or another comparable indigenous language, were to be replaced by Spanish in rural Oaxacan communities, the result would be not simply lexical replacement but substantial lexical simplification, in the ethnobiological domains, at least. The fact that a scientific terminology exists is no substitute, since that lexicon is the special province of a small number of professional botanists, who are too few to accomplish the inventory of global biodiversity we require, without the help of indigenous experts.

San Juan Mixtepec is, I believe, an "endemic culture," in the sense that the people of that village preserve a rich tradition of environmental knowledge that is a byproduct of the fact that the people of San Juan continue to provide the bulk of their own livelihood by their own labor on their own lands. Like an endemic bird or plant, their way of life is tightly bound to that small piece of the earth's surface they call home. Like an endemic mushroom or orchid, their way of life—their culture—cannot survive anywhere else. And like an endemic bat or snail, their existence enhances the quality of our lives even though we may not realize it.

CAN WE PRESERVE ENDEMIC LANGUAGES
WITH THEIR ETHNOBIOLOGICAL CONTENT?

If, as I have argued, global languages cannot as a rule replace endemic languages without substantial loss of ethnobiological content, the question of how to preserve our store of endemic languages and cultures with their rich ethnobiological content intact comes to the fore. Some scholars see the future of such endemic languages as bleak, soon doomed to extinction due to the subversive power of state-sponsored schools and of the global market and media to impose a global language (Krauss 1992). I believe there is reason for a somewhat more optimistic outlook. Survival of endemic languages may in theory take several paths: (1) survival through isolation from competing languages of wider communication; (2) survival through fusion to one degree or another with a global language, typically with massive lexical replacement (as seems to have happened to many Mexicano communities in central Mexico; see Hill and Hill 1986); and/or (3) survival through bilingual coexistence. The first option is untenable in the contemporary world and, in any case, cannot account for the fact of the survival of most Mesoamerican languages such as Zapotec. The second defeats the purpose we have identified here of preserving the rich ethnobiological content of the indigenous language, since lexical replacement would likely produce, for example, a Zapotec just as impoverished ethnobiologically as the local Spanish vernacular (cf. Hill this volume). The path of bilingual coexistence seems our best hope. What evidence is there of the long-term viability of this alternative?

I have begun a systematic comparison of the ethnobiological entries in Córdova's 1578 dictionary with the contemporary content of two closely related Zapotec languages, Mixtepec and that of Mitla in the Oaxaca Valley (Messer 1978; Stubblefield and Miller de Stubblefield 1991). Even if we cannot confidently match native terms in each of these vocabularies with their Latin equivalents, it is possible to assess the general size and shape of the taxonomies for comparative purposes and to assess the impact of Spanish on the Zapotec ethnobiological content.

My preliminary comparisons suggest that the Zapotec vocabulary of contemporary Mixtepec and Mitla Zapotec speakers is at least as elaborate as that of Córdova's sixteenth-century informants. In fact, Córdova's list of plant names is the shortest of the three, barely 300 entries out of the more than 20,000 entries in his vocabulary. By combining the results of two detailed studies of contemporary Mitla Zapotec ethnobotany (Messer 1978; Stubblefield and Miller de Stubblefield 1991) we have 477 entries, which represent just over 400 terminal taxa. My provisional inventory for San Juan Mixtepec now stands at 1349 entries representing over 699 terminal plant taxa (table 6.3). Clearly, Córdova was not trained nor motivated to produce an exhaustive account of the ethnobotanical repertoire of his consultants. Messer's Mitla research was explicitly ethnobotanical, while the Stubble-

Table 6.3   Size of Ethnobotanical Inventories in Three Zapotec
Plant Nomenclatures

	Mixtepec Zapotec	Mitla Zapotec	Córdova
Total entries[a]	1349	477	304
Total generics	435	291	—
Terminal taxa	699	404	—
Introduced plants	114	79	41
	(8.5%)	(16.6%)	(13.5%)

[a]Includes all variants, synonyms, and taxonomic ranks reported. All plants; but no fungi.

Note: Ambiguities in the interpretation of Córdova's vocabulary precluded calculating values for total generics and terminal taxa.

fields were missionary linguists who incorporated an ethnobotanical inventory as an appendix to their Zapotec dictionary and grammar. Curiously, there is no evidence that Messer and the Stubblefields compared notes.

In each case a certain number of Spanish loans are in use, either as primary terms or as modifiers of composite terms. In each case such loans are predominantly employed to name introduced plants (table 6.4). For example, peach trees (*Prunus persica*) are *yàgadurazno* in Córdova, *yàg-dràz* in San Juan Mixtepec (*durazno* 'peach'). Córdova's *-castilla* 'Castillian' is frequently employed as a modifier to distinguish introduced plants and animals from similar native species. The equivalent for Mixtepec Zapotec is *-xtîl*. For example, wheat (*Triticum aestivum*) is called "Castillian corn" in both cases: compare *xòopa castilla* (Córdova 1987:412) with Mixtepec Zapotec *zhób-xtîl*. These Spanish borrowings exhibit greater phonological modification to Zapotec norms in the contemporary lexicons than in Córdova's, reflecting four hundred years of use. But the underlying conceptual relationships the names entail is frequently precisely the same.

In sum, on the basis of my preliminary comparisons, I feel confident in concluding that in the case of Zapotec, at least, despite nearly five hundred years of intensive interaction with and subjection to Spanish colonial rule and subsequent Mexican state control, the local ethnobiological knowledge and its indigenous linguistic context have been preserved with little measurable deterioration. At present the majority of local residents of San Juan and San Pedro Mixtepec (and of a substantial number of other communities of the Sierra Sur) are bilingual as a result of local primary and secondary schooling and constant travel to Spanish-dominated areas of the state and nation for the purposes of trade and work. Yet bilingual residents of both sexes and all ages confidently provide local Zapotec names for literally hundreds of plants and all variety of animals. Children of ten can name most plants encountered in the garden and en route to the milpas and

Table 6.4 Distribution of Spanish-Derived Morphemes in Three Zapotec Plant Nomenclatures

	Mixtepec Zapotec			Mitla Zapotec			Córdova		
	I^a	Native	Total	I	Native	Total	I	Native	Total
All Z	14	934	948	14	335	349	2	253	255
Z > S	4	91	95	5	12	17			
Z = S	49	147	196	31	27	58			
Z < S	4	13	17	0	1	1			
All S	43	50	93	29	23	52			
Totals	114	1235	1349	79	398	477	44	260	304
$p =$		2.26×10^{-65}			6.63×10^{-33}				

^{a}I = plant names for post-Spanish introductions; Z, S = Zapotec and Spanish morphemes in plant names; Z > S refers to names that are composed of both Zapotec and Spanish-derived elements, but the majority of those elements are Zapotec; Z = S refers to names that are composed of both Zapotec and Spanish-derived elements, with equal numbers of elements Zapotec and Spanish; Z < S refers to names that are composed of both Zapotec and Spanish-derived elements, but the majority of those elements are of Spanish derivation

Note: Ambiguities in the interpretation of Córdova's vocabulary precluded calculating values for names combining Zapotec and Spanish-derived elements.

describe in detail how they are used. One child of four who speaks Spanish at least as well as Zapotec—learned as a result of her mother's residence in the coastal district capital of Pochutla until recently and a fact of which her grandfather is proud—nevertheless knows many Zapotec plant and animal names. Reports from other communities, such as Teponaxtla in the Cuicatlán region are less encouraging. There, despite a greater degree of physical and social isolation, the language is rarely spoken by those below the age of 50 (D. Acuca Vásquez, personal communication 1997). The difference appears to be less the simple fact of exposure to Spanish than to a local attitude of shame and embarrassment associated with the indigenous language in the case of Teponaxtla versus conscious pride in the native language in the Mixtepec area.

In conclusion, I believe that a rich ethnobiological content may be transmitted robustly via indigenous "endemic" languages, that this area of language content is extraordinarily persistent when the language retains its role as the repository of knowledge essential to subsistence production and when those who learn to speak it as their first language are proud to speak it. I do not believe the same can be said of "languages of empire." I believe "endemic" languages, like endemic species, have a critical role to play in the preservation of diversity by virtue of their endemism, their specialized attachment to and dependence on restricted local habitats. Such languages are K-selected[6] in the sense that they maintain stable populations in balance with their habitat at a relatively low total population. They are able to coexist with r-selected generalists by virtue of their superior adaptation to their specific environment, which in the cultural case implies a political-legal context that allows local communities to control how local resources are managed and an economic context that allows local residents to maintain a "modest living"[7] primarily by virtue of their sustainable use of local natural resources. These conditions still appear to hold in this part of Mexico. I suggest we seek to identify in more general terms the conditions that will allow such linguistic communities to continue to exist in intimate contact with their neighbors in our contemporary "global village."

CONCLUSIONS

I have argued that biological, linguistic, and cultural phenomena have much in common as information systems generative of behavior. But the evolutionary processes that give rise to diverse communities of biological species produce in the linguistic and cultural cases a contrary tendency. That is, cultural and linguistic diversity is progressively eliminated, a process under way since the Neolithic. Despite this historical movement toward cultural homogeneity, some "endemic cultures" strongly persist, speaking ancestral languages and conserving by that means valuable inventories of traditional environmental knowledge. Such inventories complement what Western science has learned of global biodiversity. The Zapotec languages of Oaxaca, Mexico, are a case in point. After 475 years of colonial domination, they continue to transmit to the next generation ethnobotanical knowledge comparable to that of their sixteenth-century ancestors. This is attributable to the fact that they continue to depend for their subsistence upon a community land base over which they maintain effective control (Hunn 1999). Like endemic species, endemic cultural communities represent unique long-term adaptations to highly localized environments and they deserve our vigorous support in the interest of their continued survival as examples of other ways of living on Earth.

ACKNOWLEDGMENTS

My research in San Juan Mixtepec is supported by NSF grant SBR-9515395. I am grateful to the people of San Juan and San Pedro Mixtepec for their hospitality and patient instruction and to the personnel of SERBO, A.C., for many valuable services and much useful advice. Thanks especially to Alejandro de Ávila for my initial introduction to these communities and to Robert Bye and Edelmira Linares for their generous hospitality in Mexico City. Finally, I mourn the untimely death of my colleague and partner in this research, Biol. Donato Acuca Vásquez, and dedicate this effort to his spirit.

NOTES

1. Thus over 70 percent of the bird species known from Mexico have been recorded in Oaxaca, though the state contains less than 5 percent of Mexico's land. However, no one would suggest that Oaxaca is 15 times as diverse as Mexico.

2. Measures of biodiversity are also affected by the relative arbitrariness of geographic boundaries; see, for example, the comparisons of Mexico as political unit versus "Greater Mexico" as bioregional unit in Rzedowski (1993).

3. Though a widespread species will likely exhibit greater internal diversity over its broad range, the total of its internal diversity will be less than that of an allopatric series of distinct species spread over the same range.

4. Note that bracken exhibits one type of widespread distribution, that associated with wide habitat tolerance, while the osprey exhibits a type characterized by narrow habitat tolerance.

5. The distinction between *encino,* 'live oak', and *roble,* 'deciduous oak', meaningful in Spain, has apparently dropped out of active use in Mexico today.

6. I find the traditional ecological dichotomy between K-selected and r-selected species to be a useful analogy for languages and cultures. K-selected species are those whose populations have stabilized at or near their carrying capacity (K), which is a function of their stable role in a complex community of predators and prey. Presumably, such species have adopted a reproductive strategy that is "conservative," closely adjusted to the limits of their stable environmental niche. By contrast, r-selected species pursue a reproductive strategy limited only by "r," their maximal rate of population increase. Such species are well adapted for colonizing recently disturbed sites. They reproduce very rapidly in the absence of competition from the regional specialists that have been temporarily eliminated by the disturbance, but ultimately they will be replaced by these same K-selected specialists. The allegory of the race between the tortoise and the hare is apropos. Perhaps "tribal" cultures and languages, like the tortoise, will endure after the capitalist hare dozes off short of the finish line.

7. A "modest living" is the criterion established by the U.S. Supreme Court in its review of the famed Boldt decision with respect to Indian treaty fishing rights on Puget Sound. It represents a legal distinction between subsistence-based resource use versus profit-oriented exploitation. See Cohen 1986.

REFERENCES

Barth, F. 1956. Ecologic relationships of ethnic groups in Swat, North Pakistan. *American Anthropologist* 58:1079–1089.

Berlin, B. 1992. *Ethnobiological Classification: Principles of Categorization of Plants and Animals in Traditional Societies.* Princeton, N.J.: Princeton University Press.

Cohen, F.G. 1986. *Treaties on Trial: The Continuing Controversy over Northwest Indian Fishing Rights.* Seattle: University of Washington Press.

Córdova, Fray J. de. 1987 (1578). *Vocabulario en Lengua Çapoteca.* Edición Facsimilar. México, D.F.: Ediciones Toledo.

Dawkins, R. 1976. *The Selfish Gene*. New York: Oxford University Press.

Goodenough, W.H. 1957. Cultural anthropology and linguistics. In *Report of the Seventh Annual Round Table Meeting on Linguistics and Language Study*. Pp. 167–173. Series on Language and Linguistics no. 9. Washington, D.C.: Georgetown University Press.

Heyden, D. 1983. *Mitología y Simbolismo de la Flora en el México Prehispánico*. México, D.F.: Universidad Autónoma de México, Instituto de Investigaciones Antropológicas.

Hill, J.H., and K.C. Hill. 1986. *Speaking Mexicano*. Tucson: University of Arizona Press.

Howell, S.N.G., and S. Webb. 1995. *A Guide to the Birds of Mexico and Northern Central America*. Oxford: Oxford University Press.

Hunn, E.S. 1999. The value of subsistence for the future of the world. In *Ethnoecology: Situated Knowledge/Located Lives*, ed. V. Nazarea Sandoval. Pp. 23–36. Tucson: University of Arizona Press.

Krauss, M. 1992. The world's languages in crisis. *Language* 68:4–10.

Martin, G. 1995. *Ethnobotany: A Methods Manual*. London: Chapman & Hall.

Messer, E. 1978. Zapotec plant knowledge: Classification, uses, and communication about plants in Mitla, Oaxaca. In *Prehistory and Human Ecology of the Valley of Oaxaca*, vol. 5, part 2, ed. K.V. Flannery and R.E. Blanton. Pp. 1–140. Ann Arbor: University of Michigan Press.

Pinker, S. 1994. *The Language Instinct: How the Mind Creates Language*. New York: Harper Perennial.

Plotkin, M.J. 1993. *Tales of a Shaman's Apprentice: An Ethnobotanist Searches for New Medicines in the Amazon Rain Forest*. New York: Penguin Books.

Ramamoorthy, T.P., et al., eds. 1993. *Biological Diversity of Mexico: Origins and Distribution*. Oxford: Oxford University Press.

Reeck, R. 1991. A trilingual dictionary in Zapotec, English and Spanish. M.A. thesis, Universidad de las Americas, Puebla, México.

Rendón, J.J. 1995. *Diversificación de las Lenguas Zapotecas*. Oaxaca, México: Centro de Investigaciones y Estudios Superiores en Antropología Social-Oaxaca, Instituto Oaxaqueño de las Culturas.

Rzedowski, J. 1993. Diversity and origins of the phanerogamic flora of Mexico. In *Biological Diversity of Mexico: Origins and Distribution*, ed. T. P. Ramamoorthy et al. Pp. 129–144. Oxford: Oxford University Press.

Sapir, E. 1921. *Language: An Introduction to the Study of Speech*. San Diego, Calif.: Harcourt Brace Jovanovich.

Stubblefield, M., and C. Miller de Stubblefield. 1991. *Diccionario Zapoteco de Mitla*. Vocabularios Indígenas 31. México, D.F.: Instituto Ligüístico de Verano, A.C.

Wilson, E.O. 1989. *Biodiversity*. Washington, D.C.: National Academy Press.

7

ECOLINGUISTICS, LINGUISTIC DIVERSITY, ECOLOGICAL DIVERSITY

Peter Mühlhäusler

> Coincidences of tribal boundaries to local ecology are not uncommon and imply that a given group of people may achieve stability by becoming the most efficient users of a given area and understanding its potentialities.
>
> Norman Tindale, *Aboriginal Tribes of Australia* (1974:133)

Ecolinguistics, as characterized by Fill (1996), is concerned with two main issues: (a) ecological embeddedness of human communication systems (i.e., language not being a self-contained system but an integral part of a larger ecosystem), and (b) the analysis of environmental discourses (i.e., both how people talk about the local environment, and the discourses of environmentalism). Its key concepts are those of diversity and of functional interrelationships. Its conceptual roots can be traced back to the writings of Humboldt (1836) and Whorf (1956). Both authors address the topic of the function of language diversity and suggest that different languages afford their users different perspectives on the world. The notion of functional interrelationships has a much shorter history in linguistics. It was first raised in the writings of Haugen (1972) with regard to relationships between languages and in my own and others' writings for the relationship between languages and their natural and cultural ecology (Mühlhäusler 1983, 1996a, b; Trampe 1990; Fill 1993; Brockmeier, Harré, and Mühlhäusler 1999). Meanwhile, ecolinguistics has developed rapidly and its results have begun to be taken into account by environmental scholars, though it is too early to speak of a linguistic turn of environmental studies.

LINGUISTIC APPROACHES TO DIVERSITY

Mainstream modern linguistics has concerned itself—as its other name, general linguistics, suggests—with general principles of *language*, and to be focused on language, not languages, has been its trademark. Significantly, most linguistic pronouncements in recent years have been about principles of formalization or discovery of descriptive devices capable of being applied to a wider and wider range of language data (for a detailed discussion see appendix to Mühlhäusler 1996a).

Those linguists who have concerned themselves with languages, particularly the linguistic picture of largely multilingual areas such as West Africa, the Americas, or Melanesia held views that made an ecological understanding difficult. Thus, diversity has tended to be regarded as dysfunctional, the unintended result of language splits, isolation, and lack of cooperation. This view is understandable if one takes into account the mechanistic model of human communication subscribed to by most linguists. That languages, next to transmitting information (their communicative function) also serve a large range of metacommunicative functions such as marking and sustaining group identity has become a topic only fairly recently, begun with the pioneering studies by LePage and Tabouret-Keller (1985).

A particular obstacle to an ecological view has been the working hypothesis that languages are self-contained independent systems and that the boundary between languages and their external environment and between individual languages are categorical. Where linguists have recognized the existence of linguistic diversity, this diversity has been portrayed as an inventory of named entities, related to one another like biological species on a family tree. Moreover, the documentation of human languages has been highly biased toward descriptions of single languages spoken by a group of speakers rather than the speech repertoire of a communication community or the question: What languages are employed when these speakers communicate with outsiders? As a consequence, we have a vast body of descriptions of individual vernaculars and an almost total lack of description of languages of intergroup communication. That the languages of intercultural communication in the Pacific area were perhaps as numerous as its vernaculars was shown only very recently in *Atlas of Languages of Intercultural Communication in the Pacific, Asia, and the Americas* (Wurm, Mühlhäusler and Tryon 1996).

The independence hypothesis precluded another set of research questions, including questions about adaptation of ways of speaking to specific environmental conditions (see Mühlhäusler 1996b). Such questions are now being raised by ecolinguists who propose that the result of long periods of time in which speakers develop their own interpretation of their environment has been a situation that Whorf (1956:244) characterized as follows: "Western culture has made, through language, a provisional analysis of reality and, without correctives, holds resolutely to that analysis as final. The only correctives lie in all those other tongues which by aeons of independent evolution have arrived at different, but equally logical provisional analyses." Diversity, in this view, reflects neither regressive compartmentalization nor "progress" in the Western sense, but a large number of progresses (as well as misreadings). Prehistory is full of examples of languages and cultures dying out as a result of misreading their environment.

Diversity of languages in the view I have presented here emerges as a vast repository of accumulated human knowledge and experience, or—to use a term which is becoming fashionable in many branches of knowledge—a memory. By this I mean that in a way comparable to that in which sea currents or layers of ice are "memories" of short- and long-term climatic changes and books are memories of literate cultures, human languages are memories of human inventiveness, adaptation, and survival skills. How to access and read these memories remains an awesome task, but, as Whorf once remarked (1956:215) with respect to the contribution of American Indian languages to human knowledge: "To exclude the evidence which their languages offer as to what the human mind can do is like expecting botanists to study nothing but food plants and hothouse roses and then tell us what the plant world is like!"

LINGUISTIC AND BIOLOGICAL DIVERSITY

The argument I wish to put forward is that our ability to get on with our environment is a function of our knowledge of it and that by combining specialist knowledge from many languages and by reversing the one-way flow of knowledge dominating the world's education system, solutions to our many environmental problems may be found. In particular, learning from local knowledge, such as learning from the insights and errors of traditional rainforest dwellers or desert nomads, could result in a more informed base for the sustained survival of our species. Such knowledge, I argue, is closely linked to language.

The chance of a productive symbiosis between linguistic-cultural and biological diversity is constrained by two major factors: first, the rapid disappearance of biological species, and, second, the even more rapid disappearance of linguistic diversity. Of the more than 6000 languages estimated currently to be spoken, as many as 95 percent are believed by some linguists to be on the endangered list, and their rate of extinction appears to be far greater than that of biological species. How big the former loss is to some extent must remain guesswork, as human perception of this loss presupposes knowing the names of the species that are lost. By names I do not mean just labels for single species, nor scientific labels, but native local names, as well as local names for all kinds of ecologies and, very importantly, names of parts of plants and animals of use to human beings. Let me illustrate what I mean with examples selected from the lexicon of Enga, a Papuan language of the New Guinea Highlands. The compiler of this dictionary (Lang 1975) lists a number of tree names for known species, including:

tree – breadfruit (*Ficus dammaropsis*) *kúpí, tokáka, yakáte, yongáte* (T).
  – breadfruit (wild) *yokopáti.*
  – casuarina (*Casuarina oligodon*) *kupiama, yawále.*
  – cedar (*Papuacedrus papuan* [F. Muell]) *ayápa.*
  – evergreen (*Podocarpus compactus* Wassch./*P. imbricatus*/*P. papuanus*) *páu.*
  – evergreen (*Podocarpus neriifolius*/*P. pilgeri*) *káipu.*
  – fig (*Ficus* sp.) **peke itá.*
  – mahogany (*Dysoxylum* sp.) *mamá.*
  – oak (*Lithocarpus* sp.) *lépa.*
  – palm *mulái.*

Lang (1975) also has a long list of tree names not yet described by European botanists, a list that in all likelihood further research would make considerably longer:

tree – kind of *andaita, anguana, áuki, bóna, gii, káepu, kendu, kipondu, kumú, kúngu, laikiláki, lombá, lyáka, lyakati, lyungúna, matopá, naipí, náka, nápu, opáka, pálá, patepá, péké, pelepéle, pulaka, sángú, sápo, sukú, suú, wayapé, waŋame, wano, yandále, yóké.*

It is not inconceivable that the massive logging program currently carried out in Papua New Guinea and recent bushfires triggered by poor forest management will lead to the disappearance of species whose name is known only to the peoples who used to live among them.

From an anthropocentric and utilitarian perspective, the perspective that inescapably drives the human perceptions of nature, most prominent among the Enga names for plants and plant parts are those that this particular culture has identified as being of use as foods, medicines, building materials, and so forth. The Enga dictionary contains a long list of entries naming plants that fall into this category including the following small sample (Lang 1975):

tree – (bark used as rope) light wood *ángewane* (P), *wanépa.*
  – (bark used as string) *enámbó, komau, kotále.*
  – (used in leprosy cure) *dílay.*
  – (used for throwing stick) *kongéma.*
  – (where possums are found) *miná.*
  – (seeds eaten) *ámbea mánga, kétá, tapáé, wáima, yombuta.*
  – (seeds used for hair-dye) *mílya.*
  – (wood used for spears) *mándi.*
  – (used for arrows) *mámá, yupi.*
  – (used for arrows/bows) black *plam* (?) *kupí, mimá.*
  – (used for clubs) *kulepa.*
  – (used for drums) *laíyene.*

Knowledge of these plants is under very considerable threat as the Enga are becoming dependent on foods imported in tins and containers, as their children have to attend government schools where they are expected to acquire nontraditional knowledge (which leaves little time or opportunity to acquire the full traditional knowledge), and as the habitat of much of the indigenous fauna and flora is destroyed to make way for coffee plantations and gardens in which introduced food plants are grown, as well as for roads, towns, and airstrips. Studies of many other languages of the New Guinea area point to very much the same development.

DIVERSITY, ADAPTATION, TIME

Ecologies are functional, adaptive, and dynamic and it is important to focus on these properties rather than static inventories and taxonomies. The question I would like to address is: How do languages adapt to changing environmental conditions?

This question gains importance because of the vastly increased mobility of humans as evidenced in migration, refugee movements, transmigration, tourism, and the like which take speakers of numerous languages into environments where these languages did not develop. Given the magnitude of this question, it is essential to keep the number of variables to a manageable minimum. What I mean is that a study of how Australian Aboriginal peoples adapted their languages to the conditions of a vast continent over 50,000–60,000 years is likely to tell us a great deal less than a study of recent settlements of small populations on small islands.

I have begun a comparative study of a number of such islands (Mühlhäusler 1996b) and arrived at the tentative conclusion that the development of linguistic means to talk about a new environment requires several generations. Let me briefly look at some island situations.

PITCAIRN ISLAND

The story of Pitcairn Island, one of the most isolated islands of Eastern Polynesia (where Tahiti is also found), is well known through several movies dealing with the mutiny on the *Bounty* (see Ball 1973). Pitcairn was settled in 1790 by nine British mutineers, six male and thirteen female Tahitians, with "neither party knowing more than a few words of each other's language and nothing of each other's cultural heritage" (Ross 1964:57). Pitcairn had once been inhabited:

> The mutineers found many cultivated food-trees and plants already growing in their new home: the coco-nut, bread-fruit, taro, plantain, banana and sugar-cane, as well as the [aute] the paper mulberry from which the native cloth [t^pa'], was made. Other plants, such as yams and sweet potatoes, they brought with them; and for livestock, pigs, goats and chickens. Their food supplies, including sea-birds and fish, were therefore those of Tahiti, while their cultivation methods were a blend between

Tahitian tradition and European improvisation—so far as one knows, only Brown, the botanist's assistant, had any previous knowledge of horticultural techniques. (Ross 1964:57–58)

As an uninhabited island, Pitcairn differs from such Indian Ocean islands as Mauritius, Reunion, or the Seychelles, which, in the course of European colonial expansion, were settled by Dutch, French, and English plantation owners and their African slaves. Neither group was familiar with the flora and fauna of those islands. In the case of Pitcairn, however, the larger proportion of the settlers of 1790, that is, those of Tahitian extraction, brought with them the knowledge and linguistic expressions needed to talk about their new environment. A limiting factor was that the Tahitians came from the lower strata of society and would not have had extensive knowledge of medicinal plants and life forms associated with the culture of their chiefs. Moreover, Tahitian was soon replaced by Pitkern (an English-based language with some Tahitian elements) and English, and much of the Tahitian linguistic heritage was lost. What remain are lexical items descriptive of flora, fauna, and topology but not grammatical classification systems or deeper grammar.

Over a few decades the economy changed from a subsistence economy to a market economy in which fruit and vegetables were grown to provide visiting whalers and other vessels. Introduced livestock (specially goats) and excessive use of the scarce timber supply soon caused significant environmental degradation and a number of introduced plant species soon outcompeted local ones. After supplying timber for housing, boat-building, and firewood for over a hundred and seventy years, the supply of indigenous timbers is almost nil today. An introduced plant, *Eugenia jambos,* commonly known as *rose-apple,* covers much of the unused parts of the island and, though generally considered a pest, it supplies nearly all the island's wood; indeed, its presence is the only guarantee of a sufficient supply of firewood. The degradation of Pitcairn was accelerated by the strong vegetarian beliefs its inhabitants held subsequent to their religious conversion, which brought about a decline in hunting of feral animals and a reluctance to use animal manure as fertilizer.

The naming of Pitcairn life forms did profit from the presence of the migrants from nearby Tahiti, although many of the smaller less visible life forms remain unnamed. Pitcairn after all was seen by the mutineers and their companions as a larder from which they could help themselves, not as an ecosystem in need of looking after. The naming of environmental entities bears many similarities with that encountered in other colonial contexts:

Since the flora and fauna of Pitcairn are so very different from those of England, there must, in almost all these cases, have been transfer; presumably the English settlers applied such names as best they could, guided by real or fancied similarities and, sometimes, no doubt, merely by hazy recollection of the English object. The Tahitians may have indulged in the same sort of linguistic practice, but they perhaps did so to a lesser extent by reason of the similarities between their own flora and fauna and those of the Island. Unfortunately, it is not yet possible to discuss these transfers in general. In the case of Pitcairn, an adequate flora and fauna does not yet exist. In the case of Tahiti, an adequate flora and ichthyology does exist; in the former there has been some coupling of scientific and native names, in the latter this has been rather less. The ornithology of Tahiti seems a virtually untouched field. It is, then, only in one special subject that we can make any useful comment on transfer; this is in the case of the Pitcairnese bird-names which are of English origin. The fact that we are able to do this is due to the work of Mr. Williams. I quote from a letter of his (dated July 27th, 1961). "With regard to names like sparrow, wood-pigeon, snipe, sparrow-hawk and hawk I would say that they had been applied to the nearest equivalents of the English birds. The wood-pigeon is a pigeon but you could not mistake it for the true wood-pigeon; the snipe is a shore bird but could never be mistaken for the British snipe. The sparrow does not look like a house-sparrow—or hedge-sparrow—but it was probably the only bird of the general kind of sparrow on the Island." (Ross 1964:166)

The very considerable Tahitian knowledge in using the natural environment of Pitcairn is reflected in a number of areas including coconuts and fish:

(a) coconut terminology

/a	stuff like gray cheese-cloth faded near the top of the coconut trunk at the base of the fronds
etu	sprouting coconut
hiwa	ailing to reach maturity (of coconuts or bananas)
matapele	coconut meat extracted whole from split shell
miti coconut	drinking nut
oʔoʔa	full ripe coconut
paito	the baby coconut
palu	coconut husk
taiʔro	salt water sauce made by rotting strips of green coconut meat in a bottle full of salt water

(b) fish names

fafaij	stingray
ihi	piper or garfish
kuta	barracuda
iai	St. Peter's fish
pa	fish roasted on hot coal
pa:lu:	to use ground bait to attract fish

The growing knowledge of other environmental phenomena is reflected in numerous coinings such as:

tuny-nut	a tree whose nuts have a hole at the top—a note can be produced by blowing across it
whale-bird	birds that feed on whale offal
trumpet-fish	fish with tubulous snouts
soap seed	tree with seeds that lather
shell-in-the-palm	sea shell found on the dead pandanus

That some plants and animals could be dangerous is reflected in terms such as:

dream fish	fish whose meat causes nightmares
poison trout	poisonous trout
taitaia	bad-tasting fish (from the adjective *taitai* 'tasteless')

A considerable number of animals and fish are named after the persons who first caught or used them, as in *Austin bird, Sandford* (a fish), *Frederick* (a fish), *Bernie flower* (introduced by Bernice from Mangareve), *David shell, Allan* (rock submerged at high tide), *Dorcos apple* (pineapple), yet many aspects of Pitcairn nature remain unnamed.

The conditions described by Tindale in the epigram to this chapter do not obtain where humans begin to talk about new environmental conditions, a view which is difficult to square with the misguided egalitarian view of many linguists that all languages are equally capable of expressing what their speakers need to express. I would like to argue, taking up the important points raised by Hymes (1973), that to the contrary languages differ considerably in their ability to do this and that adaptation of any language to a new environment takes several generations of speakers. In the case of Pitcairn, this was not helped by the later relocation of most of the islanders to a very different environment, Norfolk Island.

NORFOLK ISLAND

Norfolk Island lies on the Norfolk Ridge in the South Pacific. One-third the size it was when thrust from the sea bed over two and a half million years ago, the island was discovered by Captain James Cook on 10 October 1774. After circumnavigating the island, Cook landed two small boats on 11 October, finding traces of Polynesian use, but no inhabitants. Noting an abundance of pine and flax (both in great demand in British manufacturing), along with a plethora of unique flora and fauna, Cook is alleged to have recorded in his journal that the island was a "paradise" (e.g., Clarke 1986:9). An examination of his entries for 11 and 12 October 1774, shows that, while including many positive references to the "products" of Norfolk, they do not contain this epithet. He named the island in honor of the noble family Norfolk, which was one of his patrons.

The island remained uninhabited for fourteen years. On 6 March 1788, Lt. Philip Gidley King landed a party of free settlers and convicts detached from the penal settlement at Botany Bay, Australia, occupying Norfolk Island as a possession of the British Crown. But in 1814, the settlement was abandoned because the pine and flax had found been unsuitable for mast and sail making. Its buildings were either burned or demolished to discourage occupation by other nations. Whilst the government officials and a large proportion of the settlers and convicts were speakers of English, a number of other languages were represented as well (see Wright 1986:24ff).

In 1825, Norfolk Island was reoccupied as a penal settlement in which punishment was to be as severe as possible. This prison community was closed in 1855 and the island's activities shifted to whaling, trade, and the support of the Pitcairn Islanders, who were given large tracts of land on Norfolk by the British government. The Pitcairnese were transferred to Norfolk Island because of growing food shortages on Pitcairn. Despite early guarantees from the colonial administrator that they would be given free rein in dividing the lands of Norfolk Island, the Pitcairnese actually only gained about one quarter of the total land area, and the administration of land was undertaken from New South Wales, which at the time was under British colonial rule. Some Pitcairnese then returned to their place of origin and established a second colony. With the arrival of the Pitcairn Islanders on Norfolk a tradition of bilingualism began with English and Pitcairn-Norfolk both being integral parts of the islanders' speech repertoire. (The variant of the informal English-Tahitian language spoken by the Pitcairners on Norfolk gradually changed, and today Pitker and Norfolk are regarded as distinct languages by their speakers.)

In 1867, the Melanesian mission training school was transferred from Auckland, New Zealand, to Norfolk Island. It was controlled by a small number of English missionaries who educated about two hundred Melanesians at any one time. St. Barnabas chapel and training college were built away from the Pitcairnese settlement, but there were regular and friendly contacts between the two. The mission was built along the lines of a British public school and served the utopian goal of converting Melanesia to Christianity and a British way of life. The college, at its peak, consisted of six dormitories holding thirty boys each, while a much smaller number of girls were divided among the households of married missionaries. The common language of the college was a mission lingua franca, Mota, but in communicating with outsiders English and Melanesian Pidgin English appear to have been used. The mission college was closed down in the early 1920s.

The environment of Norfolk Island thus was named by three separate groups over the last two hundred years, and there has been relatively little continuity in their practices. The inhabitants of the penal colony had left when the Pitcairners arrived in 1856, and the Melanesian members of the Mota-speaking mission community lived in physical as well as almost total social isolation, with only the English missionaries communicating with the Pitcairners. Maiden (1903:715), with regard to the tree *Exocarpus phyllanthoides,* remarks that this tree, once called cherry tree by the British, became known as Isaac Wood, after Isaac Quintal from Pitcairn, who first pointed it out. "We therefore have an instance of two sets of vernaculars, the pre-Pitcairn and the post-Pitcairn." The Melanesian mission established a large number of Melanesian and English exotics around the boarding school buildings but their names disappeared when the mission left in the 1920s.

Maiden's account of Norfolk Island flora is of ecolinguistic importance on several counts:

(a) It shows that the majority of endemic and native botanical life forms do not have a local name.

(b) Most of the exotics, particularly those introduced by Pitcairners, are named.

(c) When comparing Maiden's recorded names with the most recent Norfolk dictionary (Harrison 1979), one is struck by the fact that a number of local names have disappeared (see table 7.1).

(d) The same local name often refers to plants of quite different species; for example, "sharkwood" is applied to *Coprosma pilosa* (Rubiaceae) and *Sideroxylon costatum* (Sapotadae). The term "maple" is used for two different tree species: *rauti* is recorded as referring to the Norfolk Island breadfruit (*Cordyline australis*) but also to *Lilium moderne* and *Dracaena* sp. in Harrison (1979).

(e) Maiden also comments repeatedly on the unwillingness of the Norfolk Islanders to look after their environment, as in the following two excerpts:

Making every allowance for the islanders, I still feel that they do not make adequate effort to keep the weeds in check. From all that I could gather, the islanders are something fatalists in the matter of weeds. (1903:768)

The people have so much land that at present they do not feel the deprivation of these areas which are lost to them through being rendered useless with weeds. (1903:769)

One concludes that two hundred years have not been enough for the Norfolk language to get fine-tuned to the complex ecology of Norfolk Island with its large proportion of endemic species and that in the absence of the ability to talk about their new environment the islanders unwittingly contributed to the severe environmental degradation of fauna and flora. Today 95 percent of its rainforests are gone and large areas are invaded by feral exotics.

Table 7.1 Loss of Local Names for Flora in Norfolk:
Maiden's and Harrison's Dictionaries Compared

Flora	Maiden (1903)	Harrison (1979)
waiwai	'beach'	—
home[a] *rauti*	*Cordyline terminalis*	—
neh-e	*Marathia fraxinea* 'treefern'	—

[a]*Home*, the word for 'Pitcairn' among the Pitcairn Islanders and their descendants (Maiden 1903:719).

CONCLUSIONS

There is an important aspect to any type of management: one can manage only what one knows; and a corollary: that one knows that for which one has a linguistic expression. My examples were drawn from situations where people found themselves in a new unknown environment and where they had to develop the necessary knowledge and linguistic resources to live in it. There is ample evidence that human colonization brings with it many negative consequences for the environment because of actions and practices that are in conflict with "nature." Profound changes to flora and fauna predate European colonization and occur in "traditional" societies as much as in "modern" ones. A case study such as that of the Marquesas (see Olson 1989) on the one hand reveals enormous destruction in the initial period; on the other, it also denotes that through learning processes over longer periods of time, an approximation between the contours of language and knowledge and the contours of the environment can be achieved and that, in many instances, a sustainable coexistence can be found.

The knowledge we find in older indigenous languages thus emerges as the outcome of many generations' fine-tuning of a language to local conditions, as suggested by Tindale (1974) for Australian Aboriginal societies, where 50,000 years of occupation contrast with 200 on Norfolk. Languages thus are repositories of past experience and once lost, a great deal of effort will be required to recover what has been lost with them.

REFERENCES

Ball, I.M. 1973. *Pitcairn: Children of the Bounty.* London: Gollancz.

Brockmeier, J., R. Harré, and P. Mühlhäusler. 1999. *Greenspeak.* London and Thousand Oaks: Sage.

Clarke, P. 1986. *Hell and Paradise.* Melbourne: Viking.

Fill, A. 1993. *Ökolinguistik.* Tübingen: Narr.

Fill, A. 1996. Ökologie der Linguistik—Linguistik der Ökologie. In *Sprachökologie und Ökolinguistik*, ed. A. Fill. Pp. 3–16. Tübingen: Stauffenburg.

Harrison, S. 1979. Glossary of the Norfolk Island language. M.A. thesis, Macquarie University, Sydney.

Haugen, E. 1972. *The Ecology of Language.* Selected and introduced by A.S. Dil. Stanford. Stanford University Press.

Humboldt, W. von. 1836. *Über die Verschiedenheit des menschlichen Sprachbaus.* Berlin: Königliche Akademie den Wissenschaften.

Hymes, D. 1973. Speech and language: On the origins and foundations of inequality among speakers. *Daedalus* 102(3):59–85.

Lang, A. 1975. *Enga Dictionary.* Canberra: Pacific Linguistics C-20.

LePage, R.B., and R. Tabouret-Keller. 1985. *Acts of Identity.* Cambridge: Cambridge University Press.

Maiden, J.H. 1903. The Flora of Norfolk Island. *Proceedings of the Linnean Society of New South Wales* 28:692–785.

Mühlhäusler, P. 1983. Talking about environmental issues. *Language and Communication* 3(1):71–81.

Mühlhäusler, P. 1996a. *Linguistic Ecology.* London: Routledge.

Mühlhäusler, P. 1996b. Linguistic adaptation to changed environmental conditions: Some lessons from the past. In *Sprachökologie und Ökolinguistik*, ed. A. Fill. Pp. 105–130. Tübingen: Stauffenburg.

Olson, S.L. 1989. Extinction on islands: Man as a catastrophe. In *Conservation for the Twenty-First Century*, ed. D. Western and M.C. Peal. Pp. 50–53. Oxford: Oxford University Press.

Ross, A.S.C. 1964. *The Pitcairnese Language.* London: Deutsch.

Tindale, N.B. 1974. *Aboriginal Tribes of Australia.* Berkeley: University of California Press.

Trampe, W. 1990. *Ökologische Linguistik.* Opladen: Westfälische Verlag.

Whorf, B.L. 1956. *Language, Thought, and Reality.* Cambridge, Mass.: MIT Press.

Wright, R. 1986. *The Forgotten Generation of Norfolk Island and van Diemen's Land.* Sydney: Library of Australian History.

Wurm, S.A., P. Mühlhäusler, and D.T. Tryon. 1996. *Atlas of Languages of Intercultural Communication in the Pacific, Asia, and the Americas.* Berlin: Mouton de Gruyter.

8

CULTURAL PERCEPTIONS OF ECOLOGICAL INTERACTIONS

An "Endangered People's" Contribution to the Conservation
of Biological and Linguistic Diversity

Gary P. Nabhan

Beginning perhaps with Thoreau and Rousseau, many Western philosophers have attempted to gain insights about the natural world from indigenous peoples, treating them as "native ecologists" whose traditional ecological knowledge is worthy of respect (see papers in Callicott and Nelson 1998). Nevertheless, many scholars have found this premise to be highly controversial, especially since few longitudinal studies have been conducted to confirm or contest the premise that indigenous practices based on traditional ecological knowledge positively benefit biodiversity (Denevan 1992; Diamond 1992).

That debate in its most general terms will not be resolved here, but one issue fundamental to it can be adequately addressed. I will present evidence relating to the claim that certain Native American cultures developed land management practices that enhanced biological diversity locally (Nabhan et àl. 1982; Nabhan 2000) at the same time that they developed lexicons that allowed them to describe precisely interactions among diverse plants and animals (Felger and Moser 1985; Rea 1997; Nabhan 2000).

Nevertheless, most discussions of the coevolutionary relationships between biodiversity and cultural diversity have remained general, expressed largely in terms of geographic correlations rather than causations or direct dependencies (Harmon 1996). Even though coevolutionary hotspots are now being recognized by geographers of biodiversity, few of their studies take into account that humans may be "the coevolutionary animal of them all" (Janzen, as quoted in Thompson

1999). Because such generalities make it difficult to advance any debate, I prefer to focus on recent case studies from indigenous communities in the Sonoran Desert, where it is abundantly clear that desert hunter-gatherers recognize, name, and in some cases manage ecological interactions between native plants and animals.

Although I do not wish to imply that this knowledge is valuable only if it has utility to contemporary societies, there are ways in which indigenous knowledge of ecological interactions involving threatened species could offer Western-trained scientists and resource managers hypotheses to test and strategies to integrate in their collaborative efforts to recover threatened species. Such a possibility depends, however, on how valid Western scientists believe traditional ecological knowledge to be and how compatible the protection of biodiversity and that of cultural diversity appear to be.

While conservation biologists are currently concerned with slowing the loss of rare species (Wilson 1992), language preservation activists are concerned with slowing the loss of ethnic languages (Zepeda and Hill 1991; Hale 1992). Nevertheless, neither of these groups of scholar-activists have typically proposed intervention strategies that will slow the loss of both languages and species, recognizing that they face similar if not the same pressures. The intervention strategies chosen will differ greatly in their outcomes, depending on how one defines biodiversity and linguistic diversity, how one recognizes their interactions, and how one identifies proximate and ultimate causes of these losses (Nabhan 1994).

As a shorthand among practitioners of conservation biology, biodiversity has been discussed largely in terms of "species richness." Most biologists, however, recognize the contribution of other levels of biological organization (genetic variation within populations, variability between populations, habitat heterogeneity, ecosystem diversity) that are more difficult to monitor or measure (Office of Technology Assessment 1987; Harmon 1996). As Thompson (1996, 1999) has argued, biodiversity has resulted from the diversification of species and the interactions that occur among them, but the focus of efforts to conserve biodiversity has often been primarily on species rather than on their interactions. In many cases, ignorance of biotic interactions has led to the decline of a particular plant or animal species that has lost its mutualists, even though it occurs within a formally protected area such as a national park or forest (Tewksbury et al. 1999).

Ironically, most assessments of linguistic diversity focus merely on how many extant languages there are ("language richness"), on the declining abundance of living speakers of indigenous languages ("speaker richness"), or on the erosion of idiomatic vocabularies ("lexical richness"). Relatively few studies have looked at interactions within or among languages, such as the deterioration of structural differentiation within or among languages (Hill this volume) and of the functional relationships among languages, that is, "linguistic ecologies" (Mühlhäusler this volume). In a few cases, we now know that specialized syntactical structures and their associated vocabularies have gone out of use as certain subsistence activities have been abandoned by an indigenous culture (Zepeda 1984). In most cases, however, it remains unclear whether retention of lexemes associated with the knowledge and use of biodiversity has fared any worse than overall lexical retention within an imperiled indigenous language. Likewise, we hardly know why some indigenous languages such as the Yuman dialects of the Quechan and Mohave are eroding at more rapid rates than others such as Yoeme (Yaqui and Mayo of Sonora) which originally had roughly comparable population sizes of speakers but have retained or gained speakers over the last half century (Spicer 1980; Wurm 1991; Hinton 1994).

Analogous pitfalls have tripped up the trajectory of ethnobiology, which compares cultural perceptions and traditional management practices shaping local and regional biodiversity. Concerned that both cultural traditions and life forms are rapidly disappearing, some ethnobiologists have gone into the field among imperiled cultures to "salvage" their ethnopharmacological knowledge for posterity by pressing specimens, recording names, then tallying up how many medicinal plant species remain culturally utilized. Certain projects have even hired indigenous medicine men as parataxonomists and guides for bioprospectors who promise to return economic benefits of any "discoveries" to the cultural communities to which these shamans belong (King, Carlson, and Moran 1996). But there are many shortcomings associated with this approach, as we shall see.

Most of these salvage ethnobotany missions only scratch the surface of "indigenous knowledge about the natural world" by simply recording indigenous names for plants and cataloguing their uses. Such descriptive, purely utilitarian ethnobotanical surveys hardly tell us anything about how "the natural world works" from an indigenous perspective because of the assumption of some that ethnobotanical fieldwork is no more than the elicitation of "folk taxonomies," which therefore allow correlations between indigenous names for plants and Linnaean species. In this way, ethnobotanists mirror "biodiversity systematists," who ignore ecological interactions while attempting to find regions of high species richness (Thompson 1996).

As suggested earlier, many evolutionary ecologists have become skeptical when ethnobiologists speak of indigenous peoples as "the first ecologists" because these scholars are unaware of any genuine ecological knowledge derived from ethnobiological field studies. Commonly published inventories of useful plants named in native languages typically appear to be lacking any ecological or evolutionary context.

I wish to encourage my colleagues to devote more attention to traditional ecological knowledge of plant-animal interactions. This will be essential if ethnobiological field studies are to overcome these methodological and philosophical inadequacies and to rescue this vanishing knowledge about biodiversity. I suggest that indigenous cultures retain a wealth of linguistically encoded empirical knowledge about ecological relationships among plants and animals, based on observations of interspecific interactions that may have escaped notice by field biologists. Although when one first looks at ethnoecological accounts some indigenous observations may seem irrational or counterintuitive, they may in fact be linguistically encoded means of validly explaining certain relationships between plants and animals (Anderson 1996). I will demonstrate that some indigenous hypotheses about the nature of plant-animal interactions can be tested by Western scientific means, resulting in insights of significance to ecological and evolutionary theory. I will also review the ways in which ethnoecological studies of interaction diversity can contribute to the conservation of biodiversity, particularly when the relationships that indigenous peoples recognize and describe affect endangered species. I will also mention some tangible means by which traditional ecological knowledge is being integrated into endangered species and habitat recovery programs.

BACKGROUND ON AN ENDANGERED PEOPLE

I will use examples from an endangered language spoken by one remaining ethnic group within the Sonoran Desert, the Comcáac or Seri, a subset of Hokan speakers of Sonora, Mexico. The Seri, as they are commonly called in Mexico, number less than 600 individuals residing in two permanent villages and several temporary fishing camps on the Sea of Cortés coast of Sonora near Tiburón Island, which is also part of their aboriginal territory. Roughly 550 of them still speak Cmique Iitom, a Hokan language isolate closely related to the dialects spoken by the now-extinct indigenous peoples of southern and central Baja California.

I have interviewed between 50 and 100 individuals among older generations in two tribal communities, Punta Chueca and Desemboque, as well as 30 children. A subset of 50 adults and children agreed to formal interviews regarding their participation in subsistence and ceremonial activities related to reptiles' interactions with other life-forms, including humans. Interviews were typically accomplished in Spanish and English, with native terms in Cmique Iitom used as prompts. When interviewing monolingual speakers, I was usually accompanied by bilingual relatives of the person(s), who translated and verified my understanding of the person's responses. On several occasions, sightings of the rare plant or animal elicited commentary; in most cases, however, because of the rarity of the organisms, photos and drawings of the organisms in question were utilized to elicit discussion. Folk-taxonomic information for the Seri was corroborated by consulting available linguistic and ethnographic works (Felger and Moser 1985; and materials in preparation on the Summer Institute of Linguistics' website, found at http://www.sil.org/mexico) as well as my own Seri ethnoherpetological overview (Nabhan in press).

REDEFINING ETHNOECOLOGY

Among the first definitions of ethnoecology was that published by Bye and Zigmond (1976): "the area of study that attempts to illuminate in an *ecologically* revealing fashion man's interactions with and relationship to his environment." In ethnobiological books and journal articles published since that time, scholars have sought to reveal how human cultures ecologically interact with particular environments or species, but without much appreciation of interactions in the context of evolutionary time. I wish to deepen the definition of ethnoecology to include any cultural community's means of recognizing linguistically and influencing ecologically any functional relationships among plants, animals, and their habitats through time. This assumes that indigenous management of specific ecological interactions between plant and animal species over a significant period of time may place selection pressures on these relationships. In any case, a range of indigenous knowledge of biotic interactions can be demonstrated in the following examples, taken largely from my work with colleagues in the Sonoran Desert.

Embedded in any folk taxonomy are numerous references to interactions between particular species and their habitats as well as interactions among taxa. For example, among the Seri the name given to collared lizards, *hast coof,* implies that they are similar to chuckwalla (*coof*) but favor rocky habitats (*hast*). There are at least seven reptiles in Seri territory to which they give names that have habitat markers embedded within them. Then there are thirteen plant names that make reference to reptiles, four of which refer to a reptile's direct use of a particular plant (Nabhan in press). An example of this sort of interaction is the term *xtamáaija oohit,* which means 'mud turtle's forage' among the Seri, referring collectively to three plants that grow where floodwaters accumulate in the summer.

Analysis of the folk taxonomy of the Seri (Felger and Moser 1985; Nabhan in press) indicates that 10 percent of the terminal ethnotaxa for plants have faunal referents within them. But my follow-up interviews with the Seri reveal that only 4 percent of the lexemes for these terminal ethnotaxa specifically refer to plant-animal interactions.

A word of clarification on terms may be necessary here. Keep in mind that a compound lexeme in an indigenous language may include an animal name within it, for example, *caay ixám* 'horse's gourd' (Felger and Moser 1985), but this does not necessarily mean that horses have ever eaten them. Instead, it may refer to the inferior quality of the cultivated *Cucurbita pepo*'s ornamental gourds compared to those of *xam,* the feral *Cucurbita argyrosperma* gourds that the Seri may encounter in washes in central and southern Sonora. The Seri often joke about cultivated species being poor in quality or overdomesticated by linking them linguistically with priests or livestock.

In contrast, there are at least 19 plant names used by the Seri that do refer to interactions between flowers and their pollinators, fruits and their seed dispersal agents, foliage and its herbivores or flailers, larval host plants and their larvae, brushy canopy-providers and dormant or reclusive animals, algae associated with sea turtle carapaces, and nest-providing canopies and their nesting birds. In such cases, it is reasonable to assume that plant names that recognize their faunal associates are derived from empirical observations of plant-animal interactions.

A few animal names are polysemous with the names for particular plants. For example, *hasahcápoj* is a name used to refer to the rarely seen boa constrictor that associates it with the sprawling columnar cactus, *Stenocereus alamosenis,* which has a morphology as well as northern distributional limits in Seri territory much the same as the boa's (Nabhan in press). It is clear, however, that the snake's name is derived from an archaic name for columnar cacti and not vice versa. In special cases such as this, the lexeme is polysemic for both a plant and an animal, normally when the relationship between the two is unusually robust. One such case of polysemy comes not from desert peoples, but from the Chontal Maya of the Neotropics, who use the same lexeme to name the Great Kiskadee (*Pitangus sulphuratus*) and for wild chile peppers (*Capsicum annuum*) (Vasquez-Davila 1990). The Chontales and their mestizo neighbors recognize the Great Kiskadee as an important seed dispersal agent of wild chiles in secondary growth emerging after milpa (field) abandonment. In coastal Sonora, where the Seri infrequently encounter wild chiles, these plants are associated with other birds (Nabhan 1997), so that different populations within a species may have different animal associates, which each local cultural community knows and names. Thompson (1996) has argued that it is important to recognize interactions at the population level, for each locality's or habitat's plants may be specialized to different interactions that shift in space and through time. To fully fathom a cultural community's understanding of such interactions, it is critical to go beyond mere taxonomic inquiries to interview indigenous specialists in certain plants and animals in the habitats where those species occur. Ethnobiologists should not confine themselves to taxonomic inventories but devote more time to eliciting genuine ecological knowledge from folk practitioners.

VALIDATING VARIOUS CULTURAL PERCEPTIONS
OF ECOLOGICAL INTERACTIONS

As Western-trained scientists learn of plant-animal interactions recognized and named by indigenous peoples, they can test hypotheses to elucidate the relative degree of connectivity or exclusivity in such relationships. The folk taxonomies of indigenous, Spanish- and English-speaking peoples in the Americas often distinguish wild *Capsicum* chiles from domesticated chile peppers by using a term akin to "bird pepper" but the bird in mind varies from culture to culture and place to place (Vasquez-Davila 1990; Nabhan 1997). When asked, people of any region within the range of wild *Capsicum* will name only certain birds as consumers and dispersers of these "bird peppers." Some wild chile harvesters also associate wild chiles with particular *madrinas* or *nodrizas,* 'nurse plants', which provide light-sensitive chiles with buffered microenvironments, just as coffee and chocolate plantations often utilize an overstory of madrecacao (*Gliricidia sepium*).

In a series of experiments over the last several years, I have worked with colleagues to test whether there are any peculiar relationships between red-plumaged birds and wild chile peppers as indigenous peoples suggest (Nabhan 1997; Tewksbury et al. 1999; Nabhan 2000). In particular, we wished to determine whether the roosting and foraging behavior of any resident frugivores in particular tree canopies predicted the degree of association between wild *Capsicum* shrubs and their overstory nurse plants better than other parameters could. Our results demonstrate that Northern Cardinal activity in hackberries (*Celtis* spp.) is highly correlated with wild chile presence beneath hackberries and better predicts chile distribution than do other characteristics of nurse plants or frugivorous birds. This, then, is a tangible example of how indigenous ecological knowledge can be used to guide empirical or experimental studies to learn more about plant-animal interactions.

CONTRIBUTIONS TO THE CONSERVATION BIOLOGY OF ENDANGERED BIOTA

It is clear from a number of studies, summarized in Nabhan (1992, in press), that indigenous communities are reservoirs of considerable knowledge about rare, threatened, and endemic species that has not to date been independently accumulated by Western-trained conservation biologists. What may be less obvious is that indigenous knowledge of biotic relationships with rare plants or animals can help guide the identification, management, protection, or recovery of habitats for these species. Native Americans certainly are aware of details of the diets, nesting, and refuge cover requirements of endangered species, and these details have not necessarily been recorded in the literature of conservation biology.

Take as examples the following details regarding the autoecology of four endangered animals: the desert tortoise, the Sonoran pronghorn antelope, the desert bighorn sheep, and the green sea turtle. For the desert tortoise (*Gopherus agassizii*), a key issue in its conservation management has been providing protected habitat where sufficiently diverse forages are available for its dietary use. Despite sixty years of incidental reports on desert tortoise feeding behavior and stomach contents, knowledge of the species' dietary needs remains fragmentary, especially in Mexico, where a third of the extant range of desert tortoises occur (Bury et al. in press). Despite that fact, only 2 percent of all walking surveys recording desert tortoise foraging areas have been undertaken in Mexico, and no nutritional or fecal studies have been completed south of the U.S. border.

Not surprisingly, there are six species of desert plants known for centuries to the Seri as *xtamoosni oohit* 'desert tortoise's forages'. These include four species not otherwise identified in tortoise diets in the Sonoran Desert (*Chaenactis carphoclinia; Fagonia californica* and *F. pachyacantha; Phacelia ambigua*), although another species of *Chaenactis* has been identified in Sonoran Desert tortoise diets, and *Fagonia* may be found in Mohave Desert tortoise diets (Van Devender and Schwalbe in press). A fifth species, *Chorizanthe brevicornu*, has only recently been verified as an important component of desert tortoise diets even though it is an inconspicuous, ephemeral wildflower (Van Devender and Schwalbe in press). The Seri also suggest that desert tortoises eat false purslane, *Trianthema portulacastrum*, although they give this succulent annual herb another name, different from that of other tortoise forages. Such observations may be critical to establishing indicators for good summer foraging ranges for desert tortoises, as a prelude for determining the location of future wildlife refuges.

The same is true for the Seri association of the endangered Sonoran pronghorn antelope (*Antilocapra americana sonoriensis*) with an ephemeral legume, *Phaseolus filiformis*, which they call *hamooja ihaap* 'pronghorn its-wild-bean'. Although this plant is occasionally abundant where the remnant pronghorn population lives in northwest Sonora, it is seldom abundant where the northernmost Seri bands lived, some 60–80 km south, in a poorly documented Sonoran pronghorn range. To date, this forage has not been recorded by members of the Sonoran pronghorn recovery team in their dietary studies, although it is a likely candidate.

A wild onion (*Allium haemotochiton*) is called "desert bighorn what-it-eats," referring to *Ovis canadensis mexicanus*, another threatened subspecies. Although winter-blooming onion has not been recorded in Sonoran Desert bighorn diets to date, it is also considered a good candidate by wildlife dietary ecologists (Paul Krausmann, personal communication).

Several species of algae are noted by the Seri as habitat, carapace cover, or forage for the *moosni hant koit*, the endangered green sea turtle, *Chelonia mydas: Cryptomeria obovata, Halymenia coccinea, Gracilaria textorii*, and *Rhodymenia divaricata*. The most intimate association is with the red alga *Gracilaria*, named by the Seri "sea turtle's membranes," which grows up to 30 cm tall on the carapaces of the endangered sea turtle population that overwinters, dormant, in a shallow channel of the Sea of Cortés adjacent to Seri villages (Felger and Moser 1985). Knowledge of this overwintering behavior among the sea turtle population was once unique to the Seri, but after non-Indian fishermen learned of it, they rapidly wiped out this population (Felger, Clifton, and Regal 1976). Berlin (1992) recently used the Seri as the clearest counterexample to the hypothesis that only farmers "overclassify" economically important plants and animals into "folk species," but the recognition that they also associate a particular alga with a named folk species (or distinct population) of sea turtles is even more remarkable. Unfortunately, my recent interviews with former Seri sea turtle specialists indicate that they hardly ever see individuals of this sea turtle population anymore.

THE SERI MODEL: MEANS FOR INDIGENOUS PEOPLES TO BE FULL PARTICIPANTS IN PROTECTING BIODIVERSITY

In surveys of Native Americans involved in wildlife management, hunting, fishing, and endangered species conservation, these resource managers lament that so many culturally important species have been lost from their homelands during their own lifetimes. In our region of desert, mountain, and sea, there is a long list of traditionally utilized species that native wildlife specialists feel have declined dramatically within the last fifty years: sandfood (*Pholisma sonorae*); chiltepines (*Capsicum annuum*); night-blooming cereus (*Peniocereus striatus* and *P. greggii*); ironwood (*Olneya tesota*); Hohokam agave (*Agave murpheyi*); desert bighorn (*Ovis canadensis*); Sonoran pronghorn (*Antilocapra americana*); as well as numerous birds and fish (Nabhan 1992; Rea 1997). It would be an interesting exercise to elicit from the elders of an indigenous community an entire threatened species list for their homeland and compare this "emic" or culturally based list with official federal lists for the same region.

Many of these elders are also painfully aware that their children and grandchildren have diminished exposure to these rare species and to the oral traditional knowledge about them, which is also rapidly disappearing (Rosenberg 1997; Nabhan 1998, in press). In a survey of 29 Seri individuals born before 1974 and 23 individuals born after 1974, I asked about their involvement in particular subsistence, ceremonial, or ritual activities in which traditional knowledge is conveyed from one generation to the next. These interviews, discussed in detail in Nabhan (in press), suggest a 23–30 percent decrease in participation in traditional-knowledge-renewing activities over the last generation among Seri females, and an 8–25 percent decrease among males, depending upon how one defines active participation in a community tradition.

This "extinction of experience" of rare and endemic species breaks mutually reinforcing connections between cultural and biological diversity, which have functioned over the last eight to ten millennia in the Americas, and longer elsewhere (Nabhan and St. Antoine 1993). In particular among the Seri, it appears that traditional ecological knowledge about relationships between plants and animals is being lost far more rapidly than the native names themselves for these taxa (Rosenberg 1997).

Several years ago, I made a plea to national park and wildlife refuge managers to involve Native American elders in the management not only of cultural resources but also of natural resources such as endangered species (Nabhan 1992). With the blessings and participation of the Seri council of elders and tribal governor, the Arizona-Sonora Desert Museum and Amazon Conservation Team have begun a "paraecologist" training course for young Seri men, in which elders and conservation biology experts co-teach field workshops regarding threatened flora and fauna. Each workshop has several hours of lectures by elders and by academically trained biologists, followed by field trips where the sixteen students learn about ecological interactions through hands-on exercises. Both the elders and the biologists offer questions for exams, and trainees must demonstrate competence in both realms. After receiving their diplomas, the paraecologists will be placed on field crews undertaking endangered species inventories, monitoring, and recovery in and near Seri homelands.

It is critical that Native American youth be given training opportunities for careers in endangered species and habitat conservation, particularly through courses such as this one, which values traditional ecological knowledge as complementary to Western science. If this link between cultural and biological diversity is to be in any way maintained, strengthened, or restored, it will be necessary to include indigenous peoples in the management and conservation of the world's remaining biological riches. The ethnoecological knowledge that endangered peoples such as the Seri linguistically encode in their native lexicon is a contribution to both biological and linguistic diversity. It cannot be found in the languages of neighboring peoples, nor at this moment is it in conservation biology textbooks. We can only pray that that day will come.

REFERENCES

Anderson, E.N. 1996. *Ecologies of the Heart*. New York: Oxford University Press.

Berlin, B. 1992. *Ethnobiological Classification*. Princeton: Princeton University Press.

Bury, R., et al. In press. The desert tortoise in Mexico. In *Sonoran Desert Tortoises: Their Ecology and Conservation*, ed. T. Van Devender. Tucson: University of Arizona Press.

Bye, R.A., Jr., and M.L. Zigmond. 1976. Book review: Principles of Tzetzal Plant Classification. *Human Ecology* 4(3):171–175.

Callicott, J.B., and M.P. Nelson. 1998. *The Great New Wilderness Debate*. Athens: University of Georgia Press.

Denevan, W. 1992. The pristine myth: The landscape of the Americas in 1492. *Annals of the Association of American Geographers* 82:369–385.

Diamond, J. 1992. *The Third Chimpanzee*. New York: HarperCollins.

Felger, R.S., and M.B. Moser. 1985. *People of the Desert and Sea: Ethnobotany of the Seri Indians*. Tucson: University of Arizona Press.

Felger, R.S., K. Clifton, and P. Regal. 1976. Winter dormancy in sea turtles: Independent discovery and exploitation in the Gulf of California by two local cultures. *Science* 191:283–285.

Hale, K. 1992. On endangered languages and the safeguarding of diversity. *Language* 68:1–3.

Harmon, D. 1996. The converging extinction crisis: Defining terms and understanding
trends in the loss of biological and cultural diversity. Keynote address presented April 1,
1996 at the colloquium Losing Species, Languages, and Stories: Linking Cultural and Environmental Change in the Binational Southwest, Arizona-Sonora Desert Museum,
Tucson, Arizona.

Hinton, L. 1994. *Flutes of Fire*. Berkeley, Calif.: Heyday Books.

King, S.R., T.J. Carlson, and K. Moran. 1996. Biological diversity, indigenous knowledge,
drug discovery, and intellectual property rights. In *Valuing Indigenous Knowledge*, ed.
S.B. Brush and D. Stabinsky. Pp. 167–185. Washington, D.C.: Island Press.

Nabhan, G.P. 1992. Threatened Native American plants. *Endangered Species Update* 9:1–4.

Nabhan, G.P. 1994. Proximate and ultimate threats to endangered species. *Conservation
Biology* 8:928–929.

Nabhan, G.P. 1997. Why chiles are hot. *Natural History* 106(5):24–27

Nabhan, G.P. 1998. Handing down ecological knowledge. *Orion Afield* 2(4):28–31.

Nabhan, G.P. 2000. Native American management and conservation of biodiversity in the
Sonoran Desert region: An ethnoecological perspective. In *Biodiversity and Native America*, ed. P E. Minnis and W.J. Elisens. Pp. 29–43. Norman: University of Oklahoma Press.

Nabhan, G.P. In press. *Singing the Turtles to Sea: Reptiles in Comcáac Arts and Sciences*. Berkeley: University of California Press.

Nabhan, G.P., and S. St. Antoine. 1993. The loss of floral and faunal story: The extinction
of experience—Ethnobiological perspectives on biophilia. In *The Biophilia Hypothesis*,
ed. S. Kellert and E.O. Wilson. Pp. 229–250. Washington, D.C.: Island Press.

Nabhan, G.P., et al. 1982. Papago influences on habitat and biotic diversity: Quitovac oasis
ethnoecology. *Journal of Ethnobiology* 2:124–143.

Office of Technology Assessment. 1987. *Technologies to Maintain Biological Diversity*. Washington, D.C.: U.S. Congress and U.S. Government Printing Office.

Rea, A.M. 1997. *By the Desert's Green Edge*. Tucson: University of Arizona Press.

Rosenberg, J. 1997. Documenting and Revitalizing Traditional Ecological Knowledge. M.A.
thesis, University of Arizona, Tucson.

Spicer, E. 1980. *The Yaquis: A Cultural History*. Tucson: University of Arizona Press.

Tewksbury, J.J., et al. 1999. In situ conservation of wild chiles and their biotic associates.
Conservation Biology 13(1):98–107.

Thompson, J.N. 1996. Evolutionary ecology and the conservation of biodiversity. *Trends in
Ecology and Evolution* 11(7):300–303.

Thompson, J.N. 1999. The evolution of species interactions. *Science* 284:2116–2118.

Van Devender, T.R., and C. Schwalbe. In press. Desert tortoise diets in the Arizona Uplands
of the Sonoran Desert. In *Sonoran Desert Tortoises: Their Ecology and Conservation*, ed. T.R.
Van Devender. Tucson: University of Arizona Press.

Vasquez-Davila, M.A. 1990. El Amash, El Pistoque: Un ejemplo de la etnoecología de los
Chontales de Tabasco, México. *Etnoecológica* 3:59–69.

Wilson, E.O. 1992. *The Diversity of Life*. Cambridge, Mass.: Belknap Press of Harvard University Press.

Wurm, S. 1991. Language death and disappearance: Causes and circumstances. In *Endangered Languages*, ed. R.H. Robins and E. Uhlenbeck. Pp. 1–18. New York: Berg.

Zepeda, O. 1984. *Papago Morphology*. Tucson: University of Arizona.

Zepeda, O., and J.H. Hill. 1991. The condition of Native American languages in the United States. In *Endangered Languages*, ed. R.H. Robins and E. Uhlenbeck. Pp. 135–155. New York: Berg.

9

THE VANISHING LANDSCAPE OF
THE PETÉN MAYA LOWLANDS

People, Plants, Animals, Places, Words, and Spirits

Scott Atran

Previous research has given much attention to the difficulty in managing common pool resources (Hardin 1968) and the requirements for successful management (Berkes et al. 1989). But little attention has been paid to the role of folk-ecological models in resource management. The premise of the collaborative research reported on here is that to understand people's environmental values and behavior, one must first understand the way in which people conceptualize the natural world.[1] This research explores the possibility that folk-ecological models that have worked for long periods of time, such as those of certain native peoples, may provide instruction in successful environmental accommodation. This is not to contend that all indigenous groups are equally or necessarily successful on this score; our work indicates they are not. Neither is it to deny that serious problems face any effort to scale up the local models, values, and behaviors of small-scale systems across multicultural, multinational, or global settings.

RESEARCH SETTING

Our research concerns three towns in the Municipality of San José in the north-central part of Guatemala's Department of El Petén: the native Itzaj Maya town of San José, the town of La Nueva San José, composed of immigrant Ladinos, or native Spanish-speakers, of mainly mixed European and Amerindian descent, and the immigrant Q'eqchi' Maya town of Corozal. The climate is semitropical and quasi-rainforest predominates (tropical dry or hot subtropical humid forest). Topographic and microclimatic variation allows for a dramatic range of vegetation over relatively small areas, and sustaining both this diversity and people's livelihood over the last two millennia has required correspondingly flexible agroforestry regimes (Atran 1993a).[2]

The Itzaj, who ruled the last independent Maya polity, were reduced to corvée labor after their conquest in 1697. San José was founded by the Spanish in 1708 (Soza 1970), one of a handful of "reductions" for concentrating remnants of the native Itzaj population and fragments of related groups. In 1960, the military government opened the Petén, which includes 36,000 km², or about one-third of Guatemala's territory, to immigration and colonization. In the following years, about half the forest cover of Petén was cleared. In 1990, under a "debt-for-nature" swap sponsored by the U.S. Agency for International Development and in cooperation with the United Nations, Guatemala's government included most remaining forests north of latitude 17°10' in the Maya Biosphere Reserve. San José now lies within the reserve's official "buffer zone" between that latitude and Lake Petén Itza to the south. San José has some 1500 inhabitants, most of whom still identify themselves as Itzaj, although only older adults speak the native tongue, a Lowland Mayan language related to Yukatek, Mopan, and Lakantun.

The neighboring town of La Nueva San José was established in 1978 under jurisdiction of the Municipality of San José. The vast majority of the nearly 100 households (500–600 people) are Ladinos, most of whom were born outside Petén. The town of Corozal, also under jurisdiction of the municipality, was settled at the same time by Q'eqchi' speakers, a Highland Maya group from the Guatemalan Department of Alta Vera Paz just south of Petén. Although Q'eqchi' immigration into Petén began in the early eighteenth century, massive population displacement into the area is recent. The Q'eqchi' now constitute the largest identifiable ethnic group in Petén while maintaining the smallest number of dialects and the largest percentage of monolinguals (Stewart 1980). This points to the relative isolation and suddenness of Q'eqchi' migration. Although many of the nearly 400 Q'eqchi' of Corozal understand Spanish, few willingly choose to converse in it. Q'eqchi' is not mutually intelligible with Itzaj.

All three groups live off their land, growing corn, beans, and squashes in what is generally called a "milpa." Until 1995 each household (about 5 persons) had usufruct rights on 30 manzanas (about 20 hectares) of the municipal commons (or ejido), paying a yearly rent of 2 quetzales (about 30 cents) for each manzana under cultivation. Today, new households receive only 10 manzanas, and the rent for cultivated land has doubled for everyone. All groups practice slash-and-burn agriculture and horticulture, hunt game, and often go into the rainforest to extract timber and nontimber products for sale on the market. Although all groups exhibit detailed awareness of forest life, their agroforestry practices differ appreciably. Itzaj slash and burn to prepare milpas but do not cut valuable trees and surround them with fire breaks when the rest of the patch is burned. To facilitate regrowth, they also leave trees to ring milpa plots and neither cut nor burn hill crowns (cf. Remmers and Ucan Ek' 1996 for Yukatek Maya). More often than not, Itzaj open new plots that are not contiguous with previously cultivated plots, and the bulk of an Itzaj farmer's parcel is usually left as a forest reserve.

What follows is a brief summary of each group's milpa practices (with all numerical differences mentioned between groups proving to be statistically significant). Itzaj farmers typically plant 2–3 times as many crops as Ladinos or Q'eqchi'. This results in Itzaj milpa better emulating the surrounding forest's biodiversity (cf. Nations and Nigh 1980 for Lakantun Maya). Itzaj plots tend to be small, averaging 2 manzanas per family (versus nearly 4 for a Ladino family and 6 for a Q'eqchi' family). Itzaj also let cultivated land rest fallow longer (4–5 years) than do Ladinos or Q'eqchi' (3–4 years). Like Itzaj, Ladinos cultivate the same plot for two consecutive years. Ladinos also express a need to protect hilltops and valued trees, although less consistently than do Itzaj. Unlike the Itzaj or Ladinos, the Q'eqchi' of Corozal clearcut and burn forest for new plots every year, in a contiguous pattern that snakes through the forest. The Q'eqchi' do not keep forest reserves or protect ecologically important trees and hilltops, nor do they indicate a need to do so when asked directly. The Q'eqchi', also unlike the Itzaj and Ladinos, practice milpa in corporate groups: the community of neighbors and kin help to clear and burn each household's plot, and kin groups generally seed together.

FOLK-ECOLOGICAL MODELS

For each of 6 men and 6 women from each group, we asked: "Which kinds of plants and animals are most necessary for the forest to live?" Instructions and responses were in each group's native language. When, as happened for each group, there is a cultural consensus in the formal sense (i.e., a single-factor solution in a principal components analysis), use of such restricted numbers of informants to assess groupwide patterns is statistically justified (Romney, Weller, Batchelder 1986; López et al. 1997). Among hundreds of locally known species, informants mainly nominate a small sample of some two dozen species each of plants and animals (Atran and Medin 1997; Atran and Ucan Ek' 1999). Nearly all nominated plants and animals have high cultural value, but the reasons informants give are almost always also in terms of ecological value. From these lists we formed a composite list of highly significant plants and animals known to all three populations in order to compare directly how Itzaj, Ladinos, and Q'eqchi' understand their shared environment (table 9.1). To ensure the social diversity of each sample, no persons in the sample could have immediate kinship or marriage links with one another (e.g., no grandparent, sibling, child, first cousin, spouse or in-law, or god-parenthood).

Table 9.1 Petén Plants and Animals

Ref.	Plant Name	Scientific Name	Ref.	Animal Name	Scientific Name
	FRUIT TREES			ARBOREAL ANIMALS	
P1	ramon	*Brosimum alicastrum*	A1	bat	Chiroptera
P2	chicozapote	*Manilkara achras*	A2	spider monkey	*Ateles geoffroyi*
P3	ciricote	*Cordia dodecandra*	A3	howler monkey	*Allouatta pigra*
P4	allspice	*Pimenta diocia*			*A. palliata*
P5	strangler fig	*Ficus obtusifolia*	A4	kinkajou	*Potus flavus*
		F. aurea	A5	coatimundi	*Nasua narica*
			A6	squirrel	*Sciurius deppei*
	PALMS				*S. aureogaster*
P6	guano	*Sabal mauritiiformis*		BIRDS	
P7	broom palm	*Crysophilia staurocata*			
P8	corozo	*Orbignya cohune*	A7	crested guan	*Penelope purpurascens*
		Scheelea lundelli	A8	great curassow	*Crax rubra*
P9	xate	*Chamaedorea elegans*	A9	ocellated turkey	*Meleagris ocellata*
		C. erumpens	A10	tinamou	*Tinamou major*
		C. oblongata			*Crypturellus* spp.
P10	pacaya	*Chamaedorea tepejilote*	A11	toucan	*Ramphastos sulfuratus*
P11	chapay	*Astrocaryum mexicanum*	A12	parrot	Psittacidae in part
			A13	scarlet macaw	*Ara macao*
	GRASSES/HERBS		A14	chachalaca	*Ortalis vetula*
			A15	pigeon/dove	*Columbidae*
P12	herb/underbrush	(various families)			
P13	grasses	*Cyperaceae/Poaceae*		RUMMAGERS	
	OTHER PLANTS		A16	collared peccary	*Tayassu tacaju*
			A17	white-lipped peccary	*Tayassu pecari*
P14	mahogany	*Swietania macrophylla*	A18	paca	*Cuniculus paca*
P15	cedar	*Cedrela mexicana*	A19	agouti	*Dasyprocta punctata*
P16	ceiba	*Ceiba pentandra*	A20	red-brocket deer	*Mazama americana*
P17	madrial	*Gliricidia sepium*	A21	white-tailed deer	*Odocoileus virginianus*
P18	chaltekok	*Caesalpinia velutina*	A22	tapir	*Tapirus bairdii*
P19	manchich	*Lonchocarpus castilloi*	A23	armadillo	*Dasypus novemcinctus*
P20	jabin	*Piscidia piscipula*			
P21	santamaria	*Calophyllum brasilense*		PREDATORS	
P22	amapola	*Pseudobombax ellipticum*			
		Bernoullia flammea	A24	jaguar	*Felis onca*
P23	yaxnik	*Vitex gaumeri*	A25	margay	*Felis wiedii*
P24	kanlol	*Senna racemosa*	A26	mountain lion	*Felis concolor*
P25	pukte	*Bucida buceras*	A27	boa	*Boa constrictor*
P26	water vine	*Vitis tiliifolia*	A28	fer-de-lance	*Bothrops asper*
P27	cordage vine	*Cnestidium rufescens*	A29	laughing falcon	*Herpetotheres*
P28	killer vines	(various epiphytes)			*cachinnans*

The procedure had two parts. First, participants were asked how each plant affected each animal. The task consisted of as many trials as there were plants and animals on the composite lists. On each trial, all animal picture cards were laid out and the informant was asked if any of the animals "go with" or "are companions of" the target plant, and whether the plant helps or hurts the animal. Unaffiliated animals were set aside. For each selected animal, informants were asked to explain how the plant affects the animal (P-A) and all explanations recorded. Next, informants were asked how the animal affects the plants (A-P), that is, whether each animal helps (+1), hurts (−1) or has no effect on (0) each plant (A-P). For each pair of species we also asked if disappearance of one would help, hurt, or have no effect on the other. Notice that the coding system allows each species pair to be represented in terms of one of five folk-ecological relations that can be roughly glossed as: "mutualist" (+1, +1), "commensalist" (+1, 0), "parasitic" (+1, −1), "destructive" (−1, 0 or −1, −1) or "neutral" (0, 0).

This simple coding system allows for surprisingly complex representations of interactions between clusters of folk species. For each informant, P-A and A-P relations were calibrated, and each bidirectional relation was classified as one of these five types (e.g., if the plant helped the animal but the animal hurt the plant, that pair was classified as a *parasitic* relation). For each relation type, a one-way ANOVA was performed comparing the relative frequency of occurrence for that relation across the three populations. Results are presented in table 9.2; starred relations indicate reliable difference among populations.

Significant effects were further explored using Tukey HSD post hoc tests. Itzaj perceive more *mutualist* relations than Ladinos or Q'eqchi', who do not differ from each other. Ladinos perceive more *commensalist* relations than Q'eqchi', with neither group differing from Itzaj. Itzaj and Ladinos report *parasitic* relations with comparable frequency, and both groups do so more than Q'eqchi'. No differences emerge in the frequency of *destructive* relations. Finally, Q'eqchi' report more *neutral* relations than Itzaj or Ladinos, who do not differ from each other.

The finding that Itzaj and Ladinos see comparable numbers of non-neutral relations indicates that the differences between them are more qualitative than quantitative. For example, Itzaj report that classes of animals differentially affected classes of plants, whereas Ladinos reported more universal effects. To illustrate, plant kinds were collapsed into four categories (fruit, grass/herb, palm, and other), as were animal categories (arboreal, bird, rummager, and predator). Table 9.1 shows the composition of categories; figure 9.1 shows their interaction. An ANOVA reveals a plant-by-animal interaction for Itzaj but not for Ladinos ($F_{(9,99)} = 26.04$, $MSE = .008$, $p < .0001$). On a qualitative level, although both groups acknowledge that animals have a large impact on fruit trees, Itzaj differ from Ladinos in understanding these relations. Ladinos see animals as harming plants by eating fruit. Itzaj

Table 9.2 Relative Frequencies of Plant and
Animal Relations in Petén Populations

Relation	Itzaj	Ladinos	Q'eqchi'
Mutualist*	7.2%	2.1%	0.1%
Commensalist*	8.9%	14.1%	5.7%
Parasitic*	7.4%	8.1%	0.0%
Destructive	1.2%	0.1%	0.0%
Neutral*	75.3%	75.6%	94.2%

* Indicates reliable difference among populations.

have a more nuanced appreciation of the relationship between seed properties and processing: if the seed is soft and the animal cracks the fruit casing, the animal is likely to destroy the seed and thus harm the plant, but if the seed is hard and passes through the animal's body rapidly, then the animal is apt to help the plant by dispersing and fertilizing the seed.

Our results show consistent and striking differences in the folk-ecological models of groups living in the same Lowland forest environment (Atran et al. 1999). Highland Q'eqchi' Maya immigrants see plants as passive donors to animals, and animals having no effect on plants. In sharp contrast, native Itzaj Maya have a rich, reciprocal model of animal-plant interactions, where animals can help or hurt plants. Immigrant Ladinos display a simpler and decidedly nonreciprocal model, with animals hurting plants by using them. We have similar findings in regard to Itzaj, Q'eqchi', and Ladino appreciations of the relationships between plants and people.

These striking differences in folk-ecological models also parallel the degree to which the different groups practice sustainable agroforestry. Itzaj folk ecology stresses reciprocity among plants, animals, and humans; Itzaj agricultural practices respect and preserve the forest. Q'eqchi' folk ecology reflects a belief that plants are resources to be exploited, and that animals and humans have little if any impact on them. Q'eqchi' agricultural practice is correspondingly heedless of the forest. Ladino folk ecology and agricultural practices are intermediate. Such qualitative differences may arise among people who live in the same environment via differences in conditions for learning and interest in learning. This is conceivably a function of more comprehensive cognitive frameworks through which the forest is viewed. If the forest is seen as simply a set of resources to be exploited, one might attend to plants and animals solely as noninteracting commodities. Relationships between animals and plants would only be marginally useful, as in determining where to look for certain kinds of animals. If instead, forest species and places are viewed more as a family home, the learner is invited to attend to interactions and relationships. Itzaj often describe the forest as "The Maya house" and refer to its elements and events in the con-

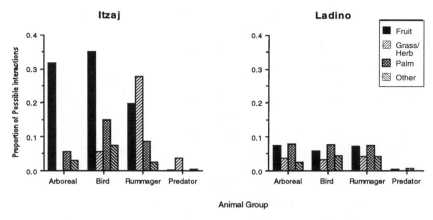

Figure 9.1. Reported impact of animal groups on plant groups for Itzaj and Ladinos.

text of stories rather than reports, which further increases attendance to interactions and relationships by imbuing those elements and events with thematic content.

The ecological consequences of the Q'eqchi' vision should be contrary to those implicit in the Itzaj vision, with short-term degeneration of the forest rather than long-term regeneration. There is some independent confirmation of this. On the one hand, recent tracking by remote sensing of Q'eqchi' expansion shows rapid and extensive deforestation along the migration routes (data summarized and mapped by Steven Sader for Conservation International and presented to the President of Guatemala in November 1996; cf. Nations this volume). On the other hand, in the last few years Itzaj have set up the only self-managed forest reserve in Petén (Atran 1993b), where there is already evidence of regeneration of plant and animal stocks in areas recently depleted by logging operations, immigrant slash-and-burn agriculture, and depredations by transient hunters. The Itzaj and Q'eqchi' patterns in this experiment (and others) are fairly clear and consistent, leading to an expectation that Itzaj are likely to act to sustain the forest and that Q'eqchi' are likely to act in ways that will degrade the forest. Expectations for Ladinos are less clear. Before turning to a fuller discussion of the issues it will help to compare social structures for these three groups.

SOCIAL NETWORK ANALYSIS

The general idea is that the methodology of social network analysis can help us to understand various socially related structures (e.g., economic behavior) as well as social change (Granovetter 1979). More specifically relevant for our purposes is the idea that "core" networks of close social ties may interact with "extended" networks involving more distant ties to forest experts to determine the community's overall ability to encounter and assimilate new information (Hammer 1983). Here we concentrate only on results that highlight overall relationships for each population between the core social network and the extended expert network (cf. Ford 1976).

The greatest overlap in the two networks occurs with the Itzaj and the least with the Q'eqchi'. For Itzaj, 14 of the most-cited social partners are among the 22 most-cited forest experts. Although the Itzaj social network is not highly central-ized, the most-cited social partner is also the second most-cited forest expert and the top forest expert is the third most-cited social partner. For Ladinos, 11 of the most-cited social partners are among the 25 most-cited forest experts. Of these 11, all are Ladino men. Ladino women tend to mention Ladino men as experts, but the top Ladino experts mostly cite the same Itzaj experts as the Itzaj themselves do, suggesting diffusion of information from Itzaj experts to a select group of so-cially well-connected Ladino men. For Q'eqchi', who have by far the most central-ized and densely connected networks, only 6 of the most-cited social partners are among the 18 most-cited forest experts. Moreover, the two forest experts far and away most cited by the Q'eqchi' are actually outside institutions (an NGO based in Washington, D.C., and the Guatemalan government agency responsible for management of the Maya Biosphere).

The three populations differ markedly in their social and expert network struc-tures; this suggests different consequences for information flow about the forest. Q'eqchi' networks suggest that information pertinent to long-term survival of the forest comes from outside organizations with little long-term experience in Petén. Moreover, what outside information there is does not seem likely to pene-trate deep into the Q'eqchi' community, because it is not conveyed by socially rel-evant actors. For Itzaj, expert information about the forest appears integrally bound to intimate patterns of social life as well as to an experiential history trace-able over many generations. For the Ladinos, expert information is also likely to be assimilated into the community. Because Ladino experts are socially well-connected, information that may come through Itzaj experts is communicable via multiple interaction pathways. These local Ladino experts also know that ulti-mate expertise lies with the Itzaj (Itzaj and Ladinos never cite government or NGO workers as forest experts). In sum, Ladinos have social connections to Itzaj expertise that may enable them to assimilate as much Itzaj knowledge as is com-patible with their own experience and understanding.

EMERGENT KNOWLEDGE STRUCTURES

Among the Itzaj, we have a society with no apparent tradition of institutions es-tablished to close off or monitor access to the commons, and without benefit of a higher authority to compel preservation of the environment for all. Yet, Itzaj plainly think and act in ways that are geared to sustaining their environment in the long run. They do more to preserve the local environment than other groups who share it, despite the fact that nearby communities have smaller populations but comparable access to common-pool resources. Itzaj do this with less reliance on kinship and densely connected social networks compared to the Q'eqchi'.

Historically, Lowland Maya readily could, and often did in hard times, leave family, friend, and village for different ones. New social attachments were relatively easy to establish (often by *compadrazgo,* or god-parenthood). Most Itzaj families, for example, can name ancestors who originated from (different) places in the Yucatán peninsula. There were also occasionally wholesale rendings of the community fabric owing to political repression, famine, or epidemics. Older Itzaj, for example, remember relatives having to flee, and some even to settle, in Soccotz on the other side of the border with British Honduras (Belize) to escape harsh corvée labor and a punishing anti-Maya language policy (effectively ending Itzaj monolingualism) during the dictatorial reign of General Jorge Ubico in the 1930s (see Schwartz 1990 for other examples).

On the face of it, this fluidity of association and the ability of a Lowland Maya nuclear family to survive on its own in the forest for seasons (and if need be for years) would seem to encourage "individualistic" rather than cooperative behavior. And, indeed, when we ask members of our three populations, "how did you learn about the forest?" as often as not Itzaj (but not Ladinos or Q'eqchi') respond: "I walk alone" (*k-in-xi'mal t-in-jun-al,* where *xi'mal* 'walk' also means 'to observe and behave appropriately,' as, for example, in the Itzaj claim that immigrants do not know how to "walk" in the forest and so are bitten by snakes).

Nowadays, Itzaj rarely coordinate hunting or planting among themselves, although in the recent past they did collectively perform rituals for these activities. Still, they have continued to behave in a collectively sustainable manner. Not until 1991 was there any institutional setup for monitoring and closing access to the forest commons. Only then was a committee of Itzaj formed to establish the Bio-Itza reserve in order to preserve the last large tree stands in the municipality from loggers who had marked them all for cutting and from immigrants who had begun to burn fields on either side of the approaches that the loggers had made into the area.[3] But the process of closing access has engendered sociopolitical division within the community (between families allied to the current municipal administration and those allied to the independently elected Bio-Itza Committee).

If, in fact, native Lowland Maya are self-sufficient, then how is it that they have been so apparently successful in collectively sustaining their society and environment over the centuries and why do they think that their fate is collectively bound to the forest? A keen student of Lowland Maya history notes:

> It is, however, no less a paradox than the emphasis placed on the autonomy and efficacy of the individual in Western urban industrial societies, where all but a few

eccentrics who subject themselves to "wilderness training" would quickly perish if deprived of the goods and services furnished by their fellows. How is it that mutual dependence is most acknowledged where least evident? (Farriss 1984:132)

The paradox is genuine, but the answers proposed are not complete, whether in terms of "emotional comfort" or "material benefits" gained in "sustained interaction with one's fellows" (ibid.), such as risk sharing in a climate that fluctuates greatly.[4] It is not so much material sharing in a society that, apart from marriages and other festivals, nowadays lacks routine villagewide commensal institutions. Rather, it is the easy sharing and feedback of *knowledge* that allows an individual to have an effective, context-sensitive evaluation of resources and responses in an environment in flux.

To begin to account for Itzaj environmental awareness and behavior toward the forest we will explore a kind of distributed *belief system* that may be deemed an *emergent knowledge structure*. An emergent knowledge structure is not a set body of knowledge or tradition that is taught or learned as shared content. Neither is it equally distributed among individual minds in any given culture or systematically distributed among the individual minds of any recognizable cultural subgroup (e.g., experts) to a reliable extent. The general idea is that one's cultural upbringing primes one to pay attention to certain observable relationships at a given level of complexity. By coaxing rather than determining adaptive behavior, this cultural upbringing also allows individual people the leeway to discover for themselves and appreciate the relationships that best fit their life circumstances. Yet, each person is also culturally attuned to the relevant discoveries of other individuals whose knowledge forms part of the emergent cultural consensus (person A can readily assimilate knowledge that person B expresses publicly in a culturally privileged way, for example, by means of a statement in Itzaj, even though A may not have direct experience with the knowledge represented in B's utterance). This allows individual knowledge and planning to develop on the basis of a much broader perspective than would be possible based exclusively on individual experience, and it does so in a way that takes into account information about other people's needs and plans.

There is both considerable overlap and variation among individuals in their appreciation of specific relationships between plants and animals (and humans). This is reflected in the statistical pattern of interinformant agreement. Although there is strong cultural consensus in the appreciation of ecological relationships (i.e., an overwhelmingly large first eigenvalue in a principal components analysis and all positive first factor scores), there are also significant individual differences in ecological appreciation (e.g., individuals A, B, and C may share most of their knowledge as do B, C, and D, but not A and D). But these differences are not systematic across individuals (i.e., little reliable correlation between observed and residual agreement). Furthermore, the pattern of variation across informants is also different for different tasks (how plants help or hurt animals and vice versa, how humans help or hurt plants and vice versa, how humans help or hurt animals and vice versa). This makes it highly unlikely that individuals are being schooled in any systematic way (e.g., if the learning pattern of individual A on task 1 overlaps more with that of individual B than of individual C, and if A overlaps more with B than with C on task 2 and more with individual D than with either B or C on task 3, then one is hard pressed to provide a coherent account of some uniform schooling process across individuals and tasks).

Emergent knowledge structures involve cognitive procedures enabling an evaluation of the future consequences of purposive behavior for one's life and for the environment in which life is embedded. The function and significance of such beliefs is not to predict an optimal ecological scenario, but to create any number of plausible and partially overlapping scenarios (with a degree of self-fulfilling involvement) to which culturally acceptable forms of human life can be fit. Such procedures resemble a theory in their ability to take particular experiences and give them general relevance; that is, they take an instance of experience and project it to a (perhaps indefinitely) larger ensemble of complexly related cases (Wisniewski and Medin 1994). Unlike theories, however, such belief systems operate as much by inciting evocation as by allowing for inference. Emergent structures do not determinately specify relationships between entities, nor do they directly imply necessary or probable consequences of actions. Rather, such structures implicate (via observable similarities, analogies, thematic or episodic contexts, etc.) wide-ranging relationships between entities that anticipate a variable and somewhat open-ended range of responses and future consequences of actions.

In the Itzaj case, there are no culturally stipulated, conventionally learned or rigorously formulated laws, standards or methods (although there are often concrete "rules of thumb" that individuals seem to recognize widely and apply intermittently, such as "birds that fully digest fruits are more likely to hurt than help the trees to which those fruits belong"). There is no *principle of reciprocity* applied to forest entities, no *rules for appropriate conduct* in the forest, and no *controlled experimental determinations* of the fitness of ecological relationships. Yet reciprocity is all-pervasive and fitness enduring. The emergent knowledge structure of the Itzaj is robustly coherent in its implications and effective in its consequences, much as in the weaving and knotting of an Itzaj hammock from the hair-thin fibers of the hennequen plant (*Agave fourcroydes*): "as in spinning a thread we twist the fiber on fiber. And the strength of the thread does not reside in the fact that some fiber runs through its whole length, but in the overlapping of many fibers" (Wittgenstein 1958).

We are only beginning to discern some (1) basic elements (e.g., generic species), (2) relational components (e.g., human-plant or plant-animal interactions) and

(3) network contours (e.g., the spirited cultural landscape of human, animal, and plant intercourse) of these knowledge structures. This latter aspect is the one that appears to be most crucial and elusive for understanding the sustainability of "Mayaland." It is to this description of the cultural landscape of Mayaland that we now briefly turn.

THE SPIRITED MAYA LANDSCAPE

Consider the relationship of the ramon tree (*Brosimum alicastrum*) to Tikal (*ti-ik'al* = 'place of the winds [spirits]'). Here, in Itzaj lore, dwell the forest spirits (*arux*) who preserve the ramon trees that grow atop the pyramids for the benefit of the Maya, the animals, and "the spirits of our great lineage (ancestors) who lived before God" (*u-pixan-oo' ki-nukuch ch'ib'-al kux-l-aj-oo' taan-il ti dyoos*). For Itzaj traditionally, Tikal was not just a place for work or visit but part of a cultural landscape: "a product of past activity that requires constant cognitive attention and behavioral intervention to preserve and reconstruct what is valued in contemporary images of the past" (Frake 1996:91).

This is still true, though Itzaj have been deprived of free access to Tikal after archaeological excavations revealed its tourist potential, and despite the fact that the forest spirits have hidden from the radios that drown out their "voices" (for these spirits are *ik'* 'wind', and their voices are heard as the sounds of the wind). Itzaj know little of what archaeologists know about Tikal (and vice versa), but they believe it to be part of themselves—part of what it means to be an Itzaj. There is a remarkable continuity in Itzaj place-names that antedates the Spanish conquest of Petén.[5] Places in the cultural landscape that Itzaj call "Mayaland" (*u-lu'um-il maayaj*) tag episodes in a person's life that Itzaj are most readily willing to communicate to others. Reference to such places "automatically" makes recounting of a personal experience culturally relevant for any other Itzaj who is listening; that is, individual and social identity are simultaneously implicated and constructed through such references. Itzaj place-names also often describe ecological associations. Not only do the places bring together social history and autobiography, they are also privileged meeting grounds for humans, plants, animals, and spirits. Nearly every place that Itzaj know and name in Petén is imbued with a sense of time-space, which is quite distinct from a chronological sequence or a spatial map. Itzaj do not locate these places only, or even primarily, in terms of spatial positioning any more than they describe their own lives in terms of temporally defined sequences of events. Rather, Itzaj refer episodes to these cultural loci that connect the different paths of individual life histories. Nor do Itzaj describe spirits or species in general terms, such as "All x are / do such-and-such." Instead, diverse episodes are interwoven into a more generalized cultural landscape where "everything has its place-and-responsibility (*kuch*) in the world" (*tulakal yan u-kuch yok'ol kab'*).

We asked the individuals interviewed in our social network study whether or not they knew stories about the forest involving spirits and, if so, to recount an example. We also asked whether or not such stories were "true." Nearly every Itzaj recounted stories they believed to be true. Many were about forest spirits, such as *arux*. For example:

> *In-ten-ej syeemprej inw-il-m-aj arux ich k'aax i wa ich kaj; in-ten t-in-cha'an-t-a(j) . . . b'ay mejen paal-ej. T-in-b'ek/t-aj kol waye' Xilil; pwes . . . y-ok'-ol in-makan-ej tu'ux yan in-näl-ej, t-uy-ok-l-aj-oo' in-makasiinoj. Ma' k-a-mootz-in-t-ik-oo' porke a'-lo' chen tun-b'ax-t-ik-oo' mak; ka' arux je'-lo' layti' chen mas yum-il ik'.*

> ("I've always seen *arux* in the forest or village; I saw him . . . like a small child. I made milpa here at Xilil [place named after an herb] . . . then, over my storage bin where I had my corn, they stole my shoes. Don't get mad at them, they're only teasing you; that *arux* he is really master of the wind.")

Younger Itzaj (who may understand but do not speak the language) also recount that *arux* play tricks on people to test the "strength" (*muk'*) of a person's "blood" (*k'ik'el*) in order to assess valor in doing what is appropriate in the forest (*jach chukul t-inw-ool im-b'et-ik* 'truly complete is my soul to do it'). The unworthy will show themselves too unsettled by such pranks and start behaving badly in the forest: cursing at the wind (for *arux* are masters of the wind), shooting too many animals, or cutting down too many trees. Then the *arux* will abandon you to hazard. But if you show respect, the *arux* will assist you, leading you to the animals when you hunt or to a stand of fruiting ramon or of chicle trees (*Manilkara achras*) full of sap. If one may be allowed an anecdotal observation: a person living by our field station was bitten by a venomous snake but refused to allow us to take him to the hospital: "I've talked to the *arux* and they will take care of me and show me what medicinal herbs (*xiw*) to take." Such spirits do not have simply a figurative or metaphorical existence but an affective presence that Itzaj will literally stake their lives on.

The Q'eqchi' knew virtually no stories about forest spirits, whereas Ladinos told stories they had heard (often believing) from the Itzaj about the ability of Itzaj sorcerers to turn themselves into animals. The Itzaj call such sorcerers *aj-waay*. Although today Itzaj consider *waay* to be demonic human souls who practice black magic, in pre-Columbian times *waay* apparently referred to the particular animal "soulmate" that made up part of each individual Maya's personal and social identity (Freidel, Schele, and Parker 1993).

The web of forest life in Mayaland is spirited with affective value that sustains reciprocity, respect, and fitness and goes beyond mere observation and consideration of the entities involved:

> It is adaptive for cognized models to engender respect for what is unknown, unpredictable, and uncontrollable, as well as for them to codify empirical knowledge. It may

be that the most appropriate cognized models, . . . from which adaptive behavior follows, are not those that simply represent ecosystemic relations in objectively "correct"
material terms, but those that invest them with significance and value beyond themselves. (Rappaport 1979:100)

The overlapping and crisscrossing network of purposive dependencies may help to
account for the historical lack of institutional arrangements to close access to the
forest commons. The Itzaj tendency to believe that when someone shows disrespect for the forest all may suffer implicates a communicative network of animals-
plants-people-spirits that is liable to making redundant direct human-human decisions to cooperate with one another in maintaining the common forest.

   If the Itzaj's lack of corporate ceremonies, institutions, and cooperative work
based on kinship poses a problem for previous analyses of the commons (cf. Ostrom 1990), then the presence of these institutions among the Q'eqchi' poses an
even more difficult problem. Ceremonial life is manifestly richer among the
Q'eqchi' than the Itzaj, including ceremonies related to agriculture. Indeed, when
Itzaj invited members of another Q'eqchi' group (from Libertad) to visit the Bio-
Itza, the Q'eqchi' expressed surprise at the lack of Itzaj ceremony and volunteered
to teach the Itzaj some of their own (Atran 1993b). Nevertheless, the Q'eqchi' do
not consider any elements or features of the Petén landscape to be sacred or to
need protection, except for the *pom* tree (*Protium copal*), whose sap provides the
incense that mediates ritualized communication with the supernatural world (but
which is not thought by the Q'eqchi' to be crucial to the survival of the Petén forest). What is surprising is that Q'eqchi' do consider the cultural landscape from
which they originated as sacred and its elements to have protected value (on protected values see Baron and Spranca 1997). Many Q'eqchi' in Petén believe, like
those who have remained in their homeland around Coban, in *tzuultaq'a,* or "the
sacred mountain valley" (Wilson 1995; cf. Schackt 1984). Belief is strongest among
elderly Catholics (less so among the Evangelical Protestant minority), just as belief about the spirited landscape of Petén is strongest among elderly Catholic Itzaj.
Mountains around Coban are named, and different towns have ancestral ties to
different mountains. Using forest land requires sacrifice to protected mountains.

   In short, the affective involvement of the Q'eqchi' with the landscape of their
homeland may resemble Itzaj involvement with Petén, but if so, little of it carries
over from Coban to Petén. As one NGO operative reported when he tried to encourage the Q'eqchi' to stress the same concern for protection of nature that he
had witnessed around Coban in order to better meet government criteria for gaining a concession in the Maya Biosphere, the Q'eqchi' responded that "in the
mountains [of Coban] we use the land with God's permission, but not in Petén,"
so that their only interest was in gaining a concession wholly given to agriculture.

   Unlike Q'eqchi', the Ladinos are learning aspects of Lowland Maya awareness, such as use and concern for the ramon tree, which the Itzaj call "the milpa of the animals" because of the many animal species that depend on it. The tree is hardly known to Highland Maya and little appreciated by Q'eqchi' in Petén. Emerging Ladino crop patterns and tree valuations also seem to reflect Itzaj attitudes and practice rather than attitudes and techniques proffered by NGOs or government agents. One wonders to what extent UN, NGO, and government activists distinguish among "locals" (cf. Arizpe, Paz, and Velázquez 1996). In fact, there is a long history in Petén of Ladinos learning Itzaj agroforestry practices. In some Petén communities that are now entirely Ladino but were formerly mixed Ladino-Maya (e.g., San Andrés, Dolores), even ancient Lowland Maya practices are readily discernible (Rice and Schwartz 1992). Some of these long-standing *petenero* Ladino communities perform variants of Maya rituals, such as the ceremony for rainstorms (*chaak*), which Itzaj have not performed in a generation. These Peteneros express and practice reciprocity in dealings with the forest (Schwartz 1995). Mayanization of Ladinos in Petén is a centuries-old process as pervasive and profound as Hispanization of the Maya, only more subtle and less forced.

ONE FINAL PUZZLE

And here we come to our final puzzle. We have suggested that Itzaj are heedful of the forest in virtue of an affective and intentional cultural commitment to the Petén landscape, just as Q'eqchi' are heedless of the Petén forest because their culture is committed to another landscape. We have also suggested that the Ladinos are learning to take heed of the forest because of their social relationship with the Itzaj. Mere proximity and exposure to Itzaj practice would not seem sufficient for learning to attend to reciprocal relationships between animals and plants (and, in other experiments, between humans and plants), much less to Itzaj forest spirits. There is, in addition, an apparent readiness among Ladinos to learn from the Itzaj, but that readiness is not apparent among the Q'eqchi'. Yet how are Ladinos able to appropriate relevant aspects of Itzaj culture without any prior understanding of its language, history, or traditions? Ladinos, in fact, claim that they are learning from and about the Itzaj; yet the Itzaj disavow teaching the Ladinos anything about the forest. Indeed, Itzaj claim contrary to fact (in 11 of 14 interviews) that the Q'eqchi' have been better able to assimilate Itzaj practice, in part because Q'eqchi' pray more like Itzaj.

How, then, could Ladinos be learning Itzaj sensibilities? The line of reasoning that follows is frankly speculative and anecdotal, but one that should motivate further research. Seeking to interview the two most-cited Itzaj experts, we found that both had gone on that particular day to the Ladino town of La Nueva. When they returned, we asked them in separate interviews if they ever teach anything about the forest to the Ladinos; both denied doing so. Then, we asked why they had gone to La Nueva and what they did there. One said that he had gone because there were no lemons to be found in San José, but he knew of some in La Nueva. He said that he had stayed so long in La Nueva after finding the lemons because he was trying to figure out with people there how it would be best to plant lemon trees. The other Itzaj said that he had gone from our field station to visit his daughter, who is married to the son of the most-cited Ladino expert. There he stayed telling stories of the barn owl (*aj-xooch' Tyto alba*) whose call augurs the death of strangers. People familiar with it cannot die from it.

A final anecdote concerns the sounds of the forest. This sensibility is not merely one of perception but of affective value. For example, Itzaj give the short-billed pigeon (*Columba nigrirostris*) the onomatopoeic name *ix-ku'uk~tz'u'uy-een*. Itzaj decompose this low, mournful sound into meaningful constituents, interpreted as follows: Pigeon was frightened of jaguar's coming. Squirrel saw this and told pigeon to leave her young with squirrel for protection. Pigeon came back to find that squirrel had eaten her young and that's why, as long as there is forest, one will hear "squirrel (*ku'uk*) tricked (*tz'u'uy* 'entangle') me (*een*)." But when we ask for identifications from Ladinos, we are sometimes told that this bird's name, Uaxactun-Uaxactun, signifies a lament for the ancient Maya spirits of Uaxactun and that is why "Itzaj named it like that." Unlike Tikal, these classic Maya ruins were given the name Uaxactun (*waxaktun* 'eight stone') in the twentieth century by an American archaeologist, Sylvanus Morley. Thus, it is hardly likely that an Itzaj elder would ever describe the pigeon's sound as these Ladinos think the Itzaj do, although some non-Itzaj speaking descendants of Itzaj speakers describe it as the Ladinos do. Yet, this misinterpretation seems to reflect a sense of what a native Maya would be obliged to heed in the forest.

Ladinos are being educated by various means—observation, story, discussion, mistranslation—to what they ought to attend to. Because the Ladinos do not have all the instruments of Itzaj culture, and are unlikely ever to own them (unlike, say, Yukatek Maya coming to live among Itzaj), there may always be doubt as to what meaning the forest holds for those in the know. The Itzaj, as a constant presence and reminder of such doubt, may be a powerful spur to attention. Perhaps, inspired by doubt and like the Ladinos, we may also be educated by a process of acquaintance and inquiry to cumulatively heed aspects of nature and knowledge that we hitherto did not imagine.

But alas, the Itzaj will tell you, there is a rub: "Vanishing is the forest, we are vanishing too, vanishing everything" (*tan u-jo'mol a-k'aax-ej, tan ki-jo'mol xan, tan u-jo'mol tulakal*).

NOTES

1. This collaborative research was supported by NSF grants SBR-9422587 and 9707761 to Scott Atran and Douglas Medin (Northwestern University). Experiments were code-signed, and data were collected and analyzed, with Norbert Ross, University of

Freiburg, Germany, Elizabeth Lynch and John Coley, Northwestern University, Edilberto Ucan Ek', Herborlaria Maya, Yucatán, Valentina Vapnarsky, Université de Paris X, and Ximena Lois, CREA-École Polytechnique, Paris. Only the more speculative interpretations are my own.

2. The pre-Columbian economy of central Petén may have been biointensive, based on agroforestry management of micro- and mesoscale environmental variation, rather than strictly on milpa as some have implied (cf. Cogwill 1962, Reina 1967). Only recently has evidence come to light at Bajo la Justa (between Yaxha and Nakun) of geointensive production in central Petén (Culbert et al. 1995). Such systems, which involve landscape modification through agroengineering works, occur more at outlying classical sites in Belize and Yucatán.

3. In March 1997, Guatemala's National Assembly recognized the Bio-Itza Committee as a moral person and legal association.

4. For example, at the height of the growing season, July rainfall in Flores went from 121 mm in 1993 to 335 mm in 1996 and in nearby Tikal from 58 mm to 137 mm; in May, when crops are first planted, there was no rainfall in Tikal in 1993 for 23 days, then 130 mm in 3 days, etc. (Guatemala government figures from INSIVUMEH.)

5. This observation is based on reports from the Franciscan missionaries who first entered Petén, which are found in the Archives of the Indies at Seville, e.g., Audiencia de Guatemala, Legajo 345, 9 October 1698, folios 302 verso to 311 recto.

REFERENCES

Arizpe, L., F. Paz, and M. Velázquez. 1996. *Culture and Global Change: Social Perceptions of Deforestation in the Lacandona Rainforest in Mexico.* Ann Arbor: University of Michigan Press.

Atran, S. 1993a. Itza Maya tropical agro-forestry. *Current Anthropology* 34:633–700.

Atran, S. 1993b. The Bio-Itza. *Anthropology Newsletter* 34(7):37.

Atran, S., and D. Medin. 1997. Knowledge and action: Cultural models of nature and resource management in Mesoamerica. In *Environment, Ethics, and Behavior,* ed. M. Bazerman, D. Messick, A. Tinbrunsel, and K. Wayde-Benzoni. Pp. 171–208. San Francisco: New Lexington Press.

Atran, S., and E. Ucan Ek'. 1999. Classification of useful plants by the Northern Peten Maya (Itzaj). In *Reconstructing Ancient Maya Diet,* ed. C. White. Pp. 19–59. Salt Lake City: University of Utah Press.

Atran, S., et al. 1999. Folkecology and commons management in the Maya Lowlands. *Proceedings of the National Academy of Sciences U.S.A.* 96:7598–7603.

Baron, J., and M. Spranca. 1997. Protected values. *Organizational Behavior and Human Decision Processes* 70:1–16.

Berkes, F., D. Feeny, B. McCay, and J. Acheson. 1989. The benefit of the commons. *Nature* 340:91–93.

Cogwill, U. 1962. An agricultural study of the southern Maya lowlands. *American Anthropologist* 64:273–286.

Culbert, T.P., L. Levi, B. McKee, and J. Kunen. 1995. Investigaciones arqueológicas en el Bajo La Justa, entre Yaxha y Nakun. Paper presented at the Simposio de Arqueología Guatemalteca.

Farriss, N. 1984. *Maya Society under Colonial Rule.* Princeton: Princeton University Press.

Ford, R. 1976. Communication networks and information hierarchies in Native American folk medicine. In *American Folk Medicine,* ed. W. Hand. Pp. 17–29. Berkeley: University of California Press.

Frake, C. 1996. A church too far near a bridge oddly placed: The cultural construction of the Norfolk countryside. In *Redefining Nature,* ed. R. Ellen and K. Fukui. Pp. 89–115. Oxford: Berg.

Freidel, D., L. Schele, and J. Parker. 1993. *Maya Cosmos: Three Thousand Years on the Shaman's Path.* New York: William Morrow.

Granovetter, M. 1979. The idea of advancement in theories of social evolution and development. *American Journal of Sociology* 85:489–515.

Hammer, M. 1983. "Core" and "extended" social networks in relation to health and illness. *Social Science Medicine* 17:405–11.

Hardin, G. 1968. The tragedy of the commons. *Science* 162:1243–1248.

López, A., S. Atran, J. Coley, D. Medin, and D. Smith. 1997. The tree of life: Universals of folkbiological taxonomies and inductions. *Cognitive Psychology* 32:251–295.

Nations, J., and R. Nigh. 1980. Evolutionary potential of Lacandon Maya sustained-yield tropical forest agriculture. *Journal of Anthropological Research* 36:1–30.

Ostrom, E. 1990. *Governing the Commons.* Cambridge: Cambridge University Press.

Rappaport, R. 1979. *Ecology, Meaning, and Religion.* Berkeley: North Atlantic Books.

Reina, R. 1967. Milpa and milperos: Implications for prehistoric times. *American Anthropologist* 69:1–20.

Remmers, G., and E. Ucan Ek'. 1996. La roza-tumba-quema maya: un sistema agroecológico tradicional frente el cambio tecnológico. *Etnoecológica* 3:97–109.

Rice, D., and S. Schwartz. 1992. Modern agricultural ecology in the Maya lowlands. Paper presented at the Fifty-seventh Annual Meeting of the Society for American Archaeology.

Romney, A.K., S. Weller, and W. Batchelder. 1986. Culture as consensus. *American Anthropologist* 88:313–338.

Schackt, J. 1984. The Tzuultak'a: Religious lore and cultural processes among the Kekchi. *Belizean Studies* 12:16–29.

Schwartz, N. 1990. *Forest Society.* Philadelphia: University of Pennsylvania Press.

Schwartz, N. 1995. Colonization, development, and deforestation in Petén, northern Guatemala. In *The Social Causes of Environmental Destruction in Latin America,* ed. M. Painter and W. Durham. Pp. 101–130. Ann Arbor: University of Michigan Press.

Soza, J.M. 1970. *Monografía del Departamento de El Petén.* 2 vols. Guatemala: Editorial José de Pineda Ibarra.

Stewart, S. 1980. *Grammatica Kekchi.* Guatemala: Editorial Academica Centro Americana.

Wilson, R. 1995. *Maya Resurgence in Guatemala: Q'eqchi' Experiences.* Norman: University of Oklahoma Press.

Wisniewski, E., and D. Medin. 1994. On the interaction of theory and data in concept learning. *Cognitive Science* 18:221–281.

Wittgenstein, L. 1958. *Philosophical Investigations.* 2d ed. Oxford: Blackwell.

10

DIMENSIONS OF ATTRITION IN LANGUAGE DEATH

Jane H. Hill

Robb (1993) estimates that the apex of human linguistic diversity may have been reached in the Neolithic, with a sociolinguistic milieu perhaps similar to that among contemporary horticulturalists in the Upper Amazon or Highland New Guinea being found planetwide. In these regions hundreds of small languages are found in relatively small areas, with speaker numbers ranging from a few hundred to perhaps 10,000. Under such circumstances most adults of normal intelligence are multilingual, and languages are not usually ranked according to enduring prestige.

This picture of egalitarian linguistic diversity must have begun to change with the rise of the earliest states. The spread of Latin in western Europe in the beginning of the common era, or of Arabic across North Africa in the Middle Ages, exemplify a shift to stratified linguistic diversity dominated by major regional languages. Yet, while language extinction is clearly nothing new, the scale of loss that we see today may be unprecedented. Krauss (1992) estimates that half of the approximately 6000 languages spoken in the 1990s will be extinct by 200. Catastrophic language loss is already well advanced in Australia and the Americas, and the linguistic diversity of Siberia, Africa, and South and Southeast Asia is also threatened. The replacing languages are of a new type: "world languages"—English and Spanish are the most important—are replacing not only small local languages, but major regional languages with speakers numbering in the millions. Linguistic diversity with its attendant multilingualism is being replaced by massive monolingualism in these world languages. Of course these languages manifest internal diversity, but this is sharply stratified into enduringly prestigious and stigmatized varieties.

We do not yet know what these processes may mean. While human languages share many universal features that are the common heritage of our species, these universals are sufficiently complex and numerous that the limits on the possibility of their combination cannot be known without a very large sample of languages (see Corbett this volume). Furthermore, small languages seem to provide for their speakers deeply embodied and very local ways of being-in-the-world, highly economical alignments of knowledge and rationality with emotional and aesthetic life that are part of a sense of belonging to a place and a community and living in these with skill and relative ease (see, for instance, Woodbury 1993). World languages may provide such a sense only for certain privileged core populations of speakers, and the kinds of knowledge favored by such populations—restricted by elaborate procedures for certification, codeable as certain kinds of text, and the like—may be qualitatively different from the kinds of local knowledge acquired within a small linguistic community. The primitive state of our understanding of linguistic diversity is not reassuring. I will attempt to sketch briefly what is known about the processes by which linguistic diversity is lost, with an eye toward interrupting these, if interruption is indeed possible.

LANGUAGE DEATH

Language death is the end of a process, sudden or gradual, that occurs when speakers abandon one language in favor of another (or others). The less favored language loses both functions—social occasions, topics, qualities of emotion—appropriate for the language, and speakers—those who are competent in it. Languages that persist in few functions, with few speakers, are said to be obsolescent. When no children speak a given language, it is said to be "moribund."

Languages that have suffered a severe loss of functions and speakers also lose structure. Sasse (1990a, b) calls this attrition of structure "language decay" and suggests that it results from the disruption of normal intergenerational transmission, such that children must learn what they know of a language under rather unfavorable cognitive and emotional circumstances. Speakers who have acquired a dying language under these unfavorable circumstances are "semi-speakers" (Dorian 1977). They exhibit impoverished structural repertoires at all levels of organization. But interrupted transmission is not the only circumstance that yields the loss of structure. I present here examples of decay in the usage of fluent speakers, who use their language in at least some functions quite frequently and for whom we have no evidence of the interruption of normal transmission of the language in childhood. These cases seem to result from the loss of contexts of life in which the lost structural possibilities were appropriate. Thus, language decay is not only instantiated in semi-speakers but may be found as well in fluent speakers of a dying language if it has retreated to a narrow functional range.

Before proceeding, I note that most of the world's population even today is multilingual, living parts of their lives in one language and parts in another, or others. We all know colleagues who conduct their home lives in, say, Lithuanian or Farsi, but who conduct their scholarly lives primarily in English. They may not know the English word for a particular kitchen implement, or the Farsi translation of a technical term in chemistry. Linguists do not see this kind of coordinate bilingualism as a sign of moribundity in either language or as a sign of impoverished competence among such speakers. Furthermore, speech communities differ widely in their evaluations of linguistic skill, with one group preferring a spare and laconic style, another preferring baroque elaboration. In some communities, a very permissive definition of "speaker of a language" appears and includes people who, for instance, may be able to count to 10, or recite the Lord's Prayer, but cannot conduct a conversation. Thus, we can judge language decay objectively only by comparing speakers with others who use the same language. Younger speakers can be compared with their elders, urban speakers with rural speakers, and speakers in immigrant communities with speakers in their country of origin. Even this kind of control may be inadequate. A community may expect younger speakers to be sparing in their talk, and older speakers to be more verbose, or vice versa. A relatively unelaborated style may mark a proud commitment to a working-class identity or the lordly nonchalance of a noble. Thus, the identification of language decay cannot be accomplished exclusively on structural grounds but requires ethnographic investigation of local ideologies and preferences, which are invariably complex and often disputed.

Speakers of dying languages should not be seen as necessarily linguistically impaired, since whatever deficiencies they may exhibit in the dying language may be amply compensated by skills in the replacing language. If we could go back and hear White Thunder, a Menominee Indian accused by no less an authority than Leonard Bloomfield (1927) of "speaking no language tolerably," we might find that his multilingual repertoire had considerable expressive power. But, as Woodbury (1993) has pointed out, the elaboration of competence in the replacing language often lends very little social advantage to the speaker of a moribund language, since the new elaboration is likely to occur within a low-status variety—indeed, its status may be low precisely because it is known to be the variety associated with an oppressed group. White Thunder may have been the victim of a value judgment in which Bloomfield unwittingly intertwined strictly linguistic judgments along with biases of class and race. Our goal as linguists is to filter out this ideological noise—both our own and that of speakers of threatened languages—and arrive at some global generalizations about what language decay is like. Good generalizations about language decay can help communities decide whether to be concerned about, for instance, the use of multilingual slang by teenagers, or the use of dominant-language expressions to connote elegance or social distance among adults. They can help communities in addressing ideological and attitudinal issues that may be implicated in language decay, such as unreasonable expectations for fluency among very young children or the idea that a language is too difficult for children and thus should only be taught in school, not in the family.

Furthermore, good generalizations about language decay will permit us to evaluate and improve the linguistic record. Today, when global linguistic diversity is rapidly decreasing (Krauss 1992), documentation of the competencies of "last living speakers" is quite properly a high priority. But linguistic data from speakers of moribund languages may be quite unlike what we would find in a fully functioning linguistic community. A better theory of the structural-functional relationships in language decay might permit us to improve our elicitation techniques to maximize the range of structures that speakers can produce. Finally, the identification of such generalizations will contribute to a general theory of language change and especially to theory about the relationship between language structure and language function.

I first briefly illustrate attrition in functional domains, and then turn to language decay proper at the levels of discourse, syntax, morphology, and lexicon.[1] I illustrate these with examples from the languages with which I am familiar, including Nahuatl or Mexicano in Mexico and Tohono O'odham in Southern Arizona. These are both languages of the Uto-Aztecan family. The common feature at all levels is that the range of structural options is reduced. Reduction of structural differentiation is, of course, also a feature of ordinary language change. However, our current understanding suggests that in ordinary language change structural reduction in one subsystem will be balanced by structural elaboration in another. For instance, the loss of the Germanic case markings in early Middle English was accompanied by the development of new syntactic complications and restrictions on word order. In language decay, in contrast, the structural reduction in the moribund language is not accompanied by any compensating elaboration. Instead, speakers elaborate their competence in the "target" language— the language toward which they are shifting.

ATTRITION ACROSS FUNCTIONAL DOMAINS

By "functional domains" I mean occasions for the production of talk and text that are defined by such dimensions as participants, topics, goals, and the like. (Probably the richest theory of domains is that suggested by Hymes in his various works on the ethnography of speaking; cf. Hymes 1972.) As an example, functional attrition in interactional domains focused on participants is seen in a continuum of practices in bilingual communities in the Malinche Volcano region of Puebla and

Tlaxcala in Central Mexico, where Nahuatl, locally called "Mexicano," is giving way to Spanish. In the most conservative communities, Mexicano is preferred if the interlocutor is identifiable as a member of the community, while Spanish is preferred with outsiders. In other towns, Mexicano is preferred only inside the domestic circle. When making a purchase in a local store or attending a community meeting, speakers are much more likely to use Spanish even if they know that their interlocutors speak Mexicano. In still other towns, people may initiate talk in Mexicano only with elderly members of their own immediate families. The result of this kind of functional splitting and attrition is that there are fewer and fewer appropriate interlocutors for talk in Mexicano.

DECAY IN DISCOURSE

I use the term "discourse" to suggest the inventory of strategies by which speakers construct coherent messages. Issues of coherence include the production of speech or text in a particular register—for instance, the expression of formality or respect versus the expression of intimacy or vulgarity—and the differentiation of genre—for instance, the difference between a ballad and a business letter—as well as such microlevel skills as the capacity to maintain a topic across several sentences. Some years ago I suggested that there were at least two major directions of register attrition: the "Latinate" pattern, where a language retreats to a complex and esoteric language of ritual, and its opposite, where a language survives only in quotidian household use (Hill 1983). But more complex and fine-grained patterns also appear.

In Tohono O'odham, there are three major ritual genres: song, oratory, and narrative. Of these, song continues to flourish, both in individual religious practice and in curing, and has inspired a program of original composition of songlike poems in bilingual classrooms. Songs are often performed on public occasions, and while certain subgenres, such as hunting songs, are being lost, the genre of power songs—songs that come to singers in dreams and that are sung to gain spiritual strength—may even be undergoing elaboration as singers draw on themes from Christian prayer and song or from popular mass media. Oratory, in contrast, has virtually disappeared. In traditional Tohono O'odham society there were many different kinds of speeches, each suitable for a different occasion. Today only a dozen elderly men know any orations, since the ritual contexts for their performance no longer occur. The situation with narrative is more complex. On several occasions in recent years individual communities have tried to arrange "winter tellings," the formal performance of the creation story over four nights. Audiences have been sparse, and the organizers were discouraged. In the last couple of years reservationwide winter tellings have been conducted at the high school in Sells, the capital of the Tohono O'odham reservation. These have attracted standing-room-only audiences, but this means that a very few storytellers serve the entire O'odham nation, rather than having many individual tellers serving local communities and even individual families. Since much traditional knowledge was encoded in oratory and ritual narrative, this is lost along with skill in the genres. Today, a few Tohono O'odham writers are beginning to gain a reputation in English or in bilingual performance, but it is difficult for this small group of people to compensate for the loss of the very substantial development of oral literature that characterized traditional O'odham culture.

DECAY IN SYNTAX

Among the many types of syntactic decay, the one that has interested me most is the loss of complex sentence types in favor of sequences of simple sentences. This phenomenon has been spotted in a number of languages since I first identified it in Cupeño and Luiseño, Uto-Aztecan languages of Southern California (Hill 1973, 1979). The early stages of such a loss are very apparent in Mexicano (see Hill 1989 for details). In Mexicano the morphological means for the construction of complex sentences, such as relative clauses, are not at all challenging. In fact, speakers have many options: they can use a conservative "headless" relative clause, that is, a relative clause with no relative pronoun, as in (1) in table 10.1, but it is quite acceptable to construct a relative clause according to Spanish-language models. Thus, speakers can use a Spanish relative pronoun like *que* (see [2] in table 10.1), or create a loan-translation of *que* using Nahuatl *tlen* 'what' as a relative pronoun, as shown in (3). I mention these several options, and emphasize their simplicity, since speakers do not fail to use relative clauses because they are "difficult" constructions. Instead, certain speakers avoid them because of a functional split between Mexicano and Spanish.

In spite of the simplicity of these constructions, some speakers use relative clauses (of any type) very infrequently when speaking Nahuatl. The decrease in frequency follows a familiar pattern. As shown in table 10.2, data from narratives reveal that women use relative clauses with equal frequency in Spanish and Nahuatl.[2] Men who identify as farmers seem to use relative clauses slightly more frequently in Spanish than in Nahuatl, but the difference is not significant. On the other hand, men who identify as wage laborers show a striking split between the two languages, with a strong preference for greater elaboration, expressed through relative clauses, when speaking Spanish.

The differences shown in table 10.2 may be the result of the intersection of speakers' construal of contexts and situations with a quality of meaning that is semiotically intrinsic to relativization. Relative clauses are the syntactic equivalent of a data-compression program that saves space on a crowded hard disk. Complex

Table 10.1   Strategies for Forming Tlaxcalan Nahuatl Relative Clauses

(1)   O:yek se: te:na:ntsi:n [o:yah Puebla]
      'There was an old lady [(who) went to Puebla]'

(2)   Okseppa o:tlayawalo:to nochi in i:peo:nes [que o:kipiaya kada asye:nda]
      'Again he went to check on all of his peons [that he used to have on each hacienda]'

(3)   Nochi [in tlen o:monesta:roaya], ni:n o:tiayah tikchi:watiweh
      'As for all [that was needed], we used to go do this'

Note: Relative clauses are in brackets.

sentences that use relative clauses, are focused on information that is "new" for the interlocutor—the information in the relative clause (e.g., "that he used to have on each hacienda" in [2] in table 10.1). This new information is in turn subordinated to still other "new" information—the information contributed by the noun that heads the clause ("his peons" in the case of [2] in table 10.1). Thus, heavily relativized text and talk is exceptionally dense with information. We would expect such density only in contexts where speakers do not share very much information. To highlight such a lack of sharedness can imply that speakers are socially distant from each other or that one is somehow superior in knowledge to the other.

Male wage laborers see their home Mexicano-speaking communities as egalitarian havens from the outside world. In that hierarchical outside world they struggle for a precarious hold in the lowest-status fringe of the Mexican market economy. In the outside world, of course, they use Spanish exclusively. The Spanish they use within the bilingual repertoire of their home communities evokes their access to this world's power. Thus, wage laborers exhibit the most advanced functional split between the two languages. They use Nahuatl primarily for the expression of egalitarian solidarity. Information-heavy speech, dense with relative clauses, is inappropriate to such solidarity and equality. Spanish, in contrast, is used in and about contexts where speakers may be hierarchically differentiated from each other, as in the workplace or in official community business. Here, an information-heavy register is appropriate because it suggests that interlocutors do not all share the same level of knowledge. It is compatible with the use of Spanish, which also suggests that the speaker may be more powerful than his interlocutor. While wage laborers can produce relative clauses in Mexicano, they do so at an extremely low frequency. We can easily imagine that their children are not getting adequate input on this kind of syntax. They are likely to grow up impaired in their ability to produce information-heavy registers in that language. As economic pressure forces more and more adults into wage labor, with the effects on their Mexicano usage illustrated above, fewer and fewer children are exposed to syntactically complex registers of Mexicano speech.

Table 10.2 Relativization in Spanish and Nahuatl

	I. Women (wives of farmers only) T-Unit Type[a]		
	Relative Clause	Other	Total
Spanish	37 (11%)	291 (89%)	328
Nahuatl	28 (12%)	201 (88%)	229
Chi square = 0.23 (NS)			

	II. Male farmers T-Unit Type		
	Relative Clause	Other	Total
Spanish	30 (17%)	142 (83%)	172
Nahuatl	44 (13%)	307 (87%)	351
Chi square = 1.90 (NS)			

	III. Male wage laborers T-Unit Type		
	Relative Clause	Other	Total
Spanish	25 (33%)	50 (67%)	75
Nahuatl	42 (7%)	546 (93%)	588
Chi square = 48.32 $p < .001$			

Note: Compared here are the numbers of relative clauses versus all other T-unit types.

[a]A "T-unit" is an analytic unit consisting of a verb and its arguments; roughly, a clause.

Farmers, in contrast, conduct many more of their daily activities within their home communities, recognizing hierarchy within them and expressing this in Nahuatl. They are likely to view performance in Spanish with suspicion, since they consider Spanish speakers to be duplicitous and insincere. Thus, in their speech we do not see the same sharp functional split, reflected in rates of relativization, that we see among wage laborers. Instead, their Spanish usage and their Mexicano usage are nearly indistinguishable in this respect.

DECAY IN MORPHOLOGY

Decay in morphology, the system for the formation of complex words, takes several forms. These include the loss of "irregular" morphology in favor of regularity and the "freezing" of complex formations as fixed lexicalizations. While regularization and freezing may be heard locally as signs of the incompetence of a speaker, they do not necessarily compromise the expressive power of a language in any obvious way. Thus, in one notable case of the freezing of complex formations, reported by Moore (1988) for Kiksht, a Chinookan language spoken along

Table 10.3   Tlaxcalan Nahuatl Noun Incorporation

(1a)   ni-tla-<u>pa:lah</u>-tsakwilia
        I-OBJ-board-block off
        'I block (the irrigation ditch) with a board'

(1b)   ni-tla-tsakwilia i:ka in pa:lah
        I-OBJ-block off with a board

(2a)   ti-<u>pu:ntah</u>-teki-ti:weh
        we-tip-cut-go to do
        'We go to cut the tips (the flowers of the maize plant)'

(2b)   ti-teki-ti:weh in pu:ntah
        we-cut-go to do the tips

(3a)   kin-<u>val</u>-inkulka:ro:-s
        Them-values-inculcate-FUT
        'One will inculcate values in them (the youth)'

(3b)   kin-inkulka:ro:-s in valores
        them-inculcate-FUT values

Note: Incorporated noun stems are underlined in type a sentences. Type b sentences have the same translations as the type a sentences, except that they are ambiguous: They can refer to particular boards, maize-plant tips, or values, as well as to generic activities.

the Columbia River in Oregon and Washington states, this has apparently been an important source for new lexical items during all the known history of the language. But some morphological decay does affect expressive power. An example is the decay of noun incorporation in Mexicano (discussed in detail in Hill and Hill 1986, Hill 1987). In noun incorporation, the noun is shifted from a position as full lexical object of the verb to a position prefixed to the verb. Examples are shown in table 10.3. Type a constructions are quite rare; out of 16,025 verbs produced by speakers on whose socioeconomic status I had good information, there are only 114 noun-incorporating verbs, and I judge 20 of these to be frozen forms. Generally, people produce expressions of type b.

I selected the examples shown in table 10.3 because the appearance of Spanish morphemes in them shows that they are not frozen ancient formulae. For each construction I show how an analytic alternative would work. Note that the type a sentences, with incorporation, always refer to a generic activity, along the lines of English "baby-sitting." One can, in fact, "baby-sit" ten-year-olds or dogs, so "baby" in this formation clearly does not mean some real baby. Instead, the incorporated noun serves as part of the label of a generic activity. Similarly, in Nahuatl the incorporated noun is nonreferential and cannot pick out any particular item in the world. In contrast, the type b sentences are ambiguous. Here, their nouns could be interpreted as referential, as specific "boards," "flowering tips," or "values" in the world.

The disappearance of noun incorporation in Nahuatl is significant because incorporation is a principal way to render a nominal element nonreferential or "generic" in order to construct "generic activity" expressions. The loss of noun incorporation represents a genuine loss of expressive potential in the language and precisely at a locus where activities that are seen as part of "culture" and "custom" are distinguished from activities seen as idiosyncratic or occasional.

DECAY IN THE LEXICON

Linguists correctly resist ideologies that see language moribundity solely in terms of "losing words" and definitions of good speakers as "people who know a lot of words." But lexical loss is important, since precisely these word-focused ideologies can erode the linguistic confidence of speakers who may have trouble remembering little-used items.

Attrition in the lexicon for local flora and fauna is of special interest. In a study of twelve O'odham and Yaqui children living in Southern Arizona, Nabhan and St. Antoine (1993) report that the children recognized on the average only 4.6 names out of 17 native species shown to them in pictures. In contrast, their grandparents averaged 11.1 names for the 17 species (Nabhan and St. Antoine 1993:244). Ofelia Zepeda and I found that substantial attrition in the biosystematic lexicon of Tohono O'odham is obvious even among people in their 60s. Between 1986 and 1990, we conducted a dialectological survey of 91 speakers of this language, all over 50 years of age. Our elicitation instrument was a scrapbook that included pictures of 50 different plants and animals. We selected these pictures in a pilot study as those most easily identifiable and most appropriate for rapid elicitation. We did not want subjects to have to pause and think about the names; instead, we wanted to collect as many lexical items as quickly as we could in order to develop a statistical base for a study of language variation. But we learned that a substantial minority of older speakers had trouble naming even these very easy stimulus pictures. These elders split into three large groups, reflected in the subsample of 33 speakers shown in table 10.4. One group not only named all of the items easily but were also reminded of other related plants and animals and gave their names as well as those of small plants or animals that were in the background of pictures intended to elicit the names of more prominently featured species. A second group (the largest) knew nearly all the names, requiring prompts on one or two items and confessing not to know perhaps one other. These speakers were probably like the "grandparents" in the Nabhan and St. Antoine (1993) sample. A third group, however, had great difficulty with the list. This group included some of the best-educated people in the sample, and two speakers who had been high tribal officers. Four of them are quite fluent and use O'odham daily in their communities; all of them speak the language at least some of the time. Three had lived off-

Table 10.4   Attrition in the Biosystematic Lexicon in Tohono O'odham

Group I:	8 elderly speakers who named all species
Group II:	19 elderly speakers who missed an average of 1.89 species (range: 1–3 missing names)
Group III:	6 speakers who missed an average of 13.5 species (range: 4–21 missing names)
	Female (age 51, schoolteacher, college degree) missed 21
	Male (age 52, high tribal officer with some college) missed 20
	Female (age 67, high school degree, San Xavier) missed 15
	Female (age 61, high school degree, San Xavier) missed 13
	Female (age 73, high school degree) missed 8
	Male (age 65, district chairperson, San Xavier) missed 4

reservation for long periods, although they were living on the reservation at the time of the study. The other three lived at San Xavier, technically a district of the reservation but for practical purposes a suburb of Tucson. Interestingly, none of these speakers used a strategy of naming with hypernyms (words like "weed," "cactus," or "bird"), even when these were available in the language. Instead, they observed that they could not remember the correct names for the species. The unknown items included even "charismatic megafauna" like big-horn sheep and bobcat, important and common dangerous animals like scorpions and gila monsters, and plants used in basketry, the main craft form for Tohono O'odham women.

This sort of lexical attrition has profound consequences. Nabhan and St. Antoine (1993) note that it cuts speakers off from their environment; they may be better able to name African megafauna on a television nature program than animals that are common in their own backyard. Furthermore, it makes speakers globally more ignorant. It is highly unlikely that the expertise about the flora and fauna of the Sonoran desert that is encoded in these names will simply be transferred to English; most English speakers, with the exception of a few desert buffs and biologists, make far fewer distinctions among the various plants and animals of the desert than do traditional Tohono O'odham, even those who are not considered particularly expert.[3] Speakers who are weak on biosystematic lexicon may, of course, have quite elaborate English vocabularies in specialized domains such as auto mechanics or needlework, but since they have relatively low prestige in "Anglo" society, this expertise may pass unacknowledged.

In a close analysis of the behavior of Tohono O'odham speakers who replied to the dialectological questionnaire, I found that people who remembered the names for plants and animals very often made comments in addition to giving the name. These comments included commonplaces from a local discourse of natural history, such as "You usually see them around water" (of *taiwig*, 'firefly'), or "They get yellow like that when they're ripe" (of *wihog*, the fruit of mesquite, *Prosopis juliflora*). They also included comments based on personal experience with the item, such as "One of those ate a colt here not too long ago" (of *mawid*, 'mountain lion', *Felis concolor*), or "My grandmother would soak those a long time and they'd come out such a pretty light color" (of the leaves of *takwui'a* 'basketry plant', *Yucca elata*). Especially interesting were spontaneous emotional reactions, such as a sound of disgust that accompanied the naming of *ko'owi*, 'rattlesnake', *Crotalus* spp., or the affectionate chuckle when speakers named *okokoi*, 'white-winged dove', *Zenaida asiatica*, a bird considered to be a charmingly greedy competitor with the O'odham for the sweet ripe fruit of the Saguaro cactus. But speakers in the third group, who could not remember many of the names, made almost no such comments, even though the items are very common locally and the speakers knew their English names. They did not even make the sounds of disgust or affection that occurred so frequently among their more fluent neighbors. Apparently, when names are lost, people lose not only encyclopedic nature knowledge; they even experience weakening of access to their own feelings and life histories. Thus, the loss of lexicon is not only a matter of the attrition of cultural knowledge but may involve the attenuation of the very selfhood of speakers insofar as memory is a part of it.[4]

CONCLUSION

In all of the cases I have reviewed, the basic process is a reduction in the range of options that speakers have available. This kind of reduction is very familiar in "normal" language change, although "normal" languages seem to maintain, overall, a sort of steady state: reduction in one area is accompanied by elaboration in others. In moribund languages, the loci of attrition seem to accumulate without compensatory complication. The language increasingly constricts into a narrow range of frozen formulae, regardless of whether these formulaic expressions are "high" (like "Give us this day our daily bread, and forgive us our debts as we forgive our debtors" in the Lord's Prayer) or "low" (like "Shut the door"). Furthermore, this decay often happens rapidly, so that the difference in competence between subgroups in the speech community may be quite apparent, yielding devastating disputes over the worth of different kinds of speech. In such communities young people frequently report that their elders are so critical of their speech in the local language that they prefer to use the national language simply to avoid the scolding. Adults may stop each other in mid-sentence to point out that a word is "not Mexicano." Visitors to a community may be subjected to "language tests" on rare or hard-to-pronounce vocabulary items. In such a charged environment, the national language is often a "safe" choice that will avoid these kinds of problems.

Some attrition of structural differentiation in the language is related to attrition of life contexts for language use. Fluent Tohono O'odham speakers lose the biosystematic lexicon because food now comes from the supermarket, an English-speaking domain, not from the desert. They lose the art of oratory because week-long communal rites like the summer wine festival are no longer possible when people must go to work for wages every day. Fluent Nahuatl speakers abandon relativization because their communities have become for them havens of egalitarianism in which elaborated speech that suggests differential levels of knowledge is inappropriate. In other cases this relationship between structure and context is not clear, and we may be better advised to turn to interrupted acquisition to account for the decay. In most dying languages many such reasons are working together, affecting not only semi-speakers, but the remaining fluent speakers as well. Such decay may have major effects on how we understand the nature of a language. For instance, it may result in a genuine typological shift, as when a split ergative case system (a system in which both ergative-absolute and nominative-accusative cases are used, but at different structural loci) is reduced to its nominative-absolutive component, which appears at all loci. This occurs in moribund Dyirbal (Schmidt 1985). When noun incorporation, the principal device of polysynthesis, is lost in Nahuatl in favor of a more analytic treatment of nonreferentiality of nouns, this disrupts our understanding of the way the polysynthesis parameter (Baker 1995) can produce a coherent organization of syntax common to many languages. The loss of complex sentence types may make it impossible for syntacticians to develop appropriate tests for detecting the details of constituent structure. Thus, dying languages are problematic as far as the linguistic record is concerned, and data from them should be used with considerable caution and awareness of these problems.

Yet even dying languages are still extremely complex systems. For the communities who have preserved them in the face of great odds, obsolescent and moribund languages are a precious heritage. Linguists must assert the right of speakers of such languages to speak and develop them, and among our most important scholarly tasks should be not only to document these languages and the social and linguistic processes that threaten them but also to help speakers recover options for expressive power that are threatened or lost.

NOTES

1. While phonological markers can have important functions in marking local identities (cf. Campbell and Muntzel 1989), space precludes attention to this dimension of attrition in the present chapter.
2. As it happens, the data on this point come only from women who were wives of farmers. The samples of Spanish narrative are small because they were collected quite by chance; Spanish usage was not part of my main research program at the time.
3. Balée (1996) has shown that the loss of ethnobotanical nomenclature does not only occur with the kind of shift from subsistence cultivation and foraging to a "grocery-

store" economy experienced by the Tohono O'odham. He presents evidence for the loss of plant names among the Guajá, a Tupian-speaking Amazonian group that probably shifted from cultivation to foraging within the last half millennium. The Guajá not only have fewer names for cultivated plants than their close linguistic relatives, the Ka'apor, who continue to practice cultivation, they also have fewer names for semidomesticates and wild plants. Hunn (this volume) notes that world languages are strikingly impoverished compared to local languages in ethnobotanical nomenclature. Comparing Hunn's data with Balée's, we might hypothesize that such lexical loss is a feature of very rapid culture change as much as it is a matter of relative detachment from the "natural" world.

4. Apart from the impact on speakers, lexical attrition is very problematic for historical and comparative linguistics. For instance, it may be impossible to find a speaker who can confirm the identification of a plant or animal mentioned in an old text, so that, even if the item was recorded at some point, it must be used in comparative work with great caution.

REFERENCES

Baker, M. 1995. *The Polysynthesis Parameter.* Oxford: Oxford University Press.

Balée, W. 1996. On the probable loss of plant names in the Guajá language (Eastern Amazonian Brazil). In *Ethnobiology in Human Welfare,* ed. S. K. Jain. Pp. 473–481. New Delhi: Deep Publications.

Bloomfield, L. 1927. Literate and illiterate speech. *American Speech* 2:432–439.

Campbell, L., and M.C. Muntzel. 1989. The structural consequences of language death. In *Investigating Obsolescence,* ed. N.C. Dorian. Pp. 197–210. Cambridge: Cambridge University Press.

Dorian, N.C. 1977. The problem of the semi-speaker in language death. *International Journal of the Sociology of Language* 12:23–32.

Hill, J.H. 1973. Subordinate clause density and language function. In *You Take the High Node and I'll Take the Low Node,* ed. C. Corum et al. Pp. 33–52. Chicago: Chicago Linguistic Society.

Hill, J.H. 1979. Language death, language contact, and language function. In *Approaches to Language,* ed. S. Wurm and W. McCormack. Pp. 44–78. The Hague: Mouton.

Hill, J.H. 1983. Language death in Uto-Aztecan. *International Journal of American Linguistics* 49(3):258–276.

Hill, J.H. 1987. Women's speech in modern Mexicano. In *Language, Gender, and Sex in Comparative Perspective,* ed. S.U. Philips, S. Steele and C. Tanz. Pp. 121–160. Cambridge: Cambridge University Press.

Hill, J.H. 1989. The social function of relativization in obsolescent and nonobsolescent languages. In *Investigating Obsolescence,* ed. N.C. Dorian. Pp. 149–164. Cambridge: Cambridge University Press.

Hill, J.H., and K.C. Hill. 1986. *Speaking Mexicano.* Tucson: University of Arizona Press.

Hymes, D.H. 1972. *Foundations of Sociolinguistics.* Philadelphia: University of Pennsylvania Press.

Krauss, M. 1992. The world's languages in crisis. *Language* 68:4–10.

Moore, R. 1988. Lexicalization versus lexical loss in Wasco-Wishram language obsolescence. *International Journal of American Linguistics* 54:453–468.

Nabhan, G.P., and S. St. Antoine. 1993. The loss of floral and faunal story: The extinction of experience. In *The Biophilia Hypothesis,* ed. S.R. Kellert and E.O. Wilson. Pp. 229–250. Washington, D.C., and Covelo, Calif.: Island Press and Shearwater Books.

Robb, J. 1993. A social prehistory of European languages. *Antiquity* 67:747–760.

Sasse, H.-J. 1990a. Theory of language death. *Arbeitspapier* Nr. 12 (Neue Folge). Institut für Sprachwissenschaft, Universität zu Köln.

Sasse, H.-J. 1990b. Language decay and contact-induced change: Similarities and differences. *Arbeitspapier* Nr. 12 (Neue Folge). Institut für Sprachwissenschaft, Universität zu Köln.

Schmidt, A. 1985. *Young People's Dyirbal.* Cambridge: Cambridge University Press.

Woodbury, A.C. 1993. A defense of the proposition, "When a language dies, a culture dies." In *SALSA I: Proceedings of the First Annual Symposium about Language and Society–Austin,* ed. R. Queen and R. Barrett. Pp. 102–130. Texas Linguistic Forum 33.

11

ACCULTURATION AND ETHNOBOTANICAL KNOWLEDGE LOSS AMONG THE PIAROA OF VENEZUELA

Demonstration of a Quantitative Method for the Empirical Study of Traditional Environmental Knowledge Change

Stanford Zent

Much has been said and written over the past twenty years about the economic, scientific, and humanitarian value of traditional environmental knowledge (TEK) and hence the need to record and preserve it for future generations (Brokensha, Warren, and Werner 1980; Williams and Baines 1993; Warren, Slikkerveer, and Brokensha 1995). Growing awareness of the fragile, eroding, and endangered status of TEK in many geographic settings has lent a sense of deep urgency to calls for its scientific documentation. Yet the specter of rapid and drastic decay of slowly accumulated, locally adapted knowledge also poses a rather puzzling question: why are so many people quickly turning away from a supposedly beneficial intellectual resource? Although some analysts have pointed rather vaguely and generally to cultural and economic globalization forces as the main reasons why TEK is disappearing at such an alarming rate, the precise determining factors, whether of local, regional, national, or international origin, and their complex interactions are still not well understood, mainly because of the dearth of empirical studies of TEK change (see Ohmagari and Berkes 1997 for a recent exception). Thus, for example, in the vast literature of ethnobotanical research it is extremely rare to find works that systematically incorporate a time dimension (excepting archaic or paleobotanical studies). As Peters (1996:242) points out, "things happen when people use plants," meaning that the distribution and abundance of local plant resources may be modified according to the type and intensity of management or exploitative techniques, and in consequence use patterns must be adjusted accordingly. It is only logical to assume that knowledge patterns will also change according to shifting environmental conditions and should be approached in the field as a dynamic phenomenon.

TOWARD A DYNAMIC EMPIRICAL PERSPECTIVE
OF TRADITIONAL ENVIRONMENTAL KNOWLEDGE

Several theoretical issues hinge on a better empirical understanding of the dynamics of TEK change. The static ethnographic treatment of TEK and our consequent failure to attend to how it is made and remade, remembered or forgotten, have unwittingly helped to foster sometimes overly optimistic and idealistic assessments of indigenous knowledge, and this distortion in turn is at the heart of current polemical and as yet unproductive debates concerning the value of scientific versus indigenous technology in development (DeWalt 1994) and the eco-political implications of leaving "ecologically noble savages" versus "ecocolonialists" in charge of biodiversity conservation (Redford 1991; Lizarralde 1992). In any event, the value and significance of TEK in the context of cataclysmic technological, economic, and social change will probably remain ambiguous until we obtain a more historically situated and empirically verifiable grasp of it, and certainly any attempt to preserve it in situ will fail unless the pertinent negative selection pressures are effectively identified and dealt with. A corollary problem is the apparent gap between theoretical and empirical formulations of TEK. On the one hand, TEK change or loss has been accorded theoretical status via the concepts of "delocalization" (Pelto and Pelto 1979), "cultural wasting" (DeWalt 1984), "monocultures of the mind" (Shiva 1993), and "extinction of experience" (Nabhan 1997), and at least conceptual awareness of this impending indigenous brain drain is no doubt behind the recent flurry of salvage research operations. On the other hand, we are still lacking empirical ethnographic studies that capture the detail and complexity of TEK change at the local level, and it is only through such careful case-by-case description and comparison that we will someday be able to distinguish the particular from the universal, the proximate from the ultimate, and the evitable from the inevitable factors driving this process.

What methodological directions are needed in order to describe empirically the evolution, or devolution, of TEK, on the ground, as it is happening? How does one operationalize the study of something that is vanishing? Given the severe paucity of previous studies that utilize an explicitly dynamic perspective of TEK, there is little methodological precedent to go on and hence a clear priority for this emerging research problem at this time is to develop descriptively and analytically adequate methodological approaches. Accordingly, one of the two main goals of this chapter is precisely to propose a coherent research strategy for the quantitative description and explanation of ethnobotanical knowledge change. This strat-

egy consists of the integrated use of four basic research methods: (1) ethnobotanical plot survey, (2) structured interview, (3) informant consensus analysis,
and (4) linear regression analysis. The second main goal is to outline some crucial
empirical features of the contemporary process of ethnobotanical knowledge
loss and acculturation among the Piaroa Indians of Venezuela.

ETHNOGRAPHIC BACKGROUND

The Piaroa (Wõthïhã) are an indigenous ethnic group numbering about 12,000
people who occupy large portions of Amazonas and Bolívar States in southern
Venezuela. Like many other contemporary native Amazonian peoples, the Piaroa
are currently undergoing dramatic cultural and ecological changes as a result of
their increasing contact with and integration to the national society. Prior to 1960,
they inhabited remote interfluvial localities and stayed generally aloof from the surrounding presence of intrusive European and neo-Venezuelan (i.e., *criollo* 'mixed
race') populations, although indirect, intermittent trading contacts with both colonizers and neighboring Indian groups have existed for centuries (see Zent 1992).
Traditional Piaroa culture displayed a strong interfluvial adaptive orientation:
small, dispersed, and mobile settlement pattern; mixed hunting-gathering and
swidden horticultural subsistence economy; well-developed crafts industry utilizing a wide range of natural products; lack of dugout watercraft; atomistic and fluid
sociopolitical structure; and pervasive mountain motifs in native cosmology.

From 1960 to 1980, a massive voluntary exodus took place and many Piaroa
families and even whole communities moved out of the hilly headwaters and
onto the downriver plains of the Orinoco and Ventuari rivers, precisely toward
the interethnic contact zones (see map 11.1). Several factors combined to break
down the former isolation of the Piaroa and draw them out of their upriver re-
doubts: improved transportation technology and infrastructure, such as road-
ways and motor vehicles, airstrips and airplanes, more affordable outboard motors; vigorous missionary activity; government-sponsored services and programs,
for example, schools, medical care, housing, financial credit, economic subsidies;
greater market and work opportunities; and eventual fading of their longstanding
terror of white and criollo people (Zent 1992:73–79). By 1980, the vast majority of
Piaroa communities and population were located in low-lying fluvial or savanna
habitats, effectively at the peripheries of their former upland territorial range.
Meanwhile, the traditional interfluvial heartland has been severely depopulated
and today only a handful of "traditional-style" communities can be found there,
mainly along the Upper Cuao River.

The geographic transition from upriver to downriver has been accompanied by
a number of significant sociocultural as well as ecological changes. The downriver communities are considerably more nucleated and sedentary than the tradi-

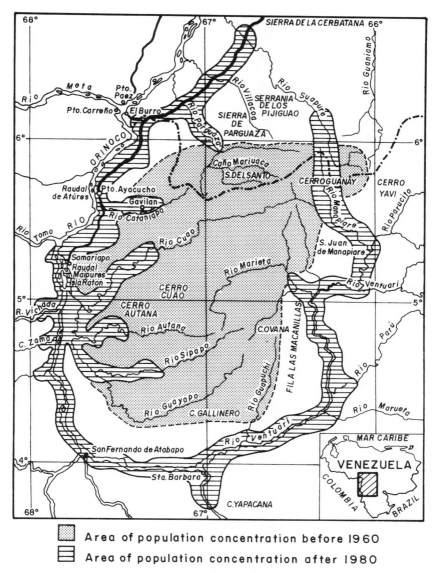

| | Area of population concentration before 1960 |
| | Area of population concentration after 1980 |

Map. 11.1. Geographic concentration of the Piaroa population before 1960 and after 1980. Reprinted from T.L. Gragson and B.G. Blount (eds.), *Ethnoecology: Knowledge, Resources, and Rights*, p. 97, copyright University of Georgia Press 1999, with permission from University of Georgia Press.

tional settlement mode. Generally speaking, the small, shifting house-village, composed socially of a tightly integrated kindred and with an average population of 25–30, has given way to permanent multihouse "towns" containing a relatively large number of socially distant family units and ranging in size from 50 to more than 500 residents. The traditional subsistence economy, based on labor and informal exchanges of food and trade items between close relatives, has been supplemented and in many cases revised with a very active commercial sector, characterized by monetary valuation of all goods and services and rational economic exchanges between individuals. In subsistence matters, there is less dependence on wild resource exploitation and greater reliance on agriculture. The major commercial activities include agricultural cash cropping, forest product extraction (e.g., rattanlike vines, forest fruits, smoked fish and game meat), gold and diamond prospecting, rattan-style furniture making, commercial artwork (e.g., ceremonial masks) and various types of wage labor. The Piaroa have become avid consumers of numerous foreign goods, such as steel tools, cooking utensils, factory-made clothing and footwear, outboard motors, radios, televisions, wristwatches, recreational equipment (e.g., bicycles, soccer balls), and processed foods. The manufacture and use of traditional craftwork has declined as a consequence of the greater dependence on Western industrial goods. Meanwhile, exposure to formal education as well as more frequent interaction with outsiders and increasingly common access to radio and television have spawned a much greater knowledge and acceptance of the values and customs of the national society. A large number of Piaroa are effective bilingual speakers of the Piaroa and Spanish languages. Perhaps more than half the tribe have nominally converted to Christianity. In sum, in the space of two decades, the Piaroa have gone from being fairly isolated interfluvial-dwelling, semisedentary, subsistence level forager-farmers who maintained their cultural autonomy and lived in close contact and interaction with the forest, to living in settled communities at the doorstep of the regional variant of the national society, strongly integrated to the regional market economy, in frequent contact with the criollo population, and well acculturated to Western culture.

Various ecological adjustments have accompanied the cultural changes mentioned above. The crowding of people and settlements in the downriver contact zones, the more sedentary lifestyle, the growing commercial economic orientation, and consequently the intensification of foraging and agricultural land use practices within more restricted home ranges has led to severe degradation of the immediate floristic and faunistic surroundings of many communities. Meanwhile, it appears that people's direct dependence on and interaction with the forest have diminished correspondingly, with the implication that their TEK is also being impacted.

PILOT STUDY OF LOSS OF ETHNOBOTANICAL KNOWLEDGE IN GAVILÁN

Given the drastic cultural and ecological changes happening to the Piaroa over the past 30 years, I set out to test the general hypothesis that TEK was in fact being lost in the acculturated habitat and attempt to identify some of the factors accounting for this process. A pilot study of the loss of ethnobotanical knowledge was made in the community of Gavilán, Río Cataniapo, Amazonas State. Ethnobotanical knowledge was chosen for the study for the following reasons: (a) it comprises a well-defined cognitive domain within TEK, and previous research amply demonstrates its deep intellectual as well as practical roots in human thought and action (cf. Hunn 1982; Atran 1990; Berlin 1992); (b) it constitutes a very important part of the traditional Piaroa subsistence economy; and (c) it is a research topic about which I have considerable previous knowledge and experience, having investigated Piaroa ethnobotany and collected approximately 1300 voucher specimens as part of my dissertation research among the Piaroa of the Upper Cuao region (Zent 1992). Data collection for the study was accomplished during two weeks of fieldwork in July–August 1994. It is important to keep in mind that the research was designed as a pilot study, aimed at testing the methodology being developed here, verifying whether the study hypothesis is consistent with empirically observable facts and thus deciding if further investigation in this cultural context is worthwhile or not, and gaining information and experience pursuant to planning a more comprehensive study of this kind at a future date. Therefore in no way do I pretend to claim that the results discussed here constitute a definitive or exhaustive examination of the research problem. Rather the objective, other than explicating a new methodological package, is to provide an introduction to an empirical understanding of the dynamics of ethnobotanical knowledge loss among the Piaroa.

STUDY SITE

The community of Gavilán was chosen as the site for the pilot study because it is a typical example of the nucleated, acculturated, and economically integrated community that dominates the Piaroa landscape today. The community is situated within the lower Cataniapo River basin, at a distance of about 30 km from Puerto Ayacucho, the capital and major urban center of the Amazonas State. A partially paved roadway connects Gavilán with Puerto Ayacucho and the journey by motor vehicle takes about 45 minutes. Daily taxi service is available and in fact most residents travel into the city at least once a week in order to carry out economic and social transactions. The present community was founded in the mid- to late 1960s, stimulated by the road, which was finished about that time, and the installation at that spot of a government-run elementary school and health clinic. In the mid-1970s, the then-territorial (now state) government donated materials for house construction and later constructed 30 cement-block, zinc-roofed houses for the local residents. Thus, boosted by government-sponsored services and the social and economic attractions of the Puerto Ayacucho urban center, Gavilán has experienced steady population growth throughout the seventies and eighties and in 1994 the population had reached approximately 350 people distributed in 50 households. The main economic activities are subsistence agriculture, cash cropping, vine extraction (used to make furniture), furniture manufacture, and wage labor. The natural vegetation of the region is tropical moist, semideciduous forest (Huber and Alarcón 1988), but because of concentrated swidden activity over the years, secondary vegetation, reflecting various stages of regrowth, now makes up a very prominent part of the floristic surroundings of the community.

METHODOLOGY

Probably the most straightforward way to determine TEK loss would be through a comparative analytical process, in which TEK data from two different time periods, for example, before and after a significant historical phase, are directly compared or reconstructed through linguistic or cultural evidence (cf. Balée 1994). But we are prevented from pursuing this technique in regards to the Piaroa for lack of previous information on their pre-1960s TEK, and the frenetic pace of change since that time makes it difficult to make any clear and simple diachronic comparisons. Thus, in the absence of comparative baseline data, the loss of cultural knowledge, like biological extinction, must be studied by indirect means, since it is virtually impossible to go out and witness the precise moment of loss or extinction of cultural or biological units under natural settings, unless one is able to do a long-term study. Previous studies by linguists and cognitive scientists, however, have taught us that synchronic variability of cultural knowledge, that is, language and cognitive categorization, foretells diachronic change of such knowledge (Casson 1981). Building on this insight, I argue that it is possible to infer historical processes of TEK change by looking at the present patterning of TEK variability and then tracing the connections between this pattern and other social variables that are indexed to temporal events or are themselves reflective of changing environmental conditions. Thus, an appropriate methodological strategy for studying TEK change would consist of two main steps: (1) to chart the pattern of knowledge variability within a Piaroa community, and (2) to study the relationship between this variability and social factors that are relevant indicators of their current situation of culture change. This basic research strategy guides the present study of the loss of ethnobotanical knowledge and acculturation among the Piaroa Indians of Venezuela and is effectively realized through the use of four essential research methods: (a) ethnobotanical plot survey, (b) structured interview, (c) informant consensus analysis, and (d) linear regression analysis. The first two concern data collection techniques, while the latter two refer to modes of data analysis.

Ethnobotanical Plot Survey

The plot survey, referring to the botanical inventory of plant species and the cultural inventory of local names and uses of these taxa within circumscribed sample plots, has become an important methodological tool in ethnobotanical research in recent years and is a key facet of the growing trend toward adopting a more quantitative approach to ethnobotany. The application of quantitative data collection techniques to ethnobotanical research, such as the plot survey, has led to qualitative advances in the analytical scope and power of this field of research through the production of more reliable, replicable, precise, and comparable databases, which are thus amenable to statistical analysis and hypothesis testing (Phillips and Gentry 1993a, b).

Sample plots have long been used by plant ecologists to measure key ecosystemic properties, such as species richness, biomass, and nutrient dynamics, whereas human ecologists and geographers have used them mostly to reveal the crop composition, successional stages, and farmer management styles of agricultural fields (Bernstein, Ellen, and Bin Antaran 1997:71–72). In the 1980s, a research team of botanists and anthropologists affiliated with the New York Botanical Garden pioneered the use of systematic botanical-ethnobotanical surveys within standard-sized one-hectare forest plots in their studies of the plant-use habits of indigenous groups of the Amazon basin (Balée 1986, 1987; Boom 1987, 1989). The adoption of this technique was initially aimed at providing a precise quantitative answer to the general question: How much of the rain forest do the Indians use? The data generated from these and subsequent studies were employed to make cross-cultural comparisons of the proportional use levels of forest trees among various native groups (Prance et al. 1987; Bennett 1992). Most recently, Bernstein, Ellen, and Bin Antaran (1997) have demonstrated the utility of plot surveys as an instrument for measuring the comprehensiveness of individual informant's ethnobotanical knowledge in their study of a Dusun community in Brunei, although the study suffers from certain methodological limitations that prevent the authors from realizing the full descriptive and analytical capacities of this technique as a tool for studying the pattern and process of ethnobotanical knowledge within a cultural community. They elicit ethnobotanical knowledge data from only two informants and do not interview them within the same sample plot, thus preventing any statistically meaningful comparison of knowledge differences between them. Consequently they are unable to relate the resulting pattern of interinformant knowledge differences to impinging social or historical factors. By contrast, in the Gavilán study, I use the plot survey as a common arena to test the knowledge of a range of informants across the same sample of plants.

A sample plot of 750 m² (5 × 150 m) was measured and marked off in a patch of high forest identified as primary vegetation by a local informant. The site of the plot is the closest sizeable patch of primary forest to the village, located a little more than two km from the village center. Primary forest was chosen for the study since this is the floristic community containing the greatest amount of overall as well as utilitarian plant diversity. A total of 48 large trees and 2 large lianas with diameter at breast height (DBH) > 10 cm were counted within the plot. Each one was marked with a number (01 to 50), the basal measurement was recorded, and herbarium specimen was collected.

Structured Interviews

Structured interviews involve asking a group of informants to respond to the same set of questions. The method is transparently quantitative in the sense that the verbatim responses can be submitted directly to statistical analysis without further coding or data manipulation (Martin 1995:96). The structured questionnaire or survey, administered in verbal or written form, has been a primary methodological tool of sociologists for many years. In anthropology, this method is associated mostly with ethnoscientists, who developed formal interviewing procedures in order to eliminate observer bias and extraneous contextual noise and therefore elicit only culturally relevant emic-type information. The formal or controlled elicitation procedure, modeled after structural linguistic methodology and adhering to a tightly controlled query-response framework for collecting data, rests on the use of standardized question frames posed in the native vernacular (e.g., What is the name of a kind of _____?). The question frame is the basic technique used by researchers of folk biology to elicit plant and animal taxonomies and use contexts (Berlin, Breedlove, and Raven 1974; Hunn 1982). The basic kinds of questions that appear most frequently in the controlled elicitation procedure may be classified as dichotomous (yes-no, true-false), multiple-choice, or fill-in-the-blank (Martin 1995:119–121).

Early ethnoscientific research was based on the administration of controlled queries to one or a few, supposedly omniscient, informant(s), without much thought being given to systematic sampling of the study population. This approach has since been upgraded, with more attention now paid to patterns of intracultural diversity and similarity of cultural knowledge as can be discerned through the application of the same query set to a representative range of informants inhabiting the same cultural space (Pelto and Pelto 1975; Gardner 1976). The pattern of knowledge distribution is configured by domain specificities, social contexts, and social use and learning situations (Boster 1987), and thus a study of this pattern can also reveal something about the organizational dynamics of knowledge within a community.

Using the marked plants contained within the forest plot at Gavilán as the sample universe for rating individual ethnobotanical knowledge, I conducted structured interviews on an individual basis among 44 male respondents, ranging from age 10 to 68. This sample represents about 40 percent of the male population of Gavilán within this demographic range. The decision to limit the interviews to males only was dictated by the severe time constraints under which the fieldwork was carried out, the general reluctance of women to participate, and the previous observation that in the traditional environment men perform the bulk of forest foraging and are more knowledgeable about high forest plants whereas women are the main agriculturalists and may be more knowledgeable about garden flora. The interviews consisted of two basic parts. In the first part, the respondent was asked to supply basic social information, about age, birthplace, residential history, family data, and education experience. This part of the interview was at least initiated in the Spanish language and, based on the respondent's performance, his bilingual ability was rated (on a scale of 0 to 3, where: 0 = no ability; 1 = knowledge of some Spanish vocabulary, but no grammatical ability; 2 = semiconversant; and 3 = reasonably fluent). The second part of the interview was conducted entirely in the native language and consisted of the use of a combination of ostension (simply pointing to the plant in question) and structured question frames regarding the name and uses of each and every one of the marked plants. The respective question frames used here were: *ta'anɨ miku piñe dau/pot'æ* 'What is the name of this tree/liana?' and *dahe heaekwæhwæthɨ piñe dau/pot'æ, kwaekwaewae/hawapo/isode adikwa/iyæsɨ/de'a rua ukwæ/kãrã iaere adikwa ka'a* 'What is this tree/liana used for? Is there food/medicine/construction/trade/animal food/other work?' The first question frame corresponds to a fill-in-the-blank type of question, while the second series correspond to a true-false format.

Informant Consensus Analysis

Informant consensus analysis refers broadly to the mathematical analysis of patterns of interinformant agreement and disagreement about selected topics of cultural interest. This technique can be used to reveal information about the cultural validity or acceptability of informant response data (e.g., modal responses, rank order) as well as the distribution of cultural knowledge within a community (e.g., degree of conformity versus idiosyncrasy, variable expertise). Various forms of consensus analysis have appeared in ethnobotanical studies of medicinal plant use (Adu-Tutu et al. 1979; Friedman et al. 1986; Johns, Kokwaro, and Kimanani 1990) and overall uses (Kainer and Duryea 1992; Phillips and Gentry 1993a, b). The recent work by Phillips and Gentry (1993a, b; Phillips 1996) provides probably the most influential example of the use of consensus analysis in ethnobotanical research, which the authors refer to as an "informant-indexing technique." They take an admittedly "plant-centric" approach that is "most appropriate for ethnobotanical research primarily oriented toward botanical, conservation, or pharmaceutical, rather than strictly anthropological, goals" (Phillips 1996:172). Their main concern is to quantify the use value of a species based on the overall average frequencies with which a group of informants state particular uses of particular species throughout a series of interviews. The interview technique used here corresponds to the fill-in-the-blank type, in which the researcher basically asks the informant to freely list the uses that he or she knows of a species on a particular day, and this exercise may be repeated once or more with the same informant for the same species at another time or place. A main weakness of this approach, in my opinion, derives from its one-sided focus on the plants instead of the people. Thus comparisons among informant knowledge are weakened by the fact that not all respondents were asked about the same set of plants; that the results may in fact be subject to bias—especially given the relative openness of the fill-in-the-blank question format in the elicitation of use data—caused by variables of the interview context, relationship between researcher and informant, and informant talkativeness (Martin 1995:168); and that the formula used for rating different knowledge levels does not take into account potentially wrong affirmative answers.

The variant of consensus analysis used in the present study is that developed by Kimball Romney and associates for the general analysis of any cultural data. (See Romney, Weller, and Batchelder 1986 and Romney, Batchelder, and Weller 1987 for details on the theoretical and mathematical postulates of this approach.) The authors define their approach as: "a formal mathematical model for the analysis of informant consensus on questionnaire data that will simultaneously provide an estimate of the cultural competence or knowledge of each informant and an estimate of the correct answer to each question asked of the informants" (Romney, Batchelder, and Weller 1987:163). Thus the consensus model, *sensu* Romney, Weller, and Batchelder, provides a mathematical estimate of the individual knowledge level of an informant, expressed as competency score, based on the extent to which his or her answers agree or disagree with the consensus choices of the entire group. Furthermore, the model provides a statistically reliable estimate of the correct answers when the answers are not known ahead of time, by measuring the degree and distribution of interinformant agreement. The correct answer usually but not always corresponds to the consensus or most popular choice. A probabilistic notion of correctness applies here and is largely a function of the dispersion of responses to a particular question (i.e., a higher degree of consensus produces a probabilistically more correct answer, while a lower consensus produces a probabilistically less correct answer), but the expertise of informants is also taken into account and may result in a majority response actually being judged as incorrect. The method is thus appropriate for the study of shared cultural knowledge but clearly not for specialist type knowledge. While based essentially on the democratic notion that the majority opinion determines cultural truth, the consensus model does, however, give more weight to the so-called cultural experts, that is, those individuals exhibiting higher overall competency scores. Although the models allow no independent confirmation of the correct answers, in the case of ethnobotanical knowledge a partial check of the results can be obtained by comparing the consensually derived folk plant taxa with their scientific counterparts as determined from plant collections.

The same model easily handles true-false, multiple-choice, and fill-in-the-blank question formats. The plant-naming portion of the interview conforms to the fill-in-the-blank type, in which the field of possible responses is essentially limitless (meaning there is less probability that one can guess at a correct answer), and an "I don't know" response is treated as a wrong answer (even when it constitutes the majority). The analysis of the plant-naming data was carried out at the basic (i.e., folk-generic) naming level (Berlin 1992). The use category portion of the interview was administered as a true-false type question, in which I asked each informant to respond affirmatively or negatively to the question of whether the plant indicated is useful for food, medicine, construction, trade, animal food, or other item work. In regards to the "other item work" use category, if the respondent answered "yes," I then prompted him with a series of dichotomous queries covering the following use subcategories: cordage and cloths, woodwork, torch-making, food storage and processing, and ornamental. This format yields less information than multiple-choice or fill-in-the-blank questions because it is easier to guess at wrong answers. This disadvantage, however, is compensated by the tighter structure and control that it imposes on the response elicitation process, thus minimizing interinformant differences due to factors other than sheer knowledge, for example, boredom, shyness, or laconism. Actual computation of the individual ethnobotanical competence scores was accomplished by using a computer program ("Consensu") expressly written for this purpose, graciously provided to me by Dr. K. A. Romney.

Linear Regression Analysis

Linear regression analysis is a very common procedure for analyzing social science data and is capable of specifying the correlation and dependency relationships between two or more sets of variables. It has been used in ethnobotany to study the relationship between folk medicines and plant families in native North America (Moerman 1991) and the relationship between age and relative knowledge of plant uses among Peruvian peasants (Phillips and Gentry 1993b). I used regression analysis to test the relationships between the spread of ethnobotanical competencies and potentially influential social factors. Both simple and multiple, linear and curvilinear regression were tried out to explore the statistical relationship between the ethnobotanical variables of plant-naming and use-value recognition competencies and the social variables of age, formal education (i.e., years of schooling completed), and bilingual ability. All of the social variables are historically relevant features of the current situation of culture change among the Piaroa. Age is a direct marker of time and can be related to key dates in the acculturation process, such as the founding of the Gavilán community. The years of education imply degree of exposure and indoctrination to a non-native knowledge form that may compete with traditional knowledge forms. Bilingual ability, or knowledge of the Spanish language, is a direct reflection of the level of contact and communication with non-Piaroa speakers.

RESULTS

Three sets of regression analysis were performed: (1) impact of the social variables of age, education, and bilingual ability on plant-naming competence score, (2) impact of plant-naming competence score on plant use-value competence score, and (3) impact of social variables on plant use-value competence score.

The regression results of the first set have been discussed in some detail elsewhere (Zent 1999) and therefore only a very brief summary of the major points will be presented here. The individual plant-naming competence scores were plotted separately against the three social variables and the respective coefficients of determination (r^2) were computed. The strongest relationship of linear regression was given by age ($r^2 = .539$), followed by education ($r^2 = .22$), and lastly by bilingual ability ($r^2 = .113$). These figures may be interpreted as indicating a fairly strong positive relationship between age and plant-naming competence, a relatively weak negative relationship between education level and plant-naming competence, and a very weak negative relationship between bilingual ability and plant-naming competence. The data were also analyzed using a polynomial or curvilinear regression model, resulting in higher coefficients of determination for all three social variables (age: .625; education: .415; bilingual ability: .209) and revealing the nonlinear or changing tendencies in the relationships described here. Thus, the competence-on-age curve (see fig. 11.1) displays a sharp rise in the lower ages (10–30 years old), then rounds off somewhat (i.e., moderate increase) in the 30–50 age interval, after which there is a sharp downward turn. The competence-on-education curve (see fig. 11.2) shows a tight distribution of relatively high competence scores at the lowest educational levels (0–1 years), a rather sharp descent

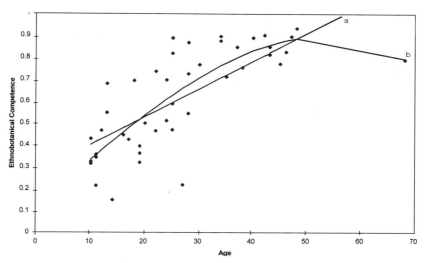

Figure 11.1. Regression of age and plant-naming competence among the Piaroa. The regression lines drawn here represent: (a) a linear or binomial model ($r^2 = .539$; $y = .2739 + .0126x$); and (b) a curvilinear or polynomial model ($r^2 = .625$; $y = 8.3118 + .4465x - 3.359\sqrt{x} - 15.7723x^2 - .0023x^3$). Reprinted from T.L. Gragson and B.G. Blount, eds., *Ethnoecology: Knowledge, Resources, and Rights*, p. 108, copyright University of Georgia Press 1999, with permission from University of Georgia Press.

throughout the middle educational levels (2–5 years), and then a mild upturn throughout the higher educational levels (7–10 years). The competence-on-bilingual ability curve (see fig. 11.3) reveals a definite although modest decline from level 0 (no bilingual ability) to level 2 (some conversational skills), followed by a modest upturn at level 3 (fluent).

The relatively strong impact of age on plant-naming competence provided the necessary stimulus for taking a deeper look at this relationship. Above it was noted that the regression of competence on age under a polynomial model deviates from strictly linear assumptions and therefore the relationship may vary according to different age groups. Furthermore, visual inspection of the scatter of data points suggests that a definite change in the relationship occurs at about the 30 year age interval. So it was decided to divide the sample of respondents into two subgroups ($<$ 30 years old and 30+ years), and compare the respective divergent regression lines between them. The results of this operation (see fig. 11.4) show a nearly flat or ever so slightly increasing level of knowledge with age among the older cohort, whereas in the younger cohort the trend is one of steep increase in competence with increasing age. Thus the regression relationship

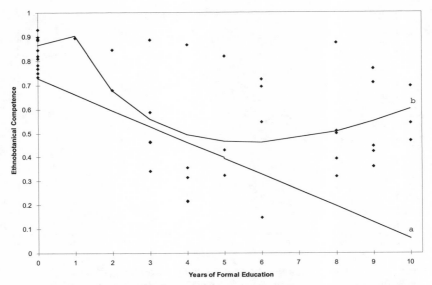

Figure 11.2. Regression of formal education and plant-naming competence among the Piaroa. The regression lines drawn here represent : (a) a linear or binomial model ($r^2 = .22$; $y = .7471 - .0307x$); and (b) a curvilinear or polynomial model ($r^2 = .415$; $y = 1.7739 + .2311x - 1.1022\sqrt{x} - .0009/x$). Reprinted from T.L. Gragson and B.G. Blount, eds., *Ethnoecology: Knowledge, Resources, and Rights*, p. 110, copyright University of Georgia Press 1999, with permission from University of Georgia Press.

between age and plant-naming competence is quite different when viewed from the contrasting perspectives of above or below the 30 year age marker; there is no apparent relationship in the former group versus a rather strong relationship in the latter group. The basic trend observed here is therefore not one of steady decrease of competence as a direct function of younger age, but rather a dramatic plunge of knowledge below the age of 30.

But is this drop of competence with age among the younger group due simply to a normal learning-with-age curve factor that reaches its culmination around the age of 30? In this regard, significance tests of the interaction of the three selected social variables in determining competence score within a multiple regression model were performed on both older and younger age groups. The tests revealed that none of the social variables are significant predictors of plant-naming competence among the older group, but both age ($p = .0001$) and bilingual ability ($p = .01$) were significant predictors, positively and negatively, respectively, of competence scores in the younger group. Such a result is also suggested by the partial contributions of the different independent variables to the combined r^2 of .527, where age accounts for .296, bilingual ability .207, and education .025 of the explained variation. Thus the knowledge differentials are not due to age alone; it

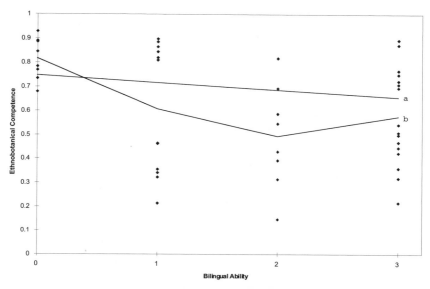

Figure 11.3. Regression of bilingual ability and plant-naming competence among the Piaroa. The regression lines drawn here represent: (a) a linear or binomial model ($r^2 = .113$; $y = .727 - .0667x$); and (b) a curvilinear or polynomial model ($r^2 = .209$; $y = .8154 - .226x + .0162x^3$). Reprinted from T.L. Gragson and B.G. Blount, eds., *Ethnoecology: Knowledge, Resources, and Rights*, p. 111, copyright University of Georgia Press 1999, with permission from University of Georgia Press.

appears that acquiring a superior command of the Spanish language also contributes significantly to an inferior knowledge of plant names. Although education was not found to be a significant predictor of plant-naming competence under the multiple regression model used here, it was found to be a very strong positive predictor ($r^2 = .709$) of greater bilingual ability, and therefore can be considered an indirect negative influence on plant-naming ability.

Plant-naming competence scores were plotted against use-value competence scores, and it was found that naming ability is a significant positive predictor of correct knowledge of use categories for all but one type of use value, food processing (see table 11.1). These results are consistent with Stross's (1973) earlier observation that correct attribute awareness is largely a function of correct name recognition of the plant among Mayan children. More important, this type of analysis gives added meaning to the previous set of regression analyses by demonstrating the practical importance of developing adequate perceptual and taxonomic categorization skills: young people who fail to learn how to identify plant taxa are likely to be unable to specify how the plant can be used effectively.

Finally, significance tests of the direct predictive impact of social variables on the respective use-value competencies were carried out. The results are summarized

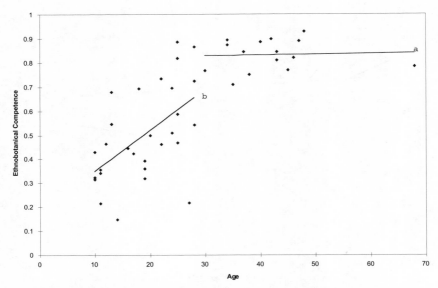

Figure 11.4. Regression of age and plant-naming competence according to age subgroups among the Piaroa. The regression lines drawn here represent: (a) a linear model for the age group 30 and above ($r^2 = .002$; $y = .8176 + .00032x$); and (b) a linear model for the age group under 30 ($r^2 = .296$; $y = .1786 + .017x$). Reprinted from T.L. Gragson and B.G. Blount, eds., *Ethnoecology: Knowledge, Resources, and Rights*, p. 109, copyright University of Georgia Press 1999, with permission from University of Georgia Press.

in table 11.2. The primary observations that can be made here are that: (a) age is a significant positive predictor of most use-value competence scores; (b) education is a significant negative predictor of a few use-value scores, such as food, medicine, trade, torch fuel, and ornament; and (c) bilingual ability is not a significant predictor of any use-value competence.

The results can be summarized as indicating: (a) age is the most important social variable reviewed here in determining the variable pattern of ethnobotanical nomenclatural and use categorization competence; (b) the strong impact of age on ethnobotanical competence is confined to the subgroup below 30 years of age, and within this subgroup the relationship can be described as one of sharply declining knowledge with younger age; (c) greater Spanish-speaking ability among the younger subgroup also conditions lower plant-naming competence; (d) years of formal education negatively affect ethnobotanical competence in an indirect sense, by exerting a strong positive influence on bilingual ability, and in a direct sense, by being associated with lesser knowledge of certain kinds of plant-use categories; and (e) knowledge of correct use value, among all age groups, is highly dependent upon a person's competence in plant identification and naming.

Table 11.1   Regression Relationships between Plant-Naming Competence and Use-Value Competence

Use Category (Dependent Variable)	Coefficient of Correlation (r)	Coefficient of Determination (r^2)	Significance Test (p)
Food	.746	.557	.000
Construction	.518	.268	.000
Medicine	.748	.559	.000
Trade	.345	.119	.022
Animal Food	.621	.385	.000
Crafts			
Cordage and cloths	.496	.246	.001
Woodwork	.42	.177	.005
Torch fuel	.72	.519	.000
Food processing	.152	.023	.326
Ornamental	.756	.572	.000

CONCLUSION

Although the results of the present study should be regarded as essentially exploratory, given the limited scope of the study, nevertheless the observed contrasts in the level and pattern of knowledge held by older versus younger generation Piaroa in Gavilán can be interpreted as consistent with the hypothesis that ethnobotanical knowledge is in fact being lost in the acculturated habitat. Admittedly, any interpretation of the cultural significance of age on ethnobotanical knowledge is necessarily complicated by the fact that age is naturally associated with the learning and accumulation of knowledge in any cultural context. But the precise pattern observed here suggests that the impact of age on ethnobotanical knowledge is a direct reflection of the prevailing process of culture change among Piaroa. The observed turning point in the linear relationship between age and ethnobotanical competence, occurring about age 30, also corresponds almost exactly with the founding of the village of Gavilán and hence the beginning of a more settled, integrated, and acculturated lifestyle for the residents. All respondents above the age of 30 spent their formative years in the interfluvial forests while a large proportion of under 30 respondents grew up in the modern-style nucleated communities in close contact with various actors and institutions of the national society. Moreover, the continued evolution and intensification of intercultural contact, economic integration, and cultural westernization during the past 30 years may indeed help explain the dramatic decline in ethnobotanical competence with age among younger people. Meanwhile, the negative correlations of

Table 11.2 Regression Relationships between Social Variables and Use-Value Competence

Independent Variable: Social Parameter	Dependent Variable: Use Category	Coefficient of Correlation (r)	Coefficient of Determination (r^2)	Significance Test (p)
Age	Food	.746	.557	.000
	Construction	.488	.238	.001
	Medicine	.701	.492	.000
	Trade	.543	.295	.000
	Animal Food	.411	.169	.006
	Crafts			
	Cordage and cloths	.399	.159	.007
	Woodwork	.436	.19	.003
	Torch fuel	.661	.437	.000
	Food processing	.158	.025	.307
	Ornamental	.683	.466	.000
Education	Food	−.39	.152	.009
	Construction	−.268	.072	.079
	Medicine	−.361	.13	.016
	Trade	−.311	.097	.04
	Animal Food	−.238	.057	.119
	Crafts			
	Cordage and cloths	−.066	.004	.670
	Woodwork	−.048	.002	.756
	Torch fuel	−.41	.168	.006
	Food processing	−.246	.06	.108
	Ornamental	−.411	.169	.006
Bilingual Ability	Food	−.206	.042	.18
	Construction	−.057	.003	.711
	Medicine	−.157	.025	.31
	Trade	−.161	.026	.296
	Animal Food	−.054	.003	.73
	Crafts			
	Cordage and cloths	−.007	.000	.966
	Woodwork	−.053	.003	.735
	Torch fuel	−.29	.084	.056
	Food processing	−.164	.027	.287
	Ornamental	−.239	.057	.118

bilingual ability and years of formal schooling with ethnobotanical knowledge seem to indicate that intrusive knowledge forms and activities are competing with and detracting from the learning of traditional environmental knowledge. Young people now spend more time in school, in the city of Puerto Ayacucho, or in the streets and playing fields of their home community than they do in the surrounding forest. Several younger individuals who were interviewed expressed informally that they hardly ever go hunting or gathering in the forest, and some older individuals admitted that they rarely take their families on camping trips away from the settlement, and when they do they do not stay very long because their children complain too much about the harsher living conditions.

A more definitive empirical description and explanation of the loss of ethnobotanical knowledge of forest plants among the Piaroa will depend of course on a more comprehensive study involving more study sites and comparing the patterning of knowledge among two or more communities displaying significant differences in type of contact and degree of acculturation. Another relevant scenario would be a restudy of the same Gavilán community at a future time. The important point from a methodological perspective is that the quantitative method described and demonstrated here is eminently capable of producing directly comparable, statistically manipulable databases of knowledge variability in different spatial or temporal contexts, and therefore offers an appropriate strategy for the empirical study of TEK change.

REFERENCES

Adu-Tutu, M., et al. 1979. Chewing stick usage in southern Ghana. *Economic Botany* 33(3): 320–328.

Atran, S. 1990. *Cognitive Foundations of Natural History.* London: Cambridge University Press.

Balée, W. 1986. Análise preliminar do inventário florestale a etnobotânica Ka'apor (Maranhão). *Boletin do Museu Goeldi, Botânica* 2:141–167.

Balée, W. 1987. A etnobotânica quantitativa dos indios Tembé (Rio Gurupi, Para). *Boletin do Museu Goeldi, Botânica* 3:29–50.

Balée, W. 1994. *Footprints of the Forest.* New York: Columbia University Press.

Berlin, B. 1992. *Ethnobiological Classification: Principles of Categorization of Plants and Animals in Traditional Societies.* Princeton, N.J.: Princeton University Press.

Berlin, B., D.E. Breedlove, and P.H. Raven. 1974. *Principles of Tzeltal Plant Classification: An Introduction to the Botanical Ethnography of a Mayan-Speaking People of Highland Chiapas.* New York: Academic Press.

Bennett, B.C. 1992. Plants and people of the Amazonian rainforests. *Bioscience* 42:599–607.

Bernstein, J.H., R. Ellen, and B. Bin Antaran. 1997. The use of plot surveys for the study of ethnobotanical knowledge: A Brunei Dusun example. *Journal of Ethnobiology* 17(1):69–96.

Boom, B. 1987. Ethnobotany of the Chácobo Indians, Beri, Bolivia. *Advances in Economic Botany* 4:1–68.

Boom, B. 1989. Useful plants of the Panare Indians of the Venezuelan Guayana. *Advances in Economic Botany* 8:57–76.

Boster, J. 1987. Introduction. *American Behavioral Scientist* 31(2):150–162.

Brokensha, D., D. Warren, and O. Werner. 1980. *Indigenous Knowledge Systems and Development*. Washington, D.C.: University Press of America.

Casson, R.W. 1981. Variability and Change. In *Language, Culture, and Cognition: Anthropological Perspectives*, ed. R.W. Casson. Pp. 269–277. New York: Macmillan Publishing Co.

DeWalt, B.R. 1984. International development paths and policies: The cultural ecology of development. *The Rural Sociologist* 4(4):255–268.

DeWalt, B.R. 1994. Using indigenous knowledge to improve agriculture and natural resource management. *Human Organization* 53(2):123–131.

Friedman, J., et al. 1986. A preliminary classification of the healing potential of medicinal plants based on a rational analysis of an ethnopharmacological field survey among Bedouins in the Negev Desert, Israel. *Journal of Ethnopharmacology* 16:275–287.

Gardner, P.M. 1976. Birds, words, and a requiem for the omniscient informant. *American Ethnologist* 3:446–468.

Huber, O., and C. Alarcón. 1988. *Mapa de vegetación de Venezuela*. 1:2.000.000. Caracas: Ministerio del Ambiente y de los Recursos Naturales Renovables and The Nature Conservancy.

Hunn, E. 1982. The utilitarian factor in folk biological classification. *American Anthropologist* 84:830–847.

Johns, T., J.O. Kokwaro, and E.K. Kimanani. 1990. Herbal remedies of the Luo of Siaya District, Kenya: Establishing quantitative criteria for consensus. *Economic Botany* 44(3):369–381.

Kainer, K.A., and M.L. Duryea. 1992. Tapping women's knowledge: Plant resource use in extractive reserves, Acre, Brazil. *Economic Botany* 46:408–425.

Lizarralde, M. 1992. Five hundred years of invasion: Eco-colonialism in indigenous Venezuela. *Kroeber Anthropological Society Papers* 75–76:62–79.

Martin, G. 1995. *Ethnobotany: A Methods Manual*. London: Chapman & Hall.

Moerman, D.E. 1991. The medicinal flora of Native North America: An analysis. *Journal of Ethnopharmacology* 31:1–42.

Nabhan, G.P. 1997. *Cultures of Habitat: On Nature, Culture, and Story*. Washington, D.C.: Counterpoint.

Ohmagari, K., and F. Berkes. 1997. Transmission of indigenous knowledge and bush skills among the Western James Bay Cree women of subarctic Canada. *Human Ecology* 25(2):197–222.

Pelto, P.J., and G.H. Pelto. 1975. Intra-cultural diversity: Some theoretical issues. *American Ethnologist* 2:1–18.

Pelto, G.H., and P.J. Pelto. 1979. *The Cultural Dimension of the Human Adventure*. New York: Macmillan Publishing Co.

Peters, C.M. 1996. Beyond nomenclature and use: A review of ecological methods for ethnobotanists. *Advances in Economic Botany* 10:241–276.

Phillips, O., and A.H. Gentry. 1993a. The useful plants of Tambopata, Peru: I. Statistical hypotheses tests with a new quantitative technique. *Economic Botany* 47(1):15–32.

Phillips, O., and A.H. Gentry. 1993b. The useful plants of Tambopata, Peru: II. Additional hypothesis testing in quantitative ethnobotany. *Economic Botany* 47(1):33–43.

Phillips, O. 1996. Some quantitative methods for analyzing ethnobotanical knowledge. *Advances in Economic Botany* 10:171–198.

Prance, G., et al. 1987. Quantitative ethnobotany and the case for conservation in Amazonia. *Conservation Biology* 1:296–310.

Redford, K.H. 1991. The ecologically noble savage. *Cultural Survival Quarterly* 13(1):46–48.

Romney, A.K., S.C. Weller, and W.H. Batchelder. 1986. Culture as consensus: A theory of culture and informant accuracy. *American Anthropologist* 88(2):313–338.

Romney, A.K., W.H. Batchelder, and S.C. Weller. 1987. Recent applications of cultural consensus theory. *American Behavioral Scientist* 31(2):163–177.

Shiva, V. 1993. *Monocultures of the Mind: Perspectives on Biodiversity and Biotechnology.* London: Zed Books.

Stross, B. 1973. Acquisition of botanical terminology by Tzeltal children. In *Meaning in Mayan Languages,* ed. M.S. Edmonson. Pp. 107–141. The Hague: Mouton.

Warren, D.M., L.J. Slikkerveer, and D. Brokensha, eds. 1995. *The Cultural Dimension of Development: Indigenous Knowledge Systems.* London: Intermediate Technology Publications.

Williams, N.M., and G. Baines, eds. 1993. *Traditional Ecological Knowledge: Wisdom for Sustainable Development.* Canberra: Centre for Resource and Environmental Studies, Australian National University.

Zent, S. 1992. Historical and ethnographic ecology of the Upper Cuao River Wõthɨhã: Clues for an interpretation of native Guianese social organization. Ph.D. diss., Columbia University, New York.

Zent, S. 1999. The quandary of conserving ethnoecological knowledge. In *Ethnoecology: Knowledge, Resources, and Rights,* ed. T. Gragson and B. Blount. Pp. 90–124. Athens: University of Georgia Press.

12

MEASURING THE EVOLUTION AND DEVOLUTION
OF FOLK-BIOLOGICAL KNOWLEDGE

Phillip Wolff and Douglas L. Medin

Although science continues to deliver new insights into the basis for life, it is hard to escape the impression that, on an individual and cultural level, knowledge about living kinds is diminishing. Anthropologists studying traditional societies often note with concern the loss of indigenous languages and a lessening of knowledge about the natural world (e.g., Nabhan and St. Antoine 1993; Wester and Yongvanit 1995; Diamond and Bishop 1999). In technologically oriented cultures, contact with biological kinds may be so minimal that researchers can demonstrate significant differences in children's biological reasoning as a function of whether they do or do not have goldfish as pets (Hatano and Inagaki 1987; Inagaki 1990).

EVOLUTION AND DEVOLUTION OF FOLK-BIOLOGICAL KNOWLEDGE

A recent survey we conducted at Northwestern University offers some index of what undergraduates know about one domain of biology, namely, trees. We provided the names of 80 trees and asked students to circle the trees they had *heard of* before, regardless of whether they knew anything about them. More than 90 percent said they had heard of birch, cedar, chestnut, fig, hickory, maple, oak, pine, and spruce. But fewer than half indicated any familiarity with alder, buckeye, catalpa, hackberry, hawthorn, honey locust, horse chestnut, larch, linden, mountain ash, sweet gum, and tulip tree—all of which are common to the Evanston, Illinois, area where Northwestern University is located. Of course, these observations by themselves do not imply a loss of knowledge. It may be that a hundred years ago Northwestern undergraduates would have proved equally unfamiliar with biological kinds. Nevertheless, such low levels of knowledge are consistent with the possibility that knowledge about trees is declining.

The Devolution Hypothesis

With modernization, it may be that knowledge about living kinds has decreased or, as we say here, *devolved*. We will refer to this possibility as the *devolution hypothesis*. Devolution might stem from two kinds of historical change. For one, the shift from rural to urban settings may result in a significant decrease in people's contact with the natural world. This reduced contact could lead to declines in knowledge, but not necessarily. The effects of reduced exposure may be offset by sufficient amounts of indirect experience with the natural world, through a culture's media, talk, and values. We will refer to this kind of exposure as *cultural support*. The idea of cultural support has to do with the degree to which a society promotes a particular area of knowledge. It has to do less with whether there are specialists who know or care about particular kinds of things than with the level to which people focus on a domain of knowledge in their everyday interactions. For example, to what extent do parents call children's attention to plants and animals? When they do, is their reference to robins, trout, and maples, or more generically to birds, fish, and trees? Declines in cultural support, like declines in exposure to the natural world, could lead to devolution.

In this chapter, we summarize our work on the devolution hypothesis with respect to the life form "trees." Trees are of special interest because they could represent a particularly strong test of the devolution hypothesis (for a fuller account, see Wolff, Medin, and Pankratz 1999). In terms of contact with the natural world, we might not expect devolution with respect to trees at all. While people in urban environments may have only limited exposure to all but a few mammals (e.g., cats, dogs, squirrels), they are likely to have seen many different kinds of trees. And trees, because of their size, are not likely to be ignored. Size is a key factor in determining those natural kinds that attract attention and get named (Hunn 1999). If the prerequisites for conceptual organization consist solely of an inherent curiosity about living kinds and a perceptual system tuned to discontinuities in nature (Berlin 1992), then even urbanized cultures should show an appreciation for different kinds of trees. But, despite our continued direct exposure to trees, it is possible that knowledge about them has devolved because cultural support for them has declined.

Measuring Cultural Support

Cultural support may take a variety of forms, many of which may be difficult to measure, especially across time. Nevertheless, we are likely to have a pretty good measure of cultural support in terms of what people write about. Are people writing about plants and animals as much as they used to? When they do so, are they writing at the life-form level (e.g., bird, tree) or at what Berlin (1992) refers to as the folk-generic (sparrow, oak) level? For English, not only are written records available, but these records are accessible in online databases that permit automated search. To the extent that there have been historical changes in the amount and specificity of discussion of biological kinds across a representative sample of sources, we have evidence for the changes in the cultural support for learning about the natural world. Note that this measure of cultural support is likely to be a conservative measure of what people may know. An author might write about noticing cottonwoods along a riverbank without being able to pick a cottonwood out of a biological lineup. The use of writing as a measure of what people know is therefore likely to overestimate the knowledge of an average citizen, hence underestimate devolution. By the same token, if changes are found, they are most likely to be historically significant.

Oxford English Dictionary

Because our interest is in a longer time span than U.S. written history affords (in terms of databases we might access), we selected a database from England for study: the *Oxford English Dictionary* (*OED*), a historical dictionary. We chose the *OED* for a variety of reasons. The *OED* seeks to capture the evolution of all words in the English language except those that became obsolete before 1150 or are intelligible only to the specialist. The first edition was published in 1933 after nearly seven decades of work. The second edition, the *OED* 2, was published in 1989. It combines the original edition, four supplemental volumes published after 1933, and results from a fourth major reading program.

The dictionary contains approximately 616,500 word forms (Murray 1989; Berg 1993). Definitions for these words are illustrated with quotations from each century of use with extra quotations provided for significant changes in meaning. The quotations were drawn from a wide range of books, with special emphasis on great literary and scientific works, but also, among other things, books of foreign travel, letters of foreign correspondents, magazines, and diaries. The total number of quotations in the *OED* 2, roughly 2.5 million, was drawn from a sample of between 5 and 6 million quotations. Given the breadth of the inquiry, we have no reason to expect that the quotations represent a biased sample with respect to the questions we aim to address. The sample may well be biased in terms of reflecting interests, values, and accessibility, but these sorts of biases are more or less orthogonal to our focus.

Recently, the entire 12-volume set was retyped into a special computer database format allowing for online searching of all definitions and quotations. The *OED* online corpus may be searched for any key words (e.g., "tree," "maple tree," "maple," etc.) and search codes may be written so that the date, source, and full quotation context will be returned.

General Predictions

Evidence for devolution may be found with two kinds of measures: (1) the number of quotations referring to trees (including kinds of trees) relative to the total number of quotations associated with a given historical period (we used 100-year blocks for our analyses), and (2) the number of sources (kinds of publications from which the quotes are drawn) relative to the total number of sources associated with a given period. Our first analysis examines the general prediction that if knowledge of trees is devolving, there should be an overall drop in the number of quotes and number of sources across time. A second major analysis examines more specific hypotheses concerning the relative usage of tree terms at different levels of taxonomic organization.

Of course, there may be historical periods of time when cultural support for biological knowledge is increasing (evolution rather than devolution). The predictions here would be more or less reversed. As we shall see, our analyses suggest both periods of evolution and devolution. Before turning to specific procedures, we state our assumptions about levels of taxonomic specificity and identify potential problems that may arise with analyses such as ours.

Levels of Specificity

In our analysis of taxonomic levels, we adopt Berlin and his colleagues' (1972, 1973, 1992) approach to taxonomic organization. According to Berlin, categories can be viewed as belonging to one of five levels of organization. At the most inclusive level, there is the unique beginner, typified by categories like *plant* and *animal*. The next level of organization, the life-form level, is commonly referred to by a single word and includes such classes as *tree, vine, grass,* and *mammal*. At the next level, the generic level, there is an explosion of categories, such as *oak, pine, catfish, perch, robin, maple tree,* or *box tree*. The generic level, according to Berlin, is the basic building block of all folk taxonomies. Among other things, it is the level most often used in describing an object, the level that is most psychologically salient, and the level first learned by children. The next two levels are the specific and varietal. Linguistically, categories at the specific level usually require two-word lexemes such as *blue spruce, white fir,* or *post oak*.

Methodological Issues

Threats to Validity

There are five general concerns associated with using text to assess change across time. One problem involves changes in spelling and naming. For example, our search revealed twenty different spellings of oak and twenty-five different spellings of tree. Spelling only became fairly uniform in the nineteenth century. Obviously, one needs to search the corpus for each of the alternative spellings. Likewise, some trees have multiple common names; for example, in England another name for linden is whitewood. The same prescription holds here.

The second concern is that the results may be affected by the particular meaning of the term being invoked in a quotation. For example, the term *pine* can be used to refer not only to a particular kind of tree, but also a particular kind of wood (e.g., pine floor), location (e.g., pine grove), activity (e.g., pine away), or proper name (e.g., the cleaning product PineSol). In the following analyses, only direct references to particular kinds of trees (the first use) were included in the analyses because it is for these uses that the devolution hypothesis makes the clearest predictions.

Our third concern is that the sources for quotes may change across time in a systematically biased manner. For example, during the age of exploration and colonization, new publications appeared (e.g., *Australian Journal*) that were devoted not to life in England but to life in the British colonies. These often include descriptions of the (novel) flora and fauna of those regions. The rise of modern science also led to technical publications. We decided to omit technical and foreign quotations and focus on what we term folk quotations.

The fourth concern is that changes between levels of specificity might be affected by the introduction of new tree terms into the language. Descriptions involving new trees may elicit more attention and favor more specific descriptions. We addressed this and some related problems by selecting a subset of 22 tree generics that were common in English from the fifteenth century to the present day. Differences between levels of specificity cannot, then, be attributed to the introduction of novel kinds.

The final concern involves possible biases in our sampling of quotations resulting from the inherent nature of the dictionary, which seeks to include all but the most specialized terms. This means that even infrequently used terms may have entries with several quotations; thus, the number of quotations within a term's entry may not reflect its actual frequency of usage. For instance, the number of quotations for low-frequency tree terms might be significantly inflated compared with their actual frequency in everyday speech. In practice, however, the *OED* does not generally include entries for tree terms at the specific level or lower (e.g., pin oak). Nevertheless, to eliminate any chance of quotation inflation, all quotations found in the entry of any tree term were eliminated from the analyses. In other words, all quotations used in these analyses came from entries of other terms.

Other Concerns

The use of the *OED* constrains our focus to England and its associated history of wars, colonialism, and increasing globalization of interests. Our task would have been more straightforward had we been able to pick a more insular culture, although the insularity of traditional cultures may be more myth than reality. This factor, however, cuts both ways. It is precisely because of its technological and global orientation that evolution or devolution of folk biology in England is of interest. Given the importance often attached to science education, it is only reasonable to ask about the cultural supports for learning about the natural world.

ANALYSIS 1: EXAMINING THE OVERALL USE OF TREE TERMS OVER TIME

The purpose of our first analysis was to test the main prediction of the devolution hypothesis: If knowledge about trees is declining, there should be an overall drop in the use of tree terms over time.

Method

The process of preparing the quotes for analysis had three main phases: (1) abstracting entries containing quotations, (2) coding the entries, and (3) correcting for uneven sampling in the *OED*.

Abstracting Entries

In the first phase, quotations containing tree terms were drawn from the *OED* using Open Text Corporation's PAT search engine. In searching for the word "tree" all alternative spellings were considered (including trau, traw, tre, tren, treo, treu, treuwum, triu, troue, trow, as well as fifteen other spellings). Alternative spellings were obtained through a word's *OED* entry. In addition to the word "tree" we also searched for 22 folk-generic level tree terms (including all associated 138 alternative spellings). The folk-generic level tree terms included alder, ash, aspen, bay, beech, birch, cypress, elm, fir, hawthorn, hazel, juniper, laurel, maple, mulberry, myrtle, oak, pine, poplar, sycamore, walnut, and willow. All of these folk-generic tree terms have been in use since the fifteenth century or earlier. The search was limited to singular forms of these terms to avoid the problem of changes in pluralization conventions over time.

Once obtained, the output from these searches was reformatted for easier coding. In the online version of the *OED,* the text contains tags that mark the start and end of entries and their associated components (e.g., definitions, quotes, sources, and dates). A program was written that removed all extraneous text and formatting markers. The resulting file contained only quotes and their associated dates and sources. Sample quotations are shown in table 12.1.

Coding Entries

The second phase of preparing quotations for analysis involved coding each entry's source, quotation, and time period. The source of the entry was coded as folk or nonfolk. An entry was considered folk if its source was neither a technical publication (e.g., *Fruit Trees, Nature, Elementary Botany, Science News, British Plants, Dictionary of Gardening*) nor a foreign one (e.g., *Jamaica, New York Times, Barbados, Journal of Upper India, Central America, Pennsylvania Archives, African Hunting*).

Each quotation was coded as either direct or indirect. Only quotes making direct references to trees were included in the analyses, as discussed earlier. Quotations were coded as indirect if they were used to modify other nouns, as in the phrase "stump of pine." In this example, the word *pine* modifies *stump,* which, of course, is not a tree. Tree terms were coded as indirect if they were used as the first term in a compound noun, as in "maple syrup." In this case the thing being referred to directly is syrup, not a tree. Finally, a tree term was coded as indirect if it referred to a substance, as in "The wall is maple." Here, *maple* refers to a kind of wood.

In addition to eliminating indirect references to trees, we also eliminated quotations that referred to something other than a tree, and quotations that included two tree terms at different levels of specificity. (The purpose of this latter restriction will be discussed in Analysis 2.) As for cases of nontree uses, quotations were excluded if the tree term was used metaphorically, as a part of speech other than a noun (e.g., verb), or as a proper name.

Coding Dates

In the preface to the *OED,* it is noted that prior to the 1400s, dialectal differences in the English language were quite pronounced. Hence, words and forms that occurred after 1500 and were dialectal were excluded from the dictionary. These factors led us to choose the late 1400s as a cutoff for our analyses. Because the most recent quotations in the *OED* were entered in 1987, we rounded this date down slightly to look at quotations from 1975 back to 1475 in 100-year intervals labeled by their median dates: 1525, 1625, 1725, 1825, and 1925.

Table 12.1   Example Quotations with Their Associated Dates and Sources

Date	Source	Quotation
1510	(Lytell Geste R Hode)	They dyde them strayt to Robyn Hode Under the grene wode tre.
1613	(Henry VIII)	We take From euery tree lop barke and part of the timber.
1785	(Henry VIII)	Eight Nuts from a tree called the Kentucke Coffee tree.
1843	(Lett)	The bunya bunya tree is noble and gigantic.
1929	(New Yorker)	The big walnut tree that was an old timer even in her day.

Correcting for Uneven Sampling in the OED

The total number of quotations across time periods varied widely. For example, the number of quotations in the 1625 time period ($N = 424{,}711$) was much higher than the number of quotations in the 1725 time period ($N = 281{,}342$). These differences were most likely due to new words entering the language and/or changes in the production of written sources. But to interpret the shifts in the number of tree quotations and sources properly, the total number of quotations and sources in the dictionary must be taken into account. That is, we need to be sure that differences in tree counts are due to factors relevant to tree terms, not sampling.

To correct for variation in number of quotes between time periods, tree counts were analyzed relative to the estimated number of folk quotations and sources in the *OED*. These estimates were obtained by taking a 1 percent random sampling of quotations and coding them for source type (i.e., folk versus nonfolk) and then multiplying them by 100. A comparable adjustment was made in evaluating number of sources.

Results and Discussion

Our search for tree terms generated a total of 22,319 quotations. Automatic coding of each quotation by source eliminated quotations from foreign and technical sources. The remaining 15,146 quotations, roughly equivalent to 900 pages of text, were further analyzed by hand according to the criteria already described. The final 6548 quotations that both made direct reference to trees and came from folk sources were roughly 29 percent of the original set of quotations.

The findings provided strong support for the main prediction of the devolution hypothesis: Cultural support for trees, as measured by the relative number of quotations and sources in the *OED*, declined markedly in the last century. Figure 12.1 shows the resulting proportions for each period of time. The confidence intervals in figure 12.1 represent ranges having a 95 percent probability of covering the true population values, assuming a binomial distribution.

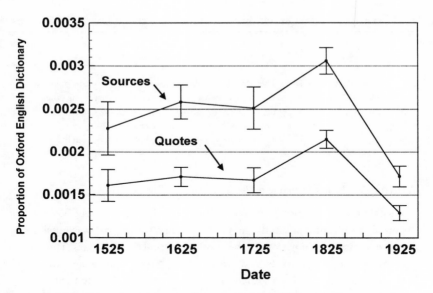

Figure 12.1. Proportion of quotations and sources in the *Oxford English Dictionary* referring to trees along with associated 95 percent confidence intervals. Reprinted from P. Wolff, D.L. Medin, and C. Pancratz, "Evolution and devolution of folkbiological knowledge," *Cognition* 73(2):187, copyright Elsevier Science 1999, with permission from Elsevier Science.

An examination of figure 12.1 shows that the proportions for quotations and sources was fairly constant through the sixteenth, seventeenth, and eighteenth centuries. In the nineteenth century, the relative number of quotations and sources increased, suggesting that knowledge of tree terms evolved during this period. However, the gains of the nineteenth century were completely lost in the twentieth century, which witnessed a striking decline in both quotations and sources using tree terms. Note that the start of the decline corresponds closely with the start of the Industrial Revolution. The confidence intervals indicate that the evolution occurring in the nineteenth century and the devolution occurring in the twentieth century are significant. The confidence intervals also indicate that the twentieth century decline was so great that writing about trees is lower now than at any other time in the history of the English language.

The only difference seems to occur between the sixteenth and seventeenth centuries: quotations indicate evolution and sources do not. Both sources and quotations increase over time until the twentieth century, but between the sixteenth and seventeenth centuries the increase proportions for sources do not exceed the confidence intervals, and, hence, they are not considered significant, whereas the increase proportions for quotations do. This difference does not change the important conclusion that we can be confident that the observed changes in quotations are not due to an overrepresentation from a particular kind or set of sources. Of limited interest, the proportions for sources were slightly higher for each time period than the proportions for quotations. These differences merely indicate that quotations containing tree terms come from a wider range of sources as compared with quotations containing other terms, on average.

In sum, the findings are perfectly consistent with the idea that there have been periods of evolution and, more recently, devolution in knowledge about trees. But the findings are also consistent with another possibility. Specifically, the overall decline in tree terms may mask important evolutionary trends at more specific levels of organization. If such countertrends are present, the overall decline in tree terms in the twentieth century may not necessarily indicate loss of knowledge. Rather it might reflect a shift from a folk-biological view to a more scientific view of trees. We will refer to this possibility as the *shift-in-knowledge hypothesis*. This hypothesis assumes the presence of two kinds of underlying change. First, the drop might reflect a tendency to use terms not covered in our searches. For example, people might talk less about particular kinds of trees and more about DNA, evolution, and the biochemical reactions associated with photosynthesis. Indeed, the concept "tree" has no status in scientific taxonomy. Short of an exhaustive analysis of all of the scientific talk in the *OED*, this change cannot be tested. Alternatively, the drop could be due to a shift toward the use of more specific terms that do refer to trees. Such a shift could occur, even with an overall decline in tree terms, assuming the overall pattern in Analysis 1 was dominated by the word "tree." This second possibility can, in fact, be tested.

The shift-in-knowledge view makes a set of predictions concerning the relative use of different levels of organization. The primary prediction is that if knowledge is increasing, generic and specific level terms should be increasing. The opposite pattern would count as evidence against the shift-in-knowledge hypothesis and for the devolution hypothesis.

ANALYSIS 2: EXAMINING TREE TERMS AT DIFFERENT LEVELS OF SPECIFICITY

The same set of quotations used in Analysis 1 was used in this analysis. In the current analysis, however, the quotations were coded according to level of organization. One of the main goals in this analysis was to understand better the observed decline in tree terms in the twentieth century. But a closer examination of the quotations could also be used to provide further insight into the apparent lack of change existing between the sixteenth and eighteenth centuries and the observed evolution of tree terms in the nineteenth century.

Method

Three levels of organization were coded. The life-form level was indicated by use of the word *tree* or one of its 22 other spellings. The generic level was indicated by quotations containing one of the 22 tree terms listed in Analysis 1. Quotations demonstrating the specific level contained one or another subset of the 22 generic tree terms.

Results and Discussion

The findings from this second analysis provide further support for the devolution hypothesis. Cultural support for trees in the twentieth century, as measured by the relative number of quotations in the *OED*, declined over time for all levels of organization. As in Analysis 1, tree counts were analyzed relative to the estimated number of folk quotations and sources in the *OED*. Figure 12.2 shows the resulting proportions for level of specificity in each period of time along with 95 percent confidence intervals. Because the proportions for sources and quotations did not differ in their overall patterning, only the proportions for quotations are displayed.

The patterns of change shown in figure 12.2 indicate both periods of evolution and devolution. Periods of evolution are indicated by the steady rise in frequency counts between the sixteenth and nineteenth centuries for both the generic and specific levels and a rise in frequency counts between the eighteenth and nineteenth centuries for the life-form level. As noted in Analysis 1, the period between the sixteenth and eighteenth centuries seemed to be devoid of change. In fact, this apparent absence belied significant shifts in use of different levels of specificity (fig. 12.2). The nineteenth century seems to represent the evolutionary climax for knowledge of trees. Talk about trees was both more frequent and at a level of greater specificity than at any other time in the history of English. All this changed in the twentieth century.

The pattern of frequency counts during the twentieth century is most consistent with devolution. It is especially relevant that the twentieth century is the only century in which frequency counts for all levels of organization declined. Thus, in contrast to the shift-in-knowledge hypothesis, an overall drop in tree terms cannot be explained as a drop in the life-form level alone, which masked increases at more specific levels of organization.

Although the overall pattern of results is consistent with devolution, at least two possible problems could be raised. The first concerns the relative resilience of the specific level to devolution, and the second concerns statistical properties of the database.

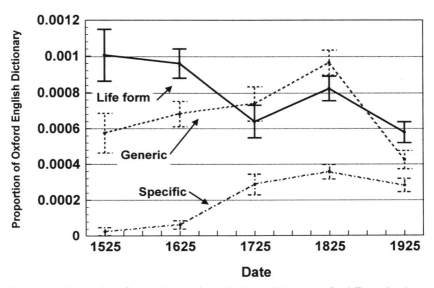

Figure 12.2. Proportion of quotations in the *Oxford English Dictionary* for different levels of specificity along with associated 95 percent confidence intervals. The figure is a variant of that appearing in P. Wolff, D.L. Medin, and C. Pancratz, "Evolution and devolution of folkbiological knowledge," *Cognition* 73(2):190, copyright Elsevier Science 1999, and is reproduced with permission from Elsevier Science.

Resilience of the Specific Level

An examination of figure 12.2 shows that the specific level declined the least among the other levels of organization during the twentieth century. That is, the specific level appears to be relatively immune to devolutionary forces, which could be taken as weak support for the shift-in-knowledge hypothesis. We believe, however, that this countercurrent is most consistent with a mixture of general devolution and a specialization of knowledge, the latter representing something of a "cognitive division of labor."

Let us take a closer look at why the specific level might be somewhat more immune to devolution than the other levels. First, because our search was limited to 22 kinds of trees at the generic level, changes in frequency counts at this level are primarily in terms of changes in tokens. But because the specific level is not limited to 22 categories of trees, changes in frequency counts at this level reflect changes in both tokens and types, which may have inflated the proportion of specific level categories relative to the other category levels. The ideal analysis would hold specific level categories constant throughout time. The primary reason why this analysis cannot be done is that there are very few terms at this level of organization that have survived across even two time periods, let alone five.

Note also that it appears it was not until the seventeenth century that the specific level even emerged.

A second and potentially more interesting rationale for expecting specific level terms to be relatively immune from devolution is based on the idea that cultures are comprised of both a general population and a small subset of specialists who act as keepers of technical knowledge. This subset may be motivated or required by the nature of their activities to operate at ranks below the generic. If the general population's knowledge is undergoing devolution in terms of how much it talks about trees, then discourse from the subset of experts or specialists will gradually comprise a greater proportion of the total amount of discussion of trees. In the limiting case, only specialists will talk about trees and name them at the specific or varietal level. In that event the number of specific and varietal terms would *increase* relative to the use of generic terms (though the absolute numbers of all three might decline).

Statistical Properties of the OED

One potential challenge to the devolution hypothesis is that the observed decline in tree terms in the twentieth century may be a statistical artifact of the OED. Assuming the twentieth century experienced an enormous explosion in new categories, it is certainly possible that talk about any one category would be diluted. Thus, the apparent decline in the twentieth century may not be due to devolution but rather to decreased talk about any one thing because there are more things to talk about. This possibility is relatively easy to check. If the twentieth-century decline is due to dilution, similar rates of decline should be observed for categories other than tree categories. If, however, the decline is due to changes in knowledge, rates of decline are likely to vary widely between the categories. To test this possibility, life-form level terms (or their equivalent) from three other domains were analyzed using the same criteria as used in Analyses 1 and 2. The specific categories analyzed were *fish, weapon,* and *bird*.

The findings provide further support for the devolution hypothesis. In contrast to the dilution hypothesis, not all the categories declined during the twentieth century. Specifically, quotations referring to the category fish steadily increased from the sixteenth century until the present. This may partially be a function of the fact that "fish" also appears in food contexts. Quotations containing the category weapon slowly declined during the sixteenth to nineteenth centuries and then asymptote during the twentieth century. Changes in the category bird mirrored those of the category tree, but not as dramatically. In sum, because declines in the twentieth century are not inevitable, we can be more confident that the observed declines in tree terms are due to changes in knowledge and not to dilution.

GENERAL DISCUSSION

The results from this research support the claim that knowledge about trees evolved during the sixteenth to nineteenth centuries and devolved during the twentieth century. We showed that the twentieth century was marked not only by a major decline in frequency in tree terms overall (Analysis 1) but at all levels of specificity (Analysis 2). These twentieth-century declines cannot be explained as simply due to an explosion of categories diluting talk about any particular kind of category. Diluting would predict that all categories should decline, but as indicated by the categories fish and weapon, decline is not inevitable.

The Relationship between Concepts and Names

One particular result from Analysis 2 deserves further discussion. Specific level categories decline less than predicted compared with the declines at other levels. As mentioned earlier, the resilience in the specific level might be due to a relatively small group of experts. There is another possible explanation. Throughout this chapter we have assumed that when people use a term, they also have some relevant knowledge about it. When a domain declines, however, this tight coupling might begin to break down. People may be able to recognize a number of tree terms, for example, birch, cedar, chestnut, fig, hickory, maple, oak, pine, or spruce, but they may not be able to pair these terms with actual referents in the world. In a sense, these terms exist as loose categories: there is no longer a correspondence between the use of a particular (specific) term and knowledge of what the term actually refers to.

What Happens When a Domain Dies?

When a domain devolves, does it reverse the order of its evolution? The answer to this question appears to be a cautious no. When a domain evolves, knowledge of the domain motivates the creation of ever more precise category labels. When a domain dies, it may be that the knowledge of the associated concepts declines faster than knowledge of specific terms. Thus, the language may preserve certain distinctions beyond the time these distinctions are still understood. It is as if knowledge builds up a terminological structure in the language, but that when knowledge declines, the structure, like an abandoned building, may remain for a while. What makes this hypothesis particularly interesting is that it seems to support recent findings assessing people's induction strategies.

A set of studies by Coley, Medin, and Atran (1997; see also Atran et al. 1997) examined the question of how knowledge of a domain might affect categorical induction. The authors investigated the induction patterns of urban Americans and Itzaj Maya for several folk-biological taxonomies including bird, fish, and tree. Coley, Medin, and Atran predicted that the Maya would treat the generic level as privileged by questioning the validity of inferences from generic to life-form categories but not varietal to specific or specific to generic. In contrast, Americans were expected to treat the life-form level as privileged, by questioning the validity of inferences from the life-form to the kingdom levels (e.g., animal, plant) but not the generic to life-form, as the Itzaj would. These predictions were based on studies showing that the Itzaj possess a great deal more knowledge about living things than do American college students (López et al. 1997) and on the assumption that the privileged level should depend on expertise. But, despite their differences in knowledge, both Itzaj and Americans treated the generic level as privileged with respect to category induction. What makes this result surprising is that Americans have been found to treat the life-form, not the generic level, as privileged on other tasks (Rosch et al. 1976; Tversky and Hemenway 1984). Coley, Medin, and Atran (1997) found a disparity between category use and knowledge, as we did, and their solution is similar to ours. When it comes to category induction, people may rely heavily on the nomenclatural patterns of their language when their knowledge for the domain is weak.

Psychological Implications of Devolution

One question not fully resolved by Coley, Medin, and Atran concerns the question of why Americans have more than one privileged (basic) level. One possibility mentioned by these authors is that the privileged level may change depending on the task. For Americans, the privileged level may be the life-form level when the tasks involve explicit knowledge or perceptual distinctions, and the generic level when the task involves category induction. What might have led to this dissociation? One possibility is suggested by the analyses in this paper, namely, the emergence of another basic level. During the nineteenth century, the basic level for different kinds of tasks may have been the same, but as folk-biological knowledge devolved, asymmetries like the ones observed by Coley, Medin, and Atran (1997) may arise.

These mismatches in degree of structure and amount of knowledge may have other psychological implications as well. For instance, they may explain why we have the intuition that knowledge about trees is dying. It may be that people sense a disparity between what they know and the terminological sophistication of the language. To end on a more positive note, it may be that while knowledge about trees has devolved, not all has been lost. Much of what has been known about trees might still be preserved in the language, albeit indirectly, and its presence there could facilitate the process of its reacquisition.

ACKNOWLEDGMENTS

The research was supported by National Science Foundation grant SBR-9511757 awarded to the second author. We thank Scott Atran, Larry Barsalou, Dedre Gentner, John Coley, Lance Rips, and the entire Folk Biology group at Northwestern University for many helpful discussions.

REFERENCES

Atran, S., et al. 1997. Generic species and basic levels: Essence and appearance in folkbiology. *Journal of Ethnobiology* 17:22–45.

Berg, D.L. 1993. *A Guide to the Oxford English Dictionary*. Oxford and New York: Oxford University Press.

Berlin, B. 1972. Speculations on the growth of ethnobotanical nomenclature. *Language in Society* 1:63–98.

Berlin, B. 1992. Ethnobiological classification. In *Cognition and Categorization*, ed. E. Rosch and B. Lloyd. Pp. 9–26. Hillsdale, N.J.: Erlbaum.

Berlin, B., D.E. Breedlove, and P.H. Raven. 1973. General principles of classification and nomenclature in folk biology. *American Anthropologist* 75:214–242.

Coley, J.D., D.L. Medin, and S. Atran. 1997. Does rank have its privilege? Inductive inferences within folkbiological taxonomies. *Cognition* 64:73–112.

Diamond, J., and D. Bishop. 1999. Ethno-ornithology of the Ketengban people, Indonesian New Guinea. In *Folkbiology*, ed. D.L. Medin and S. Atran. Pp. 17–45. Cambridge, Mass.: MIT Press.

Hatano, G., and K. Inagaki. 1987. Everyday biology and school biology: How do they interact? *The Quarterly Newsletter of the Laboratory of Comparative Human Cognition* 9:120–128.

Hunn, E. 1999. Size as limiting the recognition of biodiversity in folk biological classifications: One of four factors governing the cultural recognition of biological taxa. In *Folkbiology*, ed. D.L. Medin and S. Atran. Pp. 47–69. Cambridge, Mass.: MIT Press.

Inagaki, K. 1990. The effects of raising animals on children's biological knowledge. *British Journal of Developmental Psychology* 8:119–129.

López, A., et al. 1997. The tree of life: Universals of folkbiological taxonomies and inductions. *Cognitive Psychology* 32:251–295.

Murray, J.A.H. 1989. Preface and general explanations. In *The Oxford English Dictionary*. 2d ed. New York: Oxford University Press.

Nabhan, G.P., and S. St. Antoine. 1993. The loss of floral and faunal story: The extinction of experience. In *The Biophilia Hypothesis*, ed. S.R. Kellert and E.O. Wilson. Pp. 229–250. Washington, D.C.: Island Press.

Rosch, E., et al. 1976. Basic objects in natural categories. *Cognitive Psychology* 8:382–439.

Tversky, B., and K. Hemenway. 1984. Objects, parts, and categories. *Journal of Experimental Psychology: General* 113:169–193.

Wester, L., and S. Yongvanit. 1995. Biological diversity and community lore in Northwestern Thailand. *Journal of Ethnobiology* 15:71–87.

Wolff, P., D.L. Medin, and C. Pankratz. 1999. Evolution and devolution of folkbiological knowledge. *Cognition* 73(2):177–204.

13

SOME PROBLEMS OF DESCRIBING LINGUISTIC
AND ECOLOGICAL KNOWLEDGE

Andrew Pawley

I wish to raise some questions facing descriptive linguists and other researchers who are concerned with languages and knowledge systems: To what extent can a speech community's shared knowledge of the world properly be considered part of its language? (Let "knowledge" be understood as a subjective thing, encompassing "perceptions," "beliefs," and "understandings.") And if such shared knowledge is part of a language, how can it be effectively represented in descriptions of that language?[1]

If we accept that a great deal of cultural knowledge is part of linguistic competence, I believe we must acknowledge that conventional descriptions of languages generally do a poor job of representing such knowledge. We then need to ask why this is so and what, if anything, can be done about it.

My views on these questions have been partly formed by my experiences as a linguist taking part in two interdisciplinary field studies in the South Pacific, one in Papua New Guinea, the other in Fiji. Both projects have continued, intermittently, for some 30 years. Here, I will refer mainly to the Papua New Guinea project because it has been the more thoroughly interdisciplinary of the two and because it contains some inspirational elements.

In 1963, as a graduate student in anthropology and linguistics, I joined a project led by the late Ralph Bulmer among the Kalam- and Kobon-speaking peoples of Papua New Guinea. The Kalam (numbering about 15,000) and Kobon (about 5000) live around the junction of the Bismarck and Schrader ranges in Madang

Province on the northern margins of the central highlands. Bulmer was both an anthropologist and a naturalist. His key aim was to investigate the social organization and the perceptions and use of the natural environment of the Kalam of the Upper Kaironk Valley in the Schrader Range. These people live and make their gardens in mountainous terrain at altitudes between 1500 m and 2350 m above sea level. They also obtain wild plants and animals both from this zone and from the upper montane forests up to 2700 m, and from time to time they undertake hunting and trading expeditions to lowland regions, down to about 600 m, in the Jimi and Ramu valleys.

Like other social anthropologists, Bulmer was interested in social concepts—in kinship categories, social groupings, concepts of rank and status, supernatural beings, ritual prohibitions, notions of sacredness, and the like. But unlike most other British social anthropologists trained in the 1940s and 1950s, he was also curious about "such mundane matters as how people classify plants and animals, how they count, weigh and measure; how they conceptualize time and space; how they perceive and classify color and other physical properties which can be discriminated by the human senses" (Bulmer 1971:22). He regarded the compilation of a grammar and a dictionary of the Kalam language as a crucial part of the research and invited me to do the grammar and contribute to the dictionary.

One of Bulmer's Kalam field assistants and informants in the 1960s was a teenager called Saem Majnep, who had a deep knowledge of the wild animals and plants of his region. After some years Bulmer and Majnep began to write together. When Majnep traveled to New Zealand in the summer of 1973/74, they drafted the widely acclaimed book *Birds of My Kalam Country* (Majnep and Bulmer 1977). The text was based on Bulmer's interviews with Majnep, who organized his accounts of the various birds and their role in Kalam life into chapters using Kalam ecological and zoological criteria. By this time Majnep had learnt to write the Kalam language using the orthography developed by linguists Bruce Biggs and Andrew Pawley, and Bulmer encouraged him to write the next book on his own about the wild mammals of the Schrader Range, to be called "Animals the Ancestors Hunted." Between 1977 and 1980 Majnep tape-recorded and transcribed his accounts of Kalam knowledge, beliefs, and cultural practices concerning animals together with a number of myths relating animals and people. This material was then edited and translated into English by Bulmer, who began to add scientific and explanatory commentaries.

By 1983 they had an 800-page bilingual first draft of the book and were planning a third book, to be called "Kalam Plant Lore," dealing with the role of wild and semidomesticated plants in the life of the Kalam. Unfortunately, Bulmer died in 1988 before he had finished editing the book on Kalam hunting. The role of Majnep's editor and translator then fell to me. I edited the manuscript of the book as a series of working papers in Kalam and English, under the title *Kalam Hunting*

Traditions (Majnep and Bulmer 1990, n.d.), and together with Majnep and a New Zealand botanist, Rhys Gardner, am working on "Kalam Plant Lore." Majnep remains a farmer in the remote Kaironk Valley, supporting his family by raising crops and livestock.

KNOWLEDGE AND LANGUAGE

It is sometimes said that linguistic diversity has a value analogous to biodiversity. There are several parts to the argument. Usually the first step is to point out that each language is a store of intellectual capital. The second step is to say that no sharp boundary can be drawn between language and the rest of culture. Indeed, language is the chief means by which a speech community's knowledge and perceptions of the natural and social world, its technology, verbal arts, social values and practices, and other traditions are codified and transmitted. The third step is to say that the worldviews associated with different languages and cultures are diverse, each the product of generations of accumulated experience in different circumstances. Here are several statements along these lines, made, respectively, by a biologist, a littérateur, and two structural linguists.

> Each language is indissolubly tied up with a unique culture, literature (whether written or not) and worldview, all of which also represent the endpoint of thousands of years of human inventiveness. (Diamond 1993:84)

> It is the constructive powers of language to conceptualize the world which have been crucial to man's survival. . . . Each language maps the world differently. . . . Each tongue . . . construes a set of possible worlds and remembrances. . . . When a language dies a possible world dies with it. (Steiner 1992:xiii–xiv)

> A language is shaped by its culture, and a culture is given expression in its language, to such an extent that it is impossible to say where one ends and the other begins. . . . what can be said, and what can be talked about, may be quite different from one language-culture system to another. (Grace 1987:10)

> Of supreme significance in relation to linguistic diversity, and to local languages in particular, is the simple truth that language—in the general multi-faceted sense— embodies the intellectual wealth of the people who use it. A language and the intellectual productions of its speakers are often inseparable, in fact. Some forms of verbal art—verse, song, or chant—depend crucially on morphological and phonological, even syntactic properties of the language in which it is formed. . . . Even where the dependency is not so organic as this, an intellectual tradition may be so thoroughly a part of a people's linguistic ethnography as to be, in effect, inseparable from the language. (Hale 1992:36)

If it is true that that each language contains distinctive concepts and distinctive mechanisms for creating ideas, it makes sense to say, as Russell Bernard does, that

"Linguistic diversity is . . . at least the correlate of . . . diversity of adaptational ideas. . . . By this reasoning, any reduction of language diversity diminishes the adaptational strength of our species because it lowers the pool of knowledge from which we can draw" (Bernard 1991:2).

These statements make some strong assertions that we should not take on faith. The problem is to formulate them as precise, testable hypotheses. More than scholars from some humanist disciplines, structural linguists have tended to be fairly skeptical of the proposition that the structure of language influences thought. Recall the lively debate concerning "linguistic relativity" generated in the 1950s by the publication of Benjamin Lee Whorf's collected papers (Carroll 1956). Skepticism is healthy but in that case the difficulties of formulating Whorf's proposals in a testable form led most linguists to lose interest and to remain almost indifferent to the main issue for a generation. There is a danger that claims about the intellectual value of linguistic diversity will suffer the same fate.[2]

Ralph Bulmer was not very interested in playing elegant analytic games with Kalam terminological systems. He wanted, above all, to record what the Kalam know and believe about their world. When we came to draft entries for a dictionary, Bulmer's interests in recording as much as possible about Kalam knowledge posed a problem, one that faces every lexicographer. To what extent can knowledge of the world shared by members of a speech community be considered part of language? For example, speakers of English know many things about sheep: they are land-dwelling animals that suckle their young, they have four legs with hoofs, they eat grass and other vegetation, they are domesticated, there are many breeds, they are kept for their wool, skins, and meat, they live in herds, they are thought of as easily led and not very intelligent, and so on. Which if any of these facts belongs, in principle, in a dictionary definition of *sheep* (leaving aside whether the information is given directly under the headword or part of it in cross-referenced entries)? What information if any about Homer's works is properly part of the definition of the expression *epic struggle,* or for that matter, the expression *Homeric struggle*? Linguists and philosophers are notorious for disagreeing among themselves about questions of this sort (Haiman 1980). Is the attempt to incorporate cultural knowledge in language descriptions fundamentally misguided? Or does the generally very slight representation of such knowledge in standard dictionaries merely reflect arbitrary limitations in the practices of lexicographers arising not from theoretical principle but from practical considerations of time and money?

Now ask yourself which if any of the kinds of "ecological knowledge" referred to in the following passage are part of language.

The study of traditional ecological knowledge begins with nomenclatures [for plant and animal taxa] and ultimately proceeds to considerations of processes (functional relationships): the understandings that people have of environmental systems and the networks of cause and effect therein. A part of these perceptions involves a people's perceptions of their own roles within environmental systems: how they affect, and how they are affected by, natural processes. (Lewis 1993:9)

Obviously one would need quite specific evidence about the way ecological knowledge is encoded before one could begin to frame an answer. But the question remains unanswerable unless the construct "language" itself is defined. The fact is that the boundaries of language can be defined in various ways. There can be no one right definition, only definitions that have different purposes or values.

Two different models of language will be compared. One, which I will call the *grammar-based model,* approximates that followed by structural linguists in general (a category I take to subsume generative grammarians). It promotes linguistic descriptions that have relatively little to say about cultural knowledge. The second, which I will call the *subject matters model,* is grounded in humanist views of language. I will argue that this model, in principle, requires a great deal of cultural knowledge to be part of linguistic descriptions. Intermediate positions are possible and are held by some. It is useful, however, to compare the extremes because the key issues stand out more starkly.

Grammar-Based Views of Language

In grammar-based research the emphasis is on languages as autonomous systems of knowledge, that is, codes that can be learned, independent of particular beliefs and general "knowledge of the world." Grammarians are much concerned with languages as creative devices, with the capacity to create indefinitely many new words, phrases, or sentences. In this sense the chief creative power of a language is believed to reside in its grammar, broadly defined as a system of conventions for pairing formal signals with meaning and, specifically, with meanings that are intrinsic to the linguistic forms, not pragmatic or inferential. The core of the grammatical system consists of combinatorial rules for forming words, phrases, clauses, and complex sentences from a finite lexicon. The lexicon of any language consists of a small set of "functors" or grammatical elements plus a much larger stock of roots or words that function as nouns, verbs, and the like. Grammarians acknowledge that the meanings of most lexical units do relate to a perceived world outside of language. But insofar as the lexicon is defined as a fixed list consisting of unanalyzable or irregularly formed elements, it has traditionally been of small interest to grammarians. This is reflected in the fact that departments of linguistics normally do not offer courses in dictionary-making or in the study of lexical systems; these are regarded as low-prestige domains, entirely marginal to the central concerns of linguistics with generalizations about the nature of language.

Now to a crucial question: When can a linguistic expression be considered to be a lexical unit, something to be put in the dictionary? In the grammarian's ideal lexicon only a very small proportion of the total possible words of a language will be treated as items that belong in the dictionary. That is because most potential words are complex, derived by productive grammatical processes. The set of lexical items are the residue of exceptions—pairings that are not analyzable or not well formed. The structural approach to semantics is in the same spirit as autonomous grammar: it is much concerned with logical relations between semantic elements. This approach can throw considerable light on conceptual systems but in its purer forms is not concerned with what people know and believe about the world outside of the logic of structural semantics.

There is a certain admirable egalitarianism in this austere view of language. Grammarians generally believe in the unity of the human mind. In particular, they do not take kindly to arguments that some languages are inferior to or more limited than others. The beauty of any grammar lies precisely in its mechanical combinatorial power, which gives to the speakers of a language the capacity to generate an infinite number of sentences and the means for talking about any conceivable subject. It provides a kind of freedom, a potential means of escape from the constraints of any particular belief system.

Occasionally, one also finds a statement by a linguist that anything can be said in any language. This implies a kind of universal intertranslatability. If such intertranslatability were possible, it could only mean that each language is an expressive system that (for practical purposes) can be separated from the cultural institutions of its speakers. It is, however, clear that isomorphic or near-isomorphic translation is often impossible and that the argument that "anything that can be said in one language can be said in any other language" can only be sustained, if at all, by allowing long-winded explications and commentaries to count as "saying the same thing."

A third kind of egalitarianism in a grammar-based description is that all well-formed expressions have the same status. No well-formed phrase or sentence is singled out as more significant than any other. A much-used proverb or catchphrase has no more status in the description than any theoretically possible sentence. This is simply to say that grammarians, as grammarians, do not care about social facts or frequency of use; they are indifferent to the kinds of practical value the speech community gives to particular expressions. They are concerned with what is sayable only in the narrow sense of what is grammatically and phonologically well formed.

Humanist Views of Language

I should emphasize that I have no quarrel with reductionist models of language per se. Reductionist models are essential in science; all analyses of "reality" reduce and simplify.[3] The real issue is what a particular model is good for. The chief deficiency of grammar-based models of language is that they bypass precisely those differences among languages, and those domains of creativity, that many nonlinguists regard as particularly important. One anthropologist has written that for the purposes of translation "language appears to begin where analytical grammar leaves off" (Schrempp 1992:xvii). I have encountered similar remarks from other translators. They mean that most of what they need to know to translate texts well cannot be gleaned from grammars or indeed from conventional dictionaries. Instead, one has to spend years—or decades—gaining an idiomatic command of the various genres and styles of both the source and target languages and that means becoming steeped in knowledge of many facets of the cultures associated with each. In humanist eyes language is both the carrier of certain highly codified parts of a culture and a shaper of thoughts. Its conventions lead us to package our thoughts in a certain way, making certain views of the world available for attention by those people who care to reflect on such matters.

The force of the humanist position can be felt if you imagine yourself in the following situation. Language L has been dead for a thousand years. You are given a time machine so that you can go back in time and record L. You are also put in charge of reviving L as a spoken language serving all its former functions so when your students, the new generation of speakers, travel back in time they will be able to pass as native speakers. What would you want to include in your recordings and analysis of L, over and above an account of the rules for forming grammatical sentences and the kind of dictionary that consists of just those form-meaning units that are not predictable by rule of grammar?

To pass as a native speaker it is not enough to be able to distinguish between grammatical and ungrammatical forms. There are other domains of well-formedness. One of these is idiomaticity. One must be able to distinguish between normal (idiomatic) from odd (unidiomatic) ways of saying things; for example, in English one may tell the time in certain kinds of units but not others (thus, *It's 20 past three* and *It's a quarter past three* are idiomatic, but not *It's a third past three* or *It's three quarters to four*). In some language communities, being able to construct and recognize rhymes, puns, and metaphors is also a part of native linguistic competence. It is also important, in all communities, to know what kinds of utterances are appropriate or inappropriate in a given context (what you say when you pick up the telephone, close a meeting, or decline an invitation to dinner). Semantic well-formedness must stand beside grammaticality. As well as knowing what counts as a "clause" or a "sentence," one also needs to know what counts as a well-formed report or description of a certain kind of event or situation, in terms of the kinds and quantity of information to be mentioned or omitted, for example, when telling the time or describing the weather. (We also allow that there are specialized kinds of events and situations that not every speaker is expected to know how to describe.)

In grammar-based descriptions no technical status is given to notions such as "proverbial saying," "terminological system," or "taxonomy." Indeed, even the concepts "name (of a thing)" and "term" have no status in such descriptions. They are, however, implicitly recognized in the dictionaries of conventional lexicographers, who are chiefly concerned to record those form-meaning pairings that have the status of names or terms, without regard to their grammatical well-formedness.

CULTURAL KNOWLEDGE AND LINGUISTIC DESCRIPTIONS

The problem with much discourse about language in the humanities is that it is vague and impressionistic. People are prone to bandy about romantic but undefined notions such as "the genius of the language." The challenge is how to incorporate their insights in an analytic framework that is reasonably rigorous and that will yield readable and convincing descriptions.

I believe this can be done and has been done to some extent. I draw inspiration from ideas developed by scholars from disparate disciplines and with disparate approaches. Besides the considerable bodies of work done under the rubrics of ethnography of speaking, ethnoscience, ethnomethodology, conversational analysis, and ordinary language philosophy, I have been influenced by the writings of George Grace (1981, 1987) on "the linguistic construction of reality," Erving Goffman's analyses of social conduct in public situations (e.g., Goffman 1981), the work of Parry and Lord on Yugoslav epic poetry (Lord 1960), studies by Koenraad Kuiper and his associates on oral formulaic genres such as sports commentaries, auctioneering, and weather forecasts (e.g., Kuiper and Haggo 1984; Kuiper 1996), and research on the seemingly miraculous powers of simultaneous interpreters (Yaghi 1994).

The question of how cultural knowledge can be represented in linguistic descriptions may be put another way: How to define language in such a way that cultural knowledge will be given a prominent place? We may start with two propositions that can hardly be disputed: It is everywhere the case that communities talk about (what they perceive as) types of events, situations, and topics. And communities everywhere develop conventional means for talking about such recurrent entities, that is, each language is adapted to its uses as its speakers develop efficient and often streamlined ways of saying things. Let us therefore define a language as a system of conventions for talking about specific subject matters and for performing other culturally authorized activities.[4] Such a code is no more than an ordinary language but to distinguish it from a grammar-based language we may refer to it as a subject matters language.

What must people know to be competent in a subject matters language? They must, of course, know at least some of the things that are dealt with in grammar-based descriptions. In order to describe an event, for example, one needs to be able to construct a word picture, in the form of a clause or clauses that specify the conceptual structure of the event. But if the word picture is to be readily comprehended by an audience the speaker must also know the characteristic or normal ways of making sense of and reporting particular types of events. Grammars generate many strings of words that people would never, or hardly ever, utter. Out of the many possible grammatical ways of saying (roughly) the same thing, only a small proportion will be idiomatic. To speak idiomatically about particular subjects, speakers follow norms of discourse structure and draw on a lexicon of standard ideas with standard labels and on a phrase-book of conventional expressions that serve particular discourse functions. I will say something about each of these elements.

Names for Things or Terms

The lexicon of a subject matter language should not be confused with the lexicons posited by theoretical grammarians. The lexical items of subject matter languages correspond essentially to the headwords in conventional dictionaries and to what the man in the street calls "names for things," or "terms," or "what something is called" (Pawley 1986, 1996).

The quantity of lexical units in a lexicographer's ideal dictionary of a given language will be many times larger than in the ideal, minimalist lexicon of the grammarian. Dictionary-makers are chiefly interested in those ways of talking that are most highly codified—the repertoire of conventional expressions—and in the social evaluations associated with these expressions, as standard, formal, colloquial, vulgar, polite, literary, technical, and so on. All large general dictionaries are to some degree ethnographic, that is, as part of the definition of concepts they often include information about their significance in the culture of the speech community.

In most traditional societies, a very considerable part of the vocabulary of the language—often the largest single terminological field—will relate to plants. In Kalam there are more than 1500 terms for plant taxa, representing more than 1000 species. This amounts to 15 percent of the recorded lexicon of Kalam. (In the dictionary there are about 10,000 Kalam lexical units—not the same thing as number of headwords or entries.) It should be mentioned that in Kalam many common lexical concepts, including plant taxa, have more than one name, mainly because of name avoidance taboos. Plants are also of paramount importance in Kalam economic and social life. The Kalam eat about 170 cultivars, representing 28 species of domesticated food plants. They also eat another 40 species of wild plants. They cultivate more than 30 other species for other purposes—medicines, rituals, ornamentals, flavoring and wrapping food, and for technology. They use another 150 wild species for technological or ritual purposes and a good many more for firewood. Many other plants are important to them as the known feeding and nesting sites of different kinds of birds and animals that the Kalam hunt.

Among animal species, vertebrates are usually the most important to people in terms of economic and symbolic value. Vertebrate taxa found in Kalam territory number about 400. (The number of terms is actually much higher because of multiple synonyms for many taxa.) They eat about 200 vertebrate species. In Kalam, the largest group of vertebrate taxa is that which falls under the primary taxon or life-form *yakt,* birds (a term that also includes bats). There are about 230 *yakt* taxa, representing five species of bats and 204 species of birds, of which 140 occur in the Upper Kairial Valley. There are about 45 mammal taxa, of which 31 are categorized as *kmn* (larger game mammals), several as *as* (small mammals) three as *kopyak* ("dirty" rats). There are about 30 frog taxa, 27 snake and lizard taxa, but only 6 fish taxa (3 of them eels). In the swift mountain streams of the Kairial Valley there are no fish other than eels. There are more than 120 invertebrate taxa. At least 40 invertebrates (grubs, caterpillars, grasshoppers, spiders, snails, etc.) are eaten by the Kalam.

In conventional dictionaries the treatment of terminologies is, by scientific standards, seriously flawed (Conklin 1962; Wierzbicka 1992). More systematic treatments of terminologies appear in the ethnographic dictionaries that are now starting to appear, mainly for languages spoken by traditional societies. In entries for names of plants and animals, for instance, an ethnographic dictionary should seek to specify: (1) the meaning and reference of taxa names, matching these with scientific identifications, and taking account of sexual dimorphism, life stages, regional variants, and polymorphism; (2) the structure of names, for example, distinguishing uninomials and compounds; (3) the logical relations holding between taxa, for example, the hierarchy of generic terms vs. hyponyms, and other kinds of relations, including synonymy and overlap; (4) local knowledge about plants and animals, their roles in social and economic life, in ritual and symbolism, and their ecological contexts.

A few sample entries from the Kalam-English dictionary (Pawley and Bulmer, with Majnep, Kias, and Gi in press) follow, namely, one term for a kind of animal, one for a plant, and one for a kind of social event.

KMN[1] [kɨmɨn], *n.* 1. Game mammal. Generic taxon, including all large marsupials and rodents and some smaller arboreal forest species, and water-rats, but excluding dogs, pigs, bats and most medium sized and small rats and other small, mainly terrestrial rodents and marsupials. Synonym *sab.* Cf. *wŋbek.* Contrasts with *kopyak* (rats and mice found near homesteads), *as* (frogs and certain small, mainly terrestrial, marsupials and rodents), *yakt* (flying birds and bats), etc. Some small rodent and marsupial taxa are placed by some informants as *kmn,* by others as *as.* The crucial distinction between *kmn* and *as* appears to be in terms of culinary use and dietary prohibitions. *kmn* may be cooked ritually and may be preserved by smoking for consumption at *smi* festivals. They are not forbidden food for initiands in the *smi* or for adult male sorcerers. *as,* which are mainly collected by women, are not cooked ritually and are prohibited food for initiands between their nose-piercing and final release from restrictions at the *smi.* (G) *Kmn takn ak sugij yb aml ñapal, mey mab wog ak bteyt tblak, wog spot ak, . . . kmn ap tap okok ñbek owaknŋ, mey aml sugij ñapal.* When shooting by moonlight, they easily

find game mammals in parts of the forest where trees have been felled and in old garden areas, . . . and when the animals come there to feed they shoot them. *kmn* taxa
names include: *abpen, aklaŋ, aloñ, alks, amlgn, añ, atwak,* . . .

Ñ ŊAY [nyiŋá.y], *n.* 1. *Pueraria lobata,* a plant bearing an edible tuber, of which both
cultivated and uncultivated varieties occur in the Kalam domain. Cultivated pueraria,
though grown in relatively small quantities (up to altitudes of at least 2300 meters), is
much appreciated as a food. Vines used to tie fences, bundles of firewood etc. *ñŋay
gobŋam,* cultivated pueraria. *ñŋay aydk,* synonym *ñŋay sepeb, ñŋay sob,* uncultivated
pueraria, these terms being used indiscriminately by people living at higher altitudes
(e.g., Gobnem) for stock originating from recently cultivated plants persisting in
garden-fallow, and/or for the rather different variety with much smaller tubers which
grows wild in large patches in the *ksod* (*Themeda*) grasslands growing at 1500 meters
and below, whose tubers are used as famine food for humans and pigs.

SMI, *n.* 1. The all-night dance-festival in which a man (the *b smi* or *smi nop*) and his co-
resident extended family may hold as the climax of the ceremonial activity in which
youths and later, girls, have their nasal septa pierced (see *mluk puŋi-*), pigs are killed
and cooked, and pork, axes and shell valuables are given to affinal groups. The dance
is also attended by performers and spectators who are not kin of the hosts and who do
not stay to receive gifts and participate in the feasting on the following day. Large
quantities of taro are required to entertain guests, and the festivals are held from late
July to early December, when taro have been harvested. The *smi* is the major ceremonial institution of Kalam society involving extensive economic preparation in planting
of gardens, building up of pig stock and assembly of gift valuables and personal ornaments; the building of special dance houses (*smi kotp*); the propitiation of dead kin
(*cp kawnan*) and nature demons (*kceki*); the fulfillment, through gifts, of the obligations of kinship and affinity (*tu smen*); and providing a major index of the prosperity,
prestige and political influence of the host group and its leader.

Speech Formulas and Discourse

There are various well-tried (though far from foolproof) procedures for discovering the terms or lexical units of a language. At the most basic level, one can ask
people what the name of a thing is. But what about more complex conventions
for talking about subject matters? Here the repetitive patterns need to be uncov-
ered by close examination of texts and conversation.

Part of native command of a language involves knowing what things to say in
discourse, when and why to say them and how to say them in conventional ways.[5]
Thus, in every language there are set phrases for saying things in certain contexts,
for example, *Pleased to meet you; Excuse me; Not guilty; As you may know; If you'll par-
don the expression; There's method in my madness; Penny wise and pound foolish;* or *Politics is the art of compromise.* Lyons (1968) speaks of these as "situation-bound ex-

pressions"; I prefer to call them "speech formulas" because they are tried and true procedures for doing particular jobs in discourse.

Grammarians (and indeed some lexicographers) have seriously underestimated the number of conventional phrasal expressions for the simple reason that they have not investigated the field seriously. Speech formulas are not a small, marginal class of linguistic units. An average mature speaker of English knows tens of thousands, possibly hundreds of thousands of them. The repertoire of conventional ways of saying things goes far beyond rigidly set phrases. Productive formulas, constructions of phrase and sentence-length expressions whose content is partly specified but which contain variable elements, outnumber fixed expressions. Examples of productive speech formulas are: *I declare this meeting adjourned. X, Y, Z—you name it! X and proud of it! If it's good enough for NP_i (to do such-and-such), it's good enough for NP_j! NP_i has only PRO_i-self to blame!*[6] For telling the time, for stating people's height or weight, and for specifying other sorts of measurements there are formulas as precise and as productive as grammatical rules. An example from my own dialect is *(The time is) M to/past H,* where M stands for a number of minutes (up to 29) and H for the number of an hour (up to 12). Another is *NP is X (feet) Y (inches),* where Y stands for a number up to less than 11 plus fractions stated in terms of halves and quarters.

One might speak of a compilation of speech formulas or situation-bound expressions as a phrase book. The entries in such a phrase book will be quite lengthy. Speech formulas differ in several ways from lexemes. Lexemes are building blocks for syntactic constructions, having no discourse functions on their own. And whereas a lexeme is a form-meaning unit categorized for its possible syntactic contexts, each speech formula is typically a bundle of at least seven components: *meaning, discourse purpose, discourse context, grammatical construction,* (partly specified) *lexical form, idiomaticity constraints,* and *musical conventions* (or *prosody*), and many also exhibit an eighth component: *gesture.*

A speech formula specifies a meaning-form pairing indexed to a class of discourse contexts and discourse functions; for example, *Dear X* is a formula for beginning a letter, and *I declare this meeting adjourned* is a formula for officially opening an event. In the case of discourse that is primarily informational, as opposed to interactional and strategic, the discourse function may simply be to provide information as appropriate for the situation. This is the case when a radio commentator describes in formulaic terms details of the time of the day or the situation in a game of baseball or announces the death of a public figure: *The time is five minutes to six; The count is three and two, with the bases loaded; Mr. X, the former governor of Massachusetts, has died in Boston at the age of 92.*

The phonological component often includes special features of intonation, volume, or voice quality conventionally associated with the discourse use of the formula. For productive formulas there will be information about internal structure:

each such formula requires a minigrammar stating the constituents of the formula and defining the lexical elements that can realize each variable constituent. *Gesture* (including body language), is an obligatory or optional part (or concomitant) of many speech formulas, no less than intonation and other voice modulations. For example, the toasting formulas *To X!* and *Good health!* require the speaker to raise a glass in his or her hand and drink from it and the greeting *Hi!* calls for a smile or at least a friendly look.

NOTES ON KALAM DISCOURSE ABOUT ECOLOGY

Let us now ask whether the notions of discourse structure convention and formulaic expression are useful in describing knowledge about ecological relationships and processes, such as were referred to in the Lewis (1993) quote above.

The Kalam project has produced a large corpus of taped and written discourse about the natural environment. *Kalam Hunting Traditions,* for example, is a rich source of information about Kalam ways of talking about ecological processes. Here are English translations of a few extracts from this work to show the kind of material in Majnep's texts.[7]

> The variety of *bep* (*Rungia klossi*) that they call *bep sgaw-tmud* ("wallaby-ear *bep*") grows in the forest, including the beech zone. It comes up by itself where trees have been recently felled to make gardens, and among regrowth in garden fallow. Especially where areas have been cleared for cultivation and have afterwards been left for long periods so that the trees have grown up really tall, this *bep* will remain and the wallabies will come there to pick and eat it. (Majnep and Bulmer 1990, chap. 1, pp. 54, 56)

> However, what affects the wallabies today is not the pit-traps but the presence of dogs, and not merely the fact that people keep dogs, but that the human population has increased. Thus the wallabies have been reduced by hunting with bow and arrow and by dogs killing them, so that now there are only a few left. (Majnep and Bulmer 1990, chap. 1, p. 69)

> When the lower foliage of the fern bed is dead and dry the (sheltered spaces) underneath are like (the chambers) of a house, and many kinds of animals come here to sleep: *ymduŋ* ringtails, *atwak* cuscuses, *madaw* ground cuscuses, *mosak* (giant rat or bamboo rat, *Mallomys rothschildi*) and others. . . . In the garcinia zone there are plenty of trees, just as there are in the beech forest, but in the beech forest there are many big mossy clumps of epiphytes, whereas these are quite scarce in the garcinia forest. Thus when they hunt there for *ymduŋ* they just climb quite small trees. . . . The *ymduŋ* does not sleep like some animals do, always up in big trees. It may sleep down at the base of the tree, or in young trees, or sometimes, when it does sleep in a big tree, it may be in the hollow of a fork in the trunk or out on a lateral branch, for example in a *noman* (podocarp, *Phyllocladus* sp.). (Majnep and Bulmer 1990, chap. 2, pp. 69, 72)

It is the habit of the *gudl-ws* (the husk-shredding rat, *Anisomys imitator*) to break open and plant *gudl* (*Pandanus antaresensis*) seeds. When it collects these to eat and carry some away with its teeth it drops some of them so that they grow in places a long way from where any *gudl* was previously present. (Majnep and Bulmer 1990, chap. 3, pp. 163, 165)

People make these barriers around *kumi* pandans where *abben* (Highland Giant Tree Rats, *Uromis anak*) are doing a lot of damage to the crop, getting up the trunks and biting off the fruit-heads with devastating speed. As well as breaking off and eating the ripe fruit it removes unripe fruit and gnaws into this to drink the juice. At times when the *kumi* is flowering or newly setting fruit it gets into protective foliage and bites through the calyx so that the vapor escapes and the fruit does not develop but just goes bad. (Majnep and Bulmer 1990, chap. 4, pp. 75, 77)

The original Kalam text of passages such as these yield a good variety of terms for concepts to do with the environment and ecology. For instance, in the original text of the first quotation we find a set of terms for forest conditions: *mab wog* 'patch of forest', *kamay wog* 'patch of beech forest', *wog salm* and *wog spot* 'secondary forest' and 'old garden clearings,' and *wog tbl ymngabal* 'forest cleared for planting.' But what of speech formulas and discourse rules? There are plenty of these to be found in Majnep's texts. The topic is too large and technical to do justice to it here and only brief discussion is possible. Some Kalam discourse structure conventions and their streamlined realizations in speech formulas have been analyzed elsewhere (e.g., Pawley 1987, 1991; Lane 1991). In some respects, the rules of Kalam narrative discourse differ markedly from English. When describing an event in which someone does something to an object, a Kalam speaker does not normally simply say "A did X to O." He normally specifies, first, that the actor went to the scene of the action (if not already there), second, what he did there, third, whether he went somewhere else afterwards and took the affected object with him, and fourth what he did with the affected object at the next scene of action.

Do these discourse conventions promote excessively long-winded accounts of events? They do not. It is true that the component actions can be elaborated in separate clauses, if the narrator wishes. But each of the components can also be expressed economically by a bare verb stem within a serial verb construction (sometimes with an associated noun), along the lines of *go climb kill Giant Rat get come cook they ate (it)*. These serial verb constructions function essentially as single clauses, or rather, as the stems of single clauses, to which variable inflections may be added. The serial verb strings expressing various recurrent types of event sequences are highly conventionalized and are normally spoken as fluently as single lexemes, a point nicely demonstrated in an experiment by Talmy Givón (Givón 1990). We might say that Kalam speakers can draw on a large repertoire of "lexicalized clause stems" (Pawley and Syder 1983).

Let me now cite the first part of the original Kalam text of a memoir by Majnep

(1991), which appears (in English translation only) in a commemorative volume for Bulmer and then briefly discuss what is entailed in making sense of this text.

What the Anthropologist Did

(a)
yad	*mñi*	*Reyp*	*Bulma*	*28*	*years*	*mñab-yad*	*Kalam*	*wog*
I	now	Ralph	Bulmer	28	years	country-my	Kalam	work
ak	*nb*	*gak*						
that	such	he:did						

(b)
agngayn.
I:will:talk

(c)
Tap	*nb*	*ogok*	*tap*	*ti*	*tap*	*ti*	*nŋl,*	
thing	such	those	thing	what	thing	what	having-	
(d)
kneb	*ameb*	*owep*	*wog-wati*	*gep*	*yp,*			
sleeping	going	coming	work	doing	also			perceived

(e)
am	*okok*	*kmn*	*nen*	*pak*	*ñb*	*tagep*	*yp,*
go	wherever	game: mammal	for	kill	eat	going: about	also

(f)
am	*okok*	*yakt*	*nen*	*ñag*	*ñb*	*tagep*	*yp*
go	wherever	bird	for	shoot	eat	going: about	also

(g)
am	*bin*	*ti*	*ti*	*gl*
go	woman	what	what	having:done

(h)
tawpal,
they:exchange

(i)
agl;
having:wondered

(j)
am	*ñagep*	*pakep*	*at*	*gok*	*ti*	*ti*	*gl*
go	shooting	hitting	on	those	what	what	having:done

(k)
mdebal
they-live

(l)
agl;
having:wondered

(m)
am	*tap*	*ognap*	*kmn,*	*yakt,*	*kubap,*	*kulnok,*
go	thing	some	game: mammal	bird	green-snail-shell	dog-whelk-shell
bin-b	*penpen*	*ñeb,*				
woman-man	each:other	giving (n.)				

(n)
mñab	*pat*	*okok*	*nb*	*Kaytog,*	*Asay*	*akaŋ*
country	far	those	from	Kaironk	Asai	whether

(o)
Kaytog	*nb*	*Sbay,*	*Kaytog*	*nb*	*Malŋ,*
Kaironk	from	Simbai	Kaironk	from	Maring

(p)
tap	*ti*	*tap*	*ti*	*gl*
thing	what	thing	what	having:done

(q)
mdebal
they-live

(r)
agl;
having:wondered

(s)
wog	*nb*	*ogok*	*am*	*ngup*	*ak.*
work	such	those	go	he:observed	that

When I began to translate this extract there were several passages that I did not understand. Yet I was familiar with all the words in it and with Kalam grammar. What else does one have to know besides the words and grammar to be able to make sense of the text? Some idea can be gained if we break the material down into large sense units of two or three clauses, give a fairly free translation of each such unit and then for the more difficult cases isolate conventional expressions that are larger than a word but smaller than a sentence.

The first sense unit, consisting of clauses (a–b) is fairly straightforward:

(a–b) *I'm going to talk about Ralph Bulmer's 28 years of work in my Kalam country.*

Clauses (c) to (s) form one long sentence, with the next independent verb—a verb inflected for absolute tense, aspect, and mood and for absolute person and number of subject—not occurring until (s).

(c) *Having observed the various customs and rituals,* (d) *the everyday work and activities,*

In (c) the phrase *tap ti tap ti* 'what thing what thing' is a standard expression, meaning either 'various things', or 'the usual (or customary) things'. *kneb ameb owep wog-wati gep* 'sleeping going coming doing work' is a standard phrase roughly equivalent to 'everyday activities' or 'daily affairs'; it contains four nominalized verbs (*-eb* or *-ep* is a nominalizing suffix) each specifying a kind of everyday activity.

(e) *the hunting of game mammals,* (f) *the hunting of birds,*

Clauses (e) and (f) each contain a variant of a standard lexical formula:

am	okok	N-nen	V	ñb	tagep
go	wherever	N-after	V	eat	walking:about

where N names a category of game animal and V specifies how it is killed, for example, by striking or shooting it, that is, 'walking about going here and there after N, killing and eating'. This is a standard way of saying 'hunting N'.

(g) *the various customs associated with* (h) *obtaining wives,* (i) *he was curious about (these things),*

In (g–h) the discontinuous phrase *bin . . . tawpal,* literally 'wife buy' is modified by the conventional clause-stem *ti ti g-* 'what what do', that is, 'do the customary things'. The whole clause can be translated 'the customs associated with obtaining wives/paying bridewealth'. In (i) the verb *ag-,* which normally means 'to say, speak, make a sound', refers to inner thoughts—desires, decisions, wondering, being curious—things that, as it were, a person talks about to himself. Here *agl* may be translated 'having become curious about/interested in' all the activities mentioned in the preceding clauses.

(j) *the customs associated with warfare and other activities,* (k) *in their way of life,* (l) *he was interested in (these things),*

In (j), *am ñagep pakep g-* 'go shooting hitting do' is a lexicalized clause stem trans-latable as 'go to war', 'make war', or 'fight with lethal weapons'. In (k) *md-eb* 'live-nominalizer', that is, 'existence, living' refers to the manner of existence or way of life specified by the preceding clause. In (l) *agl* 'having wondered about' refers to being interested in the matters previously mentioned.

(m) *the trading of game mammals, feathers and green-snail shells and dog-whelk shells,*

The literal sense of (m) is 'the things people go giving each other: game mammals (i.e., furs and meat), birds (i.e., plumes), green-snail shells, dog-whelk shells.' This grouping of four names is a conventional expression which identifies a category 'valuable trade goods' by listing its most salient members.

(n) *(and the directions of trade such as) from Kaironk to Asai,* (o) *Kaironk to Simbai, and Kaironk to the Maring,*

These two clauses are straightforward.

(p) *whatever they did* (q) *in their lives,*

(p) and (q) consist of a formulaic expression which can be translated as above, or simply 'their various customary activities'.

(r) *having become interested,* (s) *these were the things he went and studied.*

(r–s) is a coda referring back to all the preceding clauses in the sentence. In (r) the generic verb of sound-making, *ag-*, again has the sense of 'wonder about, be inter-ested in', and in (s) the generic verb of cognition, *ng-* 'perceive, be aware, know', has the sense of 'investigate, study'.

CONCLUSION

Plainly, to make sense of discourse in Kalam, as in any other language, one needs to know the concepts that are common currency in the speech community—the things that people talk about—and the standard ways of expressing these. In Kalam many of the conventional concepts are complex schemas. They represent not single categories of objects and actions, but situations and events, and often sequences of events, and their standard expressions are not single words but phrases, clauses, and sequences of clauses. One also needs to know what concep-tual components are implied but conventionally omitted by the Kalam when they construct discourse. Of course these kinds of requirements are not unique to Kalam. A Kalam, Japanese, or Eskimo learning English must learn many concep-tual schemas and ways of expressing these that are conventional in English but alien to his or her own language. It is perhaps because dictionaries were first con-structed for translating between rather similar language-culture systems that lexicographers have for so long concentrated on words and paid so little attention to more complex conventional expressions.

There is no space here for a fuller account of subject matter languages but I hope I have persuaded the reader that there are other ways of viewing language besides the grammar-lexicon model of language that dominates linguistics. In fact, there is a kind of irony in the fact that some scholars should be locked into a certain view of language while remaining skeptical of the proposition that language shapes thought. We need look no further than linguistics itself, or any other science, to find illustrations of the way people construct worldviews by developing a particular set of ways of talking about a subject matter.

ACKNOWLEDGMENTS

Thanks are due to the Wenner-Gren Foundation for Anthropological Research, the New Zealand Research Grants Committee and the University of Auckland who have given me grants-in-aid for the Kalam project, and to the Papua New Guinea Biological Foundation for supporting ethnobotanical research in the Schrader Range by Ian Saem Majnep and Rhys Gardner. I am grateful to the present and former Directors of the Christensen Research Institute, Madang, Matthew Jebb and Larry Orsack, for providing a support base for Saem Majnep during his visits to Madang. I also wish to thank Nikolaus Himmelmann and Luisa Maffi for valuable comments on a draft of this paper.

NOTES

1. A third important question is: What sort of professional training and what kinds and scale of research projects are needed in order to achieve appropriately rich descriptions of languages? This question cannot be addressed here for reasons of space. Some of the relevant issues are dealt with elsewhere in this volume (e.g., in Maffi's and Moore's chapters; see also Maffi 2000).

2. See, for example, the papers in *Language* 68(1) and the responses from Ladefoged (1992) and Dorian (1993). Since writing this paper I have come across a paper by Woodbury (1994), which treats some of the same issues that I tackle here.

3. Here I draw on ideas developed in Grace (1987), which in turn are influenced by the writings of Hymes (1974) on ways of speaking.

4. But we should be careful not to reify our models. In this instance, as Grace (1981) points out, we should not equate "languages" with their grammar-lexicon descriptions.

5. Conventionalization is a matter of degree. I leave aside here the question of how one establishes the "formulaic" status of an expression. The criteria are of the same general kind as are used to assess the lexical status of complex words (Pawley 1986).

6. X, Y, and Z stand for variable expressions that fit the context; NP stands for a noun phrase, such as *the President* or *the old man;* PRO stands for a personal pronoun; and the subscripts i and j indicate whether two (or more) nominals refer to the same or a different entity.

7. If required to justify the free translation into English that has been made for each of the short extracts from Majnep's texts, one would need to give detailed arguments, just as grammarians may need to write densely argued papers to justify one grammatical analysis of a class of sentences over its competitors.

REFERENCES

Bernard, H.R. 1991. Preserving language diversity. *Human Organization* 51(1):82–89.

Bulmer, R.N.H. 1971. Science, ethnoscience, and education. *Papua New Guinea Journal of Education* 7(1):22–33.

Carroll, J.B., ed. 1956. *Language, Thought, and Reality: Selected Writings of Benjamin Lee Whorf.* New York and London: Technology Press of MIT and John Wiley.

Conklin, H. 1962. Lexicographical treatment of folk taxonomies. In *Problems in Lexicography*, ed. F.W. Householder and S. Saporta. Pp. 23–81. Bloomington: Indiana University Research Center in Anthropology, Folk Lore, and Linguistics.

Diamond, J. 1993. Speaking with a single tongue. *Discover* 14(2):78–85.

Dorian, N.C. 1993. A response to Ladefoged's other view of endangered languages. *Language* 69(3):575–579.

Givón, T. 1990. Verb serialization in Tok Pisin and Kalam: A comparative study of temporal packaging. *Studies in Language* 14:19–55.

Goffman, E. 1981. *Forms of Talk.* Philadelphia: University of Pennsylvania Press.

Grace, G. 1981. *An Essay on Language.* Columbia, N.C.: Hornbeam Press.

Grace, G. 1987. *The Linguistic Construction of Reality.* London: Croom Helm.

Haiman, J. 1980. Dictionaries and encyclopaedias. *Journal of Pragmatics* 50:329–357.

Hale, K. 1992. Language endangerment and the human value of linguistic diversity. *Language* 68(1):35–42.

Hymes, D. 1974. *Foundations in Sociolinguistics: An Ethnographic Approach.* Philadelphia: University of Pennsylvania Press.

Kuiper, K., and D. Haggo. 1984. Livestock auctions, oral poetry, and ordinary language. *Language in Society* 13:205–234.

Kuiper, K. 1996. *Smooth Talkers: The Linguistic Performance of Auctioners and Sportscasters.* Mahwah, N.J.: Lawrence Erlbaum.

Ladefoged, P. 1992. Another view of endangered languages. *Language* 68(4):809–811.

Lane, J. 1991. Kalam Serial Verb Constructions. M.A. thesis, Department of Anthropology, University of Auckland.

Lewis, H.T. 1993. Traditional ecological knowledge: Some definitions. In *Traditional Ecological Knowledge: Wisdom for Sustainable Development,* ed. N.M. Williams and G. Baines. Pp. 8–12. Canberra: Centre for Resource and Environmental Studies, Australian National University.

Lord, A. 1960. *The Singer of Tales.* Cambridge, Mass.: Harvard University Press.

Lyons, J. 1968. *Theoretical Linguistics.* Cambridge: Cambridge University Press.

Maffi, L. 2000. Language preservation vs. language maintenance and revitalization: Assessing concepts, approaches, and implications for the language sciences. *International Journal of the Sociology of Language* 142:175–190.

Majnep, I.S. 1991. What is this man up to? A Kalam view of Ralph Bulmer. In *Man and a Half: Essays in Pacific Anthropology and Ethnobiology in Honour of Ralph Bulmer,* ed. A. Pawley. Pp. 29–36. Auckland: Polynesian Society.

Majnep, I.S., and R.N.H. Bulmer. 1977. *Birds of My Kalam Country.* Auckland and Oxford: Auckland University Press and Oxford University Press.

Majnep, I.S., and R.N.H. Bulmer. 1990. *Kalam Hunting Traditions,* vols. 1–6, ed. A. Pawley. Department of Anthropology Working Papers nos. 85–90, University of Auckland.

Majnep, I.S., and R.N.H. Bulmer. n.d. *Kalam Hunting Traditions,* vols. 7–12. Printout. Department of Linguistics, Research School of Pacific and Asian Studies, Australian National University.

Pawley, A. 1986. Lexicalization. In *1985 Georgetown Round Table in Linguistics,* ed. D. Tannen and J. Alatis. Pp. 98–120. Washington, D.C.: Georgetown University Press.

Pawley, A. 1987. Encoding events in Kalam and English: Different logics for reporting experience. In *Coherence and Grounding in Discourse,* ed. R. Tomlin. Pp. 329–360. Amsterdam and Philadelphia: Benjamins.

Pawley, A. 1991. Saying things in Kalam: Reflections on language and translation. In *Man and a Half: Essays in Pacific Anthropology and Ethnobiology in Honour of Ralph Bulmer,* ed. A. Pawley. Pp. 432–444. Auckland: Polynesian Society.

Pawley, A. 1996. Grammarian's lexicon and lexicographers' lexicon: Worlds apart. In *Words: An International Symposium,* ed. Jan Svartvik. *KVHAA Konferenser* 36:189–211.

Pawley, A., and R.N.H. Bulmer, with I.S. Majnep, J. Kias, and S.P. Gi. In press. *A Dictionary of the Kalam Language of Papua New Guinea, with Ethnographic Notes.* Canberra: Pacific Linguistics.

Pawley, A., and F. Syder. 1983. Two puzzles for linguistic theory: Nativelike selection and nativelike fluency. In *Language and Communication,* ed. J. Richards and R. Schmdit. Pp. 191–225. London and New York: Longman.

Schrempp, G. 1992. *Magical Arrows: The Maori, the Greeks, and the Folklore of the Universe.* Madison: University of Wisconsin Press.

Steiner, G. 1992. *After Babel: Aspects of Language and Translation.* 2d ed. Oxford and New York: Oxford University Press.

Wierzbicka, A. 1992. Back to definitions: Cognition, semantics, and lexicography. *Lexicographica* 8:146–174.

Woodbury, A.C. 1994. A defense of the proposition, "When a language dies a culture dies." *SALSA I, Proceedings of the First Annual Symposium about Language and Society-Austin,* ed. R. Queen and R. Barrett. *Texas Linguistic Forum* 33:102–130.

Yaghi, H. 1994. A psycholinguistic model for simultaneous translation and proficiency assessment by automated acoustic analysis of discourse. Ph.D. diss., Department of Anthropology, University of Auckland.

I4

LINGUISTIC DIVERSITY AND BIODIVERSITY

Some Implications for the Language Sciences

Jeffrey Wollock

The cause of the environmental crisis is not industrial and military pollution, excessive resource extraction and harvesting, or an economic system that maximizes energy use, distorts local economic priorities, and spurs the growth of huge urban slums. These are only symptoms. *The real cause of the environmental crisis is a particular way of thinking.* The state of the world's environment is, as it were, *experimental proof* that there is something fundamentally wrong with this way of thinking, today strongly reflected in most of the world's dominant languages.

There is nothing eternally necessary about this, nor is it necessarily due to anything inherent in the *structure* of these languages. English, Spanish, French, German, Japanese, Chinese, Russian, and others all possess vast resources of expression that are now excluded from serious discourse when it comes to the major decisions that affect the lives of millions. No, the problem lies in the concrete historical evolution of rhetoric in these languages and the present expression, in them, of destructive ways of thinking (or not thinking) that guide the decisive actions of the day.[1]

LANGUAGE AND THE ENVIRONMENTAL CRISIS

A given language channels the mind in certain directions; it draws the attention in certain ways, in other ways distracts, obfuscates, and ignores. We may say it "creates a world," as long as we understand that this is simply a metaphor. What a language actually does is select from the available features of this one world and give the items in this selection certain unique slants and emotional colorations. As a speaker of a particular language, one is simply more ready to say certain things in a certain way and, indeed, to notice certain things. Among these things are the phenomena of the natural world.

Most of the world's traditional ecological knowledge, in terms of both management practices and cosmologies, is accessible only in the languages of traditional peoples who still maintain, or have until recently maintained, a sustainable way of life. It is used and preserved because it is the basis of their economic and spiritual lives. Traditional cultures have not only been built up around biodiversity, they have also helped to shape it—through technology.

If the dominant forms of modern Western technology were truly normative, the technologies of all other cultures could only be immature or privative states on this same path of development. Western technology, however, is not the natural human path of development that all others must and one day will follow. There are different technologies because different cultures have different philosophies of life.[2] It is ethnocentric to assume that, to the extent that a culture has not shaped its environment as we have, it is merely because they did not know how to.[3] This ignores the basic psychological fact that desire comes before achievement.

Each of these world-views, including a view of the relationship between humans and their environment, is institutionalized in the language used by that culture. To some extent (as claimed by the Sapir-Whorf hypothesis), this may have to do with the grammatical structure of the language (cf. Mühlhäusler 1995:160), but more obviously it is rooted in the knowledge and attitudes stored up as the content of the language and its usages, which Mühlhäusler (1995:156) refers to as "memory," and which directly influences what people are expected to think, feel, and do in relation to the environment. "Life in a particular human environment is dependent on people's ability to talk about it" (Mühlhäusler 1995:155). "Communication occurs in places, cultivates intelligible senses of those places, and thus naturally guides ways of living within them" (Carbaugh 1996:38).

Thus, *all* systems of communication, whether they are about "nature" or not—"because they occur in natural spaces [and] naturally create ways of living in those places (bodies included)"—"are affected by and carry real consequences for those places" (Carbaugh 1996:39–40). Anderson (1986:186–188) makes a related point: the study of folk natural history is as applicable to large-scale industrial society as it is to those societies that live closer to nature, not only because the discontinuous attributes (bundles of traits) of plants and animals remain the same, but also because a given culture's apathy to and explicit disinterest in particular aspects of nature are no less significant than its fascination with and expertise in others.

COLONIZING CULTURES AND THEIR EFFECT ON BIOLOGICAL
AND LINGUISTIC DIVERSITY

In developing new varieties of plants and animals, colonizing cultures do not follow the same philosophy as indigenous ones. Instead of maximum niche-diversification, they seek to exploit a few particular advantages to the maximum, altering the environment to suit a small number of uniform, profitable, high-yield biotechnological "products." The resulting systems are characterized by very high energy input and low biodiversity. In effect, such cultures trade diversity for the "big payoff" of one particular crop or livestock variety (Wollock 1994:64–65; cf. 53–55). But they are also vulnerable; they "put all their eggs in one basket." This kind of philosophy too is encoded in "what one says about things" in the languages of the particular colonizing cultures that follow it.

In the history of Western thought, such a trivialization of all other members of the ecosystem, and of the rights of traditional cultures, correlates with the gradual triumph of the nominalist strain in philosophy, which treats all universal concepts (including "nature" and "community") as arbitrary social constructs. Certain universals, of course, really are arbitrary constructs, but when these are chartered, institutionalized, and recognized by law in "developed" societies, they immediately take precedence over the the natural ones, which are considered to be no less arbitrary, but to be supported by nothing more than "mumbo-jumbo." Such diverse vernacular institutions, while they have no official sanction, are embodied in vernacular languages and dialects.

The development of nominalism in Western culture since the fourteenth century correlates closely with the triumph of revived Roman law over traditional law and with the development of nation-states and empires. Customary law was based on local ways of life, memories, and obligations; Roman law, on a centralized system of interpretation and enforcement monopolized by a professional class of lawyers and judges who together constructed a self-referential "text" that was authoritative but had no real connection with the experience of the community (Penty 1920:63–70; Vinogradoff 1929), yet to which more and more communities had to conform.

As late as the fifteenth century, however, it was still normal for nations to take pride in linguistic diversity. As Aston (1968:41–42) writes:

> The more languages, the more kings and kingdoms, so (to older ways of thinking) the more glory. It was a matter of congratulation to the English, concerned to justify the multiplicity of their "nation," as comparable to that of the French (which included Provence, Savoy, and much of Lorraine), that they could claim to possess the distinct languages of Welsh, Irish, Gascon and Cornish, "no one of which is understood by the rest."

As the centralizing tendencies implicit in making one particular language the official instrument of state began to develop, it came to be seen that alternatives would henceforth be a threat or a nuisance to the smoother running of the system. The end of the fifteenth century saw one of the first manifestations of this new consciousness. As Williams (1990:74) writes:

> Perhaps no single historical incident better illustrates the transformations occurring throughout Discovery-era Spain than Queen Isabella's acceptance of Antonio de Nebrija's Spanish *Gramática,* the first-ever grammar of any modern European language. Upon its presentation in the momentous year 1492, Isabella reportedly asked the scholar, "What is it for?" Nebrija answered Her Majesty modestly but with profound presence and insight respecting the demands of the new expansion-minded age. "Language," he reportedly stated, "is the perfect instrument of empire."[4]

There is an intimate correlation between vernacular activities, values, and language. As Illich writes:

> Traditional cultures . . . subsisted basically on vernacular values. In such societies, tools were essentially the prolongation of arms, fingers, and legs. . . . Language was drawn by each one from the cultural environment. . . . The vernacular spread just as most things and services were shared, namely, by multiple forms of mutual reciprocity . . . speech resulted from conversations embedded in everyday life, from listening to fights and lullabies, gossip, stories and dream. Even today, most people in poor countries learn all their language skills . . . in a way that nowhere compares with the self-conscious, self-important, colourless mumbling that, after a long stay in villages in South America and South-East Asia, always shocks me when I visit an American college. I feel sorrow for those students whom education has made tone deaf; they have lost the faculty for hearing the difference between the desiccated utterance of standard television English and the living speech of the unschooled. (Illich 1981:32)

Of course when languages are lost, cultural actions are never simply left to take care of themselves, or vice versa. That a language is lost necessarily means it is replaced by another one. The new language embodies another discourse, in terms of which the actions that had formerly upheld the land-management regime that maintained the traditional biodiversity may no longer make sense. If these actions and traditions are already being lost and replaced by others, it is all the more likely that the new language will be accepted. If the land is lost, the traditional actions cannot be performed, and the language itself loses much of its raison d'être. Either way (usually both processes are going on at more or less the same time), the people are left with a new language and little of their old culture. The wisdom that kept up the sustainability of the environment was encoded in that old language; in its entire, concrete usage, its proverbs, its thought patterns, its metaphors. If the language no longer exists, most of that wisdom is lost as well.

If these processes are less extensive and if they occur gradually enough, as tended to be the case in ancient times or in the early Middle Ages, people are often able to adapt to the new conditions (such as Christianity, a local market economy) and to calque a great deal of the new meanings into their own language. Conversely, if their language is lost, they may be able to calque most of its meanings into the new language; where the new language has "prestige," this process may well be voluntary. But if the change is forced and rapid, as when people suddenly lose jurisdiction over their land, or are even relocated, then culture, language, and biodiversity are all threatened.

Under such drastically altered conditions (including overpopulation, plantation monocropping on best lands, increased peasant reliance on marginal lands, etc.), the continuance of traditional activities can actually increase environmental stress, such as is seen today in the severe deforestation and advancing desertification in Africa. But it is not the traditional systems that are at fault.

The process of driving local cultural groups off their lands or at least divesting them of jurisdiction is characteristic of nineteenth- and twentieth-century imperialism. Ancient forms of imperialism demanded tribute and obedience of subject peoples but often left them their cultures and land regimes, and thus their languages. Modern imperialism takes almost everything, especially from tribal peoples, whose traditional land rights are usually nullified (through the nominalism maneuver referred to above). A uniform language is imposed and, like everything else in the new system, becomes a global commodity, which immediately puts those who do not speak it, or who speak it badly, at a grave disadvantage. The new language embodies a system of values that conceives of land and nature as arbitrary signs, objects of domination and profit, a discourse in which maintenance of biodiversity (which it equates with forgoing monetary profit) is equated with "waste." (The classic statement of this theory is John Locke's *Second Treatise on Government*, chap. 5; see Locke 1965.) As this language extends itself through compulsory education and mass media, both linguistic diversity and biodiversity decrease, leading to a condition of not only cultural but also biological simplification, which is extremely unstable and dangerous on both levels. Food production is no longer for the local economy but for "the empire"; culture is no longer vernacular but is imported and imposed. Societies are destabilized; religions become apocalyptic. Disconnected from the land, cultures homogenize and no longer uphold biodiversity.

The destruction of biodiversity and linguistic diversity have the same cause, then. Dussel (1995:32 n. 45, 35–36, 39) describes it as "the colonizer subsuming the Other as an aspect of the Same," that is, the colonized culture is viewed according to the categories of the colonizer. The mind of the colonized people is seen as a tabula rasa. What stands out is the fact that they do not speak the language of the colonizers, not that they do speak something else. If not deliberately exterminated, they are treated, at best, as wards, incompetents: until such time as they will lose their diversity and become the Same, they cannot be fit into the picture.

Their land, similarly, is a tabula rasa. Though they have shaped it according to their minds and languages, to the colonizer it is simply the potential for more of the Same (shoot all the buffalo and bring in the cows). The combined effect of indigenous peoples and a new land is too powerful not to exert an influence on the colonizer, and he soon begins to differ from his former countrymen in the Old Country; but basic categories of thought are remarkably stable, probably just as stable as the language, and in many of its characteristics the new culture only exaggerates negative tendencies that the traditions of the Old Country had kept in check.

The cause of cultural and biological destruction, then, is not migration itself, but the colonizing attitude that sometimes (but not necessarily) accompanies it, the essence of which is seeing the Other as an aspect of the Same. In the colonization of the Americas, this resulted in an unprecedented egoism (Dussel 1995: 42–47) that flowed back to the European psyche and would culminate in the birth of the Cartesian *ego cogito,* the philosophical birth of the specifically modern consciousness that separates mind from body and language from nature (1995:48; cf. Wollock 1997:xviii–xx, xxix–xxxviii).

As scientists have lately become interested in preserving biodiversity, so multinational pharmaceutical and biotechnology firms are eager to exploit it. Collecting, cataloguing, and staking a claim (i.e., pharmaceutical or genetic patenting), now referred to as bioprospecting, are being actively pursued. As information is now the key to profit, traditional knowledge regarded only a few years ago as "primitive mumbo-jumbo" is suddenly valuable and being sought by multinational pharmaceutical and biotechnology firms. Similarly, it has become important to convince multilateral loan organizations that traditional land-management systems are economically superior to the centrally dictated programs they usually favor. But the accumulated ecological knowledge that supports these systems is also a treasure chest of biological information, and language is the key to it. From this global perspective, linguistic diversity is starting to be perceived as a benefit, but *cui bono?*—for whom? Is the main goal to translate all this diversity into the one global language of science and international commerce, for the benefit of those who control it? Is linguistic diversity yet another consumable and ultimately expendable commodity for the global information industry? (Cf. Lohmann 1993.)

HOW LINGUISTIC THEORY SEES NATURE

If biodiversity and traditional cultures are mutually supporting, while colonizing cultures tend to destroy both linguistic and biological diversity, it is not surprising to find that the form of linguistic theory most favored by today's colonizing cultures has little to say about linguistic diversity (cf. Mühlhäusler 1995:154; also this volume) and either ignores or denies any connection between language and the real world.

The canonic formulation comes from Saussure's *Cours de Linguistique Générale*

(1972). In one passage, Saussure admits that linguistic signs show some trace of a natural connection with what is signified, but even if semiology should come to include modes of expression based on natural signs (pantomime gestures, for example), this would not change the fact that its main object will still be *l'arbitraire du signe*. Every social means of expression is based in principle on a collective habit or convention; those that are based on a certain natural expressivity are nonetheless fixed by rule, and it is this rule, not their intrinsic value, that requires them. (See the discussion in Norrman and Haarberg 1980:154–156.)

This is now taken as dogma by most writers on language. A typical statement is that of Bickerton (1969:38): "Meaning exists, if anywhere, only in the relationship speaker-language-hearer, not in any one of the three, and least of all in any connection with the extralinguistic universe."

Similarly, Michael Oakeshott:

Human beings are . . . composed entirely of beliefs about themselves and about the world they inhabit . . . a world composed, not of physical objects, but of occurrences which have meanings and are recognized in manners to which there are alternatives. Their contingent situations in this world are, therefore, what they understand them to be, and they respond to them by choosing to say this rather than that in relation to imagined and wished-for outcomes.

A human life is composed of performances, and each performance is a disclosure of man's beliefs about himself and the world. . . . He is what he becomes; he has a history but no "nature."

The wished-for satisfactions of human beings lie, for the most part, in the responses their utterances and actions receive from others, responses which are themselves utterances and actions related to the wished-for satisfactions of those who make them. (Oakeshott 1989:64)

According to this view, the natural world provides no meaningful input either to the speaker, the hearer, or the language. We are stuck within a self-referential semantic and pragmatic system with no touchstone to any reality outside of it. Being structurally oriented, such a theory deals with languages primarily as autonomous, self-contained systems—grammars—and when it extends to semantics, this is still a self-contained, self-referential system, a "text" (cf. Pawley this volume). Even pragmatics, the study of how linguistic systems are related to their users, remains in essence a dialectically oriented grammar of "imagined and wished-for outcomes." It does not get us beyond a "text."

Obsessed with the conception of language as a purely arbitrary system of signs, modern linguistic theory cannot accept the idea that there are really-existing systems outside of language and outside of humans, to which humans, and therefore language, bear a definite relation. Nature, if considered at all, is considered only as an abstraction, a sociolinguistic construct. To the extent that post-Cartesian linguistic thought is interested in the causes of language diversification, it forms either structural, internal hypotheses or sociopragmatic ones. It can talk about nature only in terms of subjective or socially created beliefs and desires. Here again, we see the influence of nominalism, which in the West has had a long and intimate connection with language studies (Ong 1983:145–147).

How can a culture that holds such a theory of language explain the connections between linguistic diversity and biodiversity? Biodiversity is the manifestation of the health of a mature natural ecosystem. Because language is the main guide to action, a theory of linguistics that regards social construction as the exclusive source of meaning is in itself, I believe, a threat to biodiversity. For it gives no theoretical support to the idea that the value of biodiversity is any less arbitrary than that of, say, diversity of designer jeans.

NATURE AND LANGUAGE

It is only when we shift our perspective from language as grammar (no matter how universal, generative, or otherwise), used on its own by human or artificial intelligence, to language as a guiding pattern of human action, and in itself an action of human beings with bodies as well as minds and connected with other actions within the social and natural world (cf. Wollock 1997), that it becomes possible to talk adequately about how linguistic diversity is related to biodiversity.

Many linguistic constructions of nature are possible, but some are more appropriate to nature than others. As Mühlhäusler (this volume) notes "Diversity . . . reflects neither regressive compartmentalization nor 'progress' in the Western sense, but a large number of progresses (as well as misreadings). Prehistory is full of examples of languages and cultures dying out as a result of misreading their environment." Language plays the key role in "misreading the environment." An inappropriate linguistic construct of nature will lead to inappropriate actions, like deforestation. In some cases, deforestation will naturally lead to drought. When nature yields no water, humans and animals die of thirst. *Dying of thirst is not a linguistic construct.* So a more accurate conception of the relation between nature and culture would be that of a *dialogue* between environmental nature and human nature resulting in a particular culture and a particular language that embodies the common historical experience of that community interacting in a sustainable or unsustainable way with that environment.

Linguistics connects with nature most directly through such interrelated disciplines as:

- Folk classification and ethnoscience. Cognitive ethnologists have discovered certain basic regularities throughout folk classification around the world that seem to be rooted in perceptual-cognitive regularities (e.g., Anderson 1986).
- "Ethnophysical nomenclature," that is, place-names, regional names, and

other "expressive forms that people use to render intelligible . . . [what Burke calls] 'the sheerly natural'" (Carbaugh 1996:43).

- The study of metaphor[5] as a fundamental aspect of language (Danesi 1993: 122–142). Loewenberg (1975) shows that metaphor often *requires* a natural (nonlinguistic) context for its interpretation (i.e., the recognition that it is *not* meant literally as well as the understanding of what *is* meant).
- The study of phonetic symbolism and onomatopoeia.
- The psychophysiological study of relations between perception, imagination, memory, and human action-signs, including speech (Wollock 1997).

HOW LANGUAGE AFFECTS NATURE

The positive correlation of linguistic diversity and biodiversity strongly suggests mutual influence between language and the physical environment. But because linguistic theory of the currently most accepted kind has not admitted this concept, it has not come over into natural history or geography either. For example, Gourou (1974:161, as translated in Breton 1991:20), writes:

> Languages themselves do not influence the geographic landscape (unless one wishes to talk about the geography of sound!). They are not responsible for the effectiveness, or the lack of it, of one group or another. All languages are equally capable (or equally incapable!) instruments for assuring communication; a particular language does not give one group superiority over another. The most anciently attested languages are just as complete and not less complex than those of today. Languages wrongly thought of as primitive in fact have a very highly developed degree of complexity. There is no hierarchy among languages. All languages have the same value.

This argument is instructive in how completely it misses the point. True, "languages themselves do not influence the geographic landscape," if we are speaking of languages purely as grammatical structures. But then, "languages themselves" do not influence much of anything, with the possible exception of linguists. If this were an adequate consideration of the geographic effects of language, it would be difficult to imagine why such a thing as language ever came into existence in the first place. The only perspective from which it makes any sense to ask these questions at all is that of the functioning of language in society.

To answer the argument, then: all languages are equally complex and equally effective tools of communication among their own speakers, true, and they are all theoretically intertranslatable. In reality, however, communication in a given geographic area is greatly affected by the fact that more than one language is spoken. Within a given multilingual social context, there is almost always some kind of hierarchy among the languages, mirroring social stratification, semantic registers, or at the very least, facility of use. Furthermore, ability to speak a particular language can certainly give one group social superiority over another. Finally, "languages themselves" may not influence the geographic landscape, but all languages in use are repositories of cultural memory and guides to action, and in this sense they certainly do influence the landscape, and notably, its biodiversity. These topics lie in the borderlands between sociolinguistics and geolinguistics.

Breton (1991:24) notes that geographers are quite accustomed to correlating sociohistorical and economic data with spatial data, and "especially within physical geography, the biogeographic milieu or biosphere." If he does not want to go "so far as to speak of an actual ecology of language," it is because he is reluctant to abandon the dogma that "language does not exert its influence directly on the physical world." Nevertheless, within this constraint, he manages to bring up the very problem we are now examining.

> Geographers are in a favorable position to establish how a cultural trait—language—has succeeded in spreading, maintaining itself, establishing itself in a given social setting (or even several) in a way that is the same or different from other cultural social, or economic characteristics. . . . South India, for example, which is divided into four rather different biogeographic zones, has . . . in each of them . . . a different Dravidian language: Malayalam on the narrow western coast "facing" the monsoon; Tamil in the drier plains of the east coast (domain of the winter monsoon); Kannada on the relatively cool plateaus of the northwest, in the rain shadow of the western Ghats; and Telugu on the warmer plains, hills and basins of the northeastern part of the Deccan peninsula. It is even possible to note a distribution of languages that parallels certain species of palm trees, without, of course, carrying the connection too far. (Breton 1991:24–25)

Let me "carry the connection too far," then, and suggest that these languages have to no small extent been shaped by the environments in which their speakers live and in turn have guided the shaping of environments by those who live in them. The Soviet geologist V. I. Vernadsky, who originated the concept of the biosphere, also conceptualized within it "a great geological, maybe even cosmic, force whose planetary action is usually not taken into account in conceptions of the cosmos. . . . This force is human intelligence, man's purposeful and organised will as a social being." Vernadsky termed the sum of changes wrought by human intelligence on the surface of the earth, the "noosphere" (as quoted in Novik 1968:64). Breton himself (1991:22) calls the geography of language "a geography of the human spirit, a . . . geography of the noosphere."[6]

UNITY AND DIVERSITY

Language is a motion of the soul, the noblest animate motion within the created universe, and both language and the world are elaborations of the same thing, which used to be called the Logos. Just as every people has a language, so every metaphysical tradition has such a concept, because the speech process is itself a fundamental natural symbol.

To speak in the language of Western tradition: After the disequilibrium caused by Adam's fall, the primordial unity of language remained, but it became a temptation in itself, fostering a universalist mono-mentality inspiring humans to rival God with a technological conquest of nature symbolized by the attempt to build the Tower of Babel (Genesis 11:1–9). The "confusion of tongues" that resulted from this failed attempt was not some arbitrary punishment, but a necessary corrective redirecting humans toward the endless diversity that is the reality of the planet. Yet, just as each shard of a broken mirror reflects the sun, so each language reflects a version of the primordial wisdom adequate to its own environment. Today it is interesting to see "technological man," as he again attempts to build a tower to rival the Creator, also seeking to regain that same unity of language.

Even through its "green globalism," the centralized New World Order, by attempting to integrate local communities "within large-scale economies and forms of knowledge," "undermines precisely the values it feigns to support," because it is only the existence of distinct enclaves and boundaries, "including the variations, not only in topography and natural history, but also in human history and culture, between one place and the next," that preserves biodiversity (Lohmann 1993:158–159).

Pretending to have transcended the fall and to have regained a quasi-angelic mode of perception, modern man surveys Creation with a mind that is, in reality, more arrogant than ever and artificially creates a single artificial language, ignoring the one primordial language that still lies deep within each "natural" language.

One of the most characteristic intellectual trends of the last 25 years, but especially the last decade, is what is often referred to under the rubric of "postmodernism." Essentially a response toward the continuing centralization of the world, postmodernism takes the form of a reaction against any unifying overarching system of meaning whatsoever. In this perspective, there is only diversity—unity is an illusion. Because of its emphasis on diversity, opposition to power structures, and the like, many environmental thinkers have been attracted to one or another form of postmodern thought (cf. Zimmerman 1994:91–122). I strongly agree, however, with Spretnak (1997, chap. 2) that deconstructive postmodernism is a symptom of, not a remedy for, the modern condition and in fact intensifies rather than counteracts the trends within modernism that caused the problems in the first place. Above all, it reflects the same lack of metaphysical perspective, absence of a sense of "levels" of being. Deconstructive-postmodernist ecologists recognize the falseness of the unity being imposed on the world today, but their response has been to reject the idea of unity *tout court* (cf. Harmon this volume).

Yet it is not a social construction to understand that any natural community, including the unfathomable diversity of the universe itself, shares a root unity. On whatever level you look, it is a single system, a community of communities, an order. All its members also share fundamental analogies. Every one of the world's metaphysical traditions tells us that there is a unity within this diversity. In the language of Western tradition, St. Thomas Aquinas tells us that the universe can be considered as if it were an organic body made up of parts that, though distinct and dissimilar, are connected to and supported by one another, for the conservation of the health and beauty of all (in *II Sententiarum* 1.1.1; see St. Thomas Aquinas 1929).

Diversity is recognized by all traditional cultures, including the traditional culture that existed in the West, as a fundamental property of the natural universe. As Aquinas puts it, a universe of angels and stones is a *better* universe than one consisting solely of angels (*In I Sententiarum* 44.1.2.6), because "the perfection of the universe is obtained essentially through a diversification of natures, which natures, so diversified, fill the various ranks of goodness. It is not obtained through the multiplication of the individuals in any particular one of these given natures" (ibid. 44.1.2.5).

St. Thomas gives the explanation in *Summa Contra Gentiles* III.97 (see St. Thomas Aquinas 1958): Every created substance reflects the divine goodness in some way; yet at the same time every created substance is necessarily deficient in the perfection of divine goodness. Therefore, the greatest diversity was necessary in order that the likeness of divine goodness could be more perfectly communicated. Since it could not be perfectly represented by any one of them, it would be more perfectly represented by diverse things in diverse ways. So Aquinas compares the Creation of the universe (cf. John 3:14: "In the beginning was the word, and the word was with God") to a kind of speech addressed by God to His rational creatures. Just as humans, when they cannot sufficiently express their thoughts in a word or two, multiply them in diverse and various ways to express their minds, so God used the multitude and variety of things to make more expressly clear His infinity and, at the same time, through their order, His unity. Thus Aquinas.

What is wrong with the present system is the attempt to transpose a pseudo-metaphysical, artificially imposed unity on a single global economic system led by "men of power." Not unity, but the attempt to impose unity on the wrong level, is the danger—a unity that is not built on, indeed is destructive of, maximum diversity. Like the modernists, the postmodernists cannot conceive of the coexistence of maximum diversity and unity. Like the modernists, the postmodernists reject the unity of nature, recognizing (if deploring) the existence only of arbitrary, socially constructed, pseudo-unities. Yet it is precisely in nature that we can best see the coexistence of diversity and unity, because the order of nature is the reflection, as Aquinas also said, of the supreme Unity, which, in creation, could be best reflected only through the perfect combination of maximum diversity within unity.

NOTES

1. On how contemporary Anglo-American ways of talking about the environment constrain effective awareness of problems and solutions, see Meadows Summer 1990; Bowers 1997.

2. Milton (1996:31–32, 109–114, 201–204) rightly criticizes the romantic myth that "all non-industrial ways of living" are by definition "ecologically sound." The present paper should be understood to refer, by definition, to those that *are* ecologically sound, to the extent that they are. As Mühlhäusler notes (1995:160): "there are numerous societies that have managed to ruin their natural environment with very basic technology." For that matter, it is not excluded that some sort of industrial or semi-industrial form of culture can be ecologically sound, or, with very efficient use of energy and strong pollution controls, at least a great deal more so than the present one.

3. This assumption is implied, to take one example, in the continuing discussion (largely inspired by the work of Joseph Needham) over why modern science developed in the West rather than in Muslim lands or in China, when the latter were, by the Middle Ages, far more advanced in natural philosophy and technology than the West.

4. This point is explored in greater detail by Illich (1981).

5. I agree with Lakoff (1987) and the other constructivists that metaphor is central to language. I do not agree, however, that this centrality proves the world is "constructed" by language. The centrality of metaphor, it seems to me, proves rather that language is constructed by creative imagination, which itself operates with images drawn from and rooted in nature.

   Metaphors *refer to* things in nature, they are not *definitive of* those things. In naming things in nature, they illuminate the unfamiliar by transferring meanings from what is already familiar. Take for example the word *sea-lion*. Whoever first gave the animal this name was unfamiliar with it, but noticed that it was a sharp-toothed predator that roared something like a lion and lived in the sea. The name *refers* to things that really exist in nature, *lion* and *sea*. Even though the animal is not literally a lion, it bears some natural resemblance to that animal. This resemblance is discerned through a natural psychological process in the imagination. By the same process, a mythographer might speak of a "sky-lion," and here the word does not refer to nature. This is still, however, a natural symbol, because both the sky and lions do exist in nature, and the symbolic qualities of the sky and the lion are transferred to this symbolic creature. Systems of natural metaphors may interpret nature adequately, or they may amount to parody. In either case it is from nature that they derive their power over the mind, but in the first, language will have a beneficial effect on nature, in the second, a harmful effect.

6. Of course, there are many other components to the geography of the noosphere, but my point is that language is one of them. It might also be useful to have a term for the sum of changes wrought by *human unintelligence* on the surface (and atmosphere) of the earth: we might perhaps call it the *morosphere*.

REFERENCES

Anderson, M. 1986. Folk natural history: Crossroads of language, culture, and environment. In *The Real-World Linguist: Linguistic Applications in the 1980s*, ed. P.C. Bjorkman and V. Raskin. Pp. 185–217. Norwood, N.J.: Ablex.

Aston, M. 1968. *The Fifteenth Century: The Prospect of Europe*. London: Thames & Hudson.

Bickerton, D. 1969. Prolegomena to a linguistic theory of metaphor. *Foundations of Language* 5(1):34–52.

Bowers, C.A. 1997. *The Culture of Denial: Why the Environmental Movement Needs a Strategy for Reforming Universities and Public Schools.* Albany: State University of New York Press.

Breton, R.J.-L. 1991. *Geolinguistics: Language Dynamics and Ethnolinguistic Geography.* Trans. H.F. Schiffman. Ottawa and Paris: University of Ottawa Press.

Carbaugh, D. 1996. Naturalizing communication and culture. In *The Symbolic Earth: Discourse and Our Creation of the Environment,* ed. J.G. Cantrill and C.L. Oravec. Pp. 38–57. Lexington: University Press of Kentucky.

Danesi, M. 1993. *Vico, Metaphor, and the Origin of Language.* Bloomington and Indianapolis: Indiana University Press.

Dussel, E. 1995. *The Invention of the Americas: Eclipse of "the Other" and the Myth of Modernity.* Trans. Michael D. Barber. New York: Continuum.

Gourou, P. 1974. *Pour une géographie humaine.* Paris: Flammarion.

Illich, I. 1981. Taught mother tongue and vernacular tongue. In *Multilingualism and Mother-Tongue Education,* by D.P. Pattanayak. Pp. 1–30. Delhi: Oxford University Press.

Lakoff, G. 1987. *Women, Fire, and Dangerous Things: What Categories Reveal about the Mind.* Chicago: Chicago University Press.

Locke, J. 1965. *Two Treatises on Government,* ed. P. Laslett. New York: New American Library.

Loewenberg, I. 1975. Identifying metaphors. *Foundations Of Language* 12(3):315–338.

Lohmann, L. 1993. Resisting green globalism. In *Global Ecology: A New Arena of Political Conflict,* ed. W. Sachs. Pp. 157–169. London and Atlantic Highlands, N.J.: Zed Books.

Meadows Summer, D.H. 1990. Changing the world through the informationsphere. *Gannett Center Journal* 4(3):49–61.

Milton, K. 1996. *Environmentalism and Cultural Theory: Exploring the Role of Anthropology in Environmental Discourse.* London and New York: Routledge.

Mühlhäusler, P. 1995. The interdependence of linguistic and biological diversity. In *The Politics of Multiculturalism in the Asia/Pacific,* ed. D. Myers. Pp. 154–161. Darwin, Australia: Northern Territory University Press.

Norrman, R., and J. Haarberg 1980. *Nature and Language: A Semiotic Study of Cucurbits in Literature.* London: Routledge & Kegan Paul.

Novik, I.B. 1968. Cybernetics and problems of cognition of the interrelationships of natural phenomena and of the transformation of nature. In *The Interaction of Sciences in the Study of the Earth.* Trans. V. Talmy. Pp. 64–76. Moscow: Progress Publishers.

Oakeshott, M. 1989. Education: The engagement and its frustration. In *The Voice of Liberal Learning: Michael Oakeshott in Education,* ed. T. Fuller. New Haven: Yale University Press.

Ong, W. 1983. *Ramus and the Decay of Dialogue.* Cambridge, Mass.: Harvard University Press.

Penty, A.J. 1920. *A Guildsman's Interpretation of History.* London: George Allen & Unwin.

Saussure, F. de 1972. *Cours de Linguistique Générale,* ed. C. Bally and A. Sechehaye with A. Riedlinger; critical edition by T. de Mauro. Paris: Payot.

Spretnak, C. 1997. *The Resurgence of the Real: Body, Nature, and Place in a Hypermodern World.* New York: Addison-Wesley.

Thomas Aquinas, St. 1929. *Scriptum super Libros Sententiarum,* ed. R.P. Mandonnet. 2 vols. Paris: P. Lethielleux.

Thomas Aquinas, St. 1958. *On the Truth of the Catholic Faith: Summa Contra Gentiles*. Garden City, N.Y.: Doubleday Image Books.

Vinogradoff, P. 1929. *Roman Law in Medieval Europe*. 2d ed. Oxford: Clarendon Press.

Williams, R.A., Jr. 1990. *The American Indian in Western Legal Thought: The Discourses of Conquest*. New York and Oxford: Oxford University Press.

Wollock, J. 1994. Globalizing corn: Technocracy and the Indian farmer. *Akwe:kon Journal* 11(2):53–66.

Wollock, J. 1997. *The Noblest Animate Motion: Speech, Physiology, and Medicine in Pre-Cartesian Linguistic Thought*. Studies in the History of the Language Sciences 83. Amsterdam: John Benjamins.

Zimmerman, M.E. 1994. *Contesting Earth's Future: Radical Ecology and Postmodernity*. Berkeley: University of California Press.

Part 2

BIOCULTURAL DIVERSITY PERSISTENCE AND LOSS

Case Studies

15

BIODIVERSITY AND LOSS OF INDIGENOUS LANGUAGES AND KNOWLEDGE IN SOUTH AMERICA

Manuel Lizarralde

Despite recent population increases [in America], most Indian cultures have become extinct or nearly so. Many of those groups that have survived remain threatened with extinction for much the same reasons as in the sixteenth century: disease, inhumanity, misguided "salvation," and racial and cultural mixing to the point of non-recognition.

(Denevan 1976:7)

Since Amazonian Indians are often the only ones who know both the properties of these plants and how they can best be utilized, their knowledge must be considered an essential component of all efforts to conserve and develop the Amazon. Failure to document this ethnobotanical lore would constitute a tremendous economic and scientific loss to the human race.

(Plotkin 1995:155)

These two quotes summarize a process that is occurring in South America, once a very bioculturally rich continent, hosting nearly one-third of the species of plants, birds, and fresh-water fish of the world, while having nearly one-fifth of the world's languages (Ritchie Key 1979; Toledo 1988; Wilson 1992; Wilcox and Duin 1995). Because of environmental destruction, the estimated rate of extinction of species is high and has not been properly documented. More than 35 percent of the continental biomes has been destroyed and so has 30 percent of the forest (Toledo 1988; Food and Agriculture Organization 1997). Since the arrival of

Columbus, 65 percent of the languages have also vanished. Today there are 422 known indigenous ethnolinguistic groups who speak their own language and practice their culture (Lizarralde 1993:10). These indigenous nations are survivors of the possibly 1200 indigenous groups that existed in South America in pre-Columbian times. Therefore, bioculturally, this continent has experienced and continues to experience a great and irretrievable loss.

My research with the Barí of Venezuela has demonstrated that indigenous people are capable of knowing virtually all the plants and animals that are perceptible with the naked eye (Lizarralde 1997). The case of the Barí ethnobiological knowledge and its changes exemplifies what is happening throughout the continent. As I will illustrate here with two maps showing the overlap of current locations of South American indigenous peoples with current biodiversity reserves (national parks and biosphere reserves), preservation of biodiversity is closely tied to cultural and linguistic preservation.

BIODIVERSITY AND LOSS OF NATURAL ENVIRONMENTS

South America is the richest continent in terms of biodiversity. The Amazon basin alone has one-third of the world's fresh-water fish species (Goulding 1989) and 30 percent of the bird species (Wilson 1992:196). South America has 20 percent of the world's mammal species (Emmons and Feer 1990:1) and approximately 100,000 of the estimated 320,000 species of plants (World Conservation Monitoring Centre 1992:27; cf. Ehrenfeld 1986:39). No other continent has more than 30,000 plant species (Heywood 1989).

The largest biome is the Amazonian rainforest, covering 5–6 million square km, and currently home to the greatest diversity of South American peoples (Meggers 1988:53; Sponsel, Bailey, and Headland 1996:4; see map 15.1). In this moist neotropical forest live 379 indigenous groups of people (87 percent of all groups in South America) with a total population of nearly 800,000. However, at least 16 percent (811,392 square km) of the original rainforest has already been lost (Repetto 1990; Wilson 1992; Moran 1996:156); some scholars argue that the loss approaches 30 percent (Sponsel 1985:78; cf. Toledo 1985). In any case, the loss represents an area larger than France (Moran 1996:156).

Other South American biomes have suffered a still worse fate. The most reduced forest is the Atlantic coastal forest of eastern Brazil, estimated to cover no more than 2–5 percent of its original extent (Mittermeier 1982:11). A similar scenario is being played out in the Pacific coastal forest of Colombia, where less than 25 percent is left standing, but this remainder is still being felled by timber companies (Wilson 1992:266). These two places were among the most diverse on earth in terms of number of species (Wilson 1992:261–266). At present, both Colombia and Venezuela have lost 50–60 percent of their natural vegetation (J.C. Centeno,

Map 15.1. South American indigenous peoples and types of environment.

personal communication 1995; Toledo 1988; cf. Huber and Alarcón 1988). In a frantic effort, developed nations are trying to protect the remaining pristine species-rich environments in biodiversity reserves. Now South America has 23 biodiversity reserves spread over 9 countries (McNeely, Harrison, and Dingwall 1994:22). The total protected areas, including biosphere reserves, parks, and forest reserves (not including indigenous reserves), cover 1,145,891 square km or 6.4 percent of the continent's surface (McNeely, Harrison, and Dingwall 1994:7). These areas will only protect a fraction of the biodiversity in the short run and may not be large enough to be sustainable for the long-term protection of all the remaining species (Lugo 1988; Laurance and Bierregaard 1997).

LOSS OF LANGUAGES

As a continent, South America has lost a great proportion of its indigenous population since the arrival of Europeans in 1492. Since not all the ethnolinguistic groups were recorded, no one knows precisely how many groups there were at the time of European conquest. Therefore, my figures may be an underestimation of the number of indigenous groups that once existed on the continent. Based on the best estimates, it is believed that there were perhaps some 1200 languages at European contact (see table 15.1). Čestmír Loukotka listed over 3000 different languages and dialects, but some of his names are synonyms for other tribes or subdivisions of single groups (Loukotka 1968, cited in Ritchie Key 1979:12). The number of South American indigenous groups (meaning ethnolinguistic populations) has been reduced to 35 percent of those at European contact (and their territory has been reduced to less than 6.8 percent of the South American continent). The number of surviving traditional languages and knowledge systems represents a fraction of these.

A large proportion of the current South American indigenous groups that maintain their cultures and languages are living in the same environments that are rich in biodiversity. Their current territories have been only slightly altered ecologically, in the sense that much of their natural flora and fauna is still found there. Many of these species are not well known to the developed world. This is one of the richest regions not only because of its natural resources, but also because of the biocultural knowledge of its indigenous peoples. The fact that the great majority of the South American indigenous peoples who still maintain their languages have an extensive knowledge of their environment is of incommensurable value to western science as well as crucial to the survival of South American indigenous peoples themselves.

South America still has some uncontacted indigenous groups, some of whom have recently been massacred before peaceful contact (Denny Moore, personal communication 1996). In Brazil alone, 80 different groups of indigenous people became extinct in the twentieth century. This represents an 80 percent loss of the total indigenous population (Ribeiro 1967:96; Sponsel 1985:81; Sponsel 1995:269).

Table 15.1 Indigenous Population and Ethnolinguistic
Groups in South America in Different Periods

Period	Indigenous Population	Ethnolinguistic Groups
In 1492	24,300,000[a]	1,200
In 1940	9,228,735[b]	~600
In 1988	10,129,300[c]	422
In 1996	10,028,980[d]	422

[a]Denevan 1992:370.

[b]Steward 1949:665. The number of languages existing at that time is my rough estimate.

[c]Lizarralde 1993:10.

[d]Based on Lizarralde 1993 and Ricardo 1996.

Although there is some disagreement on how many people and indigenous groups existed in South America in pre-Columbian times, there were certainly many more than was estimated earlier. The first serious attempt at a pre-Columbian population estimate was 4 million (Kroeber 1939), but recent work by Denevan (1992) makes a more reliable estimate of 24.3 million people. Today, South America has only 41 percent of the native population that existed in pre-Columbian times.

The loss of indigenous languages is a quite complex process and is at its highest rate at present. The major factors that produce language loss are economic and cultural. Almost all languages (except for Spanish, Portuguese, French, English, Dutch, and Kechua) in South America are exposed to these two influences. It is known that the number of speakers and the influence of the socioeconomic and political context on these languages determine their future. For South America, we know that 33 percent of these groups (135) are virtually lost because there are few or no remaining speakers (see fig. 15.1, Lizarralde 1993 and Ricardo 1996).

The remaining 329 groups who do speak their own languages are only 27 percent of those in South America in pre-Columbian times. Of these languages, 146 are spoken by 1000 or more speakers. These languages represent only 12 percent of pre-Columbian groups. It is not known whether 1000 speakers is enough for a language to survive. Ladefoged (1992:810) states that a language "with less than 1,000 speakers . . . [is] unlikely to remain a distinct entity." There are only 50 languages with 10,000 or more speakers, representing a total population of 7,624,778 people (mostly Kechua and Aymara). They are 2 percent of the languages that once existed in South America. These are the ethnolinguistic groups that could survive into the future if the dominant cultures of the respective countries do not inflict further pressures on their socioeconomic systems. The main problem is that, independent of the number of speakers, all these languages live under the

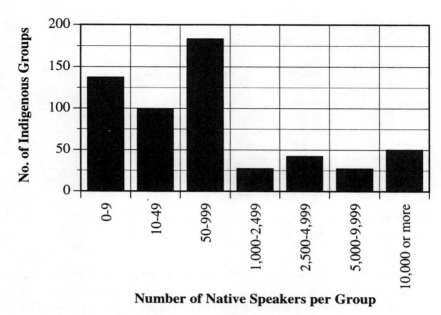

Number of Native Speakers per Group

Figure 15.1. Number of speakers per South American indigenous language.

dominance of European languages and are under heavy cultural influences that are causing their disappearance. Therefore, it is clear that a great proportion of the languages and biocultural knowledge has been lost already in South America and the rest will be lost soon if we do not act to stop these processes.

INDIGENOUS PEOPLES AND BIODIVERSITY RESERVES

Nearly three million indigenous people representing 209 ethnolinguistic groups live in indigenous reserves, natural reserves, and national parks, including 13 of the 25 biodiversity reserves (see map 15.2). These four types of area cover nearly 115 million hectares (296.5 million acres), which is only 6.6 percent of the total surface of South America (see table 15.2).

Brazilian indigenous people, representing almost half of the ethnolinguistic groups in South America, but only a bit more than 2 percent of the indigenous population, were gradually recovering their ancestral lands in recent years (Ricardo 1996). This policy was setting a good example for the rest of South America and reversing the tragedy that they experienced for centuries (cf. Hemming 1978, 1987). Brazilian Indians were able to win claims to over 17 percent of the national territory since 1988 (Paulo Pankararu, personal communication 1997). Unfortunately, this situation was reversed with the signing of the Brazilian decree no. 1775 in 1996, which will delay the demarcation of new indigenous reserves and

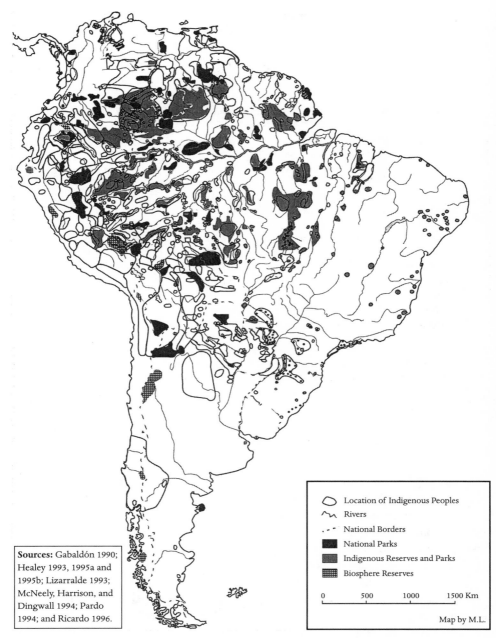

Sources: Gabaldón 1990;
Healey 1993, 1995a and
1995b; Lizarralde 1993;
McNeely, Harrison, and
Dingwall 1994; Pardo
1994; and Ricardo 1996.

Map 15.2. Indigenous reserves, biosphere reserves, and national parks in South America.

Table 15.2   South American National and Indigenous Population with
Approximate Surface Area for All Indigenous Reserves, Natural Reserves,
National Parks, and Biosphere Reserves per Country

Country	Area (Km²)	Reserves Km²	Reserves %[a]	National Population[b]	Indigenous People Groups[c]	Indigenous People Population[d]	Indigenous People %[e]
Argentina	2,776,889	30,000	1.1	34,293,000	15	140,030	0.4
Bolivia	1,098,581	60,000	5.5	7,896,000	42	2,471,340	31.3
Brazil	8,511,965	922,000	10.8	160,738,000	215	294,012	0.2
Chile	756,946	30,000	4.0	14,161,000	5	440,074	<0.1
Colombia	1,141,748	220,000	21.1	36,200,000	81	382,233	1.1
Ecuador	283,561	8,000	2.8	10,891,000	20	1,190,200	10.9
French Guiana	89,941	0	0.0	122,000	7	7,592	6.2
Guyana	214,969	0	0.0	724,000	9	28,425	3.9
Paraguay	406,752	10,000	2.5	5,358,000	18	50,697	1.0
Peru	1,285,216	100,000	7.8	24,087,000	82	4,697,348	19.5
Surinam	163,265	4,000	2.5	430,000	6	7,371	1.7
Uruguay	176,215	400	0.2	3,223,000	0	0	0.0
Venezuela	912,050	146,000	16.1	21,005,000	35	319,710	1.5
Total	17,818,098	1,530,400	8.6	319,128,000	422	10,028,980	3.2

[a]Percentage of the country's territory. The term *reserves* includes indigenous reserves, natural reserves, national parks, and biosphere reserves (Gabaldón 1990; McNeely, Harrison, and Dingwall 1994; Pardo 1994; Ricardo 1996).

[b]Based on the latest censuses given in World Bank 1995.

[c]Based on Lizarralde 1993 and Ricardo 1996.

[d]Based on Lizarralde 1993 and Ricardo 1996.

[e]Percentage of indigenous people in the total national population.

challenge the legitimacy of the claimed territories (Feferman and Borges 1996:1, 4). Decree 1775 allows nonindigenous people to have commercial and land tenure rights over the indigenous ancestral lands that were guaranteed by the 1988 Brazilian constitution. This is catastrophic, because it affects the future of 60 percent of the Amazon basin, which is the largest biome on earth and the richest bioculturally and in terms of species.

In general, indigenous peoples have been cornered in remote or marginal regions that have difficult access, although others are on private properties that are becoming ranches or plantations. Their natural environments have been largely destroyed and are quite altered if they were not converted into national parks or natural reserves. The general figures show that half of the indigenous population lives in parks, indigenous reserves, and natural reserves. The other half lives on lands owned mostly by nonindigenous people, generally of European descent.

LOSS OF KNOWLEDGE

Even though most members of the many indigenous groups speak their native language and partially practice their traditional subsistence patterns, the loss of ethnobiological knowledge is an ongoing process for many reasons. The influence of the dominant society imposing its language for political, economic, and cultural reasons is preeminent (Wurm 1991). The economic and political influences cause shifts in subsistence practices. This shift in subsistence is likely to diminish indigenous peoples' knowledge, like an object that is stored in an attic (e.g., knowledge of plants or animals that existed only in their former territory, from which they have been displaced). Eventually, this traditional subsistence knowledge is lost through disuse and with the death of the speakers.

The exposure to western goods and formal education is also a major cause of the loss of this knowledge. All these factors are present among almost all South American indigenous peoples. The loss of biocultural information is conspicuous because a great proportion of the South American indigenous cultures have experienced some degree of loss and many have lost it entirely (Denevan 1976; Lizarralde 1997; Zent 1999, this volume).

THE BARÍ PEOPLE

The loss of knowledge and language can be illustrated by the case of the Barí people of Venezuela. In fact, the Barí provide a window on a dramatic and disturbing future. The Barí have a fast-growing population of nearly 2500 people and almost all speak their language daily. Nevertheless, my research shows that there is a significant loss of Barí ethnobotanical information from one generation to the next (Lizarralde 1997). The basic problem is the shift in subsistence patterns: the introduction of a cash-crop economy that results from the increase in consumption of western goods. The need for cash is affecting their use of traditional knowledge. This shift is inducing many young people to learn the country's dominant language (Spanish) and new ways of life. How can we be certain of the impact of these changes on the maintenance of their traditional knowledge?

Using data on knowledge of forest trees by 20 Barí collaborators over nearly 17,000 naming events, I estimate that the real loss of ethnobotanical knowledge from one generation to the next may be on the order of 40–60 percent. This process has occurred over the 30 years since they were contacted. The Barí themselves acknowledge that there is considerable variation in forest knowledge. People who have lived in the area longer know more than people who have moved there relatively recently. People whose subsistence practices require more forest use (e.g., hunting versus working on ranches) know more. People with greater bilingual ability and more formal education know relatively less.

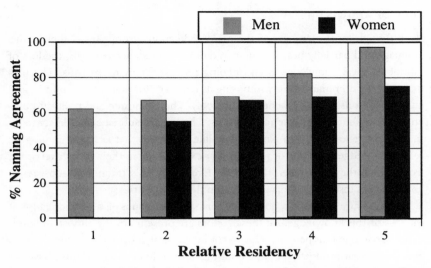

Figure 15.2. Bar graph of relative residency and mean in naming agreement among all Barí women and men in the sample ($N = 20$).

The first variable that I have observed to play an important role in knowledge of trees is the length of residency of the Barí collaborator in a specific territorial group. Most people interviewed were from the local territorial group, while some came from other groups. The length of their residency in the territory where they live is represented in figure 15.2 as follows: 1, a few years; 2, many years or recently moved to an area that has similar trees to ancestral residential area; 3, more than half of their life; 4, moved from a similar territory many years ago, or older people who moved as teenagers and grew up in the area; and 5, all their lives. There is a clear relationship between length of residency and tree knowledge as measured by percentage of agreement with accepted local taxon names (see fig. 15.2).

Statistically, the relationship between these two variables, naming agreement and relative residency, is strong, where knowledge seems to decrease with the proportional length of residency in a village territory. Therefore, any relocated tribe, put on a reservation in another territory or region will experience a significant loss of knowledge.

A second variable that causes loss of knowledge is change in subsistence practices. When asked why some people know more than others about forest trees, the Barí collaborators provided this explanation: people who spend more time in the forest and hunt more tend to know trees better. They have to know all the fruit the game animals eat and thus where the animals are more likely to be found. In order to navigate the forest and to be able to hunt animals, a Barí also needs to know all the trees as reference points to follow instructions from other hunters about the

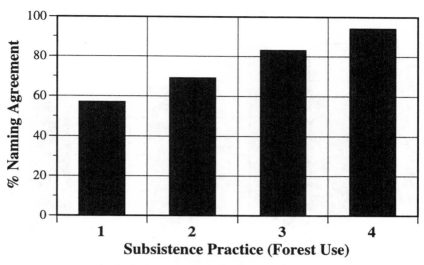

Figure 15.3. Bar graph of subsistence practice (forest use) and naming agreement among all Barí in the sample ($N = 20$).

location of certain animals seen recently. To test this emic view, I made a bar graph of the average naming agreement against uses of the forest (see fig. 15.3).

In this figure, the categories were defined based on type of subsistence pattern related to use of the forest as follows: 1, or little hunting and forest use; 2, moderate use, hunting and gathering of forest products; 3, frequent use, hunting and gathering of forest products; 4, heavy use and daily hunting. The relationship between subsistence pattern (forest use and hunting) and naming agreement among the Barí suggests that use intensity is indeed an important factor in knowledge of trees. Acculturation is causing a change in subsistence patterns for the Barí people. There is an obvious decrease in naming agreement with decreasing forest use leading to eventual loss of this knowledge.

A third variable that decreases the knowledge of forest trees (naming agreement) is the ability to speak Spanish as well as Barí (98.3 percent of the Barí speak their language; see Venezuela 1993). The negative relationship between naming agreement and bilingual ability is clear (see fig. 15.4, where "monolingual" represents people monolingual in Barí, "low" represents Barí who speak some Spanish, and "high" represents fluency in both languages). The Barí who know almost no Spanish have a higher agreement in naming trees (72 percent agreement for women and 95 percent for men) than those who speak Spanish fluently (71 percent agreement for men). The difference would be greater, but for one younger man who while fluent in Spanish knows the Barí traditional knowledge quite well. Moreover, the naming agreement rate for monolingual women is reduced be-

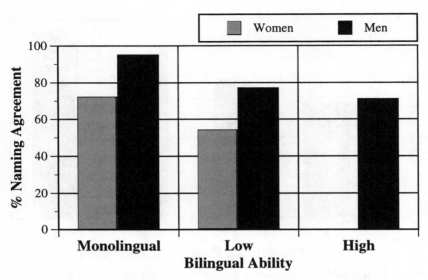

Figure 15.4. Bar graph of bilingual ability and naming agreement among all Barí women and men in the sample ($N = 20$).

cause of the poor vision of my older female collaborators. Thus, bilingual ability is a measure of acculturation and loss of traditional knowledge of trees as well as of native linguistic ability.

It is well known that exposure to Western education is a factor that decreases the cultural knowledge of a given population (Boster 1984:40; Zent 1999, this volume; Hill this volume). The more time a younger person spends studying the material of formal education, the less time she or he has to learn the traditional knowledge. If young people go to a boarding school or university, not only are they removed from the source of traditional knowledge, they may learn another cognitive-ideological system that does not value traditional knowledge. From interviews and talking with the Barí, I observed that the more educated Barí did not know the trees (even the most common ones) as well as the ones who had less formal education (see fig. 15.5).

The association between formal education and naming agreement among all men is not as strong as it should be. This weak relationship is due to the fact that I had a group of Barí who were quite knowledgeable and could excel at both Barí ethnobotanical knowledge and formal education, not a typical situation. The correlation between increasing formal education and decreasing ethnobotanical knowledge (rate of naming agreement) would have been more strongly negative if I had interviewed more "educated" Barí. However, the more educated Barí did not want to be interviewed because they were clearly aware of their ignorance of

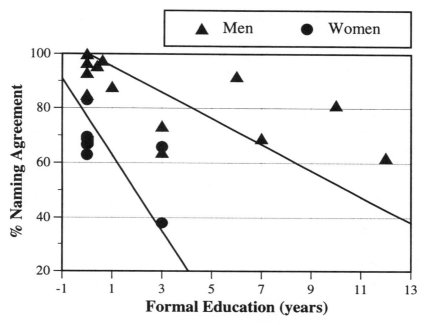

Figure 15.5. Scatter plot of formal education and naming agreement among all Barí women and men in the sample ($N = 20$).

ethnobotanical knowledge. I was able to test this informally by asking one of them to name a tree that is so easy to recognize that even an eight-year-old child could name it (*Triplaris caracasana*, Polygonaceae). This educated collaborator could not name the tree. Therefore, I believe that the rate of knowledge loss is much higher than my data reveal, perhaps 40–60 percent. This is particularly alarming since the Barí were first contacted in 1960, so my younger group of collaborators is the first generation speaking Spanish. Therefore, I think that if the Barí are representative of what is happening in the rest of South America, then the loss of knowledge is quite high in general.

CONCLUSION

At least 35–45 percent of the natural vegetation in South America has been destroyed (cf. Dourojeanni 1982:24; Toledo 1985, 1987). Countries like Colombia and Venezuela have lost 50–60 percent of their natural vegetation (J. C. Centeno, personal communication 1995; Toledo 1988; cf. Huber and Alarcon 1988). South American indigenous peoples speaking their traditional languages today amount to about 20 percent of the pre-Columbian population and are restricted to 6.8 percent of their original territory. These conditions will not support the maintenance of their traditional knowledge. Because of shifts in subsistence patterns and participation in the Western educational system, their traditional knowledge of their environments is held only in the minds of an aging population that is decreasing quite fast. Therefore, with the loss of each indigenous elder, we also lose volumes of knowledge that was accumulated through generations of practical experience with their local biodiversity.

This loss of knowledge is not only due to shifting subsistence patterns and the introduction of the Western educational system but also to reduction and destruction of natural habitat. Therefore, the grandchildren of my Barí colleagues will know as little as 10 percent of what elder Barí now know. If the other South American indigenous groups are experiencing comparable processes, the loss of traditional indigenous knowledge is occurring even faster than the extinction of species in South America. It is clear that the 6.4 percent of the continent that has been protected in reserves is utterly inadequate. Unless we take decisive action to counter these processes, both the biodiversity and the ethnobiological knowledge are inevitably condemned to disappear under these circumstances.

The involvement of indigenous peoples in the planning and management of the natural environment is essential. The Kuna Yala project, in which the World Wildlife Fund facilitated the development of a resource management plan by the Kuna people of Panama, is a good example of a successful endeavor of this sort (Ventocilla, Herrera, and Núñez 1995). There are several projects in South America that involve indigenous peoples in creating sustainable development plans based on their traditional knowledge, for instance in Ecuador, Peru, Bolivia, and Brazil (Bray and Irvine 1993; Lázaro, Pariona, and Simeone 1993; Smith 1993). By "engineering effective resource management and rural development" (Bailey 1996:316), we could stimulate the protection of the biodiversity not only in indigenous territories but those adjacent to them as well. In this way indigenous peoples' cultural knowledge and languages will be used and maintained, and their environments could be conserved.

ACKNOWLEDGMENTS

Special thanks to Luisa Maffi for inviting me to the working conference Endangered Languages, Endangered Knowledge, Endangered Environments in Berkeley, California, in 1996. I would like to thank all the Barí people for letting me conduct my research and reside in their cultural and natural environment. Funding for research to collect data in the field was provided by Wenner-Gren (Grant No. 5519), PREBELAC funds from the Institute of Economic Botany of the New York Botanical Garden and Shaman Pharmaceuticals. My former wife, Nerissa Russell, provided meticulous editorial assistance with this paper. I am indebted to Luisa Maffi for reviewing the manuscript and offering helpful suggestions.

REFERENCES

Bailey, R.C. 1996. Promoting biodiversity and empowering local people in Central African forests. In *Tropical Deforestation: The Human Dimension,* ed. L.E. Sponsel, T.N. Headland, and R.C. Bailey. Pp. 316–341. New York: Columbia University Press.

Boster, J.S. 1984. Classification, cultivation, and selection of Aguaruna cultivars of *Manihot esculenta* (Euphorbiaceae). *Advances in Economic Botany* 1:34–47.

Bray, D.B., and D. Irvine. 1993. Resource and sanctuary: Indigenous peoples, ancestral rights, and the forests of the Americas. *Cultural Survival Quaterly* 17(1):12–14.

Denevan, W.M., ed. 1976. *The Native Population of the Americas in 1492.* Madison: University of Wisconsin Press.

Denevan, W.M. 1992. The pristine myth: The landscape of the Americas in 1492. *Annals of the Association of American Geographers* 82(3):369–385.

Dourojeanni, M.J. 1982. *Recursos Naturales y Desarrollo en América Latina y el Caribe.* Lima: Universidad de Lima.

Ehrenfeld, D. 1986. Thirty million cheers for diversity. *New Scientist* 110:38–43.

Emmons, L., and F. Feer. 1990. *Neotropical Rainforest Mammals: A Field Guide.* Chicago: University of Chicago Press.

Feferman, J., and B. Borges. 1996. Brazil's Indians on alert as government hears final land rights appeals. *World Rainforest Report (Rainforest Action Network)* 13(2):1–4.

Food and Agriculture Organization. 1997. *State of the World's Forests.* Oxford, U.K.: Food and Agriculture Organization.

Gabaldón, M. 1990. *Parques Nacionales de Venezuela.* Caracas: Editorial Torino.

Goulding, M. 1989. *Amazon: The Flooded Forest.* London: BBC Books.

Gradwohl, J., and R. Greenberg. 1988. *Saving the Tropical Forests.* Washington, D.C.: Island Press.

Graves, W., ed., and J.F. Shupe. 1992. *Amazonia: A World Resource at Risk.* Washington D.C.: National Geographic Society.

Healey, K. 1993. South America: North East (map, scale 1:4,000,000). Vancouver: ITM.

Healey, K. 1995a. South America: North West (map, scale 1:4,000,000). Vancouver: ITM.

Healey, K. 1995b. South America: Southern (map, scale 1:4,000,000). Vancouver: ITM.

Hemming, J. 1978. *Red Gold: The Conquest of the Brazilian Indians.* Cambridge, Mass.: Harvard University Press.

Hemming, J. 1987. *Amazon Frontier: The Defeat of the Brazilian Indians.* Cambridge, Mass.: Harvard University Press.

Heywood, V.H. 1989. *Botanic Gardens Conservation Strategy.* Gland: World Conservation Union (IUCN).

Huber, O., and C. Alarcón. 1988. Mapa de Vegetación de Venezuela (scale 1:2,000,000). Caracas: Editorial Arte.

Kroeber, A. 1939. *Cultural and Natural Areas of Native North America.* Berkeley: University of California Press.

Ladefoged, P. 1992. Another view of endangered languages. *Language* 68(4):809–811.

Laurance, W.F., and R.O. Bierregaard, Jr. 1997. *Tropical Forest Remnants: Ecology, Management, and Conservation of Fragmented Communities.* Chicago: University of Chicago Press.

Lázaro, M., M. Pariona, and R. Simeone. 1993. A natural harvest: The Yanesha forestry co-operative in Peru and Western science and indigenous knowledge. *Cultural Survival Quarterly* 17(1):48–51.

Lizarralde, M. 1993. *Índice y Mapa de Grupos Etnolingüísticos Autóctonos de América del Sur. Antropológica,* Suplemento No. 5. Caracas: Fundación La Salle de Ciencias Naturales.

Lizarralde, M. 1997. Perception, knowledge and use of the rainforest: Ethnobotany of the Barí of Venezuela. Ph.D. diss., University of California at Berkeley.

Lugo, A.E. 1988. Estimating reductions in the diversity of tropical forest species. In *Biodiversity,* ed. E.O. Wilson. Pp. 58–70. Washington, D.C.: National Academy Press.

McNeely, J.A., J. Harrison, and P. Dingwall. 1994. *Protecting Nature: Regional Reviews of Protected Areas.* Cambridge: World Conservation Union.

Meggers, B.J. 1988. The prehistory of Amazonia. In *People of the Tropical Rain Forest,* ed. J.S. Denslow and C. Padoch. Pp. 53–62. Berkeley and Los Angeles: University of California Press.

Mittermeier, R.A. 1982. The world's endangered primates. In *Primates and the Tropical Forest,* ed. R. Mittermeier and M. Plotkin. Pp. 11–22. Pasadena: Leakey Foundation.

Moran, E.F. 1996. Deforestation in the Brazilian Amazon. In *Tropical Deforestation: The Human Dimension,* ed. L.E. Sponsel, T.N. Headland, and R.C. Bailey. Pp. 148–164. New York: Columbia University Press.

Pardo, C. 1994. South America. In *Protecting Nature: Regional Review of Protected Areas,* ed. J.A. McNeely, J. Harrison, and P. Dingwall. Pp. 348–371. Cambridge: World Conservation Union.

Plotkin, M.J. 1995. The importance of ethnobotany for tropical forest conservation. In *Ethnobotany: Evolution of a Discipline,* ed. R.E. Schultes and S. von Reis. Pp. 147–156. Portland: Dioscorides.

Repetto, R. 1990. Deforestation in the tropics. *Scientific American* 262(4):36–42.

Ribeiro, D. 1967. Indigenous cultures and languages of Brazil. In *Indians of Brazil in the Twentieth Century,* ed. J.H. Hopper. Pp. 77–166. Washington, D.C.: Institute for Cross-Cultural Research.

Ricardo, C.A. 1996. *Povos Indígenas no Brasil: 1991-1995.* São Paulo: Instituto Socioambiental.

Ritchie, Key, M. 1979. *The Grouping of South American Indian Languages.* Stuttgart: Günter Narr Verlag Tübingen.

Saffel, T. 1993. Vanishing resources, vanishing sanctuary: The remaining forest of the Americas. *Cultural Survival Quarterly* 18(4):63.

Smith, R.C. 1993. Indians, forest rights, and lumber mills. *Cultural Survival Quarterly* 17(1):52–55.

Sponsel, L.E. 1985. Ecology, anthropology, and value in Amazonia. In *Cultural Values and Human Ecology in Southeast Asia,* ed. K.L. Hutterer, A.T. Rambo, and G. Lovelace. Pp. 77–122. Ann Arbor: University of Michigan Press.

Sponsel, L.E. 1995. World system, indigenous peoples, and ecological anthropology. In *Indigenous Peoples and the Future of Amazonia: An Ecological Anthropology of an Endangered World,* ed. L.E. Sponsel. Pp. 263–293. Tucson: University of Arizona Press.

Sponsel, L.E., R.C. Bailey, and T.N. Headland. 1996. Anthropological perspectives on the causes, consequences, and solutions of deforestation. In *Tropical Deforestation: The Hu-*

man Dimension, ed. L.E. Sponsel, T.N. Headland, and R.C. Bailey. Pp. 3–52. New York: Columbia University Press.

Steward, J.H. 1949. The native population of South America. In *Handbook of South American Indians,* vol. 5: *The Comparative Ethnology of South American Indians,* ed. J.H. Steward. Pp. 655–668. Smithsonian Institution, Bureau of American Ethnology Bulletin 143. Washington, D.C.: United States Government Printing Office.

Toledo, V.M. 1985. *A Critical Evaluation of the Floristic Knowledge in Latin America and the Caribbean.* Washington, D.C.: Natural Conservancy, International Program.

Toledo, V.M. 1987. La etnobotánica en Latinoamérica: Vicisitudes, contextos, y desafios. *IV Congreso Latinoamericano de Botánica: Simposio de Etnobotánica, Bogotá, 1987,* pp. 13–34. Bogotá: Instituto Colombiano para el Fomento de la Educación Superior (ICFES).

Toledo, V.M. 1988. *The Floristic Richness of Latin America and the Caribbean as Indicated by the Botanical Inventories.* Washington, D.C., and México, D.F.: Nature Conservancy and Universidad Nacional Autónoma de México.

Ventocilla, J., H. Herrera, and V. Núñez. 1995. *Plants and Animals in the Life of the Kuna.* Austin: University of Texas Press.

Venezuela, Oficina Central de Estadística e Informática. 1993. *Censo Indígena de Venezuela 1992.* Caracas: Oficina Central de Estadística e Informática.

Wilcox, B.A., and K.N. Duin. 1995. Indigenous cultural diversity and biological utility in Latin America. *Cultural Survival Quarterly* 18(4):49–53.

Wilson, E.O. 1992. *The Diversity of Life.* New York: W.W. Norton.

World Bank. 1995. *World Development Report: Workers in an Integrating World.* Oxford: Oxford University Press.

World Conservation Monitoring Centre. 1992. *Global Biodiversity: Status of the Earth's Living Resources.* London: Chapman & Hall.

Wurm, S.A. 1991. Language death and disappearance: Causes and circumstances. In *Endangered Languages,* ed. R.H. Robins and E.M. Uhlenbeck. Pp. 1–18. Oxford and New York: Berg Publishers.

Zent, S. 1999. The quandary of conserving ethnoecological knowledge. In *Ethnoecology: Knowledge, Resources, and Rights,* ed. T. Gragson and B. Blount. Pp. 90–124. Athens: University of Georgia Press.

16

ASPECTS AND IMPLICATIONS OF ECOLOGICAL DIVERSITY IN FOREST SOCIETIES OF THE BRAZILIAN AMAZON

Katharine Milton

The vast expanse of tropical lowland forest in Amazonia is still home to a number of different indigenous societies (Rowe 1974; Clay 1993; Lizarralde 1993). Some larger societies, such as the Yanomami (> 10,000 people; Chagnon 1968), have received considerable academic attention (Chagnon 1968; Chagnon and Hames 1979; Smole 1976) and have a fair degree of public recognition as well as legal title to a large continuous tract of forest (ca. 23 million acres; de Oliveira 1993; MacDonald 1993). But many of the smaller forest societies remain largely unknown to the outside world and do not have legal title to their land.

Because these smaller societies typically number fewer than four hundred people, they lack the population density and multiple villages that could help buffer them from the changes invariably experienced with increased exposure to outside influences (Coimbra et al. 1996; Santos et al. 1997). Furthermore, because these groups are small, they also lack the numbers necessary to stage dramatic marches or to organize large meetings to bring political and media attention to their demands for land demarcation and other rights (Conklin and Graham 1995).

Information about the ecological practices of these little-known societies can serve a useful function in that it brings them into the public eye and documents their residence in and myriad uses of their forest environment. I shall present comparative data on some features of the dietary ecology of four small, forest-based indigenous societies in Brazil, as well as on the medicinal practices of two of them. As we shall see, these groups differ one from another in many aspects of their ecology—a finding that helps to emphasize the loss of cultural diversity that can occur with the demise of even one such group (Clay 1993).

BACKGROUND

In the early 1980s, having learned that there were still a number of indigenous societies living in lowland tropical (terra firme) forests of the Brazilian Amazon that had been little exposed to outside influences (at least for most of the twentieth century), I decided to visit some of them to learn more about their ecological practices. In carrying out this research, I chose to depart from the more traditional anthropological approach in which years are spent immersed in the examination of one particular society. This is a valuable and time-tested method of study, but, given the speed with which cultural practices can alter and the increasing rate of contact with these remaining forest groups, I did not think that time permitted a leisurely approach. Rather, I wanted to visit as many different societies as I could as rapidly as possible and compile the same comparative data in each case. Because all data would be collected by means of the same methodology, the end result should be a body of material that permitted valid comparisons among the ecological practices of these different forest-based groups.

In much of the earlier literature, forest societies in Amazonia were often lumped together in discussion as if they were all the same—basically static and interchangeable entities, for example, "hunter-gatherer societies," "forest-based cultures," "forest peoples," and the like—or the ecological practices of particular forest societies were compared with those of societies in other habitats such as savanna-woodland societies or riparian fisher-gardener societies. Because each environment has its own characteristics (Nelson 1973), such comparisons are likely to be forced and may often be inappropriate. Designating these forest-based groups as hunter-gatherers is, in itself, inappropriate because almost without exception such forest-based societies practice small-scale horticulture as well as hunt and gather and have done so for hundreds of years—in my opinion since well before the first Europeans ever entered the New World. For this reason, hunter-gatherer societies are extremely rare in Amazonia—one such example being the Maku of the northwestern Brazilian Amazon—and literature professing to discuss "hunter-gatherers of Amazonia" generally is discussing the behavior of hunter-gatherer-horticulturalists since small-scale cultivation has long been a feature of almost all terra firme Amazonian societies.

Three of the four groups discussed here, the Araweté, Parakaná, and Mayoruna, were visited in 1986 and 1987; the fourth group, the Matis, was visited in 1995 and 1997. The Parakaná and Araweté occupy neighboring territories in the state of Pará; each speaks a language of the Tupí-Guaraní family. The Mayoruna and Matis, who speak Panoan languages, are more distant neighbors in the state of Amazonas. Though "neighbors," traditionally each localized pair has had hostile relations with the other in the past, and overt hostilities occurred between the Araweté and Parakaná as recently as 1983 (CEDI 1981).

In this chapter, the term *in contact* is used to designate the point at which each of these societies or my particular study village of that society was brought into more or less continuous contact with members of the Brazilian Indian Bureau, Fundação Nacional do Indio (FUNAI). With such "formal contact," FUNAI personnel establish a contact post and begin living near or with the indigenous group. At the time of my initial visit, each of my study groups (or villages) had been "in contact" for only a short time.

For this reason, few individuals in any of the four societies I worked with spoke Portuguese. I did not attempt to master the language of each group but rather learned from each some more essential words and phrases. Most data were compiled through detailed observation and the systematic sampling of different activities, ecological practices, and environmental parameters rather than elicited through detailed interviews, although some interviews were possible when a reliable interpreter or translator was present.

PARAKANÁ AND ARAWETÉ OF PARÁ STATE

The Araweté live south of the town of Altamira in a single village, Ipixuna, on the Ipixuna River, while the Parakaná have three villages—two near the Tocantins River, north of the town of Marabá, and one well to the west of Marabá on the Bom Jardim, a tributary of the Xingu River. The Bom Jardim village lies to the south of the Araweté village of Ipixuna. Prior to settling on the Bom Jardim, these Parakaná had been living in the forest as nomadic hunter-gatherers for fifteen years or more. As nomads, they had been fleeing encroaching Kayapó, a populous (> 3500) Jê-speaking society whose villages traditionally were based farther to the south. The migrating Kayapó, in turn, were fleeing invasion of their traditional homeland by colonists and entrepreneurs. Because the Parakaná had only bows and arrows, they were largely defenseless when faced with Kayapó armed with rifles. Thus they fled their villages and took to the forest.

The Araweté, though they had shifted the location of their village at various times in the past, appear to have been living in the same general region, though perhaps not so far west, for hundreds of years (de Castro 1992). Like the Parakaná at Bom Jardim, the Araweté appear to have been displaced to this northwest region close to the Xingu River primarily by the movement of hostile Kayapó. The Araweté have been the subject of detailed study by E. de Castro (1988, 1992) and considerable historical and other information about them can be found in his book, *From the Enemy's Point of View.*

THE MAYORUNA AND MATIS OF AMAZONAS STATE

The Mayoruna are a fairly populous society numbering some thousand individuals in total. Most Mayoruna live in Peru, where they are known as Matses (Romanoff 1984). There are, however, some 350 Mayoruna in Brazil, living in five villages on or near the Javari River (Erikson 1996), a long, twisting tributary of the Amazon that forms a portion of the Brazil-Peru boarder. I worked with the Mayoruna of Lobo Village, which is located on the Lobo River, a small tributary east of the Javari. During the period of my research the Matis, like the Araweté, had only a single village. It is found on the west bank of the Itui River, a large tributary lying east of the Javari. Like the Araweté, the Matis have shifted the location of their villages at various times in the past but appear to have lived in the same geographical region for hundreds of years.

STAPLE CROPS

My study revealed many differences in diet among these four forest-based indigenous societies (Milton 1991). In terms of horticulture, both Pará groups devoted themselves largely to the production of a single staple cultivar, a very common tropical pattern. But each cultivated a different staple crop. The Parakaná cultivated bitter manioc, although they lacked the elaborate basketry and other equipment generally associated with its cultivation as a carbohydrate staple. The Tucanoans, for example, who are riparian fisher-gardeners found living beside large tributaries in the northwestern Brazilian Amazon, cultivate bitter manioc intensively and have a highly elaborated basketry complex, graters, large ceramic toasting pans and other equipment associated with its preparation (Milton, personal observation).

In contrast, the Parakaná place freshly harvested manioc roots into a shallow area of the river for several days. After much of the cyanogenic compounds in the manioc presumably have been leached out by water, the manioc roots are removed from the river and peeled. The water-softened pulp is wrapped in pieces of cotton cloth and squeezed by hand to further remove harmful chemicals. The squeezed root pulp is then molded into small balls that are placed on a wooden rack in the sun to dry. The dried manioc "flour" is then toasted into cakes on small clay griddles. To verify that bitter manioc was the staple cultivar of other Parakaná, I visited a second larger Parakaná village near the town of Marabá, where I found bitter manioc cultivation and the same preparation techniques.

While living as nomads in the forest, the Bom Jardim Parakaná depended heavily on nuts of the babacu palm (*Orbignya* sp.) as a carbohydrate staple. This palm species is extremely common in most forests of Pará and its density is hypothesized to relate, at least in part, to anthropogenic influences (Balée 1989; Tribin et al. 1998). At Bom Jardim, although the Parakaná cultivated bitter manioc, they also continued to utilize babacu nuts as an important food. They had sites in the forest where rocks occurred that were used to process the palm nuts, and huge heaps of shells and young palm sprouts could be found at such sites. After removal from the shell, the palm nuts were processed and used to make a type of flour to prepare a "bread" just as is done with the manioc flour.

In striking contrast to the Parakaná, the Araweté were strongly dependent on maize as a staple carbohydrate for almost all of the year (de Castro 1992). Though most forest peoples cultivate some maize, typically maize is a seasonal food, not the carbohydrate staple of the annual diet (de Castro 1992; Milton, personal observation). The Araweté had a highly elaborated technology associated with maize cultivation, storage, and utilization. It is often not appreciated that the successful exploitation of a particular crop may involve a tremendous number of human activities associated not only with the planting, tending, and harvesting of the crop, but also with the technology required to store or preserve it until it is utilized and the culinary practices required to prepare the most nourishing dishes from it (Katz, Hediger, and Valleroy 1974; E. Rozin 1983; P. Rozin 1987).[1]

Rather than having only a single staple carbohydrate crop, both the Mayoruna and Matis cultivated two: sweet manioc and plantains. Climatic and soil data show that the area of Amazonia inhabited by the Mayoruna and Matis has richer soil and receives higher and more evenly distributed annual rainfall and more hours of insolation than is the case for the area occupied by the Parakaná and Araweté (Milton 1991). These climatic and edaphic features contribute to a higher overall productivity in forests occupied by the Mayoruna and Matis. I believe it permits these groups more freedom to diversify their staple cultivars than is the case in Pará.

Data compiled on horticultural practices of the Matis some twenty years ago indicate that plantains were little in evidence (CEDI 1981). Currently, however, plantains are grown to such an extent that they are as important in the diet of the Matis as sweet manioc. My data also suggest that cultivation itself has grown more intensive among the Matis in the past twenty years. The reasons for these changes appear to relate largely, although not entirely, to recent outside influences. In order to secure a continuous flow of trade goods—now possible because of more regular contact with the outside world—recently contacted forest groups such as the Matis must have something to trade. One obvious trade good is farinha—toasted manioc flour that is consumed by most Brazilians in massive quantities. By cultivating more manioc than they require for their own nutritional needs, the Matis, who traditionally do not eat farinha but rather boiled manioc roots (and now considerable amounts of roasted plantains), can use the surplus manioc to prepare farinha for trade.

The dependence on two crops rather than one can occur, I believe, because in this more productive area of Amazonia there is freedom to diversify the diet (lesser risk if a particular crop fails and lesser likelihood of failure). The importance of plantains appears to increase in part because of increased manioc cultivation: larger fields for growing surplus manioc provide more scope for the planting of plantains and, unlike a manioc cutting, which produces only a single root crop in one or two years and then is finished, an individual plantain cutting, providing the undergrowth around the base is kept clear, will produce new shoots and stalks of fruit for a period of five or more years. Thus, plantains continue to produce far longer than manioc, and as new fields are opened each year to plant manioc, some plantains are also planted. These outlast the manioc, and slowly but surely plantain abundance continues to increase year after year.

STAPLE PREY

During my study, which involved two visits at different times of year to the Parakaná, Mayoruna, and Matis and one visit to the Araweté, I noted that the staple item of prey (here defined as the prey type with the highest number of individuals brought into the village during my periods of residence) also differed among these groups. The Araweté tended to specialize on large forest birds, the Parakaná on land tortoises, the Mayoruna on peccaries, and the Matis on woolly monkeys. This is not to imply that other prey species were not taken by each group nor that there was no overlap in many prey species between both neighboring and more distant groups.

Indeed, to a modest extent, the prey differences I documented between societies could be due to sampling error; since I was not with each society over the entire course of an annual cycle, I cannot say with certainty that the pattern I observed in each village during my visits prevailed throughout the year. De Castro (1988, 1992), whose visits to the Araweté spanned more than one annual cycle, stressed the importance of tortoises in the diet of the Araweté, but few tortoises were brought into the village during my observation period some four years later; this fact suggests that the tortoise supply in this region may now be exhausted or that there are fluctuations in the use or abundance of different prey species in different seasons or years.

But the Parakaná whose territory extends north to meet that of the Araweté consumed impressively large numbers of tortoises during my two observation periods, one of which occurred at the same time of year and same year as my visit to the Araweté. FUNAI employees who had been with the Bom Jardim Parakaná since contact stated that when the Parakaná first settled in the Bom Jardim area, tapir had comprised a very high percentage of the prey-based portion of the diet. While I was visiting the Parakaná, family groups frequently left the village and went on treks in the forest lasting ten days or more. On returning, they brought with them the remains of smoked tapirs. This is of interest because the Araweté ate no tapir while I was with them, and de Castro (1992) noted that for the Araweté, tapir is a "dangerous" food that requires special ritual preparation.

It is said that traditionally no Tupí-Guaraní-speaking societies ate monkeys.

The Bom Jardim Parakaná, who have been in contact for a shorter time than the Araweté, do not eat monkeys, nor do they kill them for any reason. In contrast, some Araweté do kill monkeys, and de Castro (1992) states that howler monkeys, though considered, like tapir, to be a "dangerous" food, are eaten by the Araweté, although I never observed this. I did, however, note that on one occasion a howler monkey was killed by an Araweté to be used as fish bait.

As the Araweté have been in formal contact for longer than the Bom Jardim Parakaná, it is tempting to suggest that their traditional food prohibition against the eating of monkeys is breaking down with increased exposure to outside food-ways. But de Castro's (1992) description of the ritual surrounding the eating of howler monkeys by the Araweté raises some questions, for it indicates that howler monkey–eating may extend back in time to well before formal contact.

The Mayoruna and Matis are avid monkey hunters and eaters, and both focus hunting principally on the same two monkey species—woolly monkeys and spider monkeys. Yet the Mayoruna, who have been in contact longer than the Matis, hunt with the bow and arrow (or shotguns if they have shells), whereas the Matis hunt monkeys exclusively with the blowgun and poisoned darts. The Matis take monkeys routinely in high numbers (on one occasion over forty woolly monkeys in one hunt), something I did not note for the Mayoruna.

In spite of considerable dietary overlap in most prey species, I also noted that prey prohibited as food in one society was often the single most important prey species in the diet of its nearest neighbor, a fact that suggests that neighboring societies might use particular prey species as cultural boundary markers (Milton 1991, 1997). Each of these forest-based societies had a very clear opinion of what constituted their hunting area, and the suspected presence of strange individuals anywhere in that hunting area elicited an immediate highly hostile response. Traditionally, relations between neighboring Amazonian indigenous societies have tended to be extremely hostile (Balée 1984; Roosevelt 1989; Milton 1991).

Items of diet are natural vehicles for symbolic meaning. Food and eating are intimate acts: foreign objects (foods) are placed in the mouth (a potentially dangerous as well as intimate gesture), and it is appreciated that there is a relationship between eating and body appearance, indicating that food is in fact incorporated into the body of the feeder. For this reason, it seems natural to view an enemy's favored prey species as "like" the enemy and to avoid eating such prey lest you take on enemy characteristics (Milton 1991, 1997). Furthermore, eschewing such prey as food shows that one is of a different people and helps to reify one's own cultural identity.

CULTURE CHANGE

Today, all Matis live together in a single village on the Itui River and have been in more or less continuous formal contact with FUNAI for about twenty years.

Though the Lobo Creek Mayoruna have been in contact for only about twenty-five years in terms of consistent interactions with FUNAI, because of their location in the Javari River area, they have actually been interacting fairly extensively with outsiders for centuries (Romanoff 1984)—something probably not the case for the Matis. The Lobo Creek Mayoruna might therefore be said to represent one level of acculturation and the Matis another, although during my two visits some Mayoruna at Lobo largely avoided contact with outsiders and still lived together in a *maloca,* a traditional multifamily longhouse constructed of wooden poles and palm thatch. Anecdotal accounts suggest that prior to around 1980 the Matis, who then lived in malocas at five different sites adjacent to the Itui River (Erikson 1996), devoted more time to life in the forest and hunting and gathering and less time to horticulture than is the case today.

In 1980, although the Matis were known to practice small-scale horticulture, they were reported to have almost no western implements associated with such cultivation and, as discussed above, plantains were stated to be little in evidence (CEDI 1981). From my observations in 1995 and 1997, it appears that in the last twelve years the Matis have become far more involved in manioc cultivation and widened their subsistence base to include the routine consumption of plantains. In addition, at present all Matis live together in a single village with each family or small extended family in an individual dwelling rather than in malocas.

The weapons each group used for hunting may also represent changes brought about through different levels of acculturation. For example, the fact that the Mayoruna do not manufacture or use the blowgun may reflect their longer period of exposure to outside influences; I was told by a Marubo informant that in the past all three Panoan-speaking groups in this region of Brazil, namely the Marubo, Mayoruna, and Matis, used the blowgun. Romanoff (1984) likewise commented that the Mayoruna used the blowgun prior to the rubber boom. My Marubo informant stated that the Marubo and Mayoruna no longer used the blowgun because "they'd forgotten how to make it" and it was "a lot of work to make it." The same informant also told me that in the past the Marubo were far more oriented toward hunting and forest collecting for subsistence than is the case today, a comment I also heard when I lived with the Mayoruna. Though the Matis still make and use blowguns intensively, especially when hunting monkeys, they currently have a new motivation for continuing to manufacture them: blowguns are very valuable items for trade with the outside world.

These comments on dietary prohibitions, horticultural practices, and weapon manufacture are provocative, for they suggest that conditions that anthropologists (speaking for myself) have taken as the "traditional" condition for a fairly isolated forest culture (and all four of these societies are quite isolated, even today)—for example, "the Mayoruna hunt with the bow and arrow while the Matis use both the bow and arrow and the blowgun" (Milton 1991)—may not, in fact, represent "traditional" reality but rather a more fluid reality based on the degree and type of exposure a given indigenous group has had with outside influences. Romanoff (1984) stressed this point in his monograph on the Matses (the Peruvian name for the Mayoruna), noting that many ecological characteristics of indigenous groups in this region of Amazonia, groups he designates "border Panoans," altered dramatically during the rubber boom (1880–1920). Such observations further emphasize the dynamic nature of these societies while heightening the need for caution in accepting current ecological practices even of very remote groups as long standing—which they may or may not be.

ILLNESS, MEDICINE, AND MAGIC

The Mayoruna have long been known as raiders for women (CEDI 1981; Romanoff 1984). Mayoruna warriors would attack other indigenous societies to steal women (Tika Mayoruna, personal communication) and also carried out extensive attacks on colonists on and along the Javari River and other nearby tributaries (CEDI 1981). In Mayoruna villages today, various women are either captives or descendants of captives. The Matis, on the other hand, do not have a history of raiding other societies for female captives nor, as far as I could tell, with one exception, was anyone in their village a non-Matis.

Romanoff (1984) has provided a vivid and detailed account of what is known of the historical background of the Mayoruna. Because of their association with the Javari River, a river with a turbulent past both because of its border status and its centrality in the rubber boom, the Mayoruna have had a highly disturbed history, one filled with strife, danger, and death (Romanoff 1984). Less is known about the past history of the Matis, but they appear sufficiently removed from the Javari region to have been less affected.

The different social ambiance I noted between the two groups—the Mayoruna appearing far more taciturn, guarded, and suspicious and the Matis more relaxed and open—seems reflected today in the respective attitudes of the two groups toward illness and its treatment. The Mayoruna at Lobo Creek showed me more than a hundred different medicinal plants that covered an extensive array of health problems including toothache, swollen testicles, diarrhea, headache, eye infection, back pain, neck pain, stomach pain, tumors, breast pain, leishmaniasis, penis pain, and even chicken pox, in addition to remedies for snake bite, ant bite, scorpion bite, and the like. Medicinal plants were routinely collected, generally by men. Most of the medicinal plants shown to me were to be used topically; leaves, for example, were to be bound onto afflicted individuals at the site of the problem or, in the case of small children, were to be used with water to bathe the child. In a few cases, however, the medicinal leaf, bark, or sap was to be taken internally or placed in the afflicted area. For example, some barks were rasped into water and the concoction drunk for headaches or toothaches or bark or sap was placed directly into a wound.

In contrast, among the Matis, actual illness and its cure, other than snake bite, did not seem to be a topic of particular interest, and in forest inventories few medicinal plants for any ailment such as those of the Mayoruna were pointed out. A few plant species were identified as useful for some ailment—toothache or scorpion bite, for example. However, almost all *remedios* (medicines) shown to me by the Matis were prophylactic, i.e., they were a remedy to prevent a condition from occurring, not to treat an existing condition. Furthermore, almost all medicinal plants shown to me pertained to the treatment of small children: the leaves of particular trees were used either to bathe a child before it got sick because of some activity of the parent—generally overeating of a particular game animal by the parent—or to ward off possible bad luck that might affect the child.

This is somewhat similar to Romanoff's (1984) description of the Matses (Mayoruna) of Peru at the time of his fifteen-month study between 1974 and 1976. He noted that the Mayoruna believed that spirits could cause illness, and for each illness there was an appropriate leaf. Leaves from forest trees were gathered, boiled, and used to bathe ill people. Romanoff (1984) also pointed out that all men knew some medicinal leaves but old men knew more. This situation is similar to the one I describe among the present-day Matis, except that I did not witness the boiling of leaves and cannot comment on beliefs in spirit illness.

In my survey of uses made of more than 350 different trees in the forest occupied by the Parakaná of Bom Jardim, only four trees in the entire inventory were stated to have medicinal use—one for malaria, one whose sap could be used to heal cuts, one "a medicine you drink," and one "medicine for the stomach." This suggests that medicinal treatments from trees did not play much role in the life of the Parakaná during my visits in 1986 and 1987.

There is debate today about the prevalence of ethnobotanical practices such as ethnomedicine among relatively uncontacted indigenous groups, as contrasted with the prevalence of such practices in societies with a longer history of acculturation (Davis and Yost 1983; Telban 1988; Balée 1994). Some data suggest that groups with little outside contact lack both extensive lists of physical ailments and extensive pharmacological inventories associated with their cure (Davis and Yost 1983; Telban 1988). Balée (1994), on the other hand, based on his intensive work with the Ka'apor of Maranhão (Brazil), takes the view that the great majority of Ka'apor medicinal plants are used to treat human diseases of pre-Columbian origin. Telban (1988) has discussed the dichotomy that exists between plants that *heal* (plants that relate to illnesses associated, for example, with the breaking of a taboo or with illness that are due to spirits or sorcery, and that are psychosocial in nature) and plants that *cure* (plants that have direct pharmacological effects and are used to treat a physical condition such as a toothache, snake bite, or malaria).

As a result of my own fieldwork in Amazonia, and independent of their work, I had reached much the same conclusions as Davis and Yost (1983) and Telban (1988), that is, I do not believe that prior to contact with and exposure to outside influences, forest-based indigenous peoples in Amazonia possessed highly elaborated pharmacopoeias of medicinal plants used to treat specific physical conditions—plants that cure. I am not suggesting that they had *no* plants of this nature, only that they had far fewer than became the case after contact. Thus, I view an interest in "curing" medicine and "curing" plants as perhaps induced to a large extent by contact with outsiders (or their diseases). Outsiders generally not only bring in new diseases that can be harmful or fatal to indigenous peoples, but they also tend to have elaborate medicine kits that contain very effective remedies since these medicines are used to treat the very diseases the outsiders have brought in— not only colds or coughs (which can kill indigenous peoples), but also serious malarias and other tropical diseases representing strains common in urban areas that tend to be far more virulent than the more benign strains in distant forests.

In 1986, I was able to talk to a Brazilian doctor involved in examining individuals in several indigenous societies in Pará at the time of first contact by the FUNAI. This doctor told me that generally at contact such forest groups were largely free of any obvious diseases. Malaria tended to be a low-grade, relatively benign forest type, not the often virulent malarial strains associated with areas of higher human habitation and large rivers. Parasite loads were also stated to be low, and, generally, in this physician's view, most individuals of all ages appeared to be in excellent health. For example, at contact the Bom Jardim Parakaná were stated to have no malaria and no ascarids, giardia, or amoebae, and little hookworm; their teeth were stated to be in beautiful condition. Individuals in excellent health are unlikely to need a large "medicine cabinet" full of medicinal plants that cure though they or their shamans might well utilize medicinal plants that heal.

Based on such reports as well as my own observations, and in keeping with Telban's (1988) remarks, I would predict a continuum of medicinal plant use vis-à-vis plants with actual pharmacological properties for treating specific physical ills, with groups farthest from outside influences showing the lowest number of "curing" medicinal plants and a gradual increase in interest in and knowledge and use of chemically active curing plants as groups experienced more contact with outside influences.

On the other hand, both Mayoruna and Matis practiced a hunting ritual involving the taking of a powerful skin secretion from the tree frog, *Phylomedusa bicolor*. In this instance, the frog secretion has dual significance (it both "cures" and "heals"); it produces actual physiological effects of great power that enhance strength and stamina as well as the acuity of the hunter's senses, while at the same time providing psychological benefits, since the secretion is also believed to bring good luck to the hunter (Daly et al. 1992; Milton 1994). It is interesting to note that this frog species is distributed across the Amazon Basin (Daly et al. 1992; Milton 1994), but with one possible exception, only Panoan-speaking peoples use the secretion. This illustrates the fact that key scientific discoveries of considerable significance may not transfer out of their original linguistic matrix, even though the plant or animal species may be found over a wide geographical area and could be used by other indigenous societies if their inhabitants were aware of its qualities (Milton 1994; Balée 1994).

On a related topic, in terms of general ethnobotanical knowledge among the Matis, a tribe so little acculturated that even in 1995 only a few individuals spoke any Portuguese, only two older (> 50 years) men and one older woman (the woman tested with only a small sample of trees) were consistently able to provide the names and uses of the majority of forest trees tagged in my sample plots. Young Matis women (15–20 years of age) appeared to know virtually no tree name unless it was a species that produced edible fruit. Though young men (15–32 years of age) claimed to know tree names, in fact they did not appear to be very certain of their information and finally told me that I needed to get one of the older men to work with me.

Younger Mayoruna, Parakaná, and Araweté appeared to have much more knowledge of forest tree identification than the Matis. Currently, I have no explanation for why younger Matis appeared so unfamiliar with the names of their forest trees. It might be that the very recent contact of the Matis with outsiders, a period beginning more or less around 1985 that would overlap with the childhood and adolescence of most individuals in the range of 15 to 30 years, was sufficiently disruptive so that it caused an actual break in the vertical transmission of ethnobotanical knowledge in this culture. This question too warrants further investigation and study.

OVERVIEW AND FUTURE PROSPECTS

Data presented here illustrate the fact that each of these indigenous societies is quite distinct in terms of many ecological and other cultural practices. Living for hundreds of years in the same geographical region and even speaking languages derived from the same linguistic stock may be reflected in some common aspects of behavior, as, for example, the taking of the frog secretion by Panoan-speaking societies, but such factors often do not appear to play an important role in the way these different societies utilize various other components of their forest environment. The biological richness of the Neotropical rainforest is reflected in the corresponding cultural richness represented by these forest peoples.

Using the recent experiences of larger indigenous groups such as the Yanomami or Kayapó as an indicator, can we predict what the future might hold for these smaller indigenous groups? Initially, of course, each society speaks only its native language. But the contacting individuals in Brazil speak Portuguese. Little by little some individuals in each village learn Portuguese, and typically these individuals are a few young men (Giannini 1996; Milton, personal observation). These young men then become the voice of the village—they negotiate, translate, explain, and in general achieve high status and considerable influence and power through their dealings with nonindigenous contacts (Giannini 1996; Milton, personal observation). They also draw attention away from the accumulated wisdom of older generations (the village elders), thereby diluting the latter's influence and disrupting the vertical transfer of traditional information from generation to generation. This syndrome is discussed in detail by Turner (1995), Giannini (1996), and others with reference to the Kayapó and Kayapó-Xikrin.

In the small forest-based societies where I work, almost everyone seems to share the view that speaking Portuguese is beneficial. I was frequently asked by both the Mayoruna of Lobo Creek and the Matis of Itui to tell FUNAI personnel in Atalaia or Brasilia to send a school teacher to their village so they could all learn to speak and read Portuguese and thus how to use currency and work with numbers. This has yet to happen for either the Matis or Mayoruna, but it shows what they perceive as their most immediate need. And, of course, when you change your language, you may change your world view—as several contributors in this volume emphasize. If you lack the vocabulary to discuss details of your forest environment, these details will no longer be important.

With human societies it is always difficult to predict what may happen. But if we take the Matis as an example, and the lack of botanical interest I noted in adolescents in 1995 continues while everyone masters Portuguese and negotiates through the initial problems of continuous contact with nonvillage people, much traditional ethnobotanical and other ecological knowledge of the forest environment may be lost—and there are no other Matis villages that can be sought out to recover it.

Yet cultural devaluation and loss may not come to pass because, as of 2000, conditions favoring the preservation of indigenous knowledge are far different and also more positive than was the case when most forest-based groups in the Brazilian Amazon were initially contacted. For example, the closest Marubo village to the Matis village is made up of a mix of far more acculturated and somewhat more acculturated individuals than is the case for the Matis. (This Marubo village is small, with only about a hundred inhabitants; it should not be confused with the large, long-missionized Marubo settlement located toward the headwaters of the Itui.) In this small Marubo village, one of the young adult men told me that he and his associates were working to ensure that all of their children learned every aspect of their traditional culture. Though these more acculturated Marubo had been living in small family dwellings, they recently had constructed two malocas and many villagers were now living in these malocas again rather than in individual houses. As expressed to me, the Marubo were happy to learn from the outside world but they were also taking steps to ensure that their children were trained in all Marubo traditions as well.

In recent years, as we know, the pendulum has swung heavily toward public respect for and strong interest in traditional knowledge and the conservation of biodiversity, including linguistic and cultural biodiversity. Indigenous groups contacted and exposed to western ways some 100, 50, or even 25 years ago in Brazil did not have this positive framework of international support and admiration. Rather, emphasis was placed on the need for indigenous peoples to divest themselves of their traditions and take on the language, behaviors, and economic practices of the wider Brazilian society. Unfortunately, there is nothing in the genome of any individual, indigenous or otherwise, about how to utilize the tropical forest environment in a sustainable manner. Rather, each generation has to learn anew the entire information set from preceding generations that hold this knowledge. And once such information vanishes, it is difficult if not impossible to recreate it.

Ironically, however, indigenous groups that have managed to stay deep in the tropical forest, far away from acculturation and outside influences, may now have the opportunity to reflect to some extent on what they want for their future rather than simply be exposed to outside customs and immediately made to feel that their own cultural creations are inferior to what the outside world has to offer. Indigenous consciousness has been raised primarily as a direct result of communication and outreach movements by more acculturated to less acculturated groups in Amazonia, and it extends even into the distant areas where I work. Observational data suggest that even in these very remote societies traditional knowledge is currently being lost. But in contrast to most other indigenous societies in Brazil, which have either disappeared or lost many of their former traditions and skills, today it would seem that indigenous peoples such as the Matis, Mayoruna, Parakaná, and Araweté may be able to determine their future for themselves—if they act now.

It would be a great tragedy if, having successfully resisted all of the many hostile and annihilating forces leveled against them over the past five centuries, forces that have resulted in the extinction of so many indigenous societies in Amazonia, these small forest-based societies could not now successfully enter the twenty-first century still in possession of their own autonomy, languages, cultural traditions, intellectual heritage, *and* their densely forested environment.

NOTE

1. The widespread incidence of pellagra in many populations (both indigenous and non-indigenous) introduced to maize (a New World cultivar) in the past few centuries was due in large part to the fact that the indigenous method of corn preparation in the

Neotropics, which involves leaching maize with an alkaline solution prior to cooking it (an activity that liberates niacin and significantly improves amino acid quality; Katz 1987) did not travel with maize seeds to their new sites of cultivation but rather was left behind—to the detriment of the new societies that tried to use maize as a dietary staple (Carpenter 1977; Katz 1987). Similarly, human groups more recently introduced to bitter manioc but not to the time-tested methods for removing its harmful cyanogenic glycosides suffered and continue to suffer nerve damage and even death as a result of the inadequate removal of these potentially lethal compounds (P. Rozin 1987).

REFERENCES

Balée, W. 1984. The ecology of ancient Tupí warfare. In *Warfare, Culture, and Environment*, ed. R. B. Ferguson. Pp. 241–265. New York: Academic Press.

Balée, W. 1989. The culture of Amazonian forests. *Advances in Economic Botany* 7:1–21.

Balée, W. 1994. *Footprints of the Forest*. New York: Columbia University Press.

Carpenter, K.J. 1977. High-cereal diets for man. *Proceedings of the Nutritional Society* 36:149–158.

CEDI. 1981. *Povos Indígenas No Brazil*, vol. 5: *Javari*. São Paulo, Brazil: Centro Ecuménico de Documentacão e Informacão.

Chagnon, N.A. 1968. *Yanomamo: The Fierce People*. New York: Holt, Rinehart & Winston.

Chagnon, N.A., and R.B. Hames. 1979. Protein deficiency and tribal warfare in Amazonia: New data. *Science* 203:910–913.

Clay, J.W. 1993. Looking back to go forward: Predicting and preventing human rights violation. In *State of the Peoples: A Global Human Rights Report on Societies in Danger*, ed. M. S. Miller. Pp. 64–71. Boston: Beacon Press.

Coimbra, C.A.E., Jr., et al. 1996. Hepatitis B, epidemiology, and cultural practices in Amerindian populations in Amazonia: The Tupí-Monde and the Xavante from Brazil. *Social Science and Medicine* 42:1738–1743.

Conklin, B.A., and L.R. Graham. 1995. The shifting middle ground: Amazonian Indians and eco-politics. *American Anthropologist* 97(4):695–710.

Daly, J.W., et al. 1992. Frog secretions and hunting magic in the Upper Amazon: Identification of a peptide interacting with adenosine receptor. *Proceedings of the National Academy of Sciences* 89:10960–10963.

Davis, E.W., and J.A. Yost. 1983. The ethnomedicine of the Waorani of Amazonian Ecuador. *Journal of Ethnopharmacology* 9(2):272–97.

De Castro, E.V. 1988. *Araweté: Os Deuses Canibais*. Rio de Janeiro: Jorge Zahar, Editor Ltda / ANPOCS.

De Castro, E.V. 1992. *From the Enemy's Point of View*. Chicago: University of Chicago Press.

Erikson, P. 1996. *La Griffe des Aieux*. SELAF no. 358. Paris: Peeters.

Giannini, I.V. 1996. The Xicrin do Catete Indigenous Area. In *Traditional Peoples and Biodiversity Conservation in Large Tropical Landscapes*, ed. K.H. Redford and J.A. Mansour. Pp. 115–36. America Verde Publications. Arlington, Virginia: Nature Conservancy, Latin America and Caribbean Division.

Katz, S.H. 1987. Fava bean consumption: A case for the coevolution of genes and culture.

In *Food and Evolution: Toward a Theory of Human Food Habits*, ed. M. Harris and E. B. Ross. Pp. 133–159. Philadelphia: Temple University Press.

Katz, S.H., M.L. Hediger, and L. Valleroy. 1974. Traditional maize processing techniques in the New Word: Anthropological and nutritional significance. *Science* 184:765–73.

Lizarralde, M. 1993. *Índice y Mapa de Grupos Etnolingüísticos Autóctonos de América del Sur. Antropológica*, Suplemento No. 5. Caracas: Fundación La Salle de Ciencias Naturales.

MacDonald, T., Jr. 1993. South America: Land and labor. In *State of the Peoples: A Global Human Rights Report on Societies in Danger,* ed. M. S. Miller. Pp. 236–238. Boston: Beacon Press.

Milton, K. 1991. Comparative aspects of diet in Amazonian forest dwellers. *Philosophical Transactions of the Royal Society,* Series B 334:253–263.

Milton, K. 1994. No pain, no game. *Natural History,* September 1994, pp. 44–51.

Milton, K. 1997. Real men don't eat deer. *Discover,* June 1997, pp. 46–53.

Nelson, R.K. 1973. *Hunters of the Northern Forest.* Chicago: University of Chicago Press.

Oliveira, J.P. de. 1993. Brazilian indigenous land policy. In *State of the Peoples: A Global Human Rights Report on Societies in Danger,* ed. M.S. Miller. Pp. 242–243. Boston: Beacon Press.

Romanoff, S.A. 1984. Matses adaptations in the Peruvian Amazon. Ph.D. diss., Columbia University, New York.

Roosevelt, A.C. 1989. Resource management in Amazonia before the conquest: Beyond ethnographic projection. *Advances in Economic Botany* 7:30–62.

Rowe, J.H. 1974. Indian Tribes of South America: A Map. Copyright J. H. Rowe 1974. No publisher.

Rozin, E. 1983. The structure of cuisine. In *The Psychobiology of Human Food Selection,* ed. L. M. Barker. Pp. 189–203. Westport, Conn.: AVI.

Rozin, P. 1987. Psychobiological perspectives on food preferences and avoidances. In *Food and Evolution: Toward a Theory of Human Food Habits,* ed. M. Harris and E.B. Ross. Pp. 181–205. Philadelphia: Temple University Press.

Santos, R.V., et al. 1997. Tapirs, tractors, and tapes: The changing economy and ecology of the Xavante Indians of central Brazil. *Human Ecology* 25(4):545–566.

Smole, W.J. 1976. *The Yanomama Indians.* Austin: University of Texas Press.

Telban, B. 1988. The role of medical ethnobotany in ethnomedicine: A New Guinea example. *Journal of Ethnobiology* 8(2):149–169.

Tribin, H.A.S., et al. 1998. La nutrición de los Nukak: Una sociedad amazónica en proceso de contacto. *Maguare* 13:117–142.

Turner, T. 1995. An indigenous peoples' struggle for socially equitable and ecologically sustainable production: The Kayapó revolt against extractivism. *Journal of Latin American Anthropology* 1:98–122.

17

ENVIRONMENT, CULTURE, AND SIRIONÓ PLANT NAMES

William L. Balée

> The differences in the culture of the various tribes cannot be explained by their living under
> different natural conditions, but only by their history.
>
> Baron Erland von Nordenskiöld (1924:3)

Amazonian environments present human languages with a formidable job: symbolic representation of a vast domain of visible, organismic minutiae. If languages adapt to people and their environments—if languages resemble viruses in their relationship to people through time (Deacon 1997:112)—the biological richness of the environment in which they occur should be partly evident in vocabulary. Eskimo words for snow, Australian Aboriginal words for sand, Hanunóo words for rice, and American English words for automobiles seem indexical of customary features of the respective environments of these languages (cf. Wierzbicka 1997:10–11). In the Amazon, words for plants should be correspondingly numerous and plant classifications complex. This is, indeed, the case, but for reasons of historical ecology there are differences of degree among Amazonian languages in terms of nomenclature and classification of plants, which to the naked eye are among the most diverse of organisms in any one habitat. Historical ecology conceives of languages and landscapes as forming rational, irreducible entities that cannot be productively analyzed in isolation from each other (Balée 1998). Language changes environments through the instrument of culture; environments change languages because languages must adapt to them in the course of migration and contact of the people who speak them. Together, these interactions constitute landscapes that can only be grasped simultaneously as rule-governed products of ecology and history acting in tandem.

This chapter concerns similarities and differences between Sirionó on the one hand and other languages of the Tupí-Guaraní language family on the other in terms of the encoding of plant names and their meanings over time. I examine how Sirionó diverges from other Tupí-Guaraní languages by degree; the divergence is understood in reference to the historical interactions, as discussed below, with landscapes of the eastern Bolivian Amazon by antecedents of the Sirionó. I propose that Sirionó ethnobotany mainly supports certain general patterns of plant nomenclature and classification already observed for the Tupí-Guaraní family, putting to rest old questions about whether Sirionó qualifies as a full member or not of the Tupí-Guaraní family, from the perspective offered by the ethnobotanical data. These data suggest that Sirionó as a language and as a culture experienced a unique but not incomparable relationship with Amazonian landscapes over time.

A HISTORY OF TREKKING IN LOWLAND BOLIVIA

The Sirionó, who speak a Tupí-Guaraní language, have lived in the biogeographic region known as the Llanos de Mojos since before the end of the seventeenth century. Ancestral Sirionó were described in 1693 as threatening to shoot arrows at a group of Jesuits and their neophytes (Anonymous 1781:104–105; Distel 1984/85:159; Block 1994:38–39; cf. Rydén 1941:23). The missionaries were seeking to make contact with the Guarayo, who were believed to be cannibals (Chávez Suárez 1986:236–237). The Guarayo have been historically located to the southeast of the Sirionó, and they also speak a Tupí-Guaraní language, perhaps most closely related to Guaraçug'wé (Pauserna) whose speakers are located to the northeast of the Sirionó along the Guaporé/Iténez River.[1] In fact, the Sirionó are similar but not identical to the Guarayo and Guaraçug'wé in language and culture, and by 1693 the Sirionó were probably already a distinct people (Anonymous 1781:105; Holmberg 1948:455).

The Guarayos had been known to the Spanish by name at the latest since 1674 (Block 1980:35) and evidently became a separate people from the Guaraçug'wé some time during the period 1525–1741 (Riester 1977:31, 33). They were largely settled in a mission town by 1794 (René-Moreno 1888:208–209). In 1696 the Sirionó were recorded as one of thirty-seven distinct Indian "nations" in the lowlands of eastern Bolivia (D'Orbigny 1946:181, cited in Chávez Suárez 1986:20–21). These Tupí-Guaraní–speaking peoples were distinguished from the Moxos Indians (Anonymous 1781), of an Arawakan tongue, who were the principal denizens of the Jesuit missions in the Llanos de Mojos (Nordenskiöld 1924:1–2; Block 1994). The historical descendants of the Moxos and the mission culture that enveloped them during the 1700s became the mestizos of the Llanos de Mojos, known today as the Benianos (Block 1994).

But the Sirionó were not mentioned as having visited a mission until 1765 (Holmberg 1948:455), and there is no record of the Sirionó having been missionized in the vast expanse of lands between the mission town of Trinidad on the Mamoré River and the Brazilian border during the Jesuit mission period from 1674–1767 and the period of mission culture, which continued into the mid 1800s (Block 1994:143). Whereas for the Moxos dwellers of the missions before 1767 the sedentary lifestyle, profusion of Western trade goods, and agricultural surpluses caused escape into the forest to be deemed "unthinkable" (Block 1994:143), the Sirionó were evidently throughout this period and until the early twentieth century quite independent of such missions and settlements. As late as 1888, the Sirionó (called then *los Sirionós bárbaros* 'the barbaric Sirionó') had yet to be missionized or otherwise settled in a contact situation (René-Moreno 1888:121). There were further attempts to missionize them early in the twentieth century, and these were partially successful then (Lunardi 1938; Rydén 1941); today most Sirionó live in a settlement founded by Protestant missionaries and most of the adults seem to classify themselves as *creyentes* 'believers.' But there is no evidence that the Sirionó were brought into contact with a mission culture speaking Língua Geral, a Tupí-Guaraní creole elsewhere in Amazonia such as the Northwest Amazon, since the Jesuit missionaries of the region would have spoken Mojos or Spanish in order to gain converts.

No substrate influence has been demonstrated for Sirionó, though doubts have persisted among some scholars that the language and people might have been Guaraní-ized (Cardús 1886:280; Nordenskiöld 1924:233; Rydén 1941:37; Susnik 1994: 86), that is, that it had been an earlier language transformed by contact with the historic migrations of Guaraní peoples from the south, in what is now Paraguay, during the fifteenth and sixteenth centuries. I argue that what happened to the Sirionó people historically, however, as with other Tupí-Guaraní peoples who lost the practice of agriculture partially or entirely, such as the Guajá, could have influenced components of the language, especially the lexicon, quite apart from any substrate influence, real or imagined. Even though it is likely that Sirionó has had foreign influences because of its highly distinctive phonology in the Tupí-Guaraní language family (Dietrich 1990:58, 97), it is, nevertheless, a full member of that family. And within that language family there are certain patterns of plant nomenclature (Balée 1989) that seem to be replicated in Sirionó, to be explored below.

The opposite point of view is derived from incomplete observations on and inferences about Sirionó culture, not language. The pioneering ethnologist Nordenskiöld stated that argument about Sirionó as follows:

> To judge from the few words we know of their language, they speak Guarani. Presumably this was not their original language, and they must have been Guarani-ized. They are the only Indians of this area who do not raise any crops, and live exclusively on the chase, fishing, and the gathering of wild fruits. They have not settled down. . . . As far as we can judge, the Siriono are the tribe in this area that promises to keep its originality longest. (1924:233)

However original and unique the Sirionó culture, it is a historical product of the gradual transformation of a more robust and diverse agricultural economy associated with primordial Tupí-Guaraní peoples. And it is now fairly clear, though perhaps it was not to Nordenskiöld, who evidently never visited a Sirionó encampment per se and only saw some who were in a mission, that the Sirionó were not entirely lacking in domesticated crops and agricultural techniques. Even the language suggests a continuous association with agriculture since proto–Tupí-Guaranian times, although some crops vanished over time.

What seems more likely than Nordenskiöld's suppositions about the origins of the language is that Sirionó is an intrusive Tupí-Guaraní language whose location in eastern Bolivia is a product of one or more Guaranian migrations alone, as with Chiriguano, Guarayo, Pauserna, and the very closely related Yuquí (Holmberg 1985:11; Stearman 1989:22; Townsend 1995:27).[2] In other words, Sirionó is as fully a member of the Tupí-Guaraní family of languages as are those mentioned above and as are Tupí-Guaraní languages of eastern Amazonia, such as Araweté, Asurini, Guajá, Ka'apor, Tembé, and Wayãpi, and it is suitable for the purpose of cross-linguistic comparison with those languages. My primary argument is that Sirionó shares features of plant nomenclature and classification previously identified for these eastern Tupí-Guaraní languages (Balée 1989; Balée and Moore 1991, 1994) that were not borrowed but are the historical-ecological product of the long-term encounter between human languages and Amazonian landscapes.

The argument that Sirionó is a language absorbed, and only imperfectly learned, by an ancient non-Tupian people conquered by the Guaraní or missionized by the Jesuits has an analogue, if not its source, in the view that Sirionó culture was depauperate in knowledge of religion, mythology and cosmology, construction and crafts, and technology related to exploitation of their habitat (see Isaac 1977:138 and Stearman 1984:630–631 for reviews; also Nordenskiöld 1924:203; Rydén 1941:8; Steward 1948:899). The Sirionó were a trekking people who did not remain in a village year round. Trekking as an environmental adaptation involves some dependence on fast producing domesticated crops combined with high mobility. It is a curious fact that it seems to be confined to lowland South America. Other groups that exhibit trekking as a mode of adaptation include some of the Maku, Yanomamö, Araweté, and Waorani. Trekking seems to be transitional between hunting-and-gathering and settled horticulture.

So the Sirionó were not in the technical sense hunter-gatherers, a technology that by the mid-twentieth century had not yet been studied in Amazonia, though it has been since for the Guajá, Yuquí, Aché-Guayaki, and Xetá.[3] Perhaps most striking in terms of technology is that the Sirionó were one of the first peoples to be described in Amazonia as lacking fire-making technology (Holmberg 1985:11). Absence of the fire-making technology of wooden drill and hearth (from just a few select tree species) is not a characteristic of any known language family in lowland South America, nor is the absence of fire-making technology more generally the case with any other human language family probably since the origin of language itself. The Sirionó share this negative feature, however, with a few other groups in the Tupí-Guaraní language family, and no doubt in other families, who are either hunters and gatherers or, like themselves, trekkers, among them being the Guajá (Balée 1999), the Yuquí (Stearman 1989), and some of the Parakanã (Fausto 1997:86). In no case is absence of fire-making technology seen with groups practicing a more sedentary horticultural mode of production in the Amazon, which has tended to include major dependence on bitter manioc among many other cultivars. Fire-making is clearly a feature that they lost (Holmberg 1985:11), not one they never had. The Sirionó have possessed domesticated crops at least since any mention was made of their agricultural technology, though these have been few and of a smaller inventory than that of proto–Tupí-Guaranian peoples, which likewise suggests a depression over time in number of domesticates, because of vagaries of migration and contact with other peoples.

The environment alone cannot account for these prehistoric phenomena. The well-drained lands of the habitat, which occur exclusively on prehistoric earthworks not built by ancestors of the Sirionó, receive about 1800 mm of rainfall per year (Townsend 1995:18) and are rich in organic soil analogous to *terra preta* of Brazil (Langstroth 1996:105, 250), factors which clearly favored dense populations dependent on agriculture in the past (Denevan 1966; Erickson 1995). The agricultural potential of this landscape was probably enhanced, not reduced, by prehistoric settlement and utilization, so the fact that the Sirionó planted few crops and had no means of fire-making cannot be attributed to it. Rather, the source of regressive changes in Sirionó agricultural technology can only be comprehended historically, through the course of migrations and contacts to which the people and their language would have been exposed. The influences of such migrations may in general be seen in the Sirionó plant lexicon, though mechanisms have yet to be identified in order to account for the specific lexical changes.

ANT TREES, CALABASHES, AND BOTTLE GOURDS

Françoise Grenand (1995) raised the point that by restricting the comparison of Tupí-Guaraní plant names to the referential level of botanical species, Balée and Moore (1991) missed showing possibly cognate forms that had wider, though related, botanical ranges of meaning. Although she did not demonstrate exactly how the referents of native terms could be restricted systematically, she did present useful objections that offer clues to language change in the tropical forest.

In the specific case of the spiny palm *Astrocaryum vulgare* Mart., Balée and Moore (1991:232) reported *tukumã'ɨ* (K), *tukumë* (T), and *awala* (W).[4] The term *awala* is a recent borrowing from Carib (Grenand 1995:31) and accounts for noncognacy with K and T here. The K and T names have their origins in a proto–Tupí-Guaraní tongue. But a closely related term has not been lost in W. That term is *tukuma,* and it is cited as a "ritual name" of the closely related, and in the Wayãpi habitat far more common, *Astrocaryum paramaca* Mart. (Grenand 1995:31–32). Grenand (1995: 23–24) argues correctly that in the course of migrations, Tupí-Guaraní words for plants have changed either internally or in their semantic range. In addition to existing plant names being adapted to new but related environmental conditions, as with the name for *Astrocaryum paramaca* Mart., plant names may be extended to new and unrelated environmental conditions where the uses of the species are very similar or the same (Grenand 1995:25–26). Some words do not need to adapt in the course of migrations because their referents are universally present and highly distinctive, like the genus *Inga* with its extra-floral nectaries. The numerous species of *Inga* are typically grouped under a cognate generic label such as *sina* (S) and *ina* (K) (of which two terms the S one is probably closer to the proto-form than the K term) in Tupí-Guaraní languages throughout their range (Grenand 1995:25). Similar processes of linguistic change and conservatism have been reported recently for British names of American birds (Brown 1992).

While a model of adaptations of plant nomenclature to new environmental conditions encountered because of the migration of Tupí-Guaraní people remains to be constructed, S also supplies evidence for such adaptations, though not all of the sort anticipated by Grenand (1995)—meaning, of course, that any such model if genuinely predictive will be necessarily complex. In part, the problem can be approached by determining how reflex forms for an original referent(s) do or do not cover the same semantic range as the original word. The terms for species of the ant-trees, *Tachigali* of the Caesalpiniaceae family and *Triplaris* of the Polygonaceae, are such an example. Balée and Moore (1991:226) showed the reflexes for *Tachigali myrmecophila* Ducke or for *Tachigali paniculata* Aubl. in four languages: *táci'i* (Ar), *taci'iwa* (As), *taši'i* (K), and *taci'iw* (T). In W, members of the same genus are called *tasi* (Grenand 1980:253; Grenand 1989:102). The ant commensals of the tree genus are *Pseudomyrmex* spp., which attack and sting intruders when any part of the tree, their home, is touched.[5] As for *Triplaris,* it harbors the same ants, but its distribution is not found in the habitats of Ar, As, K, and T, though it is seen in that of W, where it is named with marked forms: *móyu-tasi* 'anaconda-*tasi*', *móyu-tasi-pilã* 'anaconda-*tasi*-reddish', and *móyu-tasi-si* 'anaconda-*tasi*-white' (Grenand 1989:305). It is unlikely that a marking reversal would have occurred, since these species are not domesticated nor cultivated. Therefore, it can be suggested that the proto-form **taci'iβa* (my reconstruction) denoted one or more species of *Tachigali* first. In S, *tási* refers to *Triplaris americana* L. alone among plants and to the stinging pseudomyrmecine ants that inhabit its petioles and stems. The genus *Tachigali* is absent from the habitat. It can be hypothesized, therefore, that the aboriginal generic word adapted to changed environmental circumstances in the Llanos de Mojos and came to refer to a species whose name was derived from a more widespread genus of trees.

The disappearance of a name for a plant is also associated with this phenomenon. This is the case in S, where the term for calabash tree (*Crescentia cujete* L. of the Bignoniaceae) is a reflex of the term for bottle gourd (*Lagenaria siceraria* Mol. of the Polygonaceae). In other words, the reflex for calabash is φ in S, though the plant is present in the habitat. Cognate terms for calabash occur in Ar, As, K, T, and W (Balée and Moore 1991:239) and are close to *kwi* (K), whereas the term for calabash in S is *ia-í* (the fruit only is *ía;* cf. Schermair 1958:113).[6] Yet *ia-í* is in a broader sense related to *i'á* (T), *ia* (Ch), *ia* (Gy) (Dietrich 1986:354), and probably *anái* (Gw) (Horn Fitz Gibbon 1955:25), which denote the bottle gourd. Here it can be argued that somehow S *lost* the bottle gourd while keeping its name and retained the calabash while losing its name, as a result of migration and new environmental circumstances. The mechanisms that could account for such specific kinds of change remain obscure but will need to be elucidated in order to construct a systematic model for lexical change over time.

SIRIONÓ PLANT NOMENCLATURE

Sirionó nevertheless exhibits supporting evidence for definitive patterns of plant nomenclature identified earlier in the Tupí-Guaraní family (Balée 1989). First, S shares with Ar, As, Gj, K, T, and W (Balée 1989:6) the same three botanical life-form labels: *íra* (roughly, 'tree'), *kiáta* (roughly, 'herbs and grasses') and *isío* (roughly, 'vine'). As with the other Tupí-Guaraní languages studied, there is no single term for 'plant' in the taxonomic sense and there is no separate life-form term for palms in S. Two life-form labels are also polysemous, as with the other languages: *íra* also refers to 'wood' and 'objects made from wood' (Schermair 1958:133–134); *kiáta* also refers to 'forest' as well as 'bananas' (cf. Schermair 1958: 209), which are recent introductions to the Sirionó.

Second, plant names for nondomesticated plants are sometimes modeled on names for traditionally domesticated plants, but the reverse marking procedure does not seem to occur. In S, nondomesticated congenerics of the sweet potato (*Ipomoea batatas* Lam.) are called *ñiti-rémo* 'sweet potato-vine stem,' a reference to their classification as nondomesticated vines, not the domesticated sweet potato. It is in this sense cognate with numerous terms that refer to nondomesticated *Ipomoea* as 'sweet potato-similar (or false)' (Ar *yiti-rĩ,* As *yɨti-rána,* K *yɨtɨk-ran,* T *zɨtɨk-ran,* and W *yeti-lã;* Balée and Moore 1991:233), since life-form labels are semantically equivalent as markers of nondomesticated or "false" status of a marked term modeled on analogy with a domesticated plant name. Likewise, the S term *urúku-mbwè* '*Bixa orellana*-false,' which refers to nondomesticated species and varieties of annatto (*Bixa orellana* L.), is cognate with K and T *uruku-ran* (Balée 1994:277).

Third, numerous introduced domesticated treelets (including mango, coffee, tamarind, cacao, guava, and catappa nut) are named simply *íra* 'tree,' there being

Table 17.1   Sirionó Reflexes for Traditional Plant Domesticates

Sirionó Name	Names in Related Languages[a]	Botanical Referent
ibási	Ar *awači*, As *awači*, Gj *wači*, K *awaši*, W *awasi*, Gw *avásiki*	*Zea mays* L. (maize)
niñu[b]	Ar *miniyu*, As *aminiyu*, K *maneyu*, T *manizu*, W *minɨyu*, Gw *maniyu*	*Gossypium barbadense* L. (cotton)
ñíti	Ar *yiti*, As *yitíka*, K *yɨtɨk*, T *zɨtɨk*, W *yeti*, Gw *déti* or *deti*	*Ipomoea batatas* Lam. (sweet potato)
urúku	Ar *iriko'i*, K *uruku*, W *uluku* Gw *urúku*	*Bixa orellana* L. (domesticated annatto)
uúba	K *u'ɨwa*, T *u'ɨwa-a*	*Gynerium sagittatum* Beauvois (arrow cane)

[a]The data sources for the Ar, As, Gj, and K terms are found in Balée 1994 and Balée and Moore 1991. The Gw terms are from Horn Fitz Gibbon 1955:25. The S data are from my unpublished notes for the most part and Schermair 1958, 1962.

[b]It seems that several S reflexes for plant names (and no doubt other vocabulary) have been subjected to syncope. Languages in the Tupí-Guaraní family with a stress tendency toward penultimate syllable as with S are more likely to experience sycope of words than those with a tendency toward stress on the final syllable (Dietrich 1990:16), although it is true that As and W also exhibit a tendency toward penultimate stress and may show less syncope than S in the domain of plant names for unexplained reasons. In any case, the S reflex for cotton evinces syncope of the first syllable based on these examples.

no other S name for them; life-form labels in other Tupí-Guaraní languages are applied to domesticated plants evidently only when these are recent introductions (Balée 1989:12; Balée and Moore 1991:218). Likewise, bananas and plantains, as noted above, are referred to generically as "herbs."

Finally, many plant names have been lost in S through neologisms, borrowings, and ultimately peculiarities of Sirionó history and changes in environment. But names for five traditionally domesticated species have been retained in S, as shown in table 17.1. And S has retained terms for several (though not all or even many) semidomesticated plants, such as *ak^{y}ačái* (Ar *akãya'i*, As *kayuwa'ɨwa*) for edible hog plum (*Spondias mombin* L.); *imbéi* (K *ama'ɨ*) for the fibrous ambaibo (*Cecropia concolor* Willd.); *kói* (Gj *kapowa'ɨ*, K *kupa'ɨ*) for the medicinal copaíba oil tree (*Copaifera* sp.); *ñikisɨaí* (Ar *arakači'i*, Gj *arakači'a*, T *zarakači'a'ɨw*, Gw *akasía*; Horn Fitz Gibbon 1955:25) for the edible wild papaya or gargatea (*Jacaratia spinosa* [Aublet] A.DC.); *urúre* (As *murure-ete*, Gj *merere'ɨ*, and K *murure'ɨ*) for the edible Amazonian mulberry tree *Brosimum acutifolium* Huber; *sína* (K *ɨ̃ŋa* et al.) for the edible *Inga* spp.; and *tákwa* (Ar *ta'akɨ̄*, Gj *takwar*, K *takwar*) for a bamboo used in making arrow points (*Guadua* sp.). This proclivity to retain names for domesticates and semidomesticates seems to be generalized in the Tupí-Guaraní family, and S is divergent but not an exception, even if its botanical environment is unique.

CONCLUSIONS

The Sirionó language has been the subject of controversies over its origins and classification. The most extreme position has questioned the legitimacy of including Sirionó in the Tupí-Guaraní family because of a presumed profound substrate influence, but no empirical, linguistic evidence has been supplied for that. That assumption was associated with the mistaken axiom that every Tupí-Guaraní society known had a strong agricultural basis and therefore there could not have been exceptions. The Sirionó was one of the first of a series of trekking and foraging societies studied by professional ethnographers in Amazonia; several of those studied by participant observation since have proven to be of the Tupí-Guaraní language family also, such as Guajá, Yuquí, and Araweté. In terms of the linguistic evidence alone, Sirionó is a full member of the Tupí-Guaraní language, albeit in a subgroup of its own (Dietrich 1990; also see Priest 1987) or in a questionable grouping of languages that are close to it geographically but apparently distant from it in terms of vocabulary, such as Guaraçug'wé (Pauserna) and Guarayo, which are themselves close to Chiriguano. Sirionó is, nevertheless, intrusive in lowland Bolivia, and it may be that in the course of migrations, probably from the south and associated with the Guaranian invasions of lowland Bolivia in the fifteenth and sixteenth centuries, the Sirionó agricultural economy underwent changes. And so did the Sirionó language in reflecting those changes. I suggest that the early Sirionó inhabitants had lost or were in the process of losing dependence on a larger inventory of domesticated crops than they possessed in the early twentieth century and that they were becoming less sedentary. These migrations and probably contact with other cultures, perhaps of a hostile sort, led to losses of domesticates and in some cases of words for those plants. In other cases, as with the present Sirionó word for certain ant trees, the ranges of meaning of existing vocabulary changed to accommodate new conditions encountered in the new environment. Even without specifying the exact mechanisms involved in these transformations of Sirionó vocabulary, which seem intangible at present, it can be argued that Sirionó is fully a member of the Tupí-Guaraní language family in terms of plant nomenclature and classification. The Sirionó language exhibits a nomenclatural dichotomy between traditional domesticates and nondomesticated species; modeling and marking for nondomesticates based on names of traditional domesticates; and cognates for plant life forms and traditional domesticates like many other Tupí-Guaraní languages.

These similarities together with the differences that Sirionó plant names show when compared to plant names in other Tupí-Guaraní languages can fruitfully be analyzed from the perspective of historical ecology. Historical ecology considers Sirionó language change as well as language retention as products of a long-term entanglement between the speakers themselves and Amazonian landscapes. Efforts to document Sirionó and other Tupí-Guaraní languages should reflect that entanglement.

ACKNOWLEDGMENTS

Grateful acknowledgment is made to the Wenner-Gren Foundation for Anthropological Research and the Center for Latin American Studies (Tulane University) for financial underwriting of my fieldwork with the Sirionó people in 1993, 1994, and 1997. Thanks are due to the Sirionó Council and CIDDEBENI of Trinidad for logistical support and other encouragement. Responsibility for any errors herein is mine alone.

NOTES

1. In point of fact, disagreement exists on just how genetically close Guarayo and Sirionó are. There is little doubt that Guarayo and Guaraçug'we or Pauserna are very closely related linguistically and culturally (Riester 1977; Stearman 1984:640), more so than either is to Sirionó. Lemle (1971:128), Loukotka (1968:119) and Rodrigues (1984/85: 38) included them in the same proposed subgroup of the Tupí-Guaraní family on the basis of phonological criteria alone, but Wolf Dietrich (1990:115) finds Sirionó to form a subgroup of its own based on phonological, morphological, and grammatical criteria. Dietrich's arguments are persuasive, but his scheme disallows a subgroup of Wayãpi and Urubú (Ka'apor) to be formed because Ka'apor is a low-rate language in terms of retention of the criteria he uses. Yet Wayãpi seems to be more closely intelligible with Ka'apor than any other known Tupí-Guaraní language. Guarayo is understood by some Sirionó informants also, but there has been some intermarriage among them and it is not clear how much of this mutual intelligibility has been recently learned. In any case, the terms for calabash and numerous other plants are divergent between them.
2. On the basis of phonological, grammatical, and morphological evidence, Dietrich (1990:58) views Sirionó as a "peculiar case" (also see Priest 1987 on an item of phonology only); for him it is a pivotal link between Amazonian, Bolivian, and southern branches of the Tupí-Guaraní language family, forming an isolated subgroup of its own within the family. Dietrich (1986:204) also argues, without clear supporting evidence in my view, for a northern rather than southern origin for the Sirionó. Stearman (1984:640; 1989:22) suggested a southerly origin for the Yuquí and Sirionó based on linguistic, ethnohistoric, and cross-cultural evidence, and Priest (1987) showed an apparently high index of cognacy between Yuquí and Sirionó. I do agree with Dietrich (1986:204), however, that the Sirionó are intrusive in the Llanos de Mojos and not representatives of a proto-population, proto-culture, or non-Tupian proto-language that supposedly existed there before the Guaraní migrations of the 1400s and 1500s. This view does not presuppose that the region was uninhabited before they arrived.
3. Aché-Guayaki and Xetá might have been Guaraní-ized (Dietrich 1990), unlike the probable scenario with the others mentioned here.
4. The following Tupí-Guaraní language abbreviations will be used in the remainder of this chapter: Ar (Araweté), As (Asurini do Xingu), Ch (Chiriguano), Gj (Guajá), Gy (Guarayo), Gw (Guaraçug'wé), K (Ka'apor), S (Sirionó), T (Tembé), and W (Wayãpi of French Guiana).

5. The proteinaceous venom ejected by the lancets of these ants is used to reduce fever among both the Sirionó and the Ka'apor, though the same ants come from different tree species in the respective cases and though the Sirionó and Ka'apor are separated by about 2000 km of Amazon forest. The venom of a closely related species contains a polysaccharide that deactivates part of the human complement system, perhaps being useful therefore in the treatment of rheumatoid arthritis (Schultz and Arnold 1977). The similarities between the Sirionó and the Ka'apor in the use of pseudo-myrmecine ants seem to be historically parallel rather than convergent or borrowed.

6. In this regard, it is interesting to note that the bottle gourd is called *anái* (Horn Fitz Gibbon 1955:25) in Gw. Gw speakers lived near the Sirionó along the Guaporé River, which suggests a different migratory and contact history for them, in spite of possibly a common origin with S. The term is clearly noncognate if referents are restricted to botanical species (or even families, in this case).

REFERENCES

Anonymous 1781. Abrégé d'une relation Espagnole . . . In *Lettres édifiantes et curieuses, écrites des missions étrangères*, vol. 8 (orig. 1702). Paris: J.G. Merigot le jeune.

Balée, W. 1989. Nomenclatural patterns in Ka'apor ethnobotany. *Journal of Ethnobiology* 9(1):1–24.

Balée, W. 1994. *Footprints of the Forest: Ka'apor Ethnobotany—The Historical Ecology of Plant Utilization by an Amazonian People*. New York: Columbia University Press.

Balée, W. 1998. Historical ecology: Premises and postulates. In *Advances in Historical Ecology*, ed. W. Balée. Pp. 13–29. New York: Columbia University Press.

Balée, W. 1999. Mode of production and ethnobotanical vocabulary: A controlled comparison of Guajá and Ka'apor (Eastern Amazonian Brazil). In *Ethnoecology: Knowledge, Resources, and Rights*, ed. T.L. Gragson and B. Blount. Pp. 24–40. Athens: University of Georgia Press.

Balée, W., and D. Moore. 1991. Similarity and variation in plant names in five Tupí-Guaraní languages (Eastern Amazonia). *Bulletin of the Florida Museum of Natural History, Biological Sciences* 35(4):209–262.

Balée, W., and D. Moore. 1994. Language, culture, and environment: Tupí-Guaraní plant names over time. In *Amazonian Indians from Prehistory to the Present: Anthropological Perspectives*, ed. A. Roosevelt. Pp. 363–380. Tucson: University of Arizona Press.

Block, D. 1980. In search of El Dorado: Spanish entry into Moxos, a tropical frontier, 1550–1767. Ph.D. diss., University of Texas, Austin.

Block, D. 1994. *Mission Culture on the Upper Amazon: Native Tradition, Jesuit Enterprise, and Secular Policy in Moxos, 1660–1880*. Lincoln: University of Nebraska Press.

Brown, C.H. 1992. British names for American birds. *Journal of Linguistic Anthropology* 2(1):30–50.

Cardús, J. 1886. *Las Misiones Franciscanas entre los Infieles de Bolivia*. Barcelona.

Chávez Suárez, J. 1986. *Historia de Moxos*. 2d ed. La Paz: Editorial Don Bosco.

Deacon, T. 1997. *The Symbolic Species: The Co-Evolution of Language and the Brain*. New York: W.W. Norton.

Denevan, W.M. 1966. *The Aboriginal Cultural Geography of the Llanos de Mojos of Bolivia.* Berkeley: University of California Press.

Dietrich, W. 1986. *El Idioma Chiriguano: Gramática, Textos, Vocabulario.* Madrid: Instituto de Cooperación Iberoamericana.

Dietrich, W. 1990. More evidence for an internal classification of Tupi-Guarani languages. *Indiana* Supplement 12:1–136. Berlin: Gebr. Mann Verlag.

Distel, A.A.F. 1984/85. Hábitos funerarios de los Sirionó (Oriente de Bolivia). *Acta Praehistorica et Archaeologica* 16–17:159–182.

D'Orbigny, A. 1946. *Descripción geográfica, histórica y estadística de Bolivia.* La Paz.

Erickson, C.L. 1995. Archaeological methods for the study of ancient landscapes of the Llanos de Mojos in the Bolivian Amazon. In *Archaeology in the Lowland American Tropics,* ed. P.W. Stahl. Pp. 66–95. Cambridge: Cambridge University Press.

Fausto, C. 1997. A dialética da Predação e familiarização entre os Parakanã da Amazônia Oriental: Por uma teoria da guerra Ameríndia. Ph.D. diss., Museu Nacional, Rio de Janeiro.

Grenand, F. 1989. *Dictionnaire Wayãpi-Français: Lexique Français-Wayãpi (Guyane Française).* SELAF 274. Paris: Peeters and SELAF.

Grenand, F. 1995. Le voyage des mots—Logique de la nomination des plantes: Exemples dans des langues tupi du Brésil. *Revue d'Ethnolinguistique (Cahiers du LACITO)* 7:23–42.

Grenand, P. 1980. *Introduction à l'Étude de l'Univers Wayãpi: Ethnoécologie des Indiens de Haut-Oyapock (Guyane Française).* Langues et Civilisations à Tradition Orale n. 40. Paris: SELAF.

Holmberg, A. 1948. The Sirionó. In *Handbook of South American Indians,* Vol. 3: *The Tropical Forest Tribes,* ed. J.H. Steward. Pp. 455–463. Bulletin 143, Bureau of American Ethnology. Washington, D.C.: United States Government Printing Office.

Holmberg, A.R. 1985 (orig. 1950). *Nomads of the Long Bow.* Prospect Heights, Ill.: Waveland Press.

Horn Fitz Gibbon, F. von. 1955. *Breves Notas sobre la Lengua de los Indios Pausernas.* Publicaciones de la Sociedad de Estudios Geográficos e Históricos. Santa Cruz, Bolivia: Imprenta "Emília."

Isaac, B. 1977. The Siriono of eastern Bolivia: A reexamination. *Human Ecology* 5(2):137–154.

Langstroth, R.P. 1996. Forest Islands in an Amazonian Savanna of Northeastern Bolivia. Ph.D. diss., University of Wisconsin-Madison.

Lemle, M. 1971. Internal classification of the Tupi-Guarani linguistic family. In *Tupi Studies I,* ed. D. Bendor-Samuel. Pp. 107–129. Norman, Okla.: Summer Institute of Linguistics.

Loukotka, Č. 1968. *Classification of South American Indian Languages.* Los Angeles: Regents of the University of California.

Lunardi, F. 1938. I Siriòno. *Archivio per l'antropologia e l'etnologia, Firenze* 6:178–223.

Nordenskiöld, E. 1924. The ethnography of South-America seen from Mojos in Bolivia. *Comparative Ethnographical Studies* 3:1–254.

Priest, P.N. 1987. A contribution to comparative studies in the Guaraní linguistic family. *Language Sciences* 9(1):17–20.

René-Moreno, G. 1888. *Biblioteca Boliviana: Catálogo del Archivo de Mojos y Chiquitos.* Santiago: Imprenta Gutenberg.

Riester, J. 1977. *Los Guarasug'we: Crónica de Sus Ultimos Días*. La Paz: Amigos del Libro.

Rodrigues, A.D. 1984/85. Relações internas na família linguística Tupi-Guarani. *Revista de Antropologia* 27–28:33–53.

Rydén, S. 1941. *A Study of the Siriono Indians*. Gothenburg: Elanders Boktryckeri Aktiebolag.

Schermair, A. 1958. *Vocabulario Sirionó-Castellano*. Innsbruck: Innsbrucker Beiträge zur Kulturwissenschaft.

Schermair, A. 1962. *Vocabulario Castellano-Sirionó*. Innsbruck: Innsbrucker Beiträge zur Kulturwissenschaft.

Schultz, D.R., and P.I. Arnold. 1977. Venom of the ant *Pseudomyrmex* sp.: Further characterization of two factors that affect human complement proteins. *Journal of Immunology* 119(6):1690–1699.

Stearman, A.M. 1984. The Yuquí connection: Another look at Sirionó deculturation. *American Anthropologist* 86(3):630–650.

Stearman, A.M. 1989. *Yuquí: Forest Nomads in a Changing World*. New York: Holt, Rinehart & Winston.

Steward, J.H. 1948. Culture areas of the tropical forests. In *Handbook of South American Indians*, Vol. 3: *The Tropical Forest Tribes*, ed. J.H. Steward. Pp. 883–899. Bulletin 143, Bureau of American Ethnology. Washington, D.C.: United States Government Printing Office.

Susnik, B. 1994. *Interpretación Etnocultural de la Complejidad Sudamericana Antigua*, Vol. 1: *Formación y Dispersión Étnica*. Asunción: Museo Etnográfico Andrés Barbero.

Townsend, W. 1995. Living on the edge: Sirionó hunting and fishing in lowland Bolivia. Ph.D. diss., University of Florida, Gainesville.

Wierzbicka, A. 1997. *Understanding Cultures through Their Key Words: English, Russian, Polish, German, and Japanese*. New York: Oxford University Press.

18

THE ENDANGERED LANGUAGES OF AFRICA

A Case Study from Botswana

Herman M. Batibo

According to Coulmans (1983), there are about 7000 languages in the world, of which about 2500 or 31 percent are spoken in Africa. This makes Africa the continent with the highest number of languages in the world.[1] Although, on average, there would be about 50 languages per country, the distribution of languages on the continent is not so even, for there are quasi-monolingual countries[2] like Burundi, Rwanda, Somalia, Lesotho, and Swaziland, on the one hand, and multilingual and multiethnic countries, on the other. Countries with the highest linguistic diversity include Nigeria and Sudan with more than four hundred languages each, followed by Cameroon, Zaire, and Tanzania, which have more than a hundred languages each.

Although language death has been occurring in Africa for many centuries, the process appears to have been accelerated in recent years by a number of factors such as the raising of status of some lingua francas to become national media, the pressure for many small communities to integrate in wider and often more prestigious communities and the appeal of modern living associated with formal education, paid jobs, access to national-level information and knowledge, advanced technology, and urban life. Such modern living is often also associated with the more prestigious languages. Moreover, the rise and dominance of international languages like English and French have also caused threatening pressure on the small languages. As a result, many speakers of minority languages[3] choose to use the dominant languages more predominantly, even to the extent of abandoning their own mother tongues, and encourage their children to be more or exclusively proficient in the dominant languages.

It is because of this scenario that Africa is losing many of its minority languages. In fact, it is the continent where the most dramatic changes are happening. For example, in Nigeria, the most linguistically complex country in Africa, out of the more than 400 languages (the majority of them spoken in central Nigeria), 180 have under 400 speakers and are, therefore, highly threatened. Out of these, many are already moribund (Croizer and Blench 1992). The death of these languages will not only mean loss of the linguistic peculiarities of these languages but also loss of enormous cultural diversity. In Sudan, where there are also more than 400 languages, many are being pressured by the expansion of Arabic and the unstable political situation in the country. Cameroon, with its high density of languages (numbering about 240), has numerous threatened languages, several of which are already moribund.[4]

Moreover, in East Africa, with more than 230 languages, the expansion of Kiswahili as a dominant lingua franca is causing a dangerous threat to most of the other indigenous languages in the region (Batibo 1992). As rightly observed by Heine (1997), the biggest threat to the minority languages in Africa is not the presence of the colonial languages, such as English, French, or Portuguese (which normally remain the languages of the elite), but the predominance of the powerful indigenous lingua francas, which often give rise to what I have called "marked bilingualism" (Batibo 1992). According to the marked bilingualism model, language shift occurs when a bilingual community speaks two languages whose social prestige or symbolic values are not equal. The speakers will tend to abandon the language of lower social prestige (usually the minority language) in favor of the language of higher social value (usually the majority or dominant language). Hence, not only the domains of use of the minority languages diminish progressively, but also their linguistic structures and lexicon become eroded until a complete language shift to the dominant language results (cf. Hill this volume). According to some of the projections in a study carried out by Krauss (1992), as much as 90 percent of the world languages might disappear or be on the verge of extinction during the course of the twenty-first century. Many of these will be African languages.

LOSS OF LINGUISTIC, CULTURAL, AND ECOLOGICAL KNOWLEDGE

As Diamond (1993) rightly pointed out, each language is tied up with a specific culture, literary expression, and world view. Every language is the custodian of its speakers' cultural experiences, which are often the result of their many centuries of interaction with their physical milieu, inter- and intraethnic contacts, and relations with the supernatural world. Therefore, no two linguistic communities will have exactly the same past experiences. This diversity of experiences means also diversity in physical adaptations, conceptualization of ideas, ecological knowledge, and vision of the universe. Hence the existence of many language communities in Africa should not be seen as a source of linguistic or ethnic conflicts (as it has been in political spheres), but rather as an important resource residing in the wealth and diversity of zoological, botanical, and other indigenous knowledge as well as skills and methods of dealing with the physical and human worlds. Thus, if most of the languages in Africa became extinct, much of this wealth of indigenous knowledge, concepts, and skills would also disappear. This includes traditional medicine, which, in some African communities, has been found to be effective in treating complex diseases such as cancer, asthma, leprosy, and tuberculosis, as well as chronic cases of STDs, bilharzia, and anemia. Moreover, indigenous knowledge holds in-depth understanding of the ecosystem and animal behavior.

Unfortunately, it is not easy to make a thorough documentation of all this knowledge. The studies carried out by linguists, who normally focus on structural and lexical analysis, do not render much service to the presentation of this knowledge.

As a sad coincidence, the loss of linguistic and cultural diversity in Africa is occurring along with the loss of biodiversity, as a result of deforestation of the equatorial forests for timber and pulp, the clearing of the woodlands for cultivation and firewood, the overgrazing of the rich savannalands, the uncontrolled hunting of wildlife, and the pollution of the water places. As a result, Africa is now experiencing critical desertification, hostile climatic conditions, such as continued droughts, unavailability of clean water, and disappearance of certain species, including the white rhinos and elephants. Moreover, the younger generations are rapidly losing their competence in the knowledge of their ecosystem, such as names of plants and wild animals and their characteristics or uses, at the same time that they are losing interest in their folklore and traditions.

THE MULTILINGUAL AND MULTIETHNIC NATURE OF BOTSWANA

Like most countries in Africa, Botswana is a multilingual and multiethnic country; more than 30 languages are spoken as mother tongues within its borders. The languages are divided among three groups: Bantu,[5] which includes Setswana, Ikalanga, Sekgalagadi, Subiya, Thimbukushu, Shiyeyi, Sebirwa, Setswapong Lozi, Nambya (Najwa), Gciriku, Ryozi, and Otjiherero; Khoesan,[6] which includes !Xoo, #Hua, Khute, Naro, /Gwi, //Gana, Kxoe, //Ani, Ju/'hoan, #Kx'au//'ei, Deti (Teti), Kua, Tshwa, Tshasi, Buga, /Xaise, Ts'ixa, Danisi, Cara (Tshara), Hietshware, Nama (Kgothu), #Haba, Ganadi, and Shua); and, finally, Indo-European, which includes Afrikaans. Some of the Khoesan languages could, strictly speaking, be regarded as part of a dialect continuum, in view of their high intercomprehension. Botswana has the highest number of Khoesan languages in Africa.

Most of the Bantu speakers are farmers or fishermen, and they live in the relatively fertile and less dry areas in the eastern parts of the country, the Okavango Valley and Chobe District in the north. The Khoesan speakers, once commonly known as Bushmen, who are traditionally hunters and gatherers, live in the drier parts of the country, mainly in the Kalahari Desert. There are also pockets of Bushmen in the Okavango Valley, where they have become fishermen, and in the Nata area, where they are mostly pastoralists. The Afrikaans-speaking population is found largely in the Ghanzi District in the western part of the country. The speakers in this district include the descendants of the Boers, who settled in the region as farmers in 1898, and the local population that became assimilated to Afrikaans language and culture.

Setswana, demographically the most dominant language in the country, is spoken as the first language by at least 78 percent of the 1,400,000 or so Botswana inhabitants. It is, therefore, the de facto lingua franca and the country's declared national language. It is estimated that more than 15 percent of the country's population speaks it as a second language. Next in size is Ikalanga, spoken by some 150,000 people, or about 11 percent of the country's population (Andersson and Janson 1997). Some studies have put the figure even higher than 200,000 speakers, that is, about 15 percent of the country's population (Mathangwane 1996:2). In spite of its demographic importance, Ikalanga has no official status in the country. The rest of the languages are minority languages with populations of less than 40,000 speakers, as indicated in table 18.1 (after Batibo, Mathangwane, and Mosaka 1996).

English, as a former colonial language, is spoken mainly by the educated portion of the population, often urban dwellers who work in government, parastatal institutions, and the private sector. It is the declared official language of the country.

THE PATTERN OF LANGUAGE SHIFT AND DEATH IN BOTSWANA

The prevalent cases of language shift and death in Botswana could be accounted for by the marked bilingualism model that I propounded (Batibo 1992). The pattern of language shift and death appears to involve four stages.

Relative Monolingualism in the Minority Languages

Before the arrival of the dominant Bantu groups in Botswana more than 1500 years ago, the original communities in the area, the Khoesan groups, lived in relative isolation because of the small size of their communities, although some trade-related contact appears to have occurred among them. Hence, it can be assumed that most of them were largely monolingual. Their lives were based on hunting, foraging, and probably some agropastoralism.

Encroachment by the Dominant Languages

Between A.D. 1000 and 1600, significant waves of migrants settled in what is now Botswana. These groups included the Bakgalagadi, the Bakalanga, the Batswana (in their various subgroups), and later the Afrikaners. The new immigrants either

Table 18.1   Botswana Ethnic Languages and Their Estimated Speakers

Language Group	Languages	Population	Percent of Population
Bantu	Setswana	1,100,000	ca. 78.6
	Ikalanga	150,000	10.7
	Otjiherero	31,000	2.2
	Shiyeyi	20,000	1.4
	Sekgalagadi	15,000	1.1
	Thimbukushu	8,000	0.6
	Subiya	7,000	0.5
	Others	20,000	1.4
	Total (Bantu)	1,351,000	96.5
Khoesan	Shua/Tshwa/Kua	17,500	1.25
	Naro	9,000	0.6
	Ju/'hoan	3,000	0.2
	#Kx'au//'ei	2,500	0.17
	Kxoe (incl. //Ani)	2,500	0.17
	!Xoo	2,000	0.14
	//Gana	800	0.06
	/Gwi	500	0.04
	Others	2,500	0.17
	Total (Khoesan)	40,300	2.8
Indo-European	Afrikaans	ca. 3,000	ca. 0.2
Grand Total		1,394,300	99.7

absorbed or dislocated the earlier inhabitants of the area. They were not only numerically superior but also socioeconomically stronger: they owned more livestock, practiced crop farming, and used more advanced tools. The minority groups had to learn the languages of the new inhabitants in order to interact with them socially and economically.

Domination by the Encroaching Language

Although the relatively dominant languages like Ikalanga, in the northeast, Sekgalagadi, in the southwest, and Afrikaans and Naro in the Ghanzi District continued to exert influence on the smaller or more vulnerable languages in their vicinity, their impact in terms of causing language shift has not been as decisive as is the case with the Setswana language. Setswana seems to provide the most serious threat to the survival of the minority languages. Setswana, which is the de facto lingua franca, the declared national language of the country, and the undeclared alternate official language, has all the numbers in terms of prestige to induce the speakers of other languages, particularly the minority languages, to give up most, if not all, of their languages' domains of use. Setswana is the language of wider communication in the country, facilitating interethnic communication. It is the language that symbolizes national unity and identity, and it is the only indigenous language used in the first years of education. Therefore, no child can start school without reasonable proficiency in it and hope to succeed. Together with English, Setswana is the language of information and knowledge, although limited in some respects compared with English; the language of public administration, church services, and most nonofficial and technical interactions; and the presumed language of socioeconomic advancement and promotion to modern living, although in reality it is English that retains that role. Ironically, Setswana speakers in turn strive to learn English for socioeconomic advancement and promotion to modern living (Janson and Tsonope 1991). As a result, many parents in minority language groups want their children to be proficient in Setswana, even more than English, as table 18.2 indicates.

In June and July 1996, N. Mosaka and I surveyed speakers of 9 minority languages in the remote rural areas of Botswana where many people have had little or no contact with major urban centers. The aim of the survey was to find out to what extent minority language speakers were able to receive and comprehend information disseminated to them in Setswana and English and to find out how they perceived the importance of Setswana and English in the future of their children. As table 18.2 indicates, most parents in the minority language groups would like their children to be proficient in both Setswana and English. This desire is particularly unanimous among the Khoesan language groups, probably because of their socioeconomic vulnerability.

The results of the survey were very similar to those of Vossen (1988), who found that in the Ngamiland area many children of parents from different ethnic groups had Setswana as their first language and were often less proficient in the languages of their parents. The same survey revealed that the Khoesan language speakers were more vulnerable, presumably because their languages were looked down upon by the Bantu-speaking groups (Vossen 1988:27–47). Similar trends were reported by Hasselbring (1996:30) in the Ghanzi District and Batibo (1997) in the Nata area.

It is therefore evident that, given such positive attitudes toward Setswana and English, the place and role of these languages, especially Setswana, will continue to be enhanced even at the grassroots, and eventually they may replace the minority languages. One needs to remark, however, that there seems to be a hierarchy of aspirations. While the minority language speakers want to be proficient in Setswana because they believe that it is a key to modernization and better living, Setswana speakers believe rather that it is through English that they will attain modernization and better life. Hence they have a relatively low esteem of their own language, Setswana.

Table 18.2 Parents' Preference for Their
Children to Learn Setswana
and/or English

Language	Setswana (%)	English (%)
Ikalanga	82.5	80.0
Shiyeyi	88.0	82.0
Subiya	90.0	88.0
Thimbukushu	95.0	42.0
Ju/'hoan	100.0	75.0
!Xoo	100.0	100.0
Naro	100.0	100.0
#Kx'au//'ei	100.0	86.0
//Ani	100.0	100.0

Language Shift

Many cases of language shift have occurred in Botswana in recent years. Such shifts have meant that the relevant minority speakers have abandoned or are on the verge of abandoning their language in favor of a dominant or more prestigious language. Most of these cases have involved Khoesan languages (Traill, personal communication 1995): (a) Tyua, a Khoe language once spoken along the Zimbabwean border in the Sepako area; (b) S5,[7] a !Xoo variety, spoken in southwest Botswana in the Khakhea area (referred to as Masarwa by Bleek [1929]); (c) Ts'ixa, a Khoe language spoken in Mababe area; (d) Deti (or Teti), a Khoe language spoken in the north-central area. Only a few elderly speakers of Ts'ixa and Deti remain. At the moment, the most endangered languages are those that are either demographically weak (such as //Gana and /Gwi, #Hua, Tshasi, Khute), or heavily overwhelmed by dominant languages (such as /Xaise, Shua, !Xoo, Kua, and Tshwa).

Apart from the prestige factors associated with Setswana, there are a number of other factors that have precipitated the abandonment of the minority languages by their speakers. First, there has been a weakening of the social status of minority languages. This is particularly because they are not used in any domain of public life. The stigma created by their low social status has caused, in many cases, not only reduction of their domains of use but also rapid erosion or simplification of their linguistic characteristics. Second, the rapid urbanization and search for labor have forced many minority speakers to migrate into urban centers, where Setswana is the lingua franca. Third, since Botswana's independence in 1966, there has been a rapidly growing sentiment toward national unity and identity. This has created a greater inducement for the minority language speakers to adopt a majority language like Setswana. Fourth, there has also been a tendency, especially among the Khoesan language speakers, to move in scattered bands with little or no central authority and to subsist under socioeconomic uncertainty that results from the vagaries of their hunting and foraging life, especially under unfavorable climatic conditions such as times of drought. This has often made them dependent on Bantu agropastoralists and hence vulnerable to linguistic domination by other groups. And, finally, the increasing number of mixed marriages has resulted in the adoption of Setswana as the compromise language (Hasselbring 1996:28).

CONSEQUENCES OF LINGUISTIC AND CULTURAL DIVERSITY LOSS

If the present trend of language shift continues unchecked, most minority languages will disappear by the end of the twenty-first century. Such loss would be felt in both linguistic and cultural spheres.

In its linguistic diversity, Botswana not only has the highest number of Khoesan languages in the world but also has languages with the most remarkable linguistic characteristics in terms of clicks and their accompaniments, phonemic complexity, unique tone marking and peculiar morphological characteristics (Traill 1994; Andersson and Janson 1997). Moreover, because Botswana is the zone of convergence between Western and Eastern Bantu streams, the Bantu languages are unique in their inheritance of the two linguistic origins. On the other hand, the linguistic interaction between the two linguistic groups, namely Bantu and Khoesan, has given rise to very peculiar sociolinguistic situations in terms of patterns of language use, attitudes, and identity. The loss of such rich linguistic diversity would be a setback not only to Botswana, but also to humanity in general.

Botswana is also a country of great cultural diversity, as its peoples have gone through different historical and ecological experiences. Historically, each of these peoples has occupied itself with some basic socioeconomic activities varying from farming, pastoralism, fishing, to hunting and gathering. All groups have their own distinctive lifestyles, kinship and social relations, foods, traditional dress, songs and dances, literary expressions, ornaments, and crafts. Each community has also formed its own ethical values, code of conduct, and social norms. Moreover, each community has its own world view, its knowledge of the physical milieu, and its aspirations and adaptations; each has developed its own skills and appropriate tools or weaponry to deal with its physical environment and means to respond to environmental hardships. The abandonment of any of these accumulated experiences would again constitute great loss both to local people and to human heritage at large.

One typical example is that of the Khoesan people, who manifest both linguistic and cultural diversity with rich indigenous knowledge accumulated over the last 20,000 or more years of their existence in the southern African subcontinent. The Khoesan cultural experience and indigenous knowledge, as encoded also in the respective languages, embodies a sophisticated understanding of animals, plants, the soil, the weather, traditional medicine, human relations, moral values (Koch 1997:4) as well as tremendous knowledge of animal behavior, tracking, and the ecosystem (Crawhall 1997). Although these people were pushed into the hostile Kalahari desert, they have known, through adaptation, how to turn the near-barren environment into a sustainable ecosystem by creating a balanced use of the scarce animals and plants in their surroundings. Moreover, they have rich customs and traditions associated with different age groups. The ancient paintings still found on many rocks and in caves are remarkable remnants of the profound cultural diversity that was embedded in people's ways of communication, expression of their world views, and their spiritual attachment to the land and the animals they hunted. The Khoesan-speaking people also have elaborate conceptions about the divine and the origin of the universe. Their complex knowledge of the ecosystem has been described by some authors as the very origin of science as we know it today (Koch 1997:4). It would therefore be a tragedy for the Khoesan people and all of humanity if such cultural experiences and indigenous knowledge, accumulated over such a long period, should be allowed to disappear.

Another area in Botswana that abounds in both cultural and biological diversity is the Okavango Delta, which, because of its rich fauna, flora, and massive river valleys, is home to more then twelve linguistic communities of different origins. The region has a very rich history of settlement, human interaction, and socioeconomic activities, which include hunting, fishing, agriculture, and pastoralism. The various linguistic groups have also enriched their cultural experiences, adapted themselves to their ecological milieu, developed their own skills to deal with their physical world, and formed their own philosophical concepts about the universe and the supernatural world. Again, with the dominance of Setswana, most of the minority languages in the area are threatened (Vossen 1988). Their death would be the end of the area's multilingualism, cultural diversity, and tremendous wealth of indigenous knowledge. Moreover, with the dramatic increase of population, farming, and overgrazing in the region, the biodiversity in the area is also severely threatened.

EFFORTS TO PRESERVE THE MINORITY LANGUAGES AND THEIR CULTURAL EXPERIENCES

Recently, there have been concerted efforts to preserve the minority languages of Botswana. These efforts have come from researchers, missionaries, individual philanthropists, and some élites from the minority groups themselves. These groups have been concerned with the descriptions of the respective languages (orthographies, grammars, and dictionaries), the promotion of literacy, the preparation of reading materials, the formation of associations to promote the respective languages and to sensitize the speakers to be proud of their languages and cultures. Some ethnic groups, particularly the Bakalanga, the Ovaherero, and the Bayeyi, seem to be especially concerned about the maintenance of their languages and cultures.

One of the encouraging outcomes of these efforts is the creation of positive attitudes of the speakers toward their languages. Such positive attitudes would be enhanced, however, if these minority languages could have some officially recognized roles, however limited, in their respective areas. It is gratifying to note that since the 1990s there has been an apparent change of attitude in Botswana toward the minority languages, to the extent that the name of the official language council, which had been known as the National Setswana Language Council since 1981, was recently changed to the Botswana Languages Council, thus recognizing the multilingual nature of the country. Such a change in national policy might eventually trigger a change in the public attitudes toward the minority languages. This change of external setting might instill pride in the speakers themselves and desire to value and preserve their languages, while subscribing to Setswana according to a pattern of unmarked bilingualism (Batibo 1992; see below for definition).

Moreover, the recent historical decision by the Botswana Parliament that the minority languages be used at the formative level of education is a great turning point for the country, for it will lead parents to revalue their languages and encourage their children to learn them. This is because the main reason for the shift to Setswana was to facilitate primary education and integration into the wider community. It is now up to the cultural and education authorities to translate the new policy into manageable and appropriate implementation.

At present, schools treat indigenous communities as if they have no knowledge of their own. As Crawhall (1997) rightly suggests, saving the languages is part of restoring confidence and dignity in indigenous knowledge and skills within the communities. The indigenous knowledge should be used as the basis for teaching environmental sciences, geography, history, and other subjects. Developing the curriculum out of the community's experience will give the languages a new purpose and brighter future. Moreover, the "bush" terminology could be adapted to new conceptual contexts in the classroom and even in public use.

One serious dilemma that minority language speakers face, if they are not proficient in a majority language such as Setswana, is marginalization from access to vital knowledge and information concerning regional, national, and international affairs. In Botswana, where all news, information, and vital knowledge are disseminated in Setswana and English, lack of proficiency in these languages would mean not only being kept ignorant of developments at the regional, national, and international levels but also being denied the opportunity to participate in affairs at all these levels. In the survey that N. Mosaka and I carried out in remote rural areas, we discovered that a sizable percentage of the minority speakers did not understand the news or any other information disseminated to them

Table 18.3 Minority Language Speakers
Who Cannot Understand News or Other
Information Disseminated in Setswana
or English

Language	Setswana (%)	English (%)
Subiya	0.0	85.1
Ikalanga	22.5	80.3
Ju/'hoan	9.0	90.0
Seyeyi	35.2	65.8
//Ani	5.4	65.2
Naro	12.9	64.5
#Kx'au//'ei	14.5	57.1
!Xoo	10.6	85.0
Tshimbukushu	25.6	87.4

through radio, television, the newspapers, or other published material. The languages used in these cases are Setswana and English. The study was carried out by questionnaires rather than observation, and as such, the results may appear subjective to some extent; nevertheless, they suggest a pattern (see table 18.3).

Thus the minority language speakers who have limited or no proficiency in Setswana and English are unable to stay abreast of important news or information from the outside world, which may include knowledge about socioeconomic activities (e.g., agriculture, forestry, animal husbandry, hunting, fishing, mining, and water preservation), social welfare (e.g., health, education, nutrition, hygiene, literacy, and family life), diseases (e.g., the spread of AIDS, STDs, malaria, and lung disease), and spiritual life (e.g., religious teaching and morality).

CONCLUSION

The minority language groups in Botswana, just as in most other countries in Africa, have been caught in a critical dilemma: they find themselves subscribing to a double allegiance. On the one hand, they want to preserve their languages, cultures, and identity. They would like their children to push forward their linguistic and cultural heritage. On the other hand, they want to be part of the wider and modern world, and they know that the only way to have access to modern living and technology is through a language of wider communication, which, in the case of Botswana, is Setswana or English. The only way out of this dilemma of having to choose between the death of their mother tongues and marginalization is for minority speakers to have access to modern living while being able to preserve and develop their traditional values, including their languages and culture.

One solution to this dilemma is to create an atmosphere that would favor unmarked bilingualism (or multilingualism, as the case may be). Unmarked bilingualism is a state in which two languages whose functional domains may be different are accorded equal social importance by a given language community. In this case, both the minority language and the majority language have equal worth in the minds of the speakers, but their places and roles may be different. The minority language has a cultural role and is a symbol of ethnic identity, while the majority language (in this case Setswana) has a national role and is a symbol of national identity. The minority speakers would strive to be fluent in both languages and would want their children to be equally proficient in both of them. Moreover, they would cultivate positive attitudes toward both languages. Hence, while the national medium (Setswana) would be made part of the community, their mother tongues would be maintained.

This change of scenario could only be brought about by altering what Sasse (1992) calls the "external setting." According to Sasse, the external setting involves all the extralinguistic forces that favor one policy, language, or language variety over another. These extralinguistic forces include national policies, attitudes, ideologies, social values, ethnohistorical experiences, socioeconomic considerations, and political decisions. The present marked bilingual situation in Botswana could only become unmarked if the external setting became favorable to it.

In most African countries, there are local associations, NGOs, scholars (both local and foreign), government departments, and international organizations that are concerned, in one way or another, with the preservation of linguistic, cultural, and biological diversity. Their efforts are often hampered by scant resources, little or no direct government support, lack of coordination, and limited cooperation from the minority communities themselves. Usually, these organizations work separately and often have different goals and agendas. It would therefore be important to promote proper cooperation and coordination in order to enhance these efforts. Furthermore, where governments show indifference or even opposition, thinking that it would be desirable to do away with multilingualism and multiculturalism for the sake of national unity and cultural cohesion, it is essential that such governments be made to realize that one of the ways to ensure national unity and cooperation among the minority language speakers is to value their languages, cultures, and traditions. This tends to create self-confidence as well as receptive attitudes toward national development and identity.

Moreover, one needs to be mindful of the UNESCO resolution of 1952 that mother-tongue education at the formative level is most desirable and that the public use of one's language is an indisputable right (Bamgbose 1997). It should be emphasized here that the central actors in the whole enterprise of linguistic and cultural maintenance and promotion are the minority groups themselves. They must be enabled to feel proud of their languages, their cultures, and their indigenous knowledge and to feel confident that the value of any knowledge is not where it emanates from, but how it can assist humans in dealing with its complex and often hostile environment.

Lastly, there is a need to enhance international support, particularly from organizations such as UNESCO, the European Union, the Organization of African Unity, and even regional organizations such as the Southern African Development Community, in the preservation and promotion of the African languages and their respective cultures. It is evident that true development in the Third World can only be achieved if the people concerned can develop their own knowledge of their universe and skills to transform it from their own background as a base.

NOTES

1. The percentages for other regions of the world, according to Coulmans (1983), are 30 percent in Asia, 20 percent in the Pacific, 16 percent in the Americas, 1.5 percent in Europe, and 1.5 percent in the Middle East.
2. They are called here "quasi-monolingual" because, in most cases, there are small pockets of other language groups that are ignored nationally.
3. A minority language is defined here as any language group that is usually demographically smaller and often considered to be of lesser socioeconomic status vis-à-vis its more demographically dominant or sociopolitically more privileged neighbors (Batibo 1992; Mekacha 1996).
4. According to Connell (1997) there are at least six moribund languages in Mambila District of west Cameroon alone.
5. Bantu is a name that has been in use since 1962 to refer to a family of about 700 languages spoken in Africa south of the equator that have common phonological, morphological, and lexical characteristics.
6. Khoesan is a conventional name used to denote the two original language families in southern Africa, namely, San and Khoe, traditionally known as Bushmen and Hottentots. Khoesan languages are characterized by the presence of clicks. The special characters found in the names of these languages are part of the conventions used in their orthographies.
7. "S5" is the conventional name originally given to this language by Dorothy Bleek, who first worked on it. Most scholars have continued to use the convention.

REFERENCES

Andersson, L.G., and T. Janson. 1997. *Languages in Botswana*. Gaborone: Longman Botswana.

Bamgbose, A. 1997. African language use and development: Aspirations and reality. Keynote address at the Second World Congress of African Linguistics, Leipzig, Germany, 27 July–3 August 1997.

Batibo, H.M. 1992. The fate of ethnic languages in Tanzania. In *Language Death: Factual and Theoretical Explorations, with Special Reference to East Africa,* ed. M. Brenzinger. Pp. 85–98. New York: Mouton de Gruyter.

Batibo, H.M 1997. The future of the Khoesan languages in Botswana. Keynote address at the Parassession on Endangered Languages of Africa, Second World Congress of African Linguistic, Leipzig, Germany, 27 July–3 August 1997.

Batibo, H.M., J.T. Mathangwane, and N. Mosaka. 1996. Prospects for sociolinguistic research undertakings in Botswana: Priorities and strategies. In *Proceedings of the Regional Seminar on Sociolinguistic Research in Africa: Priorities and Methodologies.* Pp. 123–142. Duisburg: LICCA Publications.

Bleek, D.F. 1929. *Comparative Vocabularies of Bushman Languages.* Cambridge: Cambridge University Press.

Connell, B. 1997. Moribund languages of the Niger-Cameroon borderland. Paper presented at the Parassession on Endangered Languages of Africa, Second World Congress of African Linguistics, Leipzig, Germany, 27 July–3 August 1997.

Coulmans, F. 1983. Languages of the world. *Development Forum,* No. 91 August–September 1983:2–5.

Crawhall, N. 1997. The death of a "useless" language. *Mail and Guardian,* May 23–29, 1997:4.

Croizer, D., and R.M. Blench. 1992. *Index of Nigerian Languages.* 2d ed. Dallas: Summer Institute of Linguistics.

Diamond, J. 1993. Speaking with a single tongue. *Discover* February, 1993:78–85.

Hasselbring, S. 1996. *A Sociolinguistic Survey of the Languages of the Ghantsi District.* Botswana Language Use Project. Gaborone: Bible Society of Botswana.

Heine, B. 1997. The endangered languages of Africa: Introductory remarks. Paper presented at the Parassesssion on Endangered Languages of Africa, Second World Congress of African Linguistics, Leipzig, Germany, 27 July–3 August 1997.

Janson, T., and J. Tsonope. 1991. *Birth of a National Language: The History of Setswana.* Gaborone: Heinemann Botswana.

Koch, E. 1997. Last voice of an ancient tongue. *Mail and Guardian,* May 23–29, 1997:4.

Krauss, M. 1992. The world's languages in crisis. *Language* 68(1):4–10.

Mathangwane, J.T. 1996. Phonetics and phonology of Kalanga: A diachronic and synchronic study. Ph.D diss., University of California, Berkeley.

Mekacha, R. 1996. Language shift in the Nata area, Tanzania. *Proceedings of the Regional Seminar on Sociolinguistic Research in Africa: Priorities and Methodologies.* Pp. 57–71. Duisburg: LICCA Publications.

Sasse, H.-J. 1992. Theory of language death. In *Language Death: Factual and Theoretical Explorations, with Special Reference to East Africa,* ed. M. Brenzinger. Pp. 7–30. New York: Mouton de Gruyter.

Traill, A. 1994. *A !Xóõ. Dictionary.* Quellen zur Khoesan Forschung, 9. Cologne: Rudiger Koppe Verlag.

Vossen, R. 1988. *Patterns of Language Knowledge and Language Use in Ngamiland in Botswana.* Bayreuth African Studies, vol. 13. Bayreuth University.

19

THREATS TO INDIGENOUS KNOWLEDGE

A Case Study from Eastern Indonesia

Margaret Florey

In recent years, ecologists, anthropologists, and ethnobiologists have begun to focus attention on the issues of environmental sustainability, indigenous ecological knowledge, and threats to cultural and biological diversity among groups whose indigenous lifestyle is jeopardized by processes of social and ecological change. Linguists, too, have begun to confront the issue of the loss of linguistic diversity as awareness of the extent of language endangerment has increased. Krauss (1992:6), for example, estimates that perhaps only 10 percent of the approximately 6000 languages in the world today could be classified as "safe," while 10 percent are nearly extinct and 20 percent are moribund. Dixon (1991:230) believes the situation is likely to worsen in Asia and Oceania in the twenty-first century and stresses that a community typically does not realize its language is endangered until it is too late to reverse the process. Himmelmann (1996:2) makes a useful distinction between the symptoms and causes of language endangerment, pointing out that the latter are often less easy to discern. While earlier work in this field was often more descriptive of the symptoms of language shift or obsolescence, linguistic understanding of the theoretical issues concerning language endangerment and language obsolescence, including their causes, has been advanced through the publication of a number of key scholarly works (cf. Dorian 1989; Fishman 1991; Robins and Uhlenbeck 1991; Brenzinger 1992; Wurm 1996).

Figures such as those cited by Krauss highlight the diminished linguistic and cultural diversity and the loss of indigenous knowledge that we are facing in many parts of the world. Despite the growing research program in the field of language endangerment and the range of sociopolitical action that is taking place in many parts of the world, the languages of the Maluku region of eastern Indonesia remain neglected. The *Ethnologue* (Grimes 1996) lists 131 languages for this region, and Steinhauer (1996:36) estimates that 29 of these languages are endangered through the process of language shift, primarily toward the regional creole, Ambonese Malay, which has functioned as a lingua franca in Maluku for more than four hundred years. According to this estimate, Maluku is the most severely endangered linguistic region of Indonesia yet remains one of the least-known regions linguistically. To date only five of the estimated 56 indigenous languages of the Central Maluku region have been the subject of detailed modern linguistic research: Alune (see Florey references below), Asilulu (Collins forthcoming), Buru (Grimes 1995), Larike (Laidig and Laidig 1991), and Nuaulu (Bolton 1994).

A comprehensive long-term research program among the Alune of eastern Indonesia has been exploring language use and sociocultural and ethnobiological practices in two Alune villages. The Alune experienced rapid sociopolitical, ecological, and linguistic change throughout the twentieth century—changes that are impinging on the distribution and status of indigenous knowledge. This chapter investigates the political, religious, and economic pressures that are driving social change in these sites. As change progresses, Alune sociocultural practices are becoming devalued and a pan-Moluccan identity is emerging in place of the more local Alune identity. This development is affecting the entire complex of Alune sociocultural knowledge and is reflected in language shift to Ambonese Malay and the loss of traditional Alune ritual practices and ethnoecological knowledge. In this case study I first outline some of the key features of traditional Alune knowledge, including sociocultural practices, specialized speech registers, and some aspects of ethnobiological knowledge. I then discuss various factors implicated in the process of change, including state policies and the church, and present evidence for language shift and the loss of local knowledge of the environment. The two research sites, although historically originating from one single site, demonstrate differences in ecology, demography, recent sociopolitical history, and language use, and therefore provide a valuable basis for comparison. Finally, I examine interactions between lifestyle changes and shifts in the transmission and distribution of indigenous knowledge.

ALUNE SOCIOCULTURAL KNOWLEDGE

The Alune are an Austronesian ethnolinguistic group with a total population of approximately 10,000 people who dwell in 26 villages located in western Seram, in the eastern Indonesian province of Central Maluku. Data were gathered in three Alune villages during six fieldwork seasons totaling more than two years between 1988 and 1998. Field research was initially based in the south coastal village of Lohiatala and since 1992 primarily in the inland village of Lohiasapalewa. Some field research was also undertaken in the northern coastal village of Murnaten. Research has, at various times, been undertaken in collaboration with Prof. Chris Healey, an anthropologist, and Dr. Xenia Wolff, a botanist.

Lohiasapalewa is located in the central mountain range of west Seram, at an altitude of approximately 650 meters in submontane rainforest. Lohiasapalewa remains the most isolated Alune village and is bordered by the Alune villages of Riring, Manusa Manue, and Buria. The present-day villages of Lohiatala and Lohiasapalewa were formerly one village, located on the site of Lohiasapalewa. In 1817, conflict within Lohiasapalewa led to the departure of a breakaway group that settled in a large tract of forest approximately 20 kilometers to the south and formed the village of Lohiatala (see Makerawe and Nikolebu 1988, reproduced in Florey 1990). The historical relationship between the two villages is denoted by their retention of the name Lohia, but the addition of the name of the major river in each region, the Tala and the Sapalewa, marks their separation. A significant bond remains between Lohiasapalewa and its daughter village, but there is little contact between the villages, and the majority of villagers have not visited the other site.

Murnaten has had the longest period of contact with the non-Alune-speaking world of any Alune village. The village began to move to coastal locations at least one hundred years ago, although Murnaten people transitionally occupied several mountain sites. The mobility that marked the lifestyle of the Murnaten people in the twentieth century brought them into contact with other Seram peoples speaking Noniali, Lisabata, and Wemale, as well as with Dutch administrators and traders from as far away as Sulawesi in the west and Ternate and Tidore in the north. In 1992 the village had a population of 1075 residents in 162 households, and 10 percent of the household heads were non-Alune.

Sociocultural Practices

Traditional Alune sociocultural practices have been described in the works of Dutch missionaries, soldiers, and administrators, and Dutch and German researchers working in Maluku in the nineteenth and early twentieth centuries (cf. van Ekris 1867; Sachse 1907, 1919; Tauern 1913, 1918; Stresemann 1923; Jensen and Niggemeyer 1939; Jensen 1948; Niggemeyer 1951, 1952). Historically, the Alune people lived in extended family groups in small hamlets in the submontane rain forest. Alune social organization centers around *nuru*, patrilineal clans, and *luma*, localized lineages. Each *nuru* consists of several *luma* and recognizes a common territory from which it originated. Key leadership roles within Alune sociopolitical structures prior to the creation of the Indonesian state were held by the *'amale*, village headman, *mninu*, village chanter / town crier, *tapel upui*, guardian of the land,

and *luma matai*, the male leader of a *luma*—all of which were hereditary. Other important roles within the village included the local healers: *biane*, a woman with the right, inherited through her *luma*, to Alune midwifery knowledge, and *ma'aleru*, a healer who employs divination and incantations.

Alune cosmology focused on the placation of ancestral and local nature spirits whose goodwill was regarded as necessary for ensuring the health and vitality of the living and the productivity of the environment. Alune cosmology historically was perpetuated through ritual practices that, in part, involved the chanting of incantations to invoke the spirits of ancestors or deities who could mediate on behalf of human beings. The subject matter of incantations reveals the concerns of daily life—social control (maintaining peaceful relations with neighboring ethnolinguistic groups and within the village), the provision of an adequate food supply, the diagnosis and treatment of illness, and the protection of person and property.

Alune oral history was transmitted through origin or creation tales (*ma'lulu*) and historical narratives (*hnusu*) recounted by orators within the community who were renowned for their performance ability and linguistic skill (*ma'alulu*). Other performance genres included folktales (*tuni*), epic tales ('*apatate*), a range of song genres which are differentiated by the structure and the function of the song style and associated dance steps (including *tutu hatu, biole, mareu, maru, denu, sulite*), and riddles (*hote latu*).

Alune sociocultural life incorporated a wide range of ritual practices. Among these were the ritual held when a man's *luma* asks permission from a woman's *luma* for the couple to marry (*loa metu batai* or *sa' bina*), the ritual associated with the payment of bridewealth (*beli bata*), a ritual performed to renew intervillage alliances (*pela*), rituals held during the planting and harvesting of certain crops (particularly rice), midwifery rituals such as the ceremony performed when a baby is approximately ten to fifteen days old and is brought outside the home for the first time (*sidi 'wete belu'we*), and ceremonies during which offerings were made to the ancestors ('*oti pelate*): for example, during the blessing of a newly completed house to ensure the safety and good health of the occupants, or when making a new garden to ensure that the garden flourishes. The opening and/or closing of rituals was marked by chants (*bolu'we, nabu'we*) and often incorporated aspects of customary law, which were chanted during a formal speech (*alamanane*).

Speech Registers

A number of speech registers were associated with specialized knowledge. For example, a register utilizing metaphor (*sou mo'wai*) was employed by men to conceal their ill-intent during head-hunting expeditions. A respect relationship between a woman and her brothers-in-law or a man and his sisters-in-law was marked by the use of respect terms (*ma'mosi*) for words forming full or partial homonyms with the personal name of the person with whom one was in the respect relationship (described in Florey and Bolton 1997). A healing register (*lepate mlerude*) could be employed by healers prior to applying curative incantations to heal serious physical injury, such as a broken limb or a serious injury resulting from a fall from a tree (see Florey and Wolff 1998). The healing register involved the use of metaphorical descriptions of injuries.

Ethnobiological Knowledge

A wealth of ethnoecological knowledge reflects the importance of the interactions that Alune people traditionally have had with their physical environment.[1] Some of the more salient features of Alune ethnobotanical and ethnozoological knowledge are presented here.

During the era in which the Alune people dwelled in extended family groups in small hamlets, the forest was utilized extensively to extract materials for ritual purposes, as construction materials, to make tools, household implements, and weapons, and to harvest medicinal plants used in healing and midwifery practices. The Alune also undertook extensive harvesting of noncultivated forest products as foodstuffs, such as fruits and nuts, and plants consumed as vegetables.[2] A wide range of animals also formed a significant part of the Alune diet, including mammals (deer, pig, bats, and cuscus), birds, some reptiles, and certain invertebrates. Noncultivated foods were an essential part of the subsistence strategy of the Alune. The amount and variety of foods obtained from the forest was wide and it is clear that noncultivated plant foods provided dietary diversity. The procurement and use of noncultivated foods was supplemented by some cultivated foods, primarily tubers, which were grown in swidden fields (*mlinu*) and rice in dry rice fields (*mlinu ala*). Once crop yields became very low throughout the cultivated areas because of depletion of soil fertility, the villagers selected a new site in primary or regrown secondary forest, and reestablished the hamlet in that site. Composition of the diet was very much seasonally influenced, although rice was grown in sufficient quantities to ensure a year-round supply and sago starch was constantly available. An appreciable portion of the Alune diet was also met by casual snacking rather than during fixed meal times. In the forest, a wide variety of snacks was available, including fruits, nuts such as canari (*Canarium* sp.), shrimp and crabs, and small birds, animals, and insects, which were baked on coals.

Indigenous classification divides the entire territory owned and occupied by an Alune village in submontane and montane rainforest into eleven broad ecological zones. These zones are further delineated into named microhabitats and plant communities according to both distance from the village settlement area and specific ecological and phytogeographical characteristics. For example, there are four named zones of primary forest that delineate altitude and distance from the village settlement area. Six zones within secondary forest are named according to the stage of regeneration of the forest following the abandonment of a garden. Bamboo thicket is differentiated according to the variety of bamboo that grows in the thicket. Monoculture and multiculture groveland are distinguished, with monoculture groveland further differentiated according to whether it is owned by a *luma,* jointly by an entire village, or jointly by people from more than one village. Gardens may be terminologically distinguished according to the principal crop: for example, swidden rice (*Oryza sativa*), cassava (*Manihot esculenta*), or sweet potato (*Ipomoea batatas*). Other named zones include grassland, riverside, swampland, and floodplain, fertile, compost-rich ground, and infertile ground.

Alune terms for plant anatomy reflect both the complexity of Alune perceptions of their physical environment and the extensive uses that are made of the products of the environment. Seventy-six terms defining plant anatomy have been identified. Features that form the basis for Alune classification include the stage of development of the plant, the sex of the inflorescence, segmentation, viscosity of sap, root structure, branching structure, and the presence of thorns or spines. Monocots and dicots are differentiated in the early stages of plant development, and this is reflected in different means of propagation for these plants. The lexicon referring to plant morphology is one indicator of Alune plant taxonomy, while certain noun classifiers indicate further taxonomic principles. For example, *ai* is a classifier that occurs with fifty-six trees of the primary and secondary forest. Attributes of *ai* include woodiness, height, trunk, and branching structure, and minimal or no management of the plant (cf. Brown 1991). In contrast, *aini* can be used to specify a palm, bamboo, tree fern, or certain cultivated plants (those subject to a greater degree of management). This term denotes an individual plant other than an *ai* that can easily be distinguished from other plants, and the reference includes all parts of the plant. It is not used with grasses, creepers, or grains such as rice.

The greater cultural significance that a number of plants hold for the Alune—in particular rice (*Oryza sativa*), coconut palm (*Cocos nucifera*), betel nut palm (*Areca catechu*), and sago palm (*Metroxylon sagu*)—is apparent in the comprehensive range of information associated with these plants. For example, there are six named stages of growth for the sago palm (*pia aini*), distinguished by the development of the inflorescence and the readiness of the palm for processing as starch. An origin tale recounts the history of the Alune people's knowledge of sago as a foodstuff, including information on processing and food preparation. There is an extensive lexicon associated with processing of the palm, including styles of adze, the twenty named parts of the sago processing trough, and varieties of storage containers. A rich lexicon also exists for cooking and serving utensils and sago food products. In the respect register (*ma'mosi*), those in a respect relationship with a woman named Piai or a man named Lopia replace *pia* 'sago palm' with either of the terms *na'wa* 'sugar palm' (*Arenga pinnata*) or *nurule* 'nibung palm' (*Oncosperma tigillarium*) and replace *pia* in its secondary meaning of 'sago gelatin, staple food' with the generic term *manane* 'food'.

Alune ethnozoological knowledge reveals the importance of certain animals in the Alune world—particularly mammals such as pig (*Sus scrofa*), deer (Cervidae fam.), and cuscus (Phalangeridae fam.). For example, nine stages of growth are differentiated for pigs, primarily according to growth of the tusks and secondarily according to speed of movement of young animals. In contrast, six stages of growth are differentiated for deer, primarily according to growth of the antlers and secondarily according to ability of the young to stand upright. The cuscus (*marele*) is of greater ritual importance to the Alune than all other animals. An origin tale recounts the cuscus's role in providing women with the knowledge of midwifery skills. "Cuscus" is used metaphorically to mean a newborn male. This is a reference to men's responsibility to provide nonvegetable food, and indicates the role of cuscus as an important foodstuff—not only in mundane contexts but also on all ritual occasions. An incantation is used to ensure a successful hunt for this animal. Four named stages denote development of the cuscus in the marsupium and differentiate degrees of edibility, growth of fur, and movement outside the marsupium. The status of this animal is further indicated by the fact that certain body parts of the cuscus are distinguished terminologically from those of other mammals. For example, *molini* 'liver of cuscus'[3] contrasts terminologically with *atai* 'liver of a mammal', and *osai* 'genitalia of cuscus' contrasts terminologically with *somine* 'genitalia of a female mammal' and *tabule* 'genitalia of a male mammal'.

The term 'pregnant' is never applied to cuscus because much of this mammal's development occurs in the marsupium. But a distinction is drawn between the pregnancies of domesticated and nondomesticated animals. *Tu'une* denotes a pregnancy in humans and domesticated or commensal mammals (including dog, goat, cow, rat, civet cat) while *ntiane* denotes a pregnancy in certain other animals that give birth to live young, including nondomesticated mammals other than cuscus (deer, pig), and the Death Adder (*Acanthophis* sp.). Examples of parallelism found in Alune incantations also reflect the indigenous zoological taxonomy and reinforce local knowledge about edible and inedible foodstuffs. For example, reptiles and invertebrates are regularly paired, as are edible mammals (pig and cuscus) and inedible mammals (cat and dog).

ENDANGERED KNOWLEDGE: CONTEXTS OF CHANGE

Throughout the twentieth century, but particularly during the last fifty years in the era following Indonesian independence, the Alune have experienced rapid sociopolitical, ecological, and linguistic change. These changes have derived largely from the imposition of colonial and postcolonial political authority. The extent and pace of change is further exacerbated by change induced by church authorities following conversion from the indigenous religion to Christianity. Contemporary church and state practices have wrought wide-ranging changes that impinge on virtually all aspects of the Alune lifestyle and have affected the distribution and status of indigenous knowledge. Alongside sociopolitical changes, language shift from Alune to Ambonese Malay is also occurring.

Sociopolitical Change

At various times during the past hundred years, Alune villages have come under pressure to relocate from their mountain locations to the coast. The primary purpose of such relocation has been "pacification"—first by the Dutch colonial authorities and later by the Indonesian government—in order to make the villages more accessible to government authority and thus enable the government to exercise greater control. Some Alune villages have agreed to relocate, believing they would benefit from improved access to the facilities available in towns—a marketplace, proximity to high schools, and better health care. Other villages have strongly opposed relocation, primarily because it would have meant abandoning their land and thus losing access to the forest resources, their orchards, and hunting grounds. As the Alune are a mountain people, their territory does not extend to the coast. Relocated villages are therefore situated on land appropriated by the government from coastal villages and consequently, resources are scarce.

The people of Lohiasapalewa have twice in the past fifty years fled their village and dwelled in the forest for a period of time to resist efforts to relocate their village. The first occasion was during a guerrilla conflict fought in Maluku in the 1950s and 1960s between Indonesian military forces and the Republik Maluku Selatan (RMS), a separatist movement which strove for independence from the newly formed Indonesian Republic. Villagers were drawn into both sides of the conflict through contact with guerrillas who sought food and shelter in mountain villages and through violent reprisals from troops who attempted to minimize the support given to guerrillas. In order to remove themselves from danger, the village settlement site and associated gardens were abandoned, and the villagers reverted to their former lifestyle of living in small family groups. They built temporary shelters in the forest and moved frequently from one site to another for approximately twelve years. In 1964 the conflict ended and they were able to rebuild their village. Six years later, the regional administrator in the north coastal administrative center of Taniwel attempted to force the village to relocate to the north coast. Three village leaders who went to Taniwel to ask that the village be allowed to remain in the mountains were arrested and severely beaten but managed to escape back to Lohiasapalewa. There they ordered the villagers to abandon their village and flee to the forest while the leaders made their way to Ambon to negotiate with the governor. A period of several months elapsed before the men returned and announced they had successfully pleaded their case. In 1998, Lohiasapalewa's population was 244 residents in 32 households and there is only one non-Alune resident—the minister, who originates from the central Moluccan island of Saparua.

Unlike Lohiasapalewa, the villagers of Lohiatala were unable to resist the government's attempt to relocate them during the RMS conflict. In 1952, the villagers of Lohiatala were forced by the Indonesian military to abandon their mountain village and move to a non-Alune south coastal village, Hatusua. There they were given permission to build their own houses and to establish farms. In 1964, when the villagers were told they could return to their village, a decision was reached that a new village should be erected nearer the coast. While this decision remains controversial, it appears that the main impetus for the relocation came from younger villagers who had grown up in Hatusua and from some older villagers who wanted to remain within easier reach of high schools and market towns. The move to the new village location occurred in October 1965.

The present-day village of Lohiatala is located approximately 6.5 km inland from the south coast of western Seram, on the southern border of Lohiatala's land and some 20 km from the mountain location of the former village. The villagers retain ownership of their former village site in the mountains, and the land is still occasionally utilized, principally by elderly villagers, as hunting territory and groveland, in particular for durian (*Durio zibethinus*) and clove (*Syzygium aromaticum*). With a population in 1992 of 728 people comprising 110 households, Lohiatala is now substantially larger than its parent village and is in a much less isolated location. Lohiatala also contrasts demographically with its parent village in having a relatively high proportion of marriages with non-Alune people: 15 percent of the adult population originate from other locations. Lohiatala's nearest neighbor is the village of Waihatu, which was established in 1974 under the *Transmigrasi Nasional* program for migrants from the overpopulated islands of Java and Lombok.[4] Land was appropriated from the southernmost portion of Lohiatala's territory for the creation of this village. A short distance to the east lie the towns of Waimital and Waipirit, the former founded in 1954 under the *Transmigrasi Nasional* program and the latter established in 1965 under the *Transmigrasi Lokal* program for migrants from the nearby island of Saparua.

The indigenous system of government has also changed in Alune villages following Indonesian independence and associated restructuring of the political system. The position of 'amale has lost its hereditary status and become a government appointment with the Malay title of *kepala desa* 'village head'. In Maluku, the modern-day *kepala desa* is chosen from among the male village population. The appointee usually has spent long periods of time away from the village, often in military service or in a government job in an administrative center or in Ambon, the regional capital. Consequently, village headmen may not be fluent speakers of Alune and are often unfamiliar with village life and may be unable to perform rituals. The ritual responsibilities of the *mninu, luma matai,* and *tapel upui* have been weakened in the modern era as many of the traditional ceremonies that were associated with indigenous sociocultural beliefs have been suppressed.

Changes in the political system have also been accompanied by changes in village settlement patterns. In order to allow greater political control by local government officials, villagers now live in a fixed settlement site rather than in small hamlets throughout their village territory. This change has affected the frequency and nature of interactions within a village. For example, the *kepala desa* regularly organizes communal work projects in which all able-bodied adult men and women must participate. In Lohiasapalewa, villagers are currently involved in building a new village hall and clearing new paths within the village settlement site. In addition to these projects, the villagers of Lohiasapalewa have been granted financial assistance under the central government's program *Inpres Desa Tertinggal*[5] to assist villages identified as extremely disadvantaged. The assistance has been used to raise chickens and grow shallots for commercial sale. Adult males are required to travel to the coast to fetch building supplies and to carry harvested produce to a location from where it can be transported to the regional capital of Ambon for sale. Travel is extremely time-consuming because of the distance from the coast and men may be absent from the village for several days.

Sociopolitical change is also apparent in the education system and in employment patterns, which are contributing to changing demographics and greater mobility among young Alune villagers. While there are elementary schools that provide six years of education in all Alune villages, children attend high school in non-Alune villages. Children in Lohiatala enroll in junior and senior high school in the south coastal village of Waesamu, some 6 kilometers from Lohiatala. Young people in Lohiasapalewa leave home to attend junior and senior high school in the west Seram administrative centers of Taniwel, Piru, or Kairatu. In addition to the financial burden imposed by schooling, an important social outcome is that young people spend critical formative years in a non-Alune environment. In Lohiasapalewa it has become common practice for young women in the 15–19 age group to leave the village and seek work as servants or as shop assistants in regional centers such as Piru or Kairatu, in Ambon. In 1998, virtually no unmarried young women remained in the village. This pattern is examined in Florey and Healey 2000.

Improvements in transportation have increased the pace of change in western Seram. The World Bank has funded the building of a trans-Seram highway, which circles the coast of Seram. The central government is supporting the building of roads from the coast to interior villages by commercial enterprises. These roads also open the interior to the further settling of transmigrant communities: in 1996, a government team surveyed Lohiasapalewa's land, leading to renewed rumors that so-called unutilized land would be appropriated for settlement.

Religious Conversion

Alongside extensive sociopolitical changes in Alune society, religious change has also had a great impact on sociocultural practices. Alune villages began to convert from their ancestral religion to Christianity (Calvinism) in the earlier part of the twentieth century, a process that began at the turn of the century in coastal regions of west Seram and several decades later in the more isolated inland mountain villages. The villagers of Murnaten began to convert in 1900, while conversion began in Lohiatala in 1925 and in Lohiasapalewa in 1935. The missionaries were Malay speakers, usually from the neighboring Central Moluccan islands of Ambon and Saparua. Conversion appears to have taken place rapidly, and today all Alune villages have converted to Christianity, although a small Muslim presence exists in most coastal villages resulting from intermarriage or the establishment of small businesses. The central role the church plays in the lives of villagers is visually apparent through the large, modern church buildings that have been erected in most Alune villages. In relocated villages, the building of a permanent church is frequently cited as the principal reason preventing a permanent return to former village locations.

Religious change has resulted in the active suppression of all activities associated with pre-Christian practices. The Alune language appears to have been feared by missionaries as a vehicle for the promulgation of pre-Christian beliefs, and speaking Alune was strongly discouraged in the early days of proselytization and, on occasion, resulted in physical punishment. The laying of foundation stones of new church buildings was often accompanied by a ceremony held to mark the banning of the use of magic for any purpose. Many cultural practices were also discouraged, including traditional health care methods. Treatment of illness and injury today largely involves prayer, either as the sole healing tool or in combination with Western medicines, which are administered by a regional health practitioner (cf. Florey and Wolff 1998). The minister in Lohiatala extracted a commitment from the oldest midwife that traditional midwifery practices would cease in that site upon her death. Alune villagers are now given an Ambonese Christian personal name in place of an Alune personal name, although some covert bestowing of Alune names continues (cf. Florey and Bolton 1997).

The church, too, exerts pressure to take part in villagewide projects. In Lohiasapalewa, a large permanent church is under construction, while in Lohiatala, villagers have been constructing a new vicarage and have been carrying out repairs to the church. On communal work days in both sites, no villager is free to work in the forest or garden. Economic pressures also emanate from the church, which requests money for a variety of purposes in addition to the regular collections for the minister's salary taken at services throughout the week. For example, in both sites villagers are required to fund church building and maintenance programs. In

Lohiatala, villagers have been asked to buy new hymn books and Bibles to support the regional Protestant church authority and to donate food and money to visiting theology students.

Ecological Change

All Alune villages are experiencing some measure of ecological change. Although villages that have been able to remain in their mountain locations have undergone less ecological change than coastal villages, in all sites the lifestyle pattern, which drew heavily on forest resources, has been threatened in recent years by both external and internal changes. Lohiasapalewa occupies the most extensive lands of all Alune villages, yet land close to the village residential area is becoming scarce as the population increases. Gardens may be located up to 5 or 6 kilometers through the rain forest from the village. Greater distances to gardens make it difficult to harvest produce quickly, or to guard against the incursions of wild animals. Time constraints imposed by contemporary village life, including communal work duties, are also greatly reducing time spent in the forest hunting and gathering food. Analysis of modern agricultural practices indicates that villagers increasingly are choosing to plant root crops, primarily cassava, which previously were viewed as scarcity crops (cf. Wolff and Florey 1998). Such crops require minimal care and yield both a carbohydrate source and leafy vegetable in a relatively short time. Rice no longer constitutes a major portion of the diet, having been largely replaced by cassava and sago. Increased sedentarization and decreased reliance on noncultivated products has led to a decline in the variety of foods consumed in the everyday Alune diet. The majority of families have experienced a negative impact on their nutritional status as a result, primarily because of inadequate intake of protein.

The people of Lohiatala have undergone extensive ecological change through the move to a lowland site. A considerable number of the forest resources that were exploited in the mountain village are not found in the coastal zone. The soil is more swampy and less fertile in coastal areas and receives far less rainfall. Hence, there is less arable land and sweeping changes in horticultural practices have been made. The establishment of the neighboring transmigrant village, Waihatu, has also had ecological consequences. A branch of the lower reaches of the Nala River, which flows through Lohiatala's land, was diverted and a dam built to irrigate wet rice fields in Waihatu. The cultivation of wet rice is considered responsible for the repeated failure of swidden rice crops in Lohiatala, which is largely attributed to introduced rice pests.

A further source of ecological change is Indonesian government policies, which claim control of territory and allow commercial exploitation of natural resources. In Alune villages, these policies have resulted in commercial enterprises purchasing dammar trees (*Agathis* sp.) for sale to plywood factories, harvesting bamboo for use in paper production, and removing sago palms for packaging and sale of the starch as flour. The consequences of these actions for the Alune vary from site to site. In Lohiatala, the sale of all dammar trees has resulted in the loss of dammar resin (*'alan tuluti*) as a reliable source of income. The removal of bamboo and sago palms has also had important repercussions for the Alune people. Sago is not only a staple foodstuff, but the palm is heavily utilized for making kitchen implements and as a construction material. Bamboo is a cooking vessel, a storage container for water and wet sago starch, and a construction material. Decreased availability of these plants is one factor leading to the purchase of commercially produced kitchenwares.

Language Shift

In the Christian villages of Central Maluku, a considerable number of indigenous languages have become extinct, and the remaining languages are undergoing language shift toward the regional lingua franca, Ambonese Malay (cf. Collins 1982; Florey 1991, 1993; Grimes 1991). The introduction of the national language, Indonesian, has raised the status of Malay in this region and increased its functional domains and thus is hastening the process of language shift. Indigenous languages in Muslim villages are also undergoing language shift. Although the use of indigenous languages is reportedly maintained among all generations of speakers, there is a perceptible narrowing of the functions of the indigenous languages of these villages.

The strength of the Alune language varies among the 26 villages in which it is spoken. Language shift is well advanced in the relocated coastal villages, and Alune is virtually extinct in Kairatu, a southern Alune town that is also an administrative center. In contrast, Alune is still spoken by all generations in the mountain villages, although there is evidence for the genesis of language shift (cf. Florey 1997). A number of factors are affecting patterns of language choice. Most significant has been the pressure exerted on villagers by missionaries and ministers to cease the use of Alune. Malay is the language of Christianity—the language that demonstrates that villagers have moved out of the *masa gelap,* the "dark era," which denotes the pre-Christian period. In the modern era, Malay/Indonesian is also the language of formal education, and parents are encouraged not to speak Alune with their children in order to enhance opportunities for success in school and, later, in the workplace. Intervillage differences in language fluency can also be attributed to location, resulting in differential ease of contact and frequency of interaction with non-Alune-speaking people.

Language use in the south coastal village of Lohiatala was greatly influenced by the period the villagers of Lohiatala spent in the homes of Hatusua people during the RMS conflict. The villagers were moved from a largely monolingual Alune environment into a Malay-speaking non-Alune village in which the indigenous language is moribund. In order to examine the process of language shift in this site, observations and recordings of natural language use were supported by extensive testing of Alune language proficiency (cf. Florey 1990). A series of tests were administered to determine receptive and productive ability in Alune. The tests included recognition of lexical items, comprehension of simple sentences, comprehension of fluent Alune, and a task involving the translation of a set of sentences from Ambonese Malay to Alune. The responses of the test subjects were compared with an older fluent speaker norm. The results indicate that Ambonese Malay is the first language of all people born in Hatusua or in the new village of Lohiatala, that is, those aged approximately thirty years and younger. People in this age group no longer speak Alune, although most retain some receptive skills. They are rarely addressed in Alune and if they are, they always respond in Malay. Four age-related groups of speakers in Lohiatala were identified on the basis of the test results: near-passive bilinguals (children and young people aged eighteen or younger), imperfect speakers (aged nineteen to thirty-four), younger fluent speakers (aged thirty-five to forty-five), and older fluent speakers (aged forty-six and older).

Comparative research was later undertaken in Lohiasapalewa, Lohiatala, and Murnaten in order to determine whether the process of language shift was occurring in all sites, and, if so, how the rate of change and the process itself was affected by the different histories and differing social and cultural factors in these villages. The results demonstrated that a wide range of linguistic abilities exist between different Alune-speaking locations and that linguistic differences are paralleled by social and cultural differences (cf. Florey 1997, Florey and Kelly in press). Further, they provided evidence that Alune language proficiency is greatest in Lohiasapalewa, a relatively isolated mountain village in which knowledge and use of Alune persists through all generations. Alune language proficiency is markedly weaker in the coastal villages of Lohiatala and Murnaten, which have experienced sustained contact with other language groups, for approximately forty years in the case of Lohiatala and one hundred years in the case of Murnaten. Testing indicated that there is some retention of receptive ability by young people in Lohiatala and Murnaten, and this is confirmed by observations of linguistic interactions between older people and children. Productive ability in these two sites is, however, very limited and involves extensive mixing of Alune and Ambonese Malay.

DISCUSSION

It is apparent that the Alune world has undergone many changes in the past fifty years. This era has seen the increased sedentarization of villages, greater exposure to non-Alune peoples, language shift to Ambonese Malay, the influence of the church and its role in suppressing traditional ritual practices, decreased utilization of forest resources, limitations on the time available for agriculture resulting from heavy involvement in church and state projects, the shift to a monetary economy, and the socialization of young people away from the village during their school years.

A number of processes, then, are driving changes in the transmission and distribution of indigenous knowledge. As pre-Christian practices are increasingly devalued, older community members are no longer transmitting their knowledge of traditional Alune sociocultural practices, such as narrative genres, song styles, incantations, rituals, and specialized speech registers. Certain ethnobiological knowledge is also actively suppressed, including traditional healing methods utilizing herbal medicines. Modern horticultural practices rely to a much greater extent on the planting of a limited range of cultivated plants and to a much lesser extent on the harvesting of forest resources. Consequently, children are spending less time in the forest with their parents and are not learning about plant and animal resources, the names of species, plant management techniques, harvesting and preparation of plants as foodstuffs, or the wide range of traditional uses for plant and animal species indigenous to this environment. The lack of opportunity for exposure to the Alune ethnobiological and sociocultural knowledge and practices possessed by their parents and ancestors is exacerbated by the amount of time younger Alune villagers spend away from the village for purposes of education and employment. The extensive ethnobiological knowledge that has been documented in Lohiasapalewa is predominantly restricted to older villagers and has been acquired by only a few younger villagers in inland sites who continue to value the ways of their elders. In Lohiatala, the majority of villagers of all ages have a very limited knowledge of the ecology of interior Seram, which characterizes the traditional Alune world of their ancestors. It appears that the retention of a traditional village site and relative isolation of a village may slow the process of change but does not ensure the retention of indigenous ecological knowledge.

Change is affecting the entire complex of Alune sociocultural knowledge, which includes the ability to use the Alune language over the range of its functional domains. In Lohiasapalewa, the language is retained in its everyday functions, yet conceptual diversity is diminishing with the loss of specialized functions of Alune. The issue of the interaction between linguistic and environmental diversity remains to be addressed. Social change rather than language shift per se is driving the loss of ethnoecological and sociocultural beliefs, yet there can be little doubt that language shift and changing ecological practices are influencing each other. Decreased access to and familiarity with the physical environment diminishes both the ability and the need to talk about the natural resources and ecological practices that were encoded in Alune. Conversely, decreased fluency in the Alune language, including a lexicon increasingly restricted to fewer domains, reduces the ability to use that language to talk about natural resources. The Alune botanical and zoological taxonomies and rich lexicons for plant and animal morphology and methods of harvesting and utilizing natural resources do not seem to be transferring to the Ambonese Malay lexicon of Alune villagers. It appears, then, that ecological knowledge is being modified both through ethnobiological and sociocultural changes and through the process of language shift. There are manifest negative outcomes for environmental sustainability and indigenous ecological knowledge resulting from lifestyle changes.

Can the process of the loss of indigenous knowledge be slowed or reversed? At present, there is something of a renaissance of interest in the Alune language in Lohiasapalewa. This has, in part, arisen through the opportunities opened up by the Indonesian education department for language and cultural studies in the school curriculum. This initiative augurs well for raising the status of indigenous knowledge. At the same time, interest in language maintenance has grown through the protracted involvement of some community members in the comprehensive linguistic, ethnographic, and ethnobiological research program that my colleagues and I have undertaken. Combined, these factors appear to be influencing a revaluing of the Alune language. This process has the potential to lead to the reversing or slowing of language shift and might also lead to a revaluing of cultural knowledge.

NOTES

1. A more complete analysis of Alune ethnobiology will be presented in Florey, Healey, and Wolff, with Manakane in preparation.
2. See Wolff and Florey 1998 for a more detailed discussion of this topic.
3. Note that *moli* also means 'sacred, taboo'.
4. The *Transmigrasi Nasional* program was established by the Indonesian government in an effort to relieve the pressure on heavily overpopulated regions such as Java and Lombok by encouraging migration to less populated islands. Migrants receive assistance in the form of transportation costs, health care, housing, bedding, clothing, two hectares of land, farming tools, and seedlings. They are guaranteed an income for twelve months and are given guidance and instruction in appropriate farming techniques. The *Transmigrasi Lokal* program is organized by the regional government and concerns migration within one region or province. Limited assistance is given, including two hectares of land per family.
5. *Inpres Desa Tertinggal* is the 1993 Presidential Instruction for Assistance to Backward Villages.

REFERENCES

Bolton, R.A. 1994. A preliminary description of Nuaulu phonology and grammar. Ph.D. diss., Ann Arbor: UMI Dissertation Services.

Brenzinger, M., ed. 1992. *Language Death: Factual and Theoretical Explorations with Special Reference to East Africa.* Berlin: Mouton de Gruyter.

Brown, C.H. 1991. On the botanical life-form "tree." In *Man and a Half: Essays in Pacific Anthropology and Ethnobiology in Honour of Ralph Bulmer,* ed. A. Pawley. Pp. 72–78. Auckland: Polynesian Society.

Collins, J.T. 1982. Linguistic research in Maluku: A report of recent field work. *Oceanic Linguistics* 21(1/2):73–146.

Collins, J.T. Forthcoming. "Asilulu Dictionary."

Dixon, R.M.W. 1991. The endangered languages of Australia, Indonesia, and Oceania. In *Endangered Languages,* ed. R.H. Robins and E.M. Uhlenbeck. Pp. 229–255. Oxford: Berg.

Dorian, N., ed. 1989. *Investigating Obsolescence: Studies in Language Contraction and Death.* Cambridge: Cambridge University Press.

Ekris, A. van. 1867. Iets over het Ceramische Kakianverbond. *Tijdschrift voor Indische Taal-, Land- en Volkenkunde* 16:290–315.

Fishman, J.A. 1991. *Reversing Language Shift.* Clevedon: Multilingual Matters.

Florey, M.J. 1990. Language shift: Changing patterns of language allegiance in western Seram. Ph.D. diss., Ann Arbor: UMI Dissertation Services.

Florey, M.J. 1991. Shifting patterns of language allegiance: A generational perspective from eastern Indonesia. In *Papers in Austronesian Linguistics* 1, ed. H. Steinhauer. Pp. 39–47. Canberra: Pacific Linguistics.

Florey, M.J. 1993. The reinterpretation of knowledge and its role in the process of language obsolescence. *Oceanic Linguistics* 32(2):295–309.

Florey, M.J. 1997. Skewed performance and structural variation in the process of language obsolescence. In *Proceedings of the Seventh International Conference on Austronesian Linguistics,* ed. C. Ode and W. Stokhof. Pp. 639–660. Amsterdam and Atlanta: Editions Rodopi B.V.

Florey, M.J., and R.A. Bolton. 1997. Personal names, lexical replacement, and language shift in eastern Indonesia. *Cakalele: Maluku Research Journal* 8:27–58.

Florey, M.J., and C.J. Healey. 2000. Work well and guard your honour: Temporary labour migration and the role of adolescent women in eastern Indonesia. Asian Studies Association of Australia conference, University of Melbourne, 3–5 July.

Florey, M.J., C.J. Healey, and X.Y. Wolff, with W. Manakane. In preparation. From the edge of the forest: Alune ethnobiology in a changing world.

Florey, M.J., and B.F. Kelly. In press. Spatial reference in Alune. In *Representing Space in Oceania: Culture in Language and Mind,* ed. G. Bennardo. Canberra: Pacific Linguistics.

Florey, M.J., and X.Y. Wolff. 1998. Incantations and herbal medicines: Alune ethnomedical knowledge in a context of change. *Journal of Ethnobiology* 18(1):39–67.

Grimes, B.D. 1991. The development and use of Ambonese Malay. In *Papers on Indonesian Linguistics,* ed. H. Steinhauer. Pacific Linguistics Series A-81. Pp. 83–123. Canberra: Australian National University.

Grimes, B.F., ed. 1996. *Ethnologue: Languages of the World.* 13th edition. Dallas: Summer Institute of Linguistics.

Grimes, C.E. 1995. Buru (Masarete). In *Comparative Austronesian Dictionary: An Introduction to Austronesian Studies,* ed. D.T. Tryon. Part 1: Fascicle 1. Pp. 623–636. Berlin: Mouton de Gruyter.

Himmelmann, N.P. 1996. Language endangerment scenarios in northern Central Sulawesi. Paper presented at the International Workshop on South-East Asian Studies no. 11, Royal Institute of Linguistics and Anthropology, Leiden, 9–13 December 1996.

Jensen, A.E. 1948. *Die drei Ströme: Züge aus dem geistigen und religiösen Leben der Wemale, einem primitiven Volk in dem Molukken.* Leipzig: J.W. Goethe-Universität, Frobenius-Institut.

Jensen, A.E., and H. Niggemeyer. 1939. *Hainuwele: Volkserzählungen von der Molukken-Insel Ceram.* Frankfurt: J.W. Goethe-Universität, Forschungsinstitut für Kulturmorphologie.

Krauss, M. 1992. The world's languages in crisis. *Language* 68(1):4–10.

Laidig, W.D., and C.J. Laidig. 1991. *Larike Grammar.* Ambon: Summer Institute of Linguistics.

Niggemeyer, H. 1951. Alune-Sprache: Texte, Wörterverzeichnis und Grammatik einer Sprache West-Ceram. *Zeitschrift für Ethnologie* 76:50–69, 288–300.

Niggemeyer, H. 1952. Alune-Sprache: Texte, Wörterverzeichnis und Grammatik einer Sprache West-Ceram. *Zeitschrift für Ethnologie* 77:116–132, 238–250.

Robins, R.H., and E.M. Uhlenbeck, eds. 1991. *Endangered Languages.* Oxford: Berg.

Sachse, F.J.P. 1907. *Het Eiland Seran en Zijne Bewoners.* Leiden: Brill.

Sachse, F.J.P. 1919. *Gegevens uit de Nota Betreffende de Onderafdeeling West-Ceram.* Batavia: Encyclopaedisch Bureau.

Steinhauer, H. 1996. Endangered languages in Southeast Asia. Paper presented at the International Workshop on South-East Asian Studies no. 11, Royal Institute of Linguistics and Anthropology, Leiden, 9–13 December 1996.

Stresemann, E. 1923. Religiöse Gebräuche auf Seran. *Tijdschrift voor Indische Taal-, Land- en Volkenkunde* 62:305–424.

Tauern, O.D. 1913. Ceram. *Zeitschrift für Ethnologie* 45:162–78.

Tauern, O.D. 1918. *Patasiwa und Patalima: Vom Molukkeneiland Seran und seinen Bewohnern.* Leipzig: Voigtländer.

Wolff, X.Y., and M.J. Florey. 1998. Foraging, agricultural, and culinary practices among the Alune of west Seram, with implications for the changing significance of cultivated plants as foodstuffs. In *Old World Places, New World Problems: Exploring Resource Management Issues in Eastern Indonesia,* ed. S. Pannell and F. von Benda-Beckmann. Pp. 267–321. Canberra: Centre for Resource and Environmental Studies, Australian National University.

Wurm, S.A., ed. 1996. *Atlas of the World's Languages in Danger of Disappearing.* Paris and Canberra: UNESCO Publishing and Pacific Linguistics.

20

ON THE VALUE OF ECOLOGICAL KNOWLEDGE
TO THE KALAM OF PAPUA NEW GUINEA

An Insider's View

Ian Saem Majnep with Andrew Pawley

The Kalam people of Papua New Guinea are mountain people, living in steep-sided valleys around the junction of the Schrader and Bismarck ranges in Madang Province, on the northern fringes of the central highlands. My topic is how the Kalam think of and use their environment, especially the land and the wild plants and animals.[1] I will talk specifically about the Upper Kaironk Valley, where I live, although the main points also apply to other Kalam groups. I will say something about the work the anthropologist and ethnobiologist Ralph Bulmer and I carried out for many years on Kalam natural history and perceptions of the environment and which I am now continuing with other colleagues. And I will ask whether traditional knowledge of the wildlife of our area will still be of value to our descendants in generations to come.

Before 1959 the Kalam had no regular contact with the government or the outside world.[2] In that year a government patrol post was established in our region at the head of the Simbai Valley. Today there is still no road link to the outside world, and if we want to go to the town of Madang, we must travel by small plane from the airstrip at Simbai. But new social and economic forces are already changing our way of life and eroding traditional knowledge of wildlife, and it seems that the next generation of children won't know much about how their grandparents lived.

THE KALAM AND THEIR LAND

In the upper part of the Kaironk Valley the Kaironk River flows swiftly along at about 1500–1600 m above sea level. Many tributary streams flow down the steep slopes of the ranges, forming side valleys, before they join the Kaironk. Most people live and make their gardens on the lower slopes between the river and an altitude of about 2000 m. Like most other people in Papua New Guinea, the Kalam are farmers and the land is our chief wealth. Sweet potato is the staple crop, although taro and yams are the most important foods for ceremonial purposes and we also plant bananas, sugar cane, and various other crops. We keep pigs and, nowadays, some poultry. Until a couple of generations ago pig herds were small but now they are much larger and pigs are a very important kind of wealth for ceremonial purposes as well as an important part of our diet.

The original forest has been almost completely cleared on the lower slopes (up to 1850 m and in some places to 2000 m) except for patches in steep gullies and at streamsides. Above this level the higher slopes and mountain crest (which in some places reach more than 2700 m) are covered with primary forest. But some people clear patches of forest in order to plant and make gardens high up at the edges of the forest, or in clearings inside the forest as at Gulkm in the Aunjang Valley (at altitudes as high as 2350 m). Our custom is to move gardens around, leaving old garden land fallow for 10–20 years before we plant it again.

Unlike some other New Guinea peoples, we don't have large clans and we don't live in villages. The largest traditional social groups are small named local kin groups, usually consisting of no more than 80 or so people. People generally live in scattered homesteads or hamlets with one extended family in each hamlet. Nowadays there are also some larger clusters of houses close to the road leading to Simbai. Typically each person has rights, through his father's or mother's family, to cultivate various patches of garden land on one or both sides of the valley and also to hunt and to gather wild plants on areas of forest high up toward the crests of the mountain range. This way most people have access to land at various altitudinal levels. From time to time people travel down to the flat lowlands, as far as the Jimi Valley on the south side of the Schraders or the Ramu Valley on the north side, to visit relatives and to trade.

Nowadays the Upper Kaironk is more crowded than when I was a boy. The human population has doubled and the pig population has increased many times. We now make many more gardens to feed ourselves and our pig herds. In the 1970s we brought beef cattle in to the Upper Kaironk Valley and at one time there were more than 70. These increases in populations and introduction of new livestock have brought some problems but also people have learnt from the experience. Now we have decided to do away with the cattle because these big animals are too much trouble, taking up good gardenland for pasture and damaging the ground. Because the sweet potato and taro gardens in the main cultivation zone don't produce yields as good as they used to, people are making more gardens higher up, in places where our grandfathers cleared the primary forest. There the land is still fertile and men clear the second-growth bush and get quite good crop yields. Disputes over rights to land in the new and more crowded settlements close to the road has led some people to leave and go back to the more traditional form of separate homesteads on family lands.

Now I don't want to paint a romantic picture of life in the olden days. Some young Kalam people may imagine that their ancestors lived more or less as people do today and had quite an easy life, but this is quite wrong. On the contrary, in the olden days life was hard. Before the government came there was a lot of warfare, we only had tools of stone and wood, and there were fewer kinds of foods to eat than now. It's only quite recently that many of the crops and some of the livestock we now have reached us. The same is true of modern tools, like steel axes, and certain modern ways of doing things, like letting our pigs run free to forage during the day, instead of being kept tethered. That has become possible because steel axes make it easy to build strong fences around our gardens. In the old days only a few rich and important men owned valuable things such as high quality stone axes and pieces of greensnail and other shell valuables. People spent much of their time hunting and made only a few small garden plots for crops around their homesteads. Often food was scarce and we went hungry.

But out of this hard life came a deep knowledge of the land and the forest and the creatures that live there. I myself was brought up living on the edge of the forest and from an early age spent a lot of time hunting and gathering food there. Many older people and some young people in my area still know a great deal about the wild plants and animals of the Schrader Ranges. Much of their expertise was accumulated by the ancestors and handed on, but of course much of it was also gained by personal experience.

KALAM VIEWS OF ECOLOGICAL ZONES

Now I would like to say something about how people in my area regard the landscape and wildlife of the Schrader Ranges. Our language has various terms for particular parts of the land and the forest, in which you will find certain kinds of plants but not others, and certain kinds of animals but not others. I suppose some of these terms refer to what ecologists would call "vegetation communities," "ecological zones," and "microenvironments." I don't fully understand some of these technical terms of English, but I want to stress that in thinking about the natural world Kalam pay close attention to all sorts of details about relationships between particular plants and animals and particular conditions of climate and soil and topography. The reason we are interested in such details is because the gardens we plant, the wild plants we collect and the animals we hunt provide the essentials for our existence. It is vital for us to know where our crops will thrive, where wild plants that are important to us grow, which plants are found together, which plants are the food or sleeping places of which animals, how certain birds and other animals spread seeds, and so on.[3]

Cultivation Zone vs. Forest

In the first place, we make a broad distinction between the cultivation zone or open country (*mseŋ*) where we live and the forested upper slopes of the mountain ranges (*tluk*, or *ytk*) above about 1850 to 2000 m. We see the forested high country as a completely different world from the open country lower down, almost like the way people who live on the coast contrast the land and the sea. The *mseŋ* is an area of familiar things, it's where we build our houses, make our gardens, let our pigs run, and carry on our everyday lives. The trees in this area we call *mon yb*, common or familiar trees, because these are the ones that are most familiar and most important for building, firewood, and the like.

The high mountain forest, on the other hand, is a place full of things that grow wild—wild plants and animals and forest demons (*kceki*). It is a more dangerous and exciting world for us, a place for adventures as well as a place where we can find good things. The opposition between forest and cultivation zone is reflected in many different ways in our beliefs and customs and rituals, but that is a very complicated subject that I can't really say much about here. For men, the forest is the main place for hunting. It is also important to us for the useful plants it contains, such as vines, which are essential for building, and pandanus leaves for thatching and wild plants for food. We obtain certain plants there that are good for medicines and others that are needed in making magic.

Between the forest and the open country there are patches of secondary forest, which we call *wog salm* (or *wog spot* in some dialects of Kalam). These are areas where men have cut down the primary forest, made gardens, and then left the land fallow and where understory forest trees have grown again. *Wog salm* are neither *mseŋ* 'open country' nor *mab wog* 'primary forest' but a transition zone where we find a combination of plants and animals from both sides. There are various other local habitats that we name where particular kinds of plants tend to grow, such as the riversides (*ñg bakbak*) and the heads of streams (*ñg klam*), and landslide areas (*tp koji dp*). A garden or plantation we call simply *wog*, or *wog* plus the name of whatever crop is grown there, for example, *m wog* 'taro garden'.[4]

The Four Main Vegetation Zones

In the minds of farmers and hunters in the Kaironk Valley, however, probably the most important distinctions are those they make among four main zones according to altitude, vegetation, and climate. Let me describe these.[5]

The highest zone we call *kamay* or *kamay agn*.[6] *Kamay* is the small-leafed southern beech (*Nothofagus pullei*), a massive tree, and *kamay agn* means 'base of the southern beeches'. *Kamay* are the dominant trees on the mountain crest ridges (from about 2300 m upwards). But many other big trees grow there, such as the *molok* (a Kalam generic term applied to *Schizomeria* sp. or spp., *Spiraeopsis celebica*, *Caedcluvia* sp. or spp. and *Weinmannia* sp. or spp.), which doesn't grow right on the crests but down on the slopes, especially where the ground is damp. Three sorts of *jbl* (eugenias, *Syzygium* spp.) are present in the *kamay* area and also the cedar *sukñam* (*Papuacedrus* sp. or spp.), the podocarp *noman* (*Phyllocladus* sp.), and several other trees such as *ymges* (smaller-fruited Elaeocarpus), *agnoŋ* (*Syzygium* sp.), *maglpaceb* (*Endiandra grandifolia*), *tlum* (*Sloanea archboldiana*), and the *kodojp* and *kodlap* (*Elaeocarpus* spp.). There is a kind of wild ginger (*Alpinia* sp.) with important ritual uses that can only be found there.

This high zone is not good for gardens. The ground up on the mountain crests, where the clouds hang about, is cold and wet and unsuitable for crops. We say the cold from the wind goes into the trees and down into the ground. But other valuable things are found there. In particular, we go there to gather the nuts of the wild mountain pandan (*Pandanus brosimus/P. julianetti*). This pandan, which we call *alŋaw*, has always been a very important wild food source to us and we recognize more than 20 different kinds. Each grove of pandans has an individual owner. The *alŋaw* is propagated by a little animal, the *mug* (a forest rat, *Melomys lorentzi*), which opens the nuts and buries the kernels for food, thus scattering them around the land. The common birds in the beech forest include the Orange-billed Mountain Lory and Belford's Melidectes. The lories like to build their nests in the tall eugenias. The harpy-eagle and other large birds also nest up there in this high zone. Various important mammals that we hunt live in the trees up in the beech forest and one of them, the *atwak* (Silky Cuscus, *Phalanger sericeus*), is not found at lower altitudes. The small forest wallaby that we call *sgaw* (*Dorcopsulus vanheurni*) also lives up where the *kamay* grows, as opposed to the larger *kotwal* (the scrub wallaby or pademelon, *Thylogale browni*), which you will find only in the flat lowlands and also in the Lower Asai Valley, where some of my relatives live. Wallabies are now scarce in my area, because in recent times they have been heavily hunted with dogs as well as bows and arrows, and there are wild dogs in the forest who kill many wallabies.

As we descend the mountain slopes we come to the zone we call *sugun*. Here (from about 2300 m down to about 2100–2000 m) the typical trees are the garcinias, both the *sugun* (*Garcinia schraderi*) and the *kuam* (*Garcinia archboldiana*), and a very large tree, the *kalap db* (*Dacrydium elatum*), which resembles the rimu trees I've seen in New Zealand, and also *bbolmol* (Monimiaceae) and *ymges* (*Elaeocarpus* sp.). Other wild plants characteristic of this zone include the pandans *kumi* (a cultivar of *Pandanus brosimus/P. julianetti*, whose nuts are eaten and whose pliable leaves are used for making mats), which is like the *alŋaw*, and two other kinds of pandans that have very different leaves and fruit from these: the *gudi* (*Pandanus antaresensis*, whose leaves are used to thatch houses) and *jjak* (a wild *Pandanus* sp., similar to *gudi*, whose leaves are used to make sleeping mats). There is the bamboo called *wdn-kubsu*, used for bowstrings, fire thongs, water containers, and baskets. Several birds that are rare higher up in the beech forest are common in the garcinia zone, including the *nol* (Reichenow's Melidectes), the *gulgul* (Greater Sicklebill Bird of Paradise) and also the *ñopd* (King of Saxony Bird of Paradise).

In this warmer region sweet potato, sugar cane, and some other crops grow well, and we get good returns from new gardenland cleared from the edges of the forest. Where men have cleared the forest to make gardens, the trees that grow in the clearings first are *gupñ* (*Homalanthus* sp.), *weñgaw* (*Macaranga* sp.), and *gog* (*Saurauia* spp.). A bit later come the *wsnaŋ* (wintergreen, *Alphitonia* sp.), *kulmuŋ* (*Trema orientalis*) and *mataw* (?*Evodia* sp.).

Hunters who are used to climbing after animals up in the beech forest find it child's play to climb in the smaller trees of the garcinia zone. But if men who usually hunt in the garcinias go and search for animals in the beech forest, they find it very difficult to discover where the animals' lairs are. The reason is that in the beech forest the trees are taller and there are so many possible trees, and so many clumps of epiphytes in them, that men can climb tree after tree searching for animals, until their elbow joints feel quite weak.

Below the *sugun* region we come to the zone known as *kabi* or *sawey* (both of which refer to kinds of oak). This zone extends from about 2200 m or 2100 m down to the Kaironk River at about 1600 m. Originally most of it was covered with mixed oak forest, but much of the original forest has been cleared, especially below 1850 m. Here you will find nearly all the permanent settlements and most of our gardens. In the patches of forest that survive the dominant trees are the oaks *sawey* (*Castanopsis acuminatissima*) and *kabi* (*Lithocarpus* sp.) and also the broad-leafed southern beech (*Nothofagus carrii*), *Helicia nontana* and *Ficus augusta*. Wherever the *sawey* oaks grow, taro, yams, sweet potatoes, bananas, and sugar cane flourish. The ancestors mainly made their gardens on ridges, clearing the forest there, and later this turned into grassland. In bush fallow growing in cleared mixed oak forest, the common second-growth trees are *Trema* species, *Alphitonia, Macaranga pleiostemonia, Homalanthus populifolius, Finschia chloraxantha,* and *Piptocalyx*. Pigeons and other birds come and feed on the fruit of the *Pipturus* and *Homalanthus* trees. We can further divide the oak zone into subzones. There is still oak forest higher up and the real *mseŋ* or open country starts at about 1850 m. Nowadays we plant casuarinas all over the *mseŋ* zone, although in my grandparents' time this was not the case. The casuarinas are useful for fertilizing gardens but even more important for firewood and for providing timber for fences and houses.

Just below Womuk in the Upper Kaironk, at about 1600 m, we come to what we regard as the warm lowlands, which we call *tmen* after the Lawyer Cane (*Calamus* spp., Palmae) that grows there, or *numul*. This warm zone extends right down the Kaironk Valley to the flat lowlands of the Jimi Valley and, on the north fall of the Schraders, to the Ramu Valley (the lowlands on that side are called *cdoŋ* in Kalam). Various crops that struggle to grow in our area because it is cold grow to a large size down in the *tmen* country. In the Kaironk Valley the *tmen* country is mostly grasslands, but there you will also find the *yagad* (marita fruit pandanus, *Pandanus conoideus*), the black palm that we use for bows and some large trees and many birds and mammals and reptiles that you can't see higher up. For example, when we want to get feathers for the headdresses of dance costumes or for trading, we go to relatives down valley to trade for *wtay* (White Cockatoo), *koben* (Crown Pigeon), *kay-wl* (Hornbill), *ydam* (Pesquet's Parrot) and *yabal* (Lesser Yellow Bird of Paradise, *Paradisaea minor*).

HUNTING

Hunting is a great source of pleasure to those men and women who do it. In the dry season, when hunters climb some tree high in the forest and sit on a clump of epiphytes and scan the landscape, the distant woodlands shimmer with beauty and this is so delightful to look on that it may be hard to bring oneself to climb down and return home. On days when the skies are clear and we look up to the forests it's very frustrating to have to keep working in the gardens. We have to force ourselves to turn our eyes away, if we are to go on making fences. On fine days when we go off up into the forest we thoroughly enjoy climbing trees, looking for kapuls (game mammals). On nights when the sky is clear we also hunt by moonlight. For moonlight expeditions there are advantages to men hunting in pairs. By daylight, however, a man can hunt alone, climbing trees and getting across from one tree to another by use of a hooked pole.

These days hunting is no longer such an important source of food as it was to our ancestors. We rely on our gardens and livestock for that. But the animals people bring back from hunting serve many different kinds of purposes besides meat for the hunters and their families: game animals are needed for various rituals, initiations, sacrifices to ancestors, exchanges between kin, and for sale to other people who need them for ceremonies and exchanges.

Animals We Hunt

There are more than 150 species of birds in my area and we hunt most of them. But birds are not nearly so important to us as the animals we call *kmn*. This is the Kalam term for the wild mammals—marsupials and giant rodents—that we prize as game. Ralph Bulmer used to translate this term as *kapul,* which is its equivalent in Tok Pisin (New Guinea Pidgin English), because there is no equivalent word in standard English.[7] I will do the same here.

We hunt more than 40 kinds of *kapuls.* Although women hunt for bandicoots and other small animals on the ground and in open country, the main hunting grounds for men are the forests where the *kamay* and *sugun* grow, where lots of

animals live in the trees. Several kinds of hunting strategies are use to capture kapuls. The most important methods are:

1. Searching in trees during daylight. Men travel through the forest climbing
   trees which look as if they might have animals in them. They climb very
   tall trees and if the targeted trees don't have low branches or vines to climb
   up, a man will go up a neighboring tree and cross over by means of climbing hooks which are laid across branches and lashed at one end.
2. Searching the ground. This is the main strategy of women and men with
   disabilities.
3. Moonlight shooting in trees. This is undertaken by teams of men.
4. Trapping and digging out sets and burrows. This is also mainly done by men.

While we are out hunting kapuls in the forest, we also gather useful plants. We
collect foliage from various kinds of pandans, as I have mentioned, especially *gudi*
and *kumi* but also *alŋaw* and *jjak,* to stitch into mats and rain shields, and various
kinds of vines, for use as rope in building houses and fences. We gather edible
ferns (*Diplazium* sp.) and *kuñp* greens (the spinachlike *Oenanthe javanica*) and the
leaves of the hardy plant we call *ñepek* (*Cyrtandra* sp.), which grows on tree trunks
and tree ferns and stumps, to cook in stone ovens with the kapuls. Sometimes we
cut out the shoots of young *alŋaw* pandans and cook these, and lower down in the
garcinia forest, where *alŋaw* doesn't grow, we do the same with the shoots of *gudi.*

Hunting the Copper Ringtail and the Husk-Shredder Rat

You shouldn't think that hunting is easy. Far from it. Just as in places by the sea,
when fishermen set out they need to know where the best fishing places are, so in
my place a hunter has to have particular knowledge about which trees to climb.
Not everyone in my area is an expert hunter. In fact only a few people are truly
expert and regular hunters, and these are people who live near the edge of the
forest or in settlements inside forest clearings. To be a good hunter you must
know a lot about how the animals behave. You need various strategies for different kinds of animals. Some animals are dangerous to tackle, some are dangerous
to eat, some live in such and such places, and so on. Before he starts to climb, a
hunter will search the ground for fragments of leaves and fruit or other signs that
kapuls have been feeding, until he gets a hunch that there is a kapul curled up in
this particular tree. Hunters have to go on walking and searching until they find a
tree with a kapul lair. Visitors from some other area or from local settlements not
near the forest who accompany a hunting party get miserably hungry after climbing trees, one after another, for hours while the real hunters just carry on.

I will tell you something about one or two kapuls that we hunt. I can mention
a small part of what hunters know about these animals. The *ymduŋ* (Copper Ring-

tail, *Pseudocheirops cupreus*) is the commonest game mammal in the Upper Kaironk. It grows up to about 2 kilos in weight. One kind, which we call the beech *ymduŋ*, is found above 2300 m., up where the *kamay* or small-leafed beech trees (*Nothofagus pullei*) grow, and down to the upper part of the range of the *celed* (the larger-leafed beeches, perhaps *N. carrii*). This kapul is hard to hunt because it does not scratch claw marks on trunks and branches of trees where it has its lair; it just finds its way up without leaving any trace. Then it burrows in to make its lair, snapping off with its teeth any roots and twigs that are in its way.

The Copper Ringtail of the beech forest eats the juicy young foliage of trees such as *wask* (*Ficus ?iodotricha*), *sagal* (*?Alstonia glabriflora*), *gog* (*Saurauia* spp.), *noman* (*Phyllocladus* sp.), *majown* (*Prumnopitys amara*), *meŋñ* (*?Flindersia pimenteliana*), and *klen* (*Planchonella ?macropoda*), and the foliage of *gglɔŋ* (*Riedelia geluensis*), an epiphyte of the ginger family, *galgal* (*Alpinia* sp., also flowering gingers), *jjb* (certain epiphytic orchids), and *jsp* (small densely growing epiphytic orchids that produce a grasslike covering on tree branches). We call these little plants it eats its *bep* or 'staple greens'.

It takes quite a lot of experience for a hunter to guess reasonably accurately where the lairs of the Copper Ringtail are. When this kapul wants somewhere to sleep, it looks for a big tree with a clump of fresh succulent epiphytic moss and it sleeps under the newly grown moss. It may also sleep in bird's nest ferns, either in the large *gob-lad* fern or in the crown of the smaller *kiopi* fern, or in clusters of wild gingers, *galgal* (*Alpinia* sp.). When they get tired of sleeping higher up, these kapuls sometimes go down to streamsides and climb into hollow *beg* trees (*Perrottetia alpestris*). They will also sleep in holes in trees where branches have fallen off leaving cavities.

The Copper Ringtail does not always sleep in big trees, as some kapuls do. When men are hunting for *maygot* (the White-Tailed Highland Cuscus, *Phalanger carmelitae*) and *mosak* (the Black-Eared Giant Rat, *Mallomys rothschildi*), they have to climb really tall trees to get them, including the few that have big clumps of epiphytes, which we call *mab-yb,* such as *jbl* (*Syzygium* spp.), *kalap db* (*Dacrydium* sp./*Podocarpus* sp.), *ñŋud* (certain Lauraceae spp.), *noman* (*Phyllocladus* sp.), and *molok-kab* (*Schizomeria* sp.) But the Copper Ringtails can be found in young trees or lower down in big trees, on a lateral branch or in the hollows of a fork of the trunk. Hunters sometimes shoot *ymduŋ* by moonlight. During the day when they have noticed that the animals have been browsing on the leaf buds of a *wask* fig tree, or a *sagal* or *klen* or *noman*, they may decide to come back to that ridge when the moon is up.

Copper Ringtails and other kapuls will feed on the decayed wood of various trees, such as *noman* plancanellas, *ñŋud* laurels, *majown* (*Prumnopitys amara*), *kalap db* New Guinea podocarps, *binmuŋ* (*Ascarina subsessilis*), and *magl-paceb* (*Endiandra grandifolia*). When hunters find animals have been gnawing at one of these trees, they may make a frame of tree-fern foliage around the trunk, leaving two openings with springs set in them, and may thus catch two kapuls.

The common *ymduŋ* is found lower down, in the garcinia zone. Its lower limit is the edge of the open country where there are mixed oak forests still standing, although if there is any real forest continuing on lower down it will be found there too. Copper Ringtails at this lower altitude eat the same foods I've already mentioned for the beech *ymduŋ*. Down in the garcinia zone there are plenty of trees, but the big mossy clumps of epiphytes common in the beech forest are quite scarce there. *Ymduŋ* sometimes sleep in dry *kumi* (cultivated nut pandans). Along with many other kinds of kapuls, they also sleep in the undercroft, the natural galleries or spaces under the floor of the mountain forest which we call *abn*. Down valley they may sleep under thickets of tangle fern (*gd*). When the lower foliage of the fern bed is dead and dry, the sheltered spaces underneath are like the chambers of a house, and many kinds of animals come there to sleep.

Now I'll turn to a very different kind of kapul, the *gudi-ws* (the Husk-shredder Rat, *Anisomys imitator*). This animal has a wide distribution, being present in the beech and garcinia forests and also out in the open country in areas where trees have been felled to make gardens and in the swordgrass. It makes its nest out of foliage that it gathers together, in swordgrass beds and under the bases of tree ferns, and under fallen branches, much like the bandicoot. It is mainly caught by women, who specialize in hunting bandicoots. The *gudi-ws* is a ferocious little animal, and you have to be very careful when grabbing it.

The *gudi-ws* feeds on the nuts and fruits of *gudi* pandans, garcinias, *abok* (small-fruited breadfruit, *Artocarpus vrieseanus*), various species of fig tree, such as *wask*, *wen*, *kubap*, and *yam*, and the Alaeocarpus trees *kodojp*, *kodlap*, and *ymges*. It is the habit of this giant rat to break open and plant the seeds of *gudi* and *kumi* pandans. When it collects *gudi* seeds to eat and carries some of them away with its teeth, it drops some so that they grow in places a long way from where any such plant was previously present. It does not, however, do damage to pandans in the same way that the *abben* (Highland Giant Tree-Rat, *Uromys anak*) does, which bites off the fruit-heads and after it has eaten just a small part of these, leaves them. After the *abben* has left them broken off the *gudi-ws* and smaller bush rats come along and feed on them. The contents of the large intestine of the *gudi-ws* are not bitter but tasty, because of the sweet-tasting food it eats, and we use this to flavor greens and root vegetables before cooking them. We use the lower jaw of this animal, with incisor teeth still attached, to engrave arrow shafts and bamboo containers and to cut Lawyer Cane and orchid fiber and *Freycinettia* stems.

RECORDING KALAM NATURAL HISTORY WITH RALPH BULMER

It was in 1963, when I was about 15, that I met Ralph Bulmer, a social anthropologist who had come to work in the Upper Kaironk. He was interested in studying Kalam natural history as well as Kalam society and during his early visits to the Kaironk Valley I was one of his field assistants. A few years later he asked me to collaborate on writing a book, *Birds of My Kalam Country* (Majnep and Bulmer 1977), in which I would describe all the birds and their significance to people in a series of interviews. In preparing the book we first worked out the divisions or chapters together, then discussed the birds of each chapter mainly in Pidgin, before Ralph put my observations into English and added his comments. After that was published he asked me to write a book in my own language about either the wild animals or the wild plants of our region, which he would translate and add commentary to. So we did a very small book about some wild food plants of the Schraders and I began a larger one about the kapuls, to be called "Animals the Ancestors Hunted."

The "Animals" book was much harder to write than the "Birds" book, for many reasons. We went about the job in a completely different way. When we came to do the "Animals" book, I first taped my accounts of each animal in Kalam, then transcribed the tapes, before we worked together translating them into English. Tape-recording the chapters about the most important animals was not too difficult, because the knowledge of the trees and animals of the forest is there in my own head. But when it came to animals of the middle altitudes or the hot lowland plains I had to go around and interview many other people who had specialized knowledge that I didn't have. Ralph also asked me to record myths about animals, and I had to choose those that were suitable for the book and record them myself or get others to tell them. Transcribing the tapes, putting the words down on paper, was hard going. I had never done this kind of work before, writing in my own language. Translating the Kalam text into English was often extremely difficult. Ralph was an expert on birds, but he did not know nearly as much about kapuls and what they eat, where they sleep, and how they think. Part of the difficulty was that many Kalam words I used were not yet in our Kalam dictionary (Pawley and Bulmer in press, with I.S. Majnep, J. Kias, and S.P. Gi) and there are no Pidgin words to translate them, and indeed there are no English words to translate some of them. So when Ralph asked me what these Kalam words meant I had to stop and think, and sometimes I got really worried and began to sweat with embarrassment and ask myself what could have possessed me to take this work on! Ralph sometimes had to wait as long as five or six minutes, and I felt very ashamed. But eventually I would come up with an explanation.

There are many advantages that Bulmer and I gained from working together. If you are an outsider, such as an anthropologist or biologist or linguist, it is very hard work indeed to gain an accurate understanding of local knowledge of wildlife and the environment generally. It's much easier to record such knowledge if you are an insider. You already know the language, you already know a lot about traditional custom, and you can ask your relatives and friends about things that you yourself don't know. A foreigner will have problems with the language and often won't know when he or she is getting reliable information. But even for an insider, like me, it can be very hard work to record information about wildlife, because much of it is well known only to certain experts. Sometimes people give you inconsistent accounts. You have to check many things both by asking a range of informants and by your own observations. And when you have gathered what you believe is a body of reliable information you still have to figure out the best way to organize and present it. To present this material to an international audience you must be able to identify the animals and plant species you are talking about, for example, in terms of English and Latin names, though without making the mistake of assuming that our categories are the same as those of English speakers or those of Western science. This means you must work together with specialists from various sciences and with someone who can translate your work.

Ralph had translated the Kalam chapters of the "Animals" book into English and added some commentary, and he was working on editing the material when he died. The linguist Andrew Pawley completed the editing of the bilingual text of the "Animals" book and much of this material has been published in working paper form (Majnep and Bulmer 1990, n.d.). But Bulmer also planned to produce a streamlined English-only book version for a general readership; my friends Robin Hide and Andrew Pawley are now working on this task. Now I am working with Pawley and the botanist, Rhys Gardner, on a book about the trees and vines and other wild plants of the Kalam region (Majnep, Pawley and Bulmer, in prep.).[8]

THE FUTURE OF TRADITIONAL ECOLOGICAL KNOWLEDGE

I come now to the question: Will the knowledge and values that I have spoken of here be relevant to the lives of our grandchildren? Or are they bound to disappear?

At this stage, traditional knowledge of the natural environment is still intact in my community because there are many people around who grew up in the old days and who continue to follow most of the old customs. But during the last 30–40 years things have changed a lot. Kalam people are now being drawn into the modern world and the national life of Papua New Guinea. In the Upper Kaironk Valley most people no longer go hunting and collecting food in the forest as often as people did a generation ago. Families generally prefer to live close to the main road that runs down the north side of the valley two or three hundred meters above the river. Children nowadays must spend several years in primary school. Those who do well may go on to secondary schools in distant places, where they will live for years away from home. In order to pay school fees for their children and high bride-prices for wives men must work hard at their gardens and also plant coffee and sell some of their livestock and quite a lot of young men go to work in the towns and on the coast. Some people buy radios and play cards instead of going hunting. This story is being repeated all over Papua New Guinea.

Still, I would like to think that traditional knowledge will remain important for many Kalam. For those of us who live at home in the mountains, the land and the forest will always be the main source of wealth. Farming and hunting is in our blood and in our culture. Even though hunting kapuls and birds is no longer such an important source of food as it once was, some young people continue to hunt and to trade with hunters for other reasons: to obtain feathers, pelts, for rituals, and the like. Expert hunters are still respected. People like to live in a household where such men live, who can both bring in game and teach others how to get it. And hunting will remain an enjoyable activity for people bored with making gardens or burdened with worries. Not long ago, in my childhood, warfare and witchcraft were the main causes of anxiety and hunting was a source of happiness and freedom from worry. After days out hunting in the forest, our parents slept well and had easy minds. Now everybody has things to worry about—whether these are the state of their garden plantations or money problems or witchcraft or whatever—and it's worth remembering how our ancestors and our parents coped.

Years ago Ralph Bulmer (1971:31–32) wrote that the best way to teach children is to make them into researchers. Nowadays nearly all our children attend school where they sit at desks and study English and arithmetic and science. But learning by doing things oneself is more fun than sitting listening to a teacher or copying from a book. And in the case of biology and nature study, where better for children to start than with their own home surroundings, looking at things that they know quite a lot about? Each local school and each class could compile its own reference sources describing the plants and animals and ecology of the local environment, recording names and locations, characteristics and uses, collecting and preserving plant and insect specimens and drawing illustrations and maps. I like that idea. It should be possible for quite young children to get involved and also their parents and grandparents.

But our government needs to help. Teachers need to be trained to use such a dynamic method in place of the customary passive, spoon-feeding of information. And the Department of Education should supply schools with some simple basic materials—large sheets of paper and scrapbooks for the students, pictures of some of the animal and plant species—and perhaps encourage students by featuring some projects in exhibitions and publications and on radio and TV. After students realize that they already know a good deal about natural history, from their own experience, they will be ready to study the biology and ecology of other regions and to learn general principles. And some will appreciate that the knowledge and skills that their ancestors accumulated is still worthwhile.

NOTES

1. The main text of this paper is based partly on material written in Kalam by Majnep and partly on material from interviews with Majnep. Pawley translated and edited the material. Explanatory notes added by Pawley are enclosed in parentheses or appear as footnotes.

2. A few Kalam who were living in the foothills near Aiome in the Ramu Valley had fleeting contacts with German anthropologists before World War I.

3. Bulmer (1982:59–60) writes of the wide variety of natural resources used by the Kalam of the Upper Kaironk. In the early 1960s they cultivated some 28 species of food plants, including about 170 different cultivars. They cultivated at least 30 plant species for other purposes (flavoring and wrapping of food, medicines, rituals, as ornamentals, and for fuel). They also ate about 40 additional species that were never or seldom cultivated and used about 150 wild species more for technological or ritual purposes. Still other plants were important as the known nesting and feeding sites for different kinds of birds and animals. In all, they distinguish more than 1000 plant taxa by name. The numbers for animals are scarcely less impressive. The Kalam eat, at least occasionally, about 200 vertebrate animal species and at least 40 invertebrates. Bulmer comments that the Kalam were in no way unusual among New Guinea peoples in the range of wild plants and animals they utilize.

4. The noun *wog,* which occurs chiefly in compounds as a classifier might be glossed 'patch of vegetation, vegetation community'. In fact any patch of vegetation dominated by a particular category of plant X can be called an *X wog.*

5. The Kalam breakdown is finer-grained than the Western ecologists' classification of the altitudinal zones presented in Paijmans (1976). Paijmans contrasts three broad vegetational zones for New Guinea: lowland alluvial plains and fans (0 to 500 m); foothills and mountains below 1000 m; and lower montane (1000 to 3000 m). The Kalam make a four-way distinction within Paijmans's lower montane zone.

6. Alternatively, the Kalam often speak of zones in terms of "where (diagnostic) plant X grows," for example, "this bird is common up where the *sugun* grows."

7. Bulmer coined the term "game mammals" to translate *kmn* in a number of his papers. Smaller and less important mammals, some which are also hunted, are known as *as.*

8. The principal completed works from the collaboration are Majnep and Bulmer 1977, 1983, 1990, and n.d. See also Majnep 1982.

REFERENCES

Bulmer, R.N.H. 1971. Science, ethnoscience, and education. *Papua New Guinea Journal of Education* 7(1):22–33.

Bulmer, R.N.H. 1982. Traditional conservation practices in Papua New Guinea. In *Traditional Conservation in Papua New Guinea: Implications for Today,* ed. L. Morauta, J. Pernetta, and W. Heaney. Pp. 39–77. Monograph 16. Boroko, Papua New Guinea: Institute of Applied Social and Economic Research.

Majnep, I.S. 1982. On the importance of conserving traditional knowledge of wildlife and hunting. In *Traditional Conservation in Papua New Guinea: Implications for Today,* ed. L. Morauta, J. Pernetta, and W. Heaney. Pp. 79–82. Monograph 16. Boroko, Papua New Guinea: Institute of Applied Social and Economic Research.

Majnep, I.S., and R.N.H. Bulmer. 1977. *Birds of My Kalam Country.* Auckland and Oxford: Auckland University Press and Oxford University Press.

Majnep, I.S., and R.N.H. Bulmer. 1983. *Some Food Plants in Our Kalam Forests*. Department of Anthropology Working Paper no. 63, University of Auckland.

Majnep, I.S., and R.N.H. Bulmer. 1990. *Kalam Hunting Traditions*, vols. 1–6. Department of Anthropology Working Papers nos. 85–90, University of Auckland.

Majnep, I.S., and R.N.H. Bulmer. N.d. *Kalam Hunting Traditions*, vols. 7–12. Printout. Department of Linguistics, Research School of Pacific and Asian Studies, Australian National University.

Majnep, I.S., A. Pawley, and R.N.H. Bulmer. In preparation. Kalam plant lore.

Paijmans, K., ed. 1976. *New Guinea Vegetation*. Canberra: Australian National University Press.

Pawley, A., and R.N.H. Bulmer, with I.S. Majnep, J. Kias, and S.P. Gi. In press. *A Dictionary of the Kalam Language of Papua New Guinea, with Ethnographic Notes*. Canberra: Pacific Linguistics.

21

WA HUYA ANIA AMA VUTTI YO'ORIWA — THE WILDERNESS
WORLD IS RESPECTED GREATLY

Truth from the Yoeme Communities of the Sonoran Desert

Felipe S. Molina

Since time began for the Yoeme People of the Sonoran Desert (in the states of
Arizona, U.S., and Sonora, Mexico), the *Yoem Lutu'uria* 'Yoeme Truth' has guided
the Yoeme people spiritually, mentally, and physically into respecting all life on
Earth. The Yoeme stories, sayings, and ceremonial songs teach specifically about
understanding and respecting the *Huya Ania,* 'Wilderness World'. This traditional
knowledge is also called *Wa Yo'ora Lutu'uria,* 'the Elders' Truth'. This truth is to
help all people live in the right way on this Earth.

When a Yoeme child is born, the family and the community are happy. This
new person is another being from the Spirit World who is received and welcomed
to this Earth with much happiness and joy. The child's umbilical cord will be given
to the red ants to create a bond with the natural world. This action by the female
relatives of placing the umbilical cord on the red ant hill is a formal introduction
of the child to the insect world and all other life forms and elements of this Earth.
When the ants take the umbilical cord down into their hole, the child will be
known throughout the world by the insects and all other life. The new child will
now begin to learn how to respect humans, animals, plants, and the natural elements. The ants will not bite or hurt such a child because of the action taken by
the female relatives.

The Yoeme people have maintained many of these actions and rituals to help
one another and to give thanks to the Creator God for offering such a place as the
Earth to live on. To many Yoeme, this Earth is a brief place to live and learn to live

with a good heart and follow the teachings of the Yo'ora Lutu'uria (Elders' Truth) and the Yoem Lutu'uria (Yoeme Truth) by the elders.

Whenever we are going into the Huya Ania, the Wilderness World, we remind ourselves that we must respect the area. Anybody walking into the Huya Ania must first purify his mind, body, and soul and then ask permission out loud or mentally from the plant world, the mountains, and the animals, to be in their area. This action is extremely important for some Yoemem because we do not want to get lost or injured while we are in the Huya Ania. Yoeme people enter into the Huya Ania to gather food items and household resources or to acquire spiritual knowledge, and so it is quite necessary to do the above procedures with respect.

YOEME SONGS

The Yoeme songs teach us many of these beautiful lessons about the wilderness world. These songs can be deer songs, coyote songs or corn wine songs, and others that are no longer sung or practiced today. The following two deer songs are examples of the Yoeme Truth about the Wilderness World, Huya Ania (from Evers and Molina 1987:104, 175; see references to this chapter for other works describing Yoeme cultural and spiritual traditions).

Huya Aniwa	Wilderness World
Empo sewa yo huya aniwa	You are a flower-enchanted wilderness world
Empo yo huya aniwa	You, enchanted wilderness world
Vaewa sola voyoka huya aniwa	You lie with see-through freshness, wilderness world
Empo yo huya aniwa	You, enchanted wilderness world
Vaewa sola voyoka huya aniwa	You lie with see-through freshness, wilderness world
Empo sewa yo huya aniwa	You are a flower-enchanted wilderness world
Empo yo huya aniwa	You, enchanted wilderness world
Vaewasola voyoka huya aniwa	You lie with see-through freshness, wilderness world
Empo yo huya aniwa	You, enchanted wilderness world
Vaewa sola voyoka huya aniwa	You lie with see-through freshness, wilderness world
Empo sewa yo huya aniwa	You are a flower-enchanted wilderness world
Empo yo huya aniwa	You, enchanted wilderness world

Vaewa sola voyoka huya aniwa	You lie with see-through freshness, wilderness world
Empo yo huya aniwa	You, enchanted wilderness world
Vaewa sola voyoka huya aniwa	You lie with see-through freshness, wilderness world
Empo sewa yo huya aniwa	You are a flower-enchanted wilderness world
Empo yo huya aniwa	You, enchanted wilderness world
Vaewa sola voyoka huya aniwa	You lie with see-through freshness, wilderness world
Empo yo huya aniwa	You, enchanted wilderness world
Vaewa sola voyoka huya aniwa	You lie with see-through freshness, wilderness world
Ayamansu seyewailo huyata naisukunisu	Over there in the middle of the flower-covered wilderness world
Yo huya aniwapo	In the enchanted wilderness world
Uhyol machi hekamake uhyolisi	Beautiful with the dawn wind
Vaewa sola voyoka huya aniwa	Beautifully you lie with see-through freshness, wilderness world
Empo yo huya aniwa	You, enchanted wilderness world
Vaewa sola voyoka huya aniwa	You lie with see-through freshness, wilderness world.

Tosali Vaesevolim

White Butterflies

Tosali vaesevolimtea hepelamsum chasaka	White butterflies, they say, in a row are flying,
Tosali vaesevolimtea hepelamsum chasaka	White butterflies, they say, in a row are flying.
Tosali vaesevolimtea hepelamsum chasaka	White butterflies, they say, in a row are flying,
Tosali vaesevolimtea hepelamsum chasaka	White butterflies, they say, in a row are flying.
Tosali vaesevolimtea hepelamsum chasaka	White butterflies, they say, in a row are flying,
Tosali vaesevolimtea hepelamsum chasaka	White butterflies, they say, in a row are flying.
Tosali vaesevolimtea hepelamsum chasaka	White butterflies, they say, in a row are flying,
Tosali vaesevolimtea hepalamsum chasaka	White butterflies, they say, in a row are flying.
Ayaman ne seyewailo taa'ata yeulu weye vetana	Over there, where the flower-covered sun comes out,

yeulu katekai	they are emerging,
sime huya aniwachi	all through the wilderness world
sea hepelamsum chasaka	in a row they are flying.
Tosali vaesevolimtea	White butterflies, they say,
hepelamsum chasaka	in a row are flying.

AREAS OF CONCERN

Desert Environment

Although the Yoeme people still retain much of this rich traditional knowledge, there are problems in the communities. The Yoeme people of Arizona and Sonora are losing some of the traditional ceremonial plant resources at an alarming rate. Most of the loss is due to the rapid expansion of non-Yoeme people near Yoeme communities. Some of the plants that are talked and sung about are slowly disappearing from the desert world. Many times other plants are being used to replace the traditional plants that were commonly used in the homes and ceremonies. The mesquite and cottonwood are two examples. Mesquite trees are cut down by Yoeme and non-Yoeme for firewood, fences, shelters, art, farmland, and home sites. More and more people are moving into pristine desert lands and this, unfortunately, leads to the mass destruction of the mesquite and other desert plants important to the Yoeme. Even though there are conservation efforts within these areas, the destruction continues. Whole desert areas are bulldozed within minutes in many parts of Arizona and Sonora.

Environmentalists and other concerned citizens cry out in protest but to no avail. Hopefully, in the future, better and stricter laws will help restore and conserve the pristine areas that are being cleared for homes and businesses.

Traditional Yoeme Knowledge and Language

Another concern that the Yoeme face is the transmission of traditional knowledge and language. It is not always passed down to the young Yoeme people because of various reasons pertaining to individual families and communities. In Arizona many young Yoeme people do not speak the Yoeme language anymore so this makes it more difficult to learn about the Yoeme Truth. There are, however, small pockets of traditional families throughout the Greater Southwest who think it is important to pass on the traditional Yoeme knowledge and language.

I interviewed eight young Yoeme people from the ages of eleven to twenty-five and two adults in the Arizona Yoeme communities. The results will be explained briefly to give the reader an overview of what is happening to the Yoeme traditional knowledge and language. The questionnaire was geared toward the area of language usage in 1996 among some Yoeme families in Arizona. At the same time, valuable information was gathered on traditional Yoeme knowledge from this group of young people. I conducted the interviews one on one, asking the questions orally and receiving the response to the question orally. The responses were written in full and analyzed afterwards.

Two of the eight young people interviewed spoke the Yoeme language. The six who don't speak it expressed a strong desire to learn to speak it. All eight of the young people said that English was their dominant language. Seven of them said that they had some understanding of Spanish. All eight young people agreed that the Yoeme language must continue to be spoken by all Yoeme people. Only two said that they have two young friends each who understand a little Yoeme. All eight young people said that they would go to certain elders and to the school and public library to learn more about the Yoeme language and culture.

When I asked about the importance of maintaining the Yoeme language, I received these comments:

> "So our culture won't die and fade away. So that later we can teach our kids and keep the traditions going" (female, 20 years of age).
> "To keep the traditions going, so that more people can learn" (male, 15 years of age).
> "It is part of staying Yaqui, especially up here in the U.S. It gives a people a sense of being" (female, 24 years of age).
> "If people don't know it right off hand, it's not a way of life anymore. I prefer to hear more Yaqui . . . more a way of life" (female, 25 years of age).
> "I can speak to Dad when he becomes an elder instead of Spanish or English" (female, 12 years of age).
> "Because it's a traditional thing and I am Yaqui. . . . They have to carry it on and pass it to other people" (male, 11 years of age).
> "So I could talk to elders. What they need" (male, 17 years of age).
> "So our culture won't die" (male, 15 years of age).
> "Our language is dead. I don't feel the same anymore. When I was growing up in Marana, Arizona, everybody in the village spoke the language. We were criticized by the non-Yoeme people because we spoke a different language" (male, 42 years of age and a fluent Yoeme speaker).
> "We started to lose our language back in the sixties. We started to speak the English language. We mixed English and Yoeme together" (female, 49 years of age and a fluent Yoeme speaker).

CONCLUSIONS

For whoever is interested at this time in maintaining the native language and culture, I would recommend doing the following with tribal approval only: nature walks with our elders; invite elders to visit classrooms; videotape elders talking about cultural knowledge in the desert and in the classroom; start a Yoeme heritage school in all the communities and, finally, get the parents involved in all aspects of the students' education.

ACKNOWLEDGMENTS

The author of this chapter would like to thank the following people from Arizona: Rosario Castillo, Tomás Martinez, Francisca Martinez, Manuel Valencia (Lito), Romana Sanchez, Juana Paula Castillo, Juan Garcia, Estefana Garcia, and Dolores Miranda, as well as the following people from Sonora: Teresa Baltazar, Vicente Baltazar, Ignacio Sombra, Antonia Flores, Alfonso Leyva, Juana Romero, Juan Tampaleo Amarillas, Miki Maaso, Jesús Alvarez, José María Jaimez, and Antonia Amarillas.

The book's editor wishes to thank Steve Brown for his help with electronic scanning and editing of the original typewritten version of this chapter.

The songs "Wilderness World" and "White Butterflies" appeared originally on p. 104 and p. 175 of *Yaqui Deer Songs/Maso Bwikam: A Native American Poetry,* by Larry Evers and Felipe Molina, copyright 1987 The Arizona Board of Regents. They are reprinted here with permission of the University of Arizona Press.

REFERENCES

Evers, L., and F.S. Molina. 1981. *The South Corner of Time: Hopi, Navajo, Papago, and Yaqui Tribal Literature.* Tucson: University of Arizona Press.

Evers, L., and F.S. Molina. 1987. *Yaqui Deer Songs/Maso Bwikam: A Native American Poetry.* Tucson: University of Arizona Press.

Evers, L., and F.S. Molina. 1989. *Coyote Songs from the Yaqui Indian Bowleaders Society.* Tucson: Chax Press.

Evers, L., and F.S. Molina. 1992. Hiakim: The Yaqui homeland. *Journal of the Southwest* 34(1):1–2.

Evers, L., M. Maaso, and F.S. Molina. 1993. The elders' truth: A Yaqui sermon. *Journal of the Southwest* 35(3):225–317.

Shaul, D.L., and F.S. Molina. 1993. *A Concise Yoeme and English Dictionary.* Tucson: Tucson Unified School District.

22

RESOURCE MANAGEMENT IN AMAZONIA

Caboclo and Ribereño Traditions

Christine Padoch and Miguel Pinedo-Vasquez

Until recently, research and writing on ethnobiology and traditional resource use in Amazonia was very largely confined to studies of groups identified as indigenous peoples or Amerindians. Prior to the 1980s, with some very notable exceptions (Villarejo 1943; Galvão 1952; Wagley 1953; Sternberg 1956; San Román 1975), communities of *ribeirinhos* (or *caboclos*) of Brazil and *ribereños* of the Peruvian Amazon seemed to be invisible to scholars. Rural Amazonians were commonly assumed to belong to one of two classes: indigenous tribal groups or recently arrived colonists. In the last two decades, however, spurred largely by interest in the rubber tappers of Acre state in Brazil, as well as by several prominent research efforts carried out in Brazilian and Peruvian Amazonia, the environmental knowledge of the local but nontribal peoples of rural Amazonia has achieved some prominence in Amazonian studies.[1] But to those scholars who have focused largely on the linguistic features of Amazonian societies and of their environmental terminologies, studies of the environmental knowledge of ribeirinhos/ ribereños may still be little known.

The terms *ribeirinhos*, or *caboclos*, and *ribereños* are used to designate the rural folk of Amazonia who are neither tribally organized Amazonians nor recent immigrants. In Peruvian Amazonia and along the Solimões River and its tributaries in Brazil many caboclos and ribereños are the descendants of indigenous peoples, others in upper and lower Amazonia count both indigenous Amazonians and immigrant Europeans, Africans, and Asians as ancestors. Caboclo and ribereño populations are found throughout the basin. Their settlements and patterns of management of flora and fauna in and along the banks of the Amazon River and its many whitewater tributaries, the area known as *várzea* in Brazil, have received the greatest attention from researchers. Much of this emphasis stems from the ecological, economic, and cultural importance of rivers and their floodplains in Amazonia in both the past and the present, as well as in plans for the area's future.

MANAGING THE FLOODPLAIN

The várzea occupies only 2–3 percent of the total area of the Amazon Basin, but it has long played a very important role in the region.[2] Várzeas are often considered the most economically important of all Amazonian lands because of their good soils, their access to transport, and their relatively concentrated human populations. For centuries the floodplains have been very important areas of human settlement; research indicates that many areas of the várzea were once densely settled by highly organized indigenous peoples (Roosevelt 1989, 1991). Most indigenous várzea societies were destroyed in the early years of the European conquest of Amazonia. In contemporary Amazonia, however, most people, both rural and urban, continue to live in or close to the floodplains.

The alluvial soils of the várzea, replenished every year by floods, are fertile. But the annual floods that deposit those rich sediments also make these lands difficult to exploit by modern agricultural methods. Few modern technologies deal adequately with the risks and opportunities the diverse and changeable floodplains present. Today water buffalo ranches increasingly occupy much of the floodplain's fertile lands, particularly along its lower reaches. This extensive land use together with highly intensive and heavily fertilized vegetable production on the outskirts of a few major cities are two conspicuous modern uses. Each has an environmentally and socially questionable present and an uncertain future. Neither is economically feasible for the great majority of Amazonian farmers.

In areas like the várzea that are potentially very productive but are unsuitable for exploitation using modern agricultural technologies, it is most important to look at locally developed production systems. In the Amazon floodplains indigenous peasant populations use a great variety of agricultural and agroforestry practices that effectively exploit the diverse resources and cropping opportunities the area offers.

In this chapter we briefly discuss three examples of locally developed várzea resource management employed by ribeirinhos / ribereños. Each of these systems incorporates much environmental and technical knowledge and much production experience. We present these examples in order to emphasize the diversity, complexity, and dynamism of these locally developed Amazonian production systems with a view toward dispelling any lingering doubts about the profound environmental knowledge of rural, nontribal Amazonians.

The significance of the loss of knowledge about Amazonian forests, waters, plants, and animals that is accompanying the loss of Amazonian languages and traditions cannot be overstated. Scholars should not assume, however, that rural Amazonians whose primary or only language is a European one are ignorant of their surroundings. Nor should it be assumed that communities that are encouraged or even forced to abandon a lifestyle based largely on foraging are condemned to future ignorance of the life of forests or rivers. It is indeed tragic that few tribally organized societies continue to live in or manage the Amazon floodplains.[3] The present-day inhabitants of várzea lands, who identify themselves as ribeirinhos, caboclos, or ribereños, are estimable inheritors and transformers of indigenous floodplain management traditions.[4]

A DIVERSITY OF PATTERNS

The várzea stretches for about three thousand miles from its headwaters to the estuary. The resource use patterns employed by ribeirinho / ribereño households along this length are many, and it is difficult to generalize adequately about them. Perhaps the most important common characteristics of those local practices are their diversity, complexity, and dynamism. Most ribeirinho smallholders employ an assortment of resource management techniques that commonly include several types of farming, agroforestry, forest extraction, hunting, fishing, and occasional wage labor. Many of the strategies and techniques they employ are multistaged and complex. And despite a persistent outside view of rural folk as tradition-bound and slow to change, várzea farmers have continually shown that they are eager for change and capable of responding quickly and appropriately to both new problems and opportunities, whether environmental, economic, or political in nature.

To some degree the diversity, complexity, and dynamism of resource management is attributable to the environments that várzea dwellers manage. While the várzea is often discussed as if it were one kind of environment, we now have come to appreciate that várzea areas differ greatly. For instance, near the mouth of the Amazon, tides flood vast expanses of várzea twice daily, but seasonal variations in flood height are not large. This flood regime presents to farmers both opportunities and problems different from those of upper Amazon sites, where annual floods commonly raise water levels by ten meters or more (Denevan 1984; Padoch and Pinedo-Vasquez 1991). Great geomorphic and ecological diversity can also be found within a very small area of the várzea (cf. Denevan 1984; Hiraoka 1985a, b; Padoch and DeJong 1992). Even minor differences in elevation in the floodplain may signal crucial differences in potential for agricultural production.

The Peruvian village of Santa Rosa is a community of about 350 farmers, fishers, hunters, forest product collectors, and occasional wage laborers situated along the edge of the várzea of the Ucayali River, a large tributary of the Amazon. Santa Rosa resembles many other communities scattered along the rivers of the Peruvian Amazon.[5] All residents would identify themselves as a ribereños; they all speak Spanish, all engage in a combination of agricultural, extractive, and other activities, all are descended from families that have lived along Amazon rivers for many generations, and all are poor. There are, however, significant differences among the villagers as well. Residents of Santa Rosa trace their heritage to at least five different ethnic groups (Padoch and DeJong 1990), and they differ greatly in their life experiences, particularly in just how long they have been part of this particular Amazonian community.

This great diversity hidden under a seeming uniformity also applies to agricultural and agroforestry techniques used by the people of Santa Rosa. Exploiting the different landforms and biotopes of the floodplain, the farmers of this area of lowland Amazonia developed at least twelve distinct ways of farming. These range from the highly risky but very productive farming of rice on seasonally appearing sandbars, where production of four metric tons per hectare with no fertilization and minimal labor is not uncommon, to highly diverse, long-term, and also profitable swidden-fallow agroforestry on high levees and plots outside the floodplain. Even seemingly inhospitable environments like sandbars are used for short-term plantings of cowpeas (*Vigna* spp.), and fertile ones like levees are used for many production types including monocultures of corn and manioc, as well as large fruit plantations.

Not only do ribereño farmers use many kinds of agriculture, but they also combine these types into many different patterns. For instance, the 46 village families living in Santa Rosa combined the twelve kinds of agriculture available in 39 distinct ways. Santa Rosinos do not even agree on whether it is better to diversify resource use or to specialize in one or two types. Approximately half of Santa Rosa households used four different production types in one farming season. But three, a small but significant number, chose to employ only one production type.

Factors determining the strategy that was followed by any household were many and complex. Differences in access to some kinds of land was often important, as were differences in amount, type, and seasonality of labor available to the family. Other important variables were contrasts in access to capital and some capital resources, credit, and opportunities to work beyond the village. The family's recent farming history, including losses of crops to floods, as well as general preferences and attitudes was also significant. Some farmers were more interested in new crops, techniques, and products; others were less interested in these matters and could be considered more averse to taking risks.

COMPLEX FORMS OF MANAGING COMPLEX FORESTS

Many recent studies on ethnobotany, traditional resource management, and community-based conservation in the tropics have identified and quantified the number of species that people collect, plant, or manage in their forests (Boom 1987; DeBeer and McDermott 1989; Plotkin and Famolare 1992; and many others). These product-oriented books and articles present smallholder forest use and management in the tropics as essentially, if not exclusively, oriented toward nontimber products.

Fruits, fibers, medicinals, and other nontimber goods are indeed important to many rural Amazonians. Our (and other) studies in Amazonia, however, show that timber is an important output of forests managed for multiple outputs by Amazonian smallholders (Peluso 1993; Pinedo-Vasquez and Padoch 1996), and yet the long-term management of trees by smallholders for timber production has been generally ignored.[6] Mature forest management for multiple products including timber, doubtless like many other management practices that we have yet to recognize, has been overlooked because it is often highly complex and subtle in form and defies accepted categories of management.

Many Amazonian smallholders manage their properties for timber and are important participants in local and regional timber markets (Pinedo-Vasquez and Padoch 1996). For instance, in the Amazon estuary near the city of Macapá local sawmills are mainly supplied with timber from lands managed by very small-scale farmers and forest managers. In March, April, and May, 1994, we monitored the processing of logs at the eight sawmills located along the lower Mazagão River in the Brazilian state of Amapá. During this period each mill processed an average of 320 logs per week. The approximately fifty smallholder families that live in the river area supplied all of that production; most of the logs came from managed floodplain forests (Pinedo-Vasquez and Padoch 1996).

For over three years we studied local manipulation of the tropical moist forests of the Napo-Amazon floodplain in lowland Peru. This floodplain lies at the confluence of the Napo and Amazon rivers.[7] Amazonians have long used the area for hunting, fishing, farming, and collecting various forest products. The villagers of the region identify themselves as ribereños and, like Santa Rosinos, are largely descendants of several indigenous groups.

Of the several types of forest that are under different degrees of management along the Napo-Amazon floodplain, *capinurales,* that is, forests dominated by the tree known locally as *capinuri* (*Maquira coriacea*), are among the most important economically. Residents of the floodplain extract several products from capinurales for both domestic use and occasional sale in regional markets. These include wood for plywood and resin from the capinuri, many edible fruits, *oje* (*Ficus* spp.) latex used as a medicine, as well as a variety of other construction materials, firewood, and medicinal plants.

Capinural forests mainly occupy natural levees along the floodplain. Much of this floodplain is also farmed, and market-oriented production of rice, corn, cassava, and bananas predominates. Villagers follow a complex cropping sequence, planting a near-pure stand of rice as the first crop, followed by a near-monocultural planting of maize. Within a year the maize field tends to be succeeded by a rice-maize-cassava intercrop, a combination that may then be repeated. After one or two years, the grain crops are discontinued, and cassava usually replaces them. Cassava is often intercropped with bananas and with some fruit trees. After another two years or so cassava is phased out and bananas and other fruits remain as the only important agricultural or agroforestry crops. When bananas and tree crops take over the field, they may be more accurately referred to as agroforestry plots or managed swidden-fallows. Cultivation then becomes largely confined to occasional slash weedings and periodic harvests of the crops. Banana production rarely continues for more than eight years. Some fruit trees may live longer.

Napo-Amazon villagers do not plant capinuri trees, but seedlings of capinuri tend to become established quickly in farm fields. In older fields or swidden-fallows, capinuri juveniles thrive; and in mature capinuri forests individuals are found in all forest layers. Dense stands of capinuri, however, are formed only after a long process of management and the application of a complex of management techniques. Many of these techniques are known by names that obviously combine indigenous and Spanish words; the technologies can also be regarded as an amalgam of indigenous and exotic knowledge.

The management of capinurales can be divided into three stages. The precise number and kinds of techniques that are used varies from one stage to the other. People begin managing seedlings of capinuri and other valuable species during the first stages of agricultural production. When levee plots are cleared for rice production, individual economically valuable trees and seedlings are often spared. The clearing of land for planting grain opens up the forest canopy for light-loving tree species, encouraging their development. Care is taken by farmers that burning not destroy any valuable tree seedlings. The burn, however, actually helps promote forest development by hastening the germination of tree seeds.

Some farming and forestry operations are conveniently combined. Napo-Amazon forest managers, however, do vary the way some operations are done and apply some different labels. At this stage, two principal forest management techniques are used: *huactapeo* and *jaloneo*. *Huactapeo* is a technique that consists of three operations: selective weeding, cleaning (including the removal of the roots and stems of species that regenerate by sprouting), and controlled burning (including the elimination of regeneration materials such as the roots and stems of sprouting species by burning). The second, *jaloneo,* is a thinning technique that includes the selection of healthy, well-formed seedlings, and uprooting of seedlings with defects. By uprooting seedlings, managers reduce the probability of their resprouting, which would likely take place if offending seedlings were merely cut.

When yields of bananas decline, the fields can be better viewed as enriched fallows (*purmas*). During this stage, local people manage the juveniles of capinuri and other valuable timber species along with the planted and protected fruit and other marketable species. While selective slash weeding is applied to enhance the growth of the many types of resources in the fallows, three principal techniques, *huahuan-cheo, raleo,* and *mocheado*, are directed especially toward the management of different kinds of timbers. Each of these techniques includes sophisticated methods of selection, marking, pruning, and liberation, as well as thinning activities consisting in the removal of undesirable individuals by felling, girdling, or burning. Space and light are thus made available for growth in height and diameter of stems of capinuri, *capirona (Calycophyllum spruceanum)*, and other valuable species.

The third phase begins when the fallow has become a mature capinural forest. Two management techniques are commonly used during this phase in the Napo-Amazon: *desangrado* and *anillado*. *Anillado* involves the killing of selected stems of competitor species using girdling and fire. Its application causes the tree to die rapidly and avoids resprouting from the roots or stem. Amazonians use it to kill stems of large species that are difficult to control. *Desangrado* is a girdling technique, by which local people remove small stems of competitor species and individuals of stranglers and woody climbing vines. It involves two operations: selection, in which all individuals of vine and other species that are climbing or strangling the trunk and/or covering the canopy of valuable species are selected for removal, and girdling, which involves the removal of bark, cambium, and sapwood in a ring extending around the selected individual at the bottom of its trunk. From the ring fissure, sap, resins, and water are lost. The abundance of resin or sap attracts ants, termites and other insects that not only extract the sap or resin but also damage any new sprouts. The infestation of insects thus controls the sprouting of vines and helps kill them. These two management techniques are applied on an average of once every 6 to 8 years.

The suite of traditional management techniques outlined above is used to increase the value of capinural forests to local villagers. The application of these techniques also results in an increase in the commercial volume per hectare of timber in capinurales. The mean commercial volume of managed capinurales was 81 m³/ha for areas that had been managed as mature capinural forests for eight years, 89 m³/ha for those managed for 16 years, and 85 m³/ha for those managed for 24 years. All these values were significantly higher than the estimated mean of 54 m³/ha for the unmanaged capinural.

The application of traditional management techniques also rather surprisingly resulted in a statistically significant increase in the number of species in capinurales (Pinedo-Vasquez 1995). In contrast to conventional forestry, commercial value was raised but the floral diversity of the forest was not reduced.

NEW METHODS TO SOLVE NEW PROBLEMS

The residents of the Brazilian estuarine várzea have less direct ties to indigenous Amazonians than do most ribereños of Peru. Some of the current residents of the lower Amazon are descended from Africans who came into the area as slaves; many others are more recent immigrants from Brazil's arid northeast. They, however, share with their Peruvian counterparts a rich tradition of sophisticated resource management of várzea resources. Like the Peruvian ribereños, they actively experiment with applying techniques based on traditional environmental knowledge as well as newly acquired skills to the management of new problems.

The Brazilian village of Ipixuna and neighboring communities in the state of Amapá were until recently major exporters of bananas. Not only did these smallholders supply the urban markets of Amapá with bananas, they also exported to the major Amazonian city of Belém. In the last several years, however, banana production in the region has been almost completely wiped out by Moko disease, locally called *febre de banana*. The disease, which is common in many banana-producing areas can be controlled by a concerted and very expensive campaign of destruction of infected plants, repeated sterilization of all tools, and constant inspection of all plantings (Stover and Simmonds 1987). These control measures are not economically feasible for smallholders, and they involve the use of toxic chemicals. Ribeirinhos have instead observed and experimented locally and in the last few years developed an agroforestry system by which they manage the disease although they do not eliminate it.[8]

Those Ipixuna farmers who are most successful in maintaining adequate production of bananas do the least weeding. In order to produce bananas in an infested area, farmers permit weeds and early successional species to fill the spaces between banana plants while continuing to remove vines and other strangler species. The ribeirinhos who use this pattern, termed the *emcapoeirado* system, observed that weeding facilitates the spread of Moko disease. Some report that, as wage laborers on oil palm plantations, they first noticed that with minimal weeding, palms did well, and they decided to experiment with a similar management regime in their banana plots. Farmers who have significantly reduced the amount of weeding they do in fields that include bananas now produce relatively good crops of banana despite disease attacks; others have lost all or nearly all production. The agroforesters also get an economic return from some of the other species incorporated into their system.

Emcapoeirado is but one of many little-known resource management practices employed in the region. We are conducting further experimental studies to understand how the emcapoeirado system of banana production reduces disease infestation and to evaluate its economic viability and its ecological effects.

Emcapoeirado is a new adaptation that combines some aspects of traditional Amazonian agroforestry and forest management practices with practices learned by caboclo farmers temporarily employed as laborers on industrial plantations; and the combination has been applied to a new problem, infestation with Moko disease. It illustrates particularly well the eclectic sources and dynamic nature of locally developed Amazonian production systems.

CONCLUSIONS

The rapid and wholesale destruction of Amazonian societies and their linguistic and resource management traditions is without doubt a tragic loss. The traditions that allow innumerable caboclo and ribereño farmers to manage and farm the fickle floodplains do show, however, that some of that valuable Amazonian knowledge continues to exist and to evolve. These systems have been invisible to many of us largely because their developers and practitioners—smallholders like the ribeirinhos—have rarely been credited with sophisticated and valuable knowledge about production and resource management. The systems have perhaps also been invisible because they are confusing to us; they employ principles that many modern systems have discarded. Among these are toleration and even encouragement of biodiversity, the fusion and blurring of the most familiar resource management categories such as farming and forestry, and the successful integration of indigenous Amazonian with other resource management traditions. These new patterns, though ignored by many until now, might in their eclecticism and dynamism more accurately recall várzea management traditions from before the Conquest, in nature if not in detail, than do the practices and traditions of today's small indigenous groups. When the Amazon floodplains were home to large settled villages, organized in complex societies, and actively engaged in long-range trade, the reliance of one community on twelve forms of agriculture, the management of forests for important trade items, and the successful and rapid adaptation of agroforestry management methods to foil a new pathogen might well have not been unusual.

NOTES

1. Among the most important works on ethnicity and resource management patterns of ribeirinho/ribereño populations are Moran 1974; Parker 1985; Hiraoka 1985a, b, 1986, 1989, 1992; Padoch et al. 1985; Anderson 1988; DeJong 1988; Padoch 1988, 1992; Pinedo-Vasquez 1988, 1995; Frecchione, Posey, and da Silva 1989; Padoch and DeJong 1990, 1992; Chibnik 1991, 1994; Lima-Ayres 1992; Brondizio 1996; Brondizio and Siqueira 1997.
2. For a recent review of ecological and human ecological issues in the Amazon floodplain, see Padoch et al. 1999.
3. There are several communities of indigenous people in floodplain environments, including Miranha villages along Brazil's upper Solimões, Ticuna and Yagua villages in the Brazil-Colombia-Peru border area, and Shipibo-Conibo and Cocamilla villages located respectively along the Ucayali and Huallaga rivers in Peru.
4. For discussions of changing and problematic ethnic designations, see Lima-Chibnik 1991, 1994; Ayres 1992; Lima 1999.

5. This example is described in greater detail in Padoch and DeJong 1992. For descriptions of diversity of production in other villages in the lowland Peruvian Amazon, see Hiraoka 1985a, b; Chibnik 1994; and Pinedo-Vasquez 1995.
6. For a good overview of research on management of tropical forests by smallholders see Wiersum 1997.
7. For a more detailed discussion of resource management in the Napo-Amazon floodplain, see Pinedo-Vasquez 1995.
8. There are many discussions of Amazonian agroforestry systems, including studies of how Amazonians adapt traditional patterns to modern needs. Among these are Hecht 1982; Posey 1984; Irvine 1985; Padoch et al. 1985; Padoch and DeJong 1987, 1989, 1995; Denevan and Padoch 1988; Dufour 1990.

REFERENCES

Anderson, A.B. 1988. Use and management of native forests dominated by *acai* palm (*Euterpe oleracea* Mart.) in the Amazon estuary. In *The Palm-Tree of Life: Biology, Utilization, and Conservation,* ed. M.J. Balick. Pp. 144–154. Advances in Economic Botany 6. The Bronx: New York Botanical Garden Press.

Boom, B.M. 1987. *Ethnobotany of the Chacobo Indians, Beni, Bolivia.* Advances in Economic Botany 4. The Bronx: New York Botanical Garden Press.

Brondizio, E.S. 1996. Forest farmers: Human and landscape ecology of Caboclo populations in the Amazon estuary. Ph.D. diss., Indiana University at Bloomington.

Brondizio, E.S., and A.D. Siqueira. 1997. From extractivists to forest farmers: Changing concepts of Caboclo agroforestry in the Amazon estuary. *Research in Economic Anthropology* 18:233–279.

Chibnik, M. 1991. Quasi-ethnic groups in Amazonia. *Ethnology* 30:167–182.

Chibnik, M. 1994. *Risky Rivers.* Tucson: University of Arizona Press.

DeBeer, J.H., and M.J. McDermott. 1989. *Economic Value of Non-Timber Forest Products in Southeast Asia.* The Hague: Council for the International Union of the Conservation of Nature.

DeJong, W.A. 1988. Organización del trabajo en la Amazonía peruana: El caso de las sociedades agrícolas de Tam shiyacu. *Amazonía Indígena* 7A13):11–17.

Denevan, W.M. 1984. Ecological heterogeneity and horizontal zonation of agriculture in the Amazon floodplain. In *Frontier Expansion in Amazonia,* ed. M. Schmink and C.H. Woods. Pp. 311–366. Gainesville: University of Florida Press.

Denevan, W.M., and C. Padoch. 1988. *Swidden-Fallow Agroforestry in the Peruvian Amazon.* Advances in Economic Botany 5. The Bronx: New York Botanical Garden Press.

Dufour, D.L. 1990. Use of tropical rainforests by native Amazonians. *Bioscience* 40(9): 652–659.

Frecchione, J., D.A. Posey, and L.F. da Silva. 1989. The perception of ecological zones and natural resources in the Brazilian Amazon: An ethnoecology of Lake Coari. In *Resource Management in Amazonia,* ed. D.A. Posey and W. Balée. Pp. 260–282. Advances in Economic Botany 7. The Bronx: New York Botanical Garden Press.

Galvão, E. 1952. The religion of an Amazon community: A study in culture change. Ph.D. diss., Columbia University.

Hecht, S.B. 1982. Agroforestry in the Amazon Basin: Practice, theory and limits of a promising land use. In *Amazonia: Agriculture and Land Use Research*, ed. S.B. Hecht. Pp. 331–371. Cali, Colombia: CIAT.

Hiraoka, M. 1985a. Floodplain farming in the Peruvian Amazon. *Geographical Review of Japan* 58, Series b, 1:1–23.

Hiraoka, M. 1985b. Mestizo subsistence in riparian Amazonia. *National Ge ographic Research* 1(2):236–246.

Hiraoka, M. 1986. Zonation of mestizo farming systems in Northeast Peru. *National Geographic Research* 2(3):354–371.

Hiraoka, M. 1989. Agricultural systems in the floodplains of the Peruvian Amazon. In *Fragile Lands of Latin America*, ed. J.O. Browder. Pp. 75–99. Boulder, Colo.: Westview Press.

Hiraoka, M. 1992. Caboclo and ribereño resource management in Amazonia: A review. In *Conservation of Neotropical Forests*, ed. K. Redford and C. Padoch. Pp. 134–157. New York: Columbia University Press.

Irvine, D. 1985. Succession management and resource distribution in an Amazonian rainforest. Paper presented at the Annual Meeting of the American Anthropological Association, Washington, D.C.

Lima-Ayres, D. de M. 1992. History, social organization, identity, and outsiders' social classification of the rural population of the Amazonian region. Ph.D. diss., King's College, Cambridge University.

Lima-Ayres, D. de M. 1999. Equity, sustainable development, and biodiversity preservation: Some questions about ecological partnerships in the Brazilian Amazon. In *Várzea: Diversity, Development, and Conservation of Amazonia's Whitewater Floodplain*, ed. C. Padoch et al. Pp. 247–266. Advances in Economic Botany 13. The Bronx: New York Botanical Garden Press.

Moran, E.F. 1974. The adaptive system of the Amazonian caboclo. In *Man in the Amazon*, ed. C. Wagley. Pp. 136–159. Gainesville, Fla.: University Presses of Florida.

Padoch, C. 1988. People of the floodplain and forest. In *People of the Tropical Rain Forest*, ed. J.S. Denslow and C. Padoch. Pp. 127–140. Berkeley, Calif.: University of California Press.

Padoch, C. 1992. Marketing of non-timber forest products in western Amazonia: General observations and research priorities. In *Non-Timber Forest Product Extraction from Tropical Forests: Evaluation of a Conservation and Development Strategy*, ed. D.C. Nepstad and S. Schwartzman. Pp. 43–50. Advances in Economic Botany 9. The Bronx: New York Botanical Garden Press.

Padoch, C., and W. DeJong. 1987. Traditional agroforestry practices of native and ribereño farmers in the lowland Peruvian Amazon. In *Agroforestry: Realities, Possibilities, and Potentials*, ed. H.L. Gholz. Pp. 179–195. Dordrecht, The Netherlands: Martinus Nijhoff and Dr. W. Junk Publishers.

Padoch, C., and W. DeJong. 1989. Production and profit in agroforestry: An example from the Peruvian Amazon. In *Fragile Lands of Latin America*, ed. J. Browder. Pp. 102–114. Boulder, Colo.: Westview Press.

Padoch, C., and W. DeJong. 1990. Santa Rosa: The impact of the minor forest products trade on an Amazonian place and population. In *New Directions in the Study of Plants and People,* ed. G.T. Prance and M.J. Balick. Pp. 151–158. Advances in Economic Botany 8. The Bronx: New York Botanical Garden Press.

Padoch, C., and W. DeJong. 1992. Diversity, variation, and change in ribereño agriculture. In *Conservation of Neotropical Forests: Working from Traditional Resource Use,* ed. K. Redford and C. Padoch. Pp. 158–174. New York: Columbia University Press.

Padoch, C., and W. DeJong. 1995. Subsistence- and market-oriented agroforestry in the Peruvian Amazon. In *The Fragile Tropics: Sustainable Management of Changing Environments,* ed. T. Nishizawa and J.I. Uitto. Pp. 226–237. Hong Kong: United Nations University Press.

Padoch, C., and M. Pinedo-Vasquez. 1991. Floodtime on the Ucayali. *Natural History* May, pp. 48–57.

Padoch, C., et al. 1985. Amazonian agroforestry: A market-oriented system in Peru. *Agroforestry Systems* 3:47–58.

Padoch, C., et al. 1999. *Várzea: Diversity, Development, and Conservation of Amazonia's Whitewater Floodplain.* Advances in Economic Botany 13. The Bronx: New York Botanical Garden Press.

Parker, E. 1985. Caboclization: The transformation of the Amerindian in Amazonia, 1615–1800. *Studies in Third World Societies* 32:1–49.

Peluso, N. 1993. *The impacts of social and environmental changes on indigenous people's forest management in West Kalimantan, Indonesia.* Forest, Trees, and People Monograph Series. Rome: United Nations Food and Agriculure Organization.

Pinedo-Vasquez, M. 1988. The river people of Maynas. In *People of the Tropical Rain Forest,* ed. J.S. Denslow and C. Padoch. Pp. 141–142. Berkeley, Calif.: University of California Press.

Pinedo-Vasquez, M. 1995. Human impact on Várzea ecosystems in the Napo-Amazon, Peru. Ph.D. diss., Yale School of Forestry and Environmental Studies.

Pinedo-Vasquez, M., and C. Padoch. 1996. Managing forest remnants and forest gardens in Peru and Indonesia. In *Forest Remnants in Tropical Landscapes,* ed. J. Schelhas and R. Greenberg. Pp. 327–341. Washington, D.C.: Island Press.

Plotkin, M.J., and L.M. Famolare. 1992. *Sustainable Harvest and Marketing of Rain Forest Products.* Washington, D.C.: Island Press.

Posey, D.A. 1984. A preliminary report on diversified management of tropical forest by the Kayapo Indians of the Brazilian Amazon. In *Ethnobotany in the Neotropics,* ed. G.T. Prance and J. Kallunki. Pp. 112–126. Advances in Economic Botany 1. The Bronx: New York Botanical Garden Press.

Roosevelt, A.C. 1989. Resource management in Amazonia before the conquest: Beyond ethnographic projection. In *Resource Management in Amazonia: Indigenous and Folk Strategies,* ed. D. Posey and W. Balée. Pp 30–62. Advances in Economic Botany 7. The Bronx: New York Botanical Garden Press.

Roosevelt, A.C. 1991. *Moundbuilders of the Amazon: Geophysical Archaeology on Marajo Island, Brazil.* San Diego: Academic Press.

San Román, J. 1975. *Perfiles históricos de la Amazonía peruana*. Lima, Peru: Paulinas-Centro de Estudios Teológicos de la Amazonía.

Sternberg, H.O. 1956. *A água e o homem na várzea do Careiro*. Rio de Janeiro: Universidade do Brasil.

Stover, R.H., and N.W. Simmonds. 1987. *Bananas*. Harlow, England: Longman Scientific and Technical.

Villarejo, A. 1943. *Así es la Selva*. Iquitos, Peru: CETA.

Wagley, C. 1953. *Amazon Town: A Study of Man in the Tropics*. New York: Macmillan.

Wiersum, K.F. 1997. Indigenous exploitation and management of tropical forest resources: An evolutionary continuum in forest-people interactions. *Agriculture, Ecosystems and Environment*. 63:1–16.

Part 3

PERPETUATING THE WORLD'S BIOCULTURAL DIVERSITY

Agenda for Action

23

BIOLOGICAL AND CULTURAL DIVERSITY

The Inextricable, Linked by Language and Politics

Darrell A. Posey

It is estimated that there are currently at least 300 million people world wide who are indigenous (Gray 1999). They occupy a wide geographical range from the Polar regions to the deserts, savannas, and forests of the tropical zone. The fact that 4000 to 5000 of the more than 6000 languages in the world are indigenous strongly implies that Indigenous peoples constitute most of the world's cultural diversity (UNESCO 1993). They include groups as disparate as the Quechua descendants of the Inca civilization in Bolivia, Ecuador, and Peru, who collectively number more than 10 million, and the Gurumalum band of Papua New Guinea, who number fewer than ten. In New Guinea alone, more than 600 languages are spoken among a population of only 6 million, and in India there are more than 1600 languages (Durning 1992).

Yet human cultural diversity is threatened on an unprecedented scale. It has been estimated that up to 90 percent of the world's languages—the storehouses of peoples' intellectual heritages and the framework for their unique understandings of life—will disappear within a century (Krauss 1992). According to the United Nations Educational, Scientific and Cultural Organisation (UNESCO 1993), nearly 2500 languages are in immediate danger of extinction.

Many of the areas of highest biological diversity on the planet are inhabited by Indigenous peoples, providing what the Declaration of Belém (from the First International Congress of Ethnobiology of the International Society of Ethnobiology, 1988) calls an "inextricable link" between biological and cultural diversity

(Posey and Dutfield 1996a:2; and see Appendix 1). In fact, of the nine countries that together harbor 60 percent of human languages, six are both centers of cultural diversity and megadiversity countries with exceptional numbers of unique plant and animal species (Durning 1992; Harmon 1996).

The links between biological and cultural diversity are inherently recognized in the 1992 Convention on Biological Diversity (CBD). Indigenous peoples and local communities are recognized as playing important roles in in situ conservation, for they have often sustainably managed the natural resources of fragile, biologically rich ecosystems for millennia. This is specifically noted in the Preamble, which recognizes the

> close and traditional dependence of many indigenous and local communities embodying traditional lifestyles on biological resources, and the desirability of sharing equitably benefits arising from the use of traditional knowledge, innovations and practices relevant to the conservation of biological diversity and the sustainable use of its components.

Indigenous peoples themselves frequently emphasize that conserving biological diversity requires respect and recognition of their rights, as in the 1992 Indigenous Peoples' Earth Charter (IWGIA 1992b):

> Recognising indigenous peoples' harmonious relationship with Nature, indigenous sustainable development strategies and cultural values must be respected as distinct and vital sources of knowledge. (Clause 67)

Similarly, the Final Statement from the 1995 Consultation on Indigenous Peoples' Knowledge and Intellectual Property Rights in Suva, Fiji (UNDP 1995):

> We assert that in situ conservation by indigenous peoples is the best method to conserve and protect biological diversity and indigenous knowledge, and encourage its implementation by indigenous communities and all relevant bodies. (Clause 2.2)

According to the 1992 Charter of the Indigenous-Tribal Peoples of the Tropical Forests (IWGIA 1992a):

> The best guarantee of the conservation of biodiversity is that those who promote it should uphold our rights to the use, administration, management and control of our territories. We assert that guardianship of the different ecosystems should be entrusted to us, indigenous peoples, given that we have inhabited them for thousands of years and our very survival depends on them. (Article 42)

Indigenous and local communities are not just fighting to maintain biodiversity: they are fighting for their very survival. Thus, attempts to conserve the world's biological diversity require the immediate strengthening and enhancement of Indigenous peoples and local communities. They also require respect for traditional knowledge and creative application of this knowledge—with equitable sharing of benefits. Such a view is implicit in Article 8(j) of the CBD, which states that each contracting party must:

> Subject to its national legislation, respect, preserve and maintain knowledge, innovations and practices of indigenous and local communities embodying traditional lifestyles relevant for the conservation and sustainable use of biological diversity and promote the wider application with the approval and involvement of the holders of such knowledge, innovations and practices and encourage the equitable sharing of the benefits arising from the utilisation of such knowledge, innovations and practices.

Glowka, Burhenne-Guilmin, and Synge, in collaboration with McNeely and Gundling (1994), note that use of the word *traditional* could be understood to imply that those not embodying traditional lifestyles are excluded from the CBD. Thus, it is extremely important that the word *traditional* should be defined in such a way that it conveys its living reality: a filter through which *innovation* occurs. Pereira and Gupta (1993) argue that "it is the traditional methods of research and application, not always particular pieces of knowledge" that persist, and they call this a "tradition of invention and innovation." Thus, technological changes do not simply lead to modernization and loss of traditional practice, but rather can provide additional inputs into vibrant, adaptive and adapting, holistic systems of knowledge, innovation, and practice.

As a report to the Secretariat of the CBD by the Four Directions Council of Canada explains (Four Directions Council 1996):

> What is "traditional" about traditional knowledge is not its antiquity, but *the way it is acquired and used*. In other words, the social process of learning and sharing knowledge, which is unique to each Indigenous culture, lies at the very heart of its "traditionality". Much of this knowledge is actually quite new, but it has a social meaning, and legal character, entirely unlike the knowledge indigenous people acquire from settlers and industrialised societies.

Thus, traditional livelihood systems are constantly adapting to changing social and environmental conditions. There are as many adaptive strategies as there are local communities, and it is that diversity of adaptations that is the background of survival and sustainability.

SUSTAINABILITY, KNOWLEDGE, AND CULTURAL DIVERSITY

The concept of sustainability is embodied in indigenous and traditional livelihood systems. Historical evidence exists that demonstrates the sustained productivity of indigenous systems, in some cases for thousands of years on the same land. Indigenous peoples and traditional communities often possess a conservation ethic developed from living in particular ecosystems. This ethic cannot be regarded as universal, but indigenous systems do tend to emphasize the following specific values and features:

- Cooperation
- Family bonding and cross-generational communication, including links with ancestors
- Concern for the well-being of future generations
- Local-scale, self-sufficiency, and reliance on locally available natural resources
- Rights to lands, territories, and resources that tend to be collective and inalienable rather than individual and alienable
- Restraint in resource exploitation and respect for nature, especially for sacred sites

The "traditional knowledge, innovations and practices" of "indigenous and local communities embodying traditional lifestyles" mentioned in the CBD are often referred to by scientists as traditional ecological knowledge (TEK)—defined by Gadgil, Berkes, and Folke (1993:151) as "A cumulative body of knowledge and beliefs handed down through generations by cultural transmission about the relationship of living beings (including humans) with one another and with their environment." TEK is far more than a simple compilation of facts. It is the basis for local-level decision making in areas of contemporary life, including natural resource management, nutrition, food preparation, health, education, and community and social organization. TEK is holistic, inherently dynamic, constantly evolving through experimentation and innovation, fresh insight, and external stimuli.

Scientists are becoming increasingly aware of the sophistication of TEK among many indigenous and local communities. For example, the Shuar people of Ecuador's Amazonian lowlands use 800 species of plants for medicine, food, animal fodder, fuel, construction, and fishing and hunting supplies. Traditional healers in Southeast Asia rely on as many as 6500 medicinal plants, and shifting cultivators throughout the tropics frequently sow more than 100 crops in their forest farms (Durning 1992). Much of the world's crop diversity is in the custody of farmers who follow age-old farming and land use practices that conserve biodiversity and provide other local benefits. Among such benefits are the promotion of indigenous diet diversity, income generation, production stability, minimization of risk, reduced insect and disease incidence, efficient use of labor, intensification of production with limited resources, and maximization of returns under low levels of technology. These ecologically complex agricultural systems associated with centers of crop genetic diversity include traditional cultivars, or landraces, that constitute an essential part of our world crop genetic heritage and nondomesticated plant and animal species that serve humanity as biological resources.

Many ancient indigenous agricultural and other sustainability systems remained intact until the colonial period. Classic examples are the raised bed systems used for millennia by traditional farmers of tropical America, Asia, and Africa. Known in Mesoamerica as *chinampas, waru waru,* or *tablones,* these were extremely effective for irrigation, drainage, soil fertility maintenance, frost control, and plant disease management. Pre-Conquest Aztecs at Lake Texcoco had 10,000 hectares of land under *chinampas* feeding a population of 100,000 people (Willett 1993). There is also evidence that farming in India has continued on the same land for more than two thousand years without a drop in yields and remained remarkably free of pests. Indian peasants grew as many as forty-one different crops annually in certain localities and possessed the ability to vary seeds according to the needs of the soil and the season. Comparable levels of productivity today "can be obtained only in the best of the Green Revolution areas of the country, with the most advanced, highly expensive, and often environmentally ruinous technologies" (cited in Willett 1993).

Most traditional peoples who inhabit forests or areas close to forests rely extensively upon hunted, collected, or gathered foods and resources, a significant portion of which are influenced by humans to meet their needs. These semidomesticates or nondomesticated resources (Posey 1994, 1996) form the basis for a vast treasury of useful species that have systematically been undervalued and overlooked by science, yet provide food and medicinal security for local communities around the world.

Failure to understand that "wild" landscapes often are actually modified by humans has also blinded outsiders to the management practices of Indigenous peoples and communities. Many "pristine" landscapes are, in fact, *cultural landscapes,* either created by humans or modified by human activity, such as natural forest management, cultivation, and the use of fire. According to the Four Directions Council (1996), "the territories in which Indigenous peoples traditionally live are *shaped environments,* with biodiversity as a priority goal, notwithstanding the fact that the modifications may be subtle and can be confused with the natural evolution of the landscape."

Some recently "discovered" cultural landscapes include those of Australian Aboriginal peoples who, 100,000 years before the term "sustainable development" was coined, were storing and trading seeds, as well as dividing tubers to propagate domesticated and nondomesticated plant species. Sacred sites act as conservation areas for vital water sources and plant and animal species by restricting access and behavior. Traditional technologies, such as fire use, are now recognized as extremely sophisticated systems that shape and maintain the balance of vegetation and provide habitats for hunted animals. The decline of fire management that occurred when Aboriginal peoples were centralized into settlements is one reason given for the rapid decline of mammals throughout the arid regions (Sultan, Craig, and Ross 1996).

Another example of a cultural landscape is the *apêtê* 'forest islands' of the Kayapó Indians of Brazil (Posey 1996). Kayapó practices of planting and transplanting within and between many ecological zones demonstrate the degree to which the indigenous population has modified Amazonia. The existence of extensive plantations of fruit and nut trees, as well as *apêtê,* has forced scientists to reevaluate what have erroneously been considered "natural" Amazonian landscapes. The Kayapó techniques of constructing *apêtê* in savanna show the degree to which this Amazon group can create and manipulate microenvironments within and between ecozones to increase biological diversity. Such ecological engineering requires a detailed knowledge of soil fertility, microclimate, and plant varieties, as well as of the interrelationships between components of a human-modified ecological community. Successful *apêtê* are dependent not just on the cultivator's knowledge of their immediate properties but also of long-term successional relationships that change as the forest islands mature and grow, and their complexity greatly increases as many plants are specifically grown to attract useful animals. *Apêtê* are managed both as agroforestry units and game reserves. Kayapó knowledge of *apêtê* formation and succession offers invaluable insights into processes of forestation in savanna and reforestation in denuded areas.

Kagore Shona people of Zimbabwe have sacred sites, burial grounds, and other sites with special historical significance embedded in the landscape (Matowanyika 1996). Outsiders often cannot recognize them and may attempt to impose their own values on the land, for example, during land use planning exercises. In societies where there is no written language, and where there are no large edifices such as European cathedrals or similar displays of a people's heritage, sacred spaces may be hard to discern, hidden in hills, mountains, and valleys, but this does not make them less significant.

Cultural landscapes and their link with the conservation of biological diversity are now recognized under the 1972 UNESCO Convention Concerning the Protection of the World Cultural and Natural Heritage, commonly known as The World Heritage Convention. A new category of World Heritage Site, the Cultural Landscape, recognizes "the complex interrelationships between man and nature in the construction, formation and evolution of landscapes." The first cultural landscape World Heritage Site was Tongariro National Park, a sacred region for the Maori people of New Zealand.

Indigenous peoples and a growing number of scientists believe that it is no longer acceptable simply to assume that just because landscapes and species appear to outsiders to be "natural," they are therefore "wild." The Aboriginal Resolution from the 1995 Ecopolitics IX Conference held in Darwin, Australia, states:

> The term "wilderness", as it is popularly used, and related concepts as "wild resources", "wild foods", etc., [are unacceptable]. These terms have connotations of *terra nullius* [empty or unowned land and resources] and, as such, all concerned people and organisations should look for alternative terminology which does not exclude Indigenous history and meaning.

Shifts in terminology are more than just semantic. By declaring landscapes as *wildernesses* or resources as *wild*, scientists have effectively placed these into the public domain—thereby ignoring historical interactions and dispossessing local communities of their rights to traditional territories and resources (Posey 1999). This is an example of how politically dominant scientific terminology can be misused to undermine cultural diversity.

Even though there is still insufficient recognition of the role of Indigenous peoples and local communities in sustainable management, such recognition is growing. For example, most conservation organizations and development agencies now believe that their strategies are more likely to succeed if they support indigenous and traditional resource management practices. As the Four Directions Council (1996) points out, this is because:

1. Indigenous peoples tend to use, and therefore to value and protect, a much *larger variety of species* than their nonindigenous neighbors.
2. Indigenous peoples tend to make modifications in their environment with the aim of *increasing* the diversity of ecological niches and associated species.
3. The traditional knowledge of Indigenous peoples is not only particularly rich in local ecological detail and predictive power but continues to be *updated and supplemented* as long as people continue to use the ecosystem and study its behaviour.

All of these principles are enmeshed in the languages that transmit TEK. As Maffi (in Maffi, Skutnabb-Kangas, and Andrianarivo 1999) emphasizes, languages are marked by diversity not just between language groups but within the same language, and even the same community. There can be considerable variations in knowledge systems between individuals, kin groups, clans, age grades, and gender groups; all of these combine to form the rich store of linguistically encoded knowledge of any language community.

THE CONCEPT OF STEWARDSHIP

Indigenous peoples are—and frequently view themselves as—stewards of nature, of some of the richest and most diverse terrestrial as well as coastal and marine biomes on Earth. They conserve biological diversity and, in some cases, provide other environmental benefits, for example, soil and water conservation, soil fertility enhancement, and management of game and fisheries. But the ultimate goal of linking indigenous systems and sustainability should be to harness the totality, rather than the components, of TEK systems in sustainability strategies, so that the quality of indigenous management can benefit the wider society. As expressed in the 1992 Kari-Oca Declaration (also known as Indigenous Peoples' Earth Charter):

the creator has placed us, the indigenous peoples, upon our Mother the Earth. We cannot be removed from our lands. We the Indigenous Peoples, are connected by the circle of life to our land and environments. (Posey and Dutfield 1996a:189)

Although indigenous knowledge is highly pragmatic and practical, Indigenous peoples generally view this knowledge as emanating from a *spiritual* base: all Creation is sacred, the sacred and secular are inseparable, spirituality is the highest form of consciousness, and spiritual consciousness is the highest form of awareness. In this sense, a dimension of indigenous knowledge is not *local* knowledge, but knowledge of the *universal* as expressed in the local.

Knowledge of the environment depends not only on the relationship between humans and nature, but also on that between the visible world and the invisible spirit world. According to Opoku (1978, as quoted in Posey and Dutfield 1996b:37), the distinctive feature of traditional African religion is that it is:

A way of life, [with] the purpose of . . . order[ing] our relationship with our fellow men and with our environment, both spiritual and physical. At the root of it is a quest for harmony between man, the spirit world, nature, and society.

In indigenous experience, states Opoku, the unseen is as much a part of reality as that which is seen. The spiritual is as much a part of reality as the material, and there is a complementary relationship between the two, *with the spiritual being more powerful than the material.* The community is of the dead as well as the living. And in nature, behind visible objects lie essences, or powers, which constitute the true nature of those objects.

Harmony and equilibrium among components of the cosmos are central concepts in most indigenous cosmologies. Agriculture, for example, can provide "balance for well-being" through relationships not only among people but also nature and deities. In this concept, the blessing of a new field represents not mere spectacle but an inseparable part of life where the highest value is harmony with the Earth. Most indigenous knowledge traditions recognize linkages between health, diet, properties of different foods and medicinal plants, and horticultural and natural resource management practices—all within a highly articulated cosmological and social context.

Indigenous knowledge embraces information about location, movements, and other factors explaining spatial patterns and timing in the ecosystem, including sequences of events, cycles, and trends. For Indigenous peoples, the main significance of their knowledge systems is that their connection to land and the relationships and obligations that arise from that connection are the core of their identity. These connections are expressed in song, dance, and other ceremonial activities. They are often expressed in myths and stories that encode in highly symbolic forms the intricate understanding that local communities have about land, life, and the cosmos. When languages are lost, these treasures of knowledge also disappear.

EQUITABLE PARTNERSHIPS AND RIGHTS

Historically, relationships between Indigenous peoples and states have been called "partnerships." Unfortunately, "partnerships" have been notably skewed in favor of governments, causing Indigenous peoples to seek their own community-controlled sustainability strategies in accordance with their own priorities and criteria. The role of external support or intervention has become that of creating more space for *indigenous control,* based on certain rights, including self-determination, territorial and land rights, the right to development, collective rights, community empowerment, prior informed consent and privacy, control of access, religious rights and religious freedom, and the right to a unique language.

Self-Determination

Self-determination includes the right and authority of Indigenous peoples to negotiate with states, on an equal basis, the standards and mechanisms that will govern relationships between them. The principle of self-determination is set forth in Article 55 of the Charter of the United Nations (1945) and in Article 1 of the 1966 International Covenant on Economic, Social, and Cultural Rights. The latter states that by virtue of the right to self-determination all peoples can freely determine their political status and freely pursue their economic, social, and cultural development.

Territorial and Land Rights

The right of ownership and control over traditional lands and resources is central to self-determination and, therefore, is considered a fundamental right. Article 14 of the 1989 International Labor Organization's Indigenous and Tribal Peoples Convention (No. 169) states:

> The rights of ownership and possession of the peoples concerned over the lands which they traditionally occupy shall be recognised. In addition measures shall be taken in appropriate cases to safeguard the right of the peoples concerned to use lands not exclusively occupied by them, but to which they have traditionally had access for their subsistence and traditional activities. Particular attention shall be paid to the situation of nomadic peoples and shifting cultivators in this respect.

The 1994 Draft UN Declaration on the Rights of Indigenous Peoples, although still far from being ratified by the United Nations General Assembly, provides one of the major "soft law" documents to guide international law and practice concerning Indigenous peoples. It states that:

> Indigenous peoples have the right to maintain and strengthen their distinctive spiritual and material relationships with the lands, territories, waters and coastal seas and other resources which they have traditionally owned or otherwise occupied or used, and to uphold their responsibilities to future generations in this regard. (Article 25)

The Right to Development

The right to development for Indigenous peoples includes (a) the right of access to resources on their territories and (b) the right to seek development on their own terms. The need for Indigenous peoples to secure rights to their subsistence base, enforced by the state and backed by international law, is a precondition for fostering and maintaining traditional systems of ecosystem management and in situ biodiversity conservation.

According to the UN Declaration on the Right to Development, adopted by the General Assembly in 1986:

> The right to development is an inalienable human right by virtue of which every human person and all peoples are entitled to participate in, and enjoy economic, social, cultural and political development. (Article 1)

The principle of the right to development is enshrined in international law in two United Nations human rights covenants, the 1966 International Covenants on Civil and Political Rights (ICCPR) and on Economic, Social, and Cultural Rights (ICESCR), and also in the International Labor Organization Convention 169, which states:

> The peoples concerned shall have the right to decide their own priorities for the process of development as it affects their lives, beliefs, institutions and spiritual well-being and the lands they occupy or otherwise use, and to exercise control, to the extent possible, over their own economic, social and cultural development. In addition, they shall participate in the formulation, implementation and evaluation of plans and programmes for national and regional development which may affect them directly. (7.1)

> The rights of the peoples concerned to the natural resources pertaining to their lands shall be specially safeguarded. These rights include the right of these peoples to participate in the use, management and conservation of these resources. (15.1)

Collective Rights

The recognition of collective rights is a critical aspect of self-determination. Communally shared concepts and communally owned property are fundamental aspects of indigenous societies. Indigenous proprietary systems are often highly complex. Generally, Indigenous peoples have a collective responsibility for their own territory in that individuals and families may hold lands and resources for their own use, but ownership is subject to customary law and practice and based on the collective consent of the community. According to the Basic Points of Agreement on Intellectual Property Rights and Biodiversity drafted at the 1994 Regional Meeting of the Coordinating Body of Indigenous Organizations of the Amazon Basin (COICA):

> For members of indigenous peoples, knowledge and determination of the use of resources are collective and intergenerational. No indigenous population, whether of individuals or communities, nor the Government, can sell or transfer ownership of resources which are the property of the people and which each generation has an obligation to safeguard for the next. (Point 7)

The ICCPR and ICESCR are the two main international human rights legal instruments that uphold collective rights. Article 1(2) of both these documents calls for the recognition of collective human rights:

> all peoples may, for their own ends, freely dispose of their natural wealth and resources without prejudice to any obligations arising out of international economic cooperation, based upon the principle of mutual benefit, and international law. In no case may a people be deprived of its own means of subsistence.

Community Empowerment

For Indigenous peoples, community-controlled sustainable development and community empowerment should be mutually reinforcing. Agenda 21 (the program for action in international environmental policy agreed upon by governments at the UN Conference on Environment and Development in 1992) clearly articulates international recognition that community empowerment is necessary for sustainable development. Chapter 26 includes, as measures to fulfill, the objective of establishing a process to empower Indigenous peoples and their communities (Article 26.3):

> (iii) Recognition of their values, traditional knowledge and resource-management practices with a view to promoting environmentally sound and sustainable development;
>
> (vi) Support for alternative environmentally sound means of production to ensure a range of choices on how to improve their quality of life so that they effectively participate in sustainable development.

Community empowerment also requires recognition that:

> the lands of indigenous people and their communities should be protected from activities that are environmentally unsound or that the indigenous people concerned consider to be socially and culturally inappropriate. (Article 26.3 [ii])

And that:

> traditional and direct dependence on renewable resources and ecosystems, including sustainable harvesting [of wildlife], continues to be essential to the cultural, economic, and physical well-being of indigenous people and their communities. (Article 26.3 [iv])

Elders, women, and youth play significant roles in indigenous societies. Often, their contributions are ignored as transmitters of culture and language, producers of food, providers of medicines, and collectors of fuel, fodder, and water. Thus, community empowerment must respect these groups that are frequently the most vulnerable to environmental and social change.

Prior Informed Consent and Privacy

An important tool in the transformation from "partnership" to equitable relationships is prior informed consent. Such consent is recognized in international law, including the Convention on Biological Diversity. Article 15, clause 5 of the CBD states: "Access to genetic resources shall be subject to prior informed consent of the Contracting Party providing such resources, unless otherwise determined by that Party."

Prior informed consent reflects full acceptance of an activity. It implies the right to stop the activity from proceeding, and/or to halt it if already underway. The types of activity that should be subject to the consent condition need to include the following:

1. All projects affecting local communities, such as construction works, colonization schemes, and protected areas.
2. The extraction of biological material and minerals from local communities or the territories of traditional communities, whether or not the communities have legal title to these lands.
3. The acquisition of knowledge from a person, group, or people.

Prior informed consent assumes the right to privacy, meaning the right to be free from intrusion and unwanted public attention. Indigenous peoples are vulnerable to such privacy invasions as the public disclosure and use of sacred or secret knowledge, the taking of images and other sensitive information, and unwanted intrusions by tourists seeking photographs. The right to privacy is recognized as a basic human right in international law. Article 17 of the ICCPR states:

> No one shall be subjected to arbitrary or unlawful interference with his privacy, family, home or correspondence, nor to unlawful attacks on his honour and reputation.
>
> Everyone has the right to the protection of the law against such interference and attacks.

Most indigenous groups see bioprospecting as a breach of their privacy, as they do unwanted and authorized surveys carried out by governments, universities and scientific institutions.

Control of Access

Community control over access to traditional knowledge and resources is seen as a basic right and is supported by a number of international agreements and conventions. The Draft UN Declaration on the Rights of Indigenous Peoples states:

> Indigenous peoples are entitled to the recognition of the full ownership, control and protection of their cultural and intellectual property.
> They have the right to special measures to control, develop and protect their sciences, technologies and cultural manifestations, including human and other genetic resources, seeds, medicines, knowledge of the properties of fauna and flora, oral traditions, literatures, designs, and visual and performing arts. (Article 29)

Nonindigenous users of indigenous knowledge and resources need to assist traditional communities to establish and strengthen local control by respecting their rights to deny access, veto any projects that affect them, and ensure that authorization has been secured through prior informed consent.

The extreme expression of denial of access comes through a growing moratorium movement. Many indigenous groups are now calling for a complete stop to additional research, collections, or bioprospecting until adequate mechanisms for protection and benefit-sharing are provided by national laws.

For example, the Mataatua Declaration, the statement from the First International Conference on the Cultural and Intellectual Property Rights of Indigenous Peoples, which took place in Whakatane, Aotearoa/New Zealand, in 1993, states:

> A moratorium on any further commercialisation of Indigenous medicinal plants and human genetic materials must be declared until Indigenous communities have developed appropriate protection measures. (Recommendation 2.8)

The Final Statement from the United Nations Development Program's Pacific Regional Consultation on Indigenous Peoples' Knowledge and Intellectual Property Rights, Suva, Fiji (UNDP 1995), echoes this call and urges Indigenous peoples "not to co-operate in bioprospecting activities until appropriate protection measures are in place" (Article 2).

The movement will probably spread to nonindigenous groups in the near future but is already effectively restricting access on indigenous lands and territories in Canada, Ecuador, Panama, New Zealand, Australia, Fiji, and throughout the Pacific.

Religious Rights and Religious Freedom

Religious rights are very important for Indigenous peoples, especially considering the spiritual nature of indigenous knowledge systems. In a seminar on intellectual property rights at the United Nations Human Rights Convention in Vienna, June 1993, Ray Apoaka of the North American Indian Congress suggested that intellectual property rights are essentially a question of religious freedom for Indigenous peoples:

> Much of what they want to commercialise is sacred to us. We see intellectual property as part of our culture—it cannot be separated into categories as [Western] lawyers would want. (Apoaka, quoted in Posey 1994:234.)

Pauline Tangiora, a Maori leader, agrees:

> Indigenous peoples do not limit their religion to buildings, but rather see the sacred in all life. (Tangiora, quoted in Posey 1994:234)

Therefore, in extracting the components of biodiversity, it is often forgotten that what may be of commercial value to outsiders has deep religious significance for the local community. According to the Universal Declaration of Human Rights (1948):

> Everyone has the right to freedom of thought, conscience and religion . . . and freedom, either alone or in community with others and in public or private, to manifest his religion or belief in teaching, practice, worship and observance. (Article 18)

Right to a Unique Language

Language has always played a major role in the indigenous rights movement. The right to speak one's language and to transmit culture through language is never overlooked in indigenous statements. For example, the Declaration of Principles of the World Council of Indigenous Peoples (1982) states that:

> All Indigenous Peoples have the rights to be educated in their own language and to establish their own educational institutions. Indigenous Peoples' languages shall be respected by nation-states in all dealings between them on the basis of equality and non-discrimination. (Principle 15)

The Kari-Oca Declaration (Indigenous Peoples' Earth Charter) demands that:

> Indigenous peoples should have the right to their own knowledge, language, and culturally appropriate education, including bicultural and bilingual education. Through recognizing both formal and informal ways, the participation of family and community is guaranteed. (Paragraph 25)
>
> The use of existing indigenous languages is our right. These languages must be protected. (Paragraph 90)
>
> States that have outlawed indigenous languages and their alphabets should be censured by the United Nations. (Paragraph 91)

The Charter of the Indigenous-Tribal Peoples of the Tropical Forests (Penang, Malaysia, 15 February 1992) calls for:

> Establishment of systems of bilingual and intercultural education. These must revalidate our beliefs, religious traditions, customs, and knowledge; allowing our control over these programmes, by the provision of suitable training, in accordance with our cultures; in order to achieve technical and scientific advances for our peoples, in tune with our own cosmo-visions, and as a contribution to the world community. (Article 32)

International agreements negotiated within the United Nations also guarantee linguistic rights. The principal legally binding agreement is found in the ICCPR:

> In those States in which ethnic, religious or linguistic minorities exist persons belonging to such minorities shall not be denied the right, in community with the other members of their group, to enjoy their own culture, to profess and practice their own religion, or to use their own language. (Article 27)

The 1989 Convention on the Rights of the Child also calls on states to agree that education of the child be directed to:

> The development of respect for the child's parents, his or her own cultural identity, language and values, for the national values of the country in which the child is living; the country from which he or she may originate, and for civilizations different from his or her own. (Article 29)

> In those States in which ethnic, religious or linguistic minorities or persons of indigenous origin exist, a child belonging to such a minority or who is indigenous shall not be denied the right, in community with other members of his or her group, to enjoy his or her own culture, to profess and practise his or her own religion, or to use his or her own language. (Article 30)

The 1993 Vienna Declaration on Human Rights makes the following provisions:

> Considering the importance of the promotion and protection of the rights of persons belonging to minorities and the contribution of such promotion and protection to the political and social stability of the States in which such persons live.
>
> The World Conference on Human Rights reaffirms the obligation of States to ensure that persons belonging to minorities may exercise fully and effectively all human rights and fundamental freedoms without any discrimination and in full equality before the law in accordance with the Declaration on the Rights of Persons Belonging to National or Ethnic, Religious and Linguistic Minorities.
>
> The persons belonging to minorities have the right to enjoy their own culture, to profess and practice their own religion and to use their own language in private and in public, freely and without interference or any form of discrimination. (Item 19)

The Draft UN Declaration on the Rights of Indigenous Peoples states:

Indigenous peoples have the right to revitalize, use, develop and transmit to future generation their histories, languages, oral traditions, philosophies, writing systems and literatures, and to designate and retain their own names for communities, places and persons.

States shall take effective measures, whenever any right of indigenous peoples may be threatened, to ensure this right is protected and also to ensure that they can understand and be understood in political, legal and administrative proceedings, where necessary through the provision of interpretation or by any other appropriate means. (Item 14)

Indigenous children have the right to all levels and forms of education of the State. All indigenous peoples also have this right and the right to establish and control their educational systems and institutions providing education in their own languages, in a manner appropriate to their cultural methods of teaching and learning.

Indigenous children living outside their communities have the right to be provided access to education in their own culture and language.

States shall take effective measures to provide appropriate resources for these purposes. (Item 15)

Indigenous peoples have the right to establish their own media in their own languages. They also have the right to equal access to all forms of non-indigenous media.

States shall take effective measures to ensure that State-owned media duly reflect indigenous cultural diversity. (Item 17)

The June 1996 World Conference on Linguistic Rights in Barcelona, Spain, developed a Draft Universal Declaration of Linguistic Rights, which identifies a strong link between linguistic rights and the collective rights, not of states but of linguistic communities. According to Article 3.2, the collective rights of language groups may include *inter alia*:

- The right for their own language and culture to be taught
- The right of access to cultural services
- The right to an equitable presence of their language and culture in the communications media
- The right to receive attention in their own language from government bodies and in socioeconomic relations.

The declaration ends by recommending that the United Nations General Assembly create a Council of Languages within the UN system. (For further details on linguistic human rights, see Skutnabb-Kangas this volume.)

CONCLUSION

The Convention on Biological Diversity honors the inextricable link between in situ conservation of biodiversity and the enhancement of local communities through the use and application of their knowledge, innovations, and practices. It does not, however, expressly recognize the importance of language and languages in the transmission of traditional ecological knowledge. Like most international agreements, it is also vague about how basic rights will be secured for the lands, resources, and cultures of indigenous and local communities. It is increasingly clear, however, that without favorable conditions for the flourishing of language diversity, transmission of knowledge will break down—and with it the intricate management and livelihood systems that mold and maintain local biodiversity.

Although some local knowledge will be retained even in the face of language loss or deterioration, the richness of traditional ecological knowledge can be conserved only through the maintenance of language diversity. This implies favorable conditions not only for the flourishing of language groups, but also the continuation of robust internal diversity expressed through linguistic variation of individual specialists, and in gender, age, and inheritance groups.

The integral (holistic) nature of knowledge systems has been shown to be linked to land and territory. Thus, it is impossible to discuss conservation of cultural and linguistic diversity without discussing the basic rights of local peoples and their self-determination and control over their own lands and resources. This, of course, makes future activities for linguists, anthropologists, environmentalists, and others working with indigenous and local communities a profoundly political matter. And it implies that continued research into language and cultural diversity requires a more collaborative approach in which equitable partnerships evolve from mutual interest between researchers and local communities. The days of "our" studying "them" (with the added barb of "before they become extinct") must be replaced with collaboration to conserve the biological, linguistic, and cultural diversity of the planet—before we *all* become extinct.

REFERENCES

Durning, A.T. 1992. *Guardians of the Land: Indigenous Peoples and the Health of the Earth.* Worldwatch Paper 112. Washington, D.C.: Worldwatch.

Four Directions Council. 1996. *Forests, Indigenous Peoples, and Biodiversity: Contribution of the Four Directions Council to the Secretariat of the Convention on Biological Diversity.* Lethbridge, FDC.

Gadgil, M., F. Berkes, and C. Folke. 1993. Indigenous knowledge for biodiversity conservation. *Ambio* 22(2/3):151–156.

Glowka, L., F. Burhenne-Guilmin, and H. Synge, with J.A. McNeely and L. Gundling. 1994. *A Guide to the Convention on Biological Diversity.* IUCN Environmental Policy and Law Paper No. 30. Gland: IUCN.

Gray, A. 1999. Indigenous peoples, their environments and territories: Introduction. In *Cultural and Spiritual Values of Biodiversity,* ed. D.A. Posey. Pp. 61–66. London and Nairobi: Intermediate Technology Publications and UNEP.

Harmon, D. 1996. Losing species, losing languages: Connections between biological and linguistic diversity. *Southwest Journal of Linguistics* 15:89–108.

International Society of Ethnobiology. 1988. *Declaration of Belém*. First International Congress of Ethnobiology, Belém, Brazil.

IWGIA 1992a. *IWGIA Newsletter* 2:21–24.

IWGIA 1992b. *IWGIA Newsletter* 4:57–61.

Krauss, M. 1992. The world's languages in crisis. *Language* 68(1):4–10.

Maffi, L., T. Skutnabb-Kangas, and J. Andrianarivo. 1999. Linguistic diversity. In *Cultural and Spiritual Values of Biodiversity*, ed. D.A. Posey. Pp. 21–57. London and Nairobi: Intermediate Technology Publications and UNEP.

Matowanyika, J.Z.Z. 1996. Resource management and the Shona people in rural Zimbabwe. In *Indigenous Peoples and Sustainability: Cases and Actions*, ed. D.A. Posey and G. Dutfield. Pp. 257–266. Gland and Utrecht: IUCN and International Books.

Opoku, K.A. 1978. *West African Traditional Religions*. Lagos: FEP International.

Pereira, W., and A.K. Gupta. 1993. A dialogue on indigenous knowledge. *Honey Bee* 4(4):6–10.

Posey, D.A. 1994. International agreements and intellectual property right protection for indigenous peoples. In *Intellectual Property Rights for Indigenous Peoples: A Sourcebook*, ed. T. Greaves. Pp. 223–251. Oklahoma City: Society for Applied Anthropology.

Posey, D.A. 1996. Indigenous knowledge, biodiversity, and international rights: Learning about forests from the Kayapó Indians of the Brazilian Amazon. In *The Commonwealth Forestry Review* 76(1) [ed. A.J. Grayson]:53–60. Oxford: Commonwealth Forestry Association.

Posey, D.A., ed. 1999. *Cultural and Spiritual Values of Biodiversity*. London and Nairobi: Intermediate Technology Publications and UNEP.

Posey, D.A., and G. Dutfield. 1996a. *Beyond Intellectual Property: Toward Traditional Resource Rights for Indigenous Peoples and Local Communities*. Ottawa: International Development Research Centre.

Posey, D.A., and G. Dutfield, eds. 1996b. *Indigenous Peoples and Sustainability: Cases and Actions*. Gland and Utrecht: IUCN and International Books.

Sultan, R., D. Craig, and H. Ross. 1996. Aboriginal joint management of Australian national parks: Uluru-Kata Tjuta. In *Indigenous Peoples and Sustainability: Cases and Actions*, ed. D.A. Posey and G. Dutfield. Pp. 326–338. Gland and Utrecht: IUCN and International Books.

UNDP. 1995. *Statements of the Regional Meetings of Indigenous Representatives on the Conservation and Protection of Indigenous Knowledge*. New York: United Nations Development Programme.

UNESCO. 1993. *Amendment to the Draft Programme and Budget for 1994–1995 (27 C/5), Item 5 of the Provisional Agenda (27 C/DR.321)*. Paris: UNESCO.

United Nations. 1945. *Charter of the United Nations and Statute of the International Court of Justice* New York: UN Department of Public Information.

Vienna Declaration. 1993. *The Vienna Declaration and Programme of Action*. World Conference on Human Rights, June 1993, Vienna, Austria.

Willett, A.B.J. 1993. Indigenous knowledge and its implications for agricultural development and agricultural extension: A case study of the Vedic tradition in Nepal. Ph.D. diss., Iowa State University.

24

LINGUISTIC HUMAN RIGHTS IN EDUCATION
FOR LANGUAGE MAINTENANCE

Tove Skutnabb-Kangas

They sound like . . . the clucking of hens or the clucking of turkeys.

When they speak they fart with their tongues in their mouths.

They seem to resound always with the very nature, the poetic character of the lands where they were used. The cadences of the wild, of water and earth, rock and grass, roll onomatopoetically along the tongue. Khoikhoi words . . . crack and softly rustle, and click. The sand and dry heat and empty distance of the semi-arid lands where the Khoikhoi originated are embedded in them. But so is softness, greenness. They run together like the very passage of their olden days.

Quoted in Koch and Maslamoney 1997:28)

These descriptions of the Khoekhoe "click" languages of South Africa and Namibia, all by outsiders—the first two by some of the first colonizers of the Khoekhoe areas, the third by historian Noel Mostert—reflect a possible change in attitudes, which might, in the best case, have some positive consequences for the maintenance of Khoekhoe languages, already thought to be dead in South Africa. The Khoekhoe have been called Hottentots by outsiders, and "to this day the Shorter Oxford English Dictionary notes that the word Hottentot is used to describe 'a person of inferior intellect and culture'" (Koch and Maslamoney 1997:28). It is easier to rob the lands and kill the languages of peoples if they are construed as "inferior." This is exactly what has happened and is still happening in the world, only now with more sophisticated methods than centuries ago. One of the results is physical, linguistic, and cultural genocide and disappearance of diversities. Loss of biodiversity, erosion of traditional cultures, and subtractive learning of dominant languages leading to loss of diversity of human languages, all these are happening at an accelerating pace (see Krauss 1992; Harmon 1995; Maffi, Skutnabb-Kangas, and Andrianarivo 1999).

As numerous resolutions and actions in recent decades show, indigenous peoples see their languages and cultural traditions—and the fight to protect or restore their linguistic rights—as essential elements in their struggle for survival and self-determination. Language and land are considered by most of them as equally constitutive of their identity. Likewise, the Organisation for Security and Cooperation in Europe (OSCE) High Commissioner on National Minorities, Max van der Stoel says that the minorities he has encountered in his work (which is diplomatic "conflict prevention in situations of ethnic tension"—Rothenberger 1997:3) have two main demands: self-determination and the right to education through the medium of their own language. Land rights are part of self-determination, but they are only a part of the solution and an either-or approach is not enough. Land and language need to go together. Both need maintenance, development, and—in many cases—revitalization.

Says Hendrik Stuurman, a Khoikhoi Nama who came back to South Africa in the mid-1990s from Namibia, where some of his people had gone after a forced removal from their ancestral lands during apartheid (quoted in Koch and Maslamoney 1997):

> Now we have our land back. We are doing well with houses, schools and clinics. We thought that with our land we would be able to heal the culture that is the soul of our people. But we find instead that our language is dying. At least while we were in exile we were able to read and speak our language (Nama is recognised in Namibia). We now realise that, in the act of regaining our land, we may have destroyed our culture.

As for Stuurman himself, he already belongs to the generation in which the intergenerational transmission was incomplete; his main language is Afrikaans. This is how he feels about it:

> a feeling deep inside me that something is wrong . . . that I have drunk the milk of a strange woman, that I grew up alongside another person. I feel like this because I do not speak my mother's language.

Centrality of language for Stuurman and others reflects the key role that language plays in all aspects of human life everywhere. Language is central to our conceptualization of the world and for interpreting, understanding, and changing it. Language supports us in organizing our world and frees our energy for other tasks. Words for concepts are like pegs on which we hang the meanings that we store in the storehouse of our mind. They are the framework that binds together the details into a totality, a meaningful whole. Verbalizing helps us remember and reproduce meaning and thus make sense of reality. Through the verbal socialization process we also learn much of our own culture's ethics. Together with the words for objects and phenomena, we learn our culture's connotations, associations, emotions, and value judgments. The definition and construction of our ecosocial world, including group identity, status, and world view, are reflected in, reflect on, and are realized through language.

The particular social and ecological circumstances in which different human groups develop over time—the specific relationships each group establishes among its members and with other people around, as well as with the place in which they live—lead to different and historically changing ways of defining, understanding, and interpreting the world via language. The diversity of languages (and cultures) around the world has arisen through these complex and dynamic processes.

The linguistic human rights of both indigenous peoples and linguistic minorities, especially educational language rights, play a decisive role in maintaining and revitalizing languages and in supporting linguistic and cultural diversity and, through them, also biological diversity on Earth. (See Skutnabb-Kangas 1998a, b, 2000, and Skutnabb-Kangas in Maffi, Skutnabb-Kangas, and Andrianarivo 1999 for a more extensive treatment of the topic.) Before introducing the main international linguistic rights instruments, I discuss some of the ideological attitudes and policy measures that most seriously threaten the respect and protection of linguistic diversity.

LINGUICISM AND ETHNICISM: DIVISION OF POWER BASED ON LANGUAGE AND CULTURE

The fate of languages is of utmost and growing importance. The struggle over the power and resources of the world is conducted increasingly through ideological means, and ideas are mainly mediated through language. This also partially explains the spread of numerically large languages (English, Spanish, Russian, Mandarin Chinese, etc.) at the cost of the smaller ones. The ideas of the power-holders cannot be spread, nationally or internationally, unless those with less power understand the power-holders' language (e.g., "international English" worldwide, or standard Italian in Italy). Mass media and formal education are used for this language learning, which is mostly organized subtractively (the "major" language is learned at the cost of one's own language, not in addition to it). Mass media, formal education, and religions also form an important part of the consciousness industry through which the content of the hegemonic ideas of power-holders is spread.

Language and culture are in the process of replacing "race" as bases for discrimination. Access to material resources and structural power is increasingly determined not on the basis of skin color, or "race" (as biologically argued racism has it), but on the basis of ethnicity and language(s) (mother tongue and competence, or lack thereof, in official and/or "international" languages). Linguistically argued racism, *linguicism,* and culturally argued racism, *ethnicism* or *culturism,* can be defined as "ideologies, structures and practices which are used to legitimate, effectuate and reproduce an unequal division of power and (both material and non-material) resources between groups which are defined on the basis of language (linguicism) and culture or ethnicity (ethnicism/culturism)" (Skutnabb-Kangas 1988:13).

LINGUISTIC GENOCIDE REFLECTS THE MONOLINGUAL REDUCTIONISM OF "NATION-STATES"

Given the growing importance of language worldwide, state resistance to smaller languages is to be expected. Cobarrubias (1983) provides the following taxonomy of possible state policies vis-à-vis indigenous or minority languages: (1) attempting to kill a language; (2) letting a language die; (3) unsupported coexistence; (4) partial support of specific language functions; (5) adoption as an official language. The division of power and resources in the world partially follows linguistic lines along which more accrues to speakers of "big" languages, because they can use their languages for most official purposes (case 5 or at least 4 above). Speakers of smaller languages are often forced to learn the big languages subtractively (at the expense of their own languages) rather than additively (in addition to their own languages), because the latter are not used for official purposes, including education; that is, the state has adopted one or other of policies 1–3 above.

In understanding reduction versus maintenance of linguistic diversity, it is useful to compare the concept of *linguicide* (linguistic genocide) with that of *language death.* The notion of "language death" does not necessarily imply a causal agent but is seen as a natural, inevitable result of social change and "modernization," leading toward the development of a unified world with a world language, possibly coexisting with national languages with a reduced role. Within this paradigm, language death is interpreted as the result of voluntary language shift by each speaker.

Linguicide, by contrast, implies *agency involved in causing the death of languages.* The agents can be *active* ("attempting to kill a language") or *passive* ("letting a language die" or "unsupported coexistence"). The *causes* of linguicide and linguicism must be analyzed from both structural and ideological angles, covering the struggle for structural power and material resources, on the one hand, and on the other the legitimization, instantiation, and reproduction of the unequal division of power and resources between groups based on language. The *agents* of linguicide/linguicism can also be structural (a state, an institution, laws and regulations, budgets, etc.) or ideological (norms and values ascribed to different languages and their speakers). Thus there is nothing "natural" in language death. Language death has causes, which can be identified and analyzed.

In preparation for the 1948 United Nations International Convention for the Prevention and Punishment of the Crime of Genocide (E 793, 1948), linguistic and cultural genocide were discussed alongside physical genocide as serious crimes against humanity (Capotorti 1979). In the UN General Assembly, however, Article 3 covering linguistic and cultural genocide was voted down and is thus not part of the final convention. What remains is a definition of linguistic genocide that was acceptable to most states then in the UN:

> Prohibiting the use of the language of the group in daily intercourse or in schools, or the printing and circulation of publications in the language of the group. (Article 3.1.)

Linguistic genocide in this form is practiced throughout the world. The use of an indigenous or minority language can be prohibited overtly and directly, through laws, imprisonment, torture, killings, and threats (as, e.g., Turkey does vis-à-vis Kurdish; see Skutnabb-Kangas and Bucak 1994). It can also be prohibited covertly, more indirectly, via ideological and structural means, such as in educational systems. Every time there are indigenous or minority children in day-care centers and schools with no bilingual teachers authorized to use the languages of the children as the main teaching and child care media, this is tantamount to prohibiting the use of minority languages "in daily intercourse or in schools." This is the situation for most indigenous First Nations as well as immigrant and refugee minority children.

Linguicism is a major factor in determining whether speakers of particular languages are allowed to enjoy their linguistic human rights. Lack of these rights, for instance the absence of these languages from school time-tables, makes indigenous and minority languages invisible. Alternatively, minority mother tongues are construed and presented as nonresources, as handicaps that are said to prevent indigenous or minority children from acquiring the majority language—portrayed as the only valued linguistic resource—so that minority children should get rid of them in their own interest. At the same time, many minorities, especially minority children, are in fact prevented from fully acquiring majority resources, and especially majority languages, through disabling educational structures in which instruction is organized through the medium of the majority languages in ways that contradict most scientific evidence on how bilingual education should be structured. (See Cummins 1996, 1998 and Skutnabb-Kangas 1984, 1990, 2000 for references.)

The processes reducing the linguistic and cultural diversity of the world are symptomatic of an ideology of monolingual reductionism, which consists of several myths. The first four include the beliefs that monolingualism, at both individual and societal levels, is *normal;* that it is *unavoidable* ("it is a pity but you cannot make people cling to small languages which are not useful; they want to shift"); that it is *sufficient* to know a "big" language, especially English ("everything important is in English, or, if it is important enough and has been written in another language, it will be translated into English"; "the same things are being said in all languages so why bother?") and *desirable* ("you learn more if you can use all your energies on one language instead of needing to learn many"; "monolingual countries are richer and more developed"; "it is cheaper and more efficient to have just one language"). In fact all these myths can be easily refuted—they are rather fallacies (Skutnabb-Kangas 1996a, b).

The fifth myth is that the granting of linguistic and cultural human rights inevitably leads to the disintegration of present states. At this level, monolingual reductionism can be characterized as an ideology used to rationalize linguistic genocide (especially in education) by states calling themselves "nation-states," which treat the existence of (unassimilated) minorities as a threat leading to the state's potential disintegration. The principle of the territorial integrity and political sovereignty of contemporary states is often presented as being in conflict with another fundamental human rights principle, that of self-determination. Minorities that enjoy linguistic human rights are portrayed as aiming to demand first internal self-determination, for example, cultural and other autonomy, and then independent status, that is, external self-determination. By denying them linguistic rights and by bringing about homogenization through linguistic and cultural genocide in education and elsewhere, states claim that they are seeking to eliminate the threat of groups that may eventually demand self-determination. To deny minorities those human rights that are most central to their reproducing themselves as distinctive groups—namely, linguistic and cultural human rights, and especially educational language rights—while observing (or appearing to observe) several of the basic human rights for all its citizens, including minorities, is a covert way for a state to make languages disappear at the same time that it retains its legitimacy in the eyes of (most of) its citizens and the international community. Covert linguicide of this type appears to be extremely effective. It is often far more difficult to struggle against covert violence, against the colonization of the mind, where short-term "benefits" may obscure longer-term losses, than it is to fight physical violence and oppression.

Contra these myths, however, there are strong reasons why states should support, rather than try to eliminate, linguistic and cultural diversity and grant linguistic human rights. Some states might indeed break up in the process, but this should be acceptable if the human right of self-determination is to be upheld (Clark and Williamson 1996). By and large, though, granting linguistic and cultural human rights to minorities reduces rather than creates the potential for "ethnic" conflict, prevents the disintegration of states, and may avoid anarchy in which even the rights of the élites will be severely curtailed by conditions that increasingly resemble civil war, especially in inner cities. Linguistic and cultural identity are at the core of the cultures of most ethnic groups (Smolicz 1979).

When threatened, these identities can have a very strong potential to mobilize groups: "attempts to artificially suppress minority languages through policies of assimilation, devaluation, reduction to a state of illiteracy, expulsion or genocide are not only degrading of human dignity and morally unacceptable, but they are also an *invitation* to separatism and an *incitement* to fragmentation into mini-states" (Smolicz 1986:96; emphasis added). Thus, fostering diversity by granting linguistic human rights can actually promote a state's self-interest.

As Asbjørn Eide (1995:29–30) of the UN Human Rights Commission points out, cultural rights have received little attention both in human rights theory and in practice, despite the fact that today "ethnic conflict" and "ethnic tension" are seen as the most important potential causes of unrest, conflict, and violence in the world. Absence or denial of linguistic and cultural rights are today effective ways of promoting, not curbing, this "ethnic" conflict and violence. Linguicide is ineffective as a strategy for preventing the disintegration of present-day states. "Preservation of the linguistic and cultural heritage of humankind" (one of UNESCO's declared goals) presupposes preventing linguicide. Linguistic diversity at local levels is not only a necessary counterweight to the hegemony of a few "international" languages but represents a recognition of the fact that all individuals and groups have basic linguistic human rights and is a necessity for the survival of the planet. The perpetuation of linguistic diversity is a necessary component of any discourse on and strategy for the maintenance of biological and cultural diversity on the planet.

HUMAN RIGHTS INSTRUMENTS AND LANGUAGE RIGHTS IN EDUCATION

In many international, regional, and multilateral human rights instruments, language is mentioned in the preamble and general clauses (e.g., United Nations Charter, Article 13; Universal Declaration of Human Rights [1948], Article 2; International Covenant on Civil and Political Rights [ICCPR; 1966, in force since 1972], Article 2.1) as one of the characteristics on the basis of which discrimination is forbidden, together with "race, colour, sex, religion, political or other opinion, national or social origin, property, birth or other status" (ICCPR Article 2.1). This suggests that language has been seen as one of the most important characteristics of humans in terms of human rights issues in the key documents that have pioneered the UN effort after 1945. Yet, the most important linguistic human rights, especially in education, are still absent from human rights instruments. Language gets much poorer treatment in human rights law than other important human attributes, like gender, "race" or religion.

For the maintenance and development of linguistic and cultural diversity on our planet, educational language rights are not only vital but the most important linguistic human rights. Intergenerational transmission of languages is the most vital factor for their maintenance. If children do not get the opportunity to learn their parents' idiom fully and properly so that they become at least as proficient as their parents, the language cannot survive. As more children get access to formal education, much of the language learning that earlier happened in the community must happen in schools. Beyond the nonbinding preambles, however, in the educational clauses of human rights instruments, two phenomena can be observed. One is that language disappears completely, as, for instance, in the Universal Declaration of Human Rights, where the paragraph on education (26) does not refer to language at all. Similarly, the ICCPR, having mentioned language on a par with race, color, sex, religion, and the like in Article 2.1, refers to "racial, ethnic or religious groups" but omits reference to language or linguistic groups in its educational Article 13. Second, if language-related rights are specified, the articles dealing with these rights are typically so weak as to be de facto meaningless. For example, in the UN Declaration on the Rights of Persons Belonging to National or Ethnic, Religious and Linguistic Minorities (adopted in 1992), most of the articles use the obligating formulation "shall" and introduce few opt-out clauses or alternatives— except where linguistic rights in education are concerned:

> 1.1. States *shall protect* the existence and the national or ethnic, cultural, religious and linguistic identity of minorities within their respective territories, and *shall encourage* conditions for the *promotion* of that identity.
>
> 1.2. States *shall* adopt *appropriate* legislative *and other* measures *to achieve those ends.*
>
> 4.3. States *should* take *appropriate* measures so that, *wherever possible,* persons belonging to minorities have *adequate* opportunities to learn their mother tongue *or* to have instruction in their mother tongue. (Emphasis added.)

Similarly, in the European Charter for Regional or Minority Languages (1992), the formulations in the educational articles include a range of modifications including "as far as possible," "relevant," "appropriate," "where necessary," "pupils who so wish in a number considered sufficient," "if the number of users of a regional or minority language justifies it," as well as a number of alternatives, as in "to allow, encourage *or* provide teaching in *or* of the regional or minority language at all the appropriate stages of education" (emphasis added). Writing binding formulations that are sensitive to local conditions presents unquestionably real problems. But opt-out clauses and alternatives permit a reluctant state to meet the requirements in a minimalist way by claiming that a provision was not "possible" or "appropriate," or that numbers were not "sufficient" or did not "justify" a provision, or that it "allowed" the minority to organize teaching of their language as a subject, at their own cost.

The new Council of Europe Framework Convention for the Protection of National Minorities was adopted by the Committee of Ministers of the Council of Europe in 1994. Again the article covering medium of education is much more heavily qualified than any other:

In areas inhabited by persons belonging to national minorities traditionally or in substantial numbers, *if there is sufficient demand,* the parties shall *endeavour* to ensure, *as far as possible* and *within the framework of their education systems,* that persons belonging to those minorities have *adequate* opportunities for being taught in the minority language *or* for receiving instruction in this language. (Article 14.2; emphasis added.)

Thus, the situation is not improving despite new instruments in which language rights are mentioned and even treated in detail. Even where linguists have participated in drafting new or planned instruments, as in the case of the Draft Universal Declaration of Linguistic Rights, the results are far from perfect.

DRAFT UNIVERSAL DECLARATION OF LINGUISTIC RIGHTS

The Draft Universal Declaration of Linguistic Rights, handed over to UNESCO in June 1996, is the first attempt at formulating a universal document about language rights exclusively. It grants rights to three different entities: *individuals* (i.e., "everyone"), *language groups,* and *language communities.* No dutyholders are specified anywhere.

Language communities have more rights in the declaration than the other two categories, including in education (Article 23), but the rights granted to them are formulated in such a way that the whole declaration runs the risk of being seen as full of pious, unrealistic wishes that cannot be taken seriously.

For *language groups,* collective rights to one's own language are not seen as inalienable. For those *individuals* who speak a language other than the language of the territory, education in their own language is not a positive right. In addition, the declaration grants members of language communities extensive rights to any *foreign* language in the world, whereas the rights granted to "everyone" include only the (negative "does *not exclude*") right to "oral and written knowledge" of *one's own* language.

All language communities are entitled to an education which will enable their members to acquire a full command of their own language, including the different abilities relating to all the usual spheres of use, as well as *the most extensive possible command of any other language* they may wish to know. (Article 26 on rights of language communities; emphasis added)

1. Everyone is entitled to receive an education *in the language specific to the territory where s/he resides.*

2. This right does not exclude the right to acquire *oral and written knowledge of any language* which may be of use to him/her as an instrument of communication with other *language communities.* (Article 29 on rights of "everyone"; emphasis added)

The declaration would force everybody who is not defined as a member of a language community to assimilate. This also makes the declaration vulnerable in several respects, especially in relation to states that claim they do not have minority language communities. In addition, even the language rights for language communities are formulated in such a way as to make them completely unrealistic for anybody except, maybe, a few hundred of the world's language communities, most of them dominant linguistic majorities. For most African, Asian, and Latin American countries, the rights in the declaration are at present practically, economically, and even politically impossible to realize, as was clearly expressed at the first UNESCO meeting where the declaration was discussed. It therefore seems extremely unlikely that the declaration will be accepted in its present form.

The Draft UN Declaration on Rights of Indigenous Peoples formulates language rights strongly, especially in education. If these rights were to be granted in their present form, some 60 to 80 percent of the world's oral languages would have decent legal support. But according to the chair of the Draft Declaration's Working Group, Erica-Irene Daes (1995), there is little chance of the declaration being accepted in the present form, and even if it were, implementation by UN member states is, of course, a completely different matter.

The Draft Universal Declaration of Linguistic Rights, clearly less than ideal in its present form, represents the first attempt at formulating language rights at a universal level to reach a stage that permits serious international discussion to start. From the point of view of maintaining the planet's linguistic diversity, the immediate fate of the Draft UN Declaration on Rights of Indigenous Peoples is probably more important, however, because it has at least some chance of being accepted, signed, and ratified, even if in a form that reduces the rights granted in the present draft.

STARTING POINTS FOR POSITIVE DEVELOPMENTS

Some recent instruments could be a starting point for more positive developments. One is the reinterpretation of UN International Covenant on Civil and Political Rights, Article 27:

> In those states in which ethnic, religious or linguistic minorities exist, persons belonging to such minorities shall not be denied the right, in community with other members of their group, to enjoy their own culture, to profess and practise their own religion, or to use their own language.

In the customary reading of Article 27, rights were only granted to individuals, not collectivities. And "persons belonging to . . . minorities" only had these rights in states that accepted their existence. This has not helped immigrant minorities because they have not been seen as minorities in the legal sense by the states in which they live. More recently (6 April 1994), the UN Human Rights Commission adopted a General Comment on Article 27 that interprets it in a substantially broader and more positive way than earlier.

In customary reading, the article was interpreted as

- excluding immigrants or migrants, who have not been seen as minorities;
- excluding groups, even if they are citizens, that are not recognized by the state (in the same way as the European Charter does) either as minorities or as "indigenous," a formulation that has been added to Article 30 in the 1989 UN Convention on the Rights of the Child, which is otherwise identical to ICCPR Article 27;
- conferring only some protection from discrimination, that is, "negative rights," but not a positive right to maintain or even use one's language; and
- not imposing any obligations on the states.

The UN Human Rights Commission sees the article as

- protecting all individuals on the state's territory or under its jurisdiction, that is, immigrants and refugees as well, irrespective of whether they belong to the minorities specified in the article or not;
- stating that the existence of a minority does not depend on a decision by the state but requires to be established by objective criteria;
- recognizing the existence of a "right" rather than only a nondiscrimination prescription); and
- imposing positive obligations on the states.

It remains to be seen to what extent this General Comment will influence the state parties in relation to linguistic human rights of speakers of smaller languages. It depends on the extent to which the committee's interpretation ("soft law") will become the general norm observed by the countries where indigenous peoples and migrant and refugee minorities live.

The second positive development is the new educational guidelines issued by the Foundation on Inter-Ethnic Relations for the OSCE High Commissioner on National Minorities, Max van der Stoel (The Hague Recommendations Regarding the Education Rights of National Minorities and Explanatory Note, October 1996). These guidelines were elaborated by a small group of experts on human rights and education, of which I was a member. They "attempt to clarify in relatively straight-forward language the content of minority education rights"; international human rights standards "have been interpreted in such a way as to ensure their coherence in application" (p. 3). In the section "The Spirit of International Instruments," bilingualism is set as a right and responsibility for persons belonging to national minorities (Article 1), and states are reminded not to interpret their obligations in a restrictive manner (Article 3):

> 1) The right of persons belonging to national minorities to maintain their identity can only be fully realised if they acquire a proper knowledge of their mother tongue

during the educational process. At the same time, persons belonging to national minorities have a responsibility to integrate into the wider national society through the acquisition of a proper knowledge of the State language. . . .

3) It should be borne in mind that the relevant international obligations and commitments constitute international minimum standards. It would be contrary to their spirit and intent to interpret these obligations and commitments in a restrictive manner.

In the section "Minority Education at Primary and Secondary Levels," education through the medium of the mother tongue is recommended at all levels, including the teaching of the dominant language as a second language by bilingual teachers (Articles 11–13). Teacher training is made a duty on the state (Article 14):

11) The first years of education are of pivotal importance in a child's development. Educational research suggests that the medium of teaching at pre-school and kindergarten levels should ideally be the child's language. Wherever possible, States should create conditions enabling parents to avail themselves of this option.

12) Research also indicates that in primary school the curriculum should ideally be taught in the minority language. The minority language should be taught as a subject on a regular basis. The State language should also be taught as a subject on a regular basis preferably by bilingual teachers who have a good understanding of the children's cultural and linguistic background. Towards the end of this period, a few practical or non-theoretical subjects should be taught through the medium of the State language. Wherever possible, States should create conditions enabling parents to avail themselves of this option.

13) In secondary school a substantial part of the curriculum should be taught through the medium of the minority language. The minority language should be taught as a subject on a regular basis. The State language should also be taught as a subject on a regular basis preferably by bilingual teachers who have a good understanding of the children's cultural and linguistic background. Throughout this period, the number of subjects taught in the State language, should gradually be increased. Research findings suggest that the more gradual the increase, the better for the child.

14) The maintenance of the primary and secondary levels of minority education depends a great deal on the availability of teachers trained in all disciplines in the mother tongue. Therefore, ensuing from the obligation to provide adequate opportunities for minority language education, States should provide adequate facilities for the appropriate training of teachers and should facilitate access to such training.

Finally, the Explanatory Note states that

submersion-type approaches whereby the curriculum is taught exclusively through the medium of the State language and minority children are entirely integrated into classes with children of the majority are not in line with international standards (p. 5).

This means that the children to whom the recommendations apply might be granted some of the central educational linguistic human rights. The question now is to what extent the 55 OSCE countries will apply the recommendations and how they will interpret their scope. The recommendations could in principle apply to all minorities, even the "everyone" with very few rights in the Draft Universal Declaration on Linguistic Rights. And since indigenous peoples should have at least all the rights that minorities do, these recommendations might provide them also with a tool, while the Draft UN Declaration on the Rights of Indigenous Peoples is still under discussion.

Finally, I wish to suggest what a universal convention of linguistic human rights should guarantee at an individual level, especially in relation to the important educational language rights (see Skutnabb-Kangas 1998a) and for linguistic diversity to be maintained. In a civilized state, there should be no need to debate the right for indigenous peoples and minorities to exist, to decide about their own affairs (self-determination) and to reproduce themselves as distinct groups, with their own languages and cultures. This includes the right to ownership and guardianship of their own lands and material (natural and other) resources as a prerequisite for the maintenance of nonmaterial resources (cf. Posey this volume). It is a self-evident, fundamental collective human right. There should be no need to debate the right to identify with, to maintain, and to fully develop one's mother tongue(s) (the language[s] a person has learned first and/or identifies with). It is a self-evident, fundamental individual linguistic human right.

Necessary individual linguistic rights have to do with access to the mother tongue and an official language in a situation of stable bilingualism and with language-related access to formal primary education. At an *individual level,* a universal convention of linguistic human rights should guarantee, first, in relation to the mother tongue, that everybody can:

- identify with their mother tongue and have this identification accepted and respected by others;
- learn the mother tongue fully, orally (when physiologically possible) and in writing (which presupposes that minorities are educated through the medium of their mother tongue, within the state-financed educational system); and
- use the mother tongue in most official situations, including day care, schools, courts, emergency situations of all kinds, health care, hospitals, and many governmental and other offices.

Second, in relation to other languages:

- that everybody whose mother tongue is not an official language in the country of residence can become bilingual (or trilingual, if a person has two mother tongues) in the mother tongues and (one of) the official language(s) according to one's own choice;
- that suitably trained bilingual teachers are available; and

- that parents know enough about research results when they make their educational choices; for example, minority parents must know that good mother-tongue medium teaching leads to better proficiency both in the mother tongue and in the dominant language, than dominant-language medium submersion teaching.

Third, concerning the relationship between languages:

- that any change of mother tongue is not imposed, but voluntary, that is, it includes knowledge of alternatives and of long-term consequences of choices.

And fourth, concerning profit from education:

- that everybody can profit from education, regardless of what one's mother tongue is ("profit" being defined in terms of educational equal outcome, not just of equal opportunity).

A universal convention of linguistic human rights must also make states duty-holders, in a firm and detailed way, that is, it must provide enforceable rights. If these rights are not granted and implemented, it seems likely that the present pessimistic prognoses of some 90 percent of the world's oral languages not being around anymore by 2100 (Krauss 1992) may actually turn out to have been optimistic. Languages that are not used as main media of instruction will cease to be passed on to children at the latest when we reach the fourth generation of groups in which everybody goes to school—and many languages may be killed much earlier. There is still much work to be done for education through the medium of the mother tongue to be recognized as a human right. Yet this is what is most urgently needed to ensure that indigenous and minority peoples will be able to maintain and develop their languages and perpetuate linguistic diversity on Earth.

REFERENCES

Capotorti, F. 1979. *Study of the Rights of Persons Belonging to Ethnic, Religious, and Linguistic Minorities.* New York: United Nations.

Clark, D., and R. Williamson, eds. 1996. *Self-Determination: International Perspectives.* London: Macmillan.

Cobarrubias, J. 1983. Ethical issues in status planning. In *Progress in Language Planning: International Perspectives,* ed. J. Cobarrubias and J.A. Fishman. Pp. 41–85. Berlin: Mouton.

Cummins, J. 1996. *Negotiating Identities: Education for Empowerment in a Diverse Society.* Ontario, Calif.: California Association for Bilingual Education.

Cummins, J., ed. 1998. *Bilingual Education: The Encyclopedia of Language and Education.* Dordrecht: Kluwer Academic.

Daes, E.-I. 1995. Redressing the balance: The struggle to be heard. Paper presented at the Global Cultural Diversity Conference, Sydney, 26–28 April 1995.

Eide, A. 1995. Economic, social, and cultural rights as human rights. In *Economic, Social, and*

Cultural Rights: A Textbook, ed. A. Eide, C. Krause, and A. Rosas. Pp. 21–40. Dordrecht, Boston, and London: Martinus Nijhoff Publishers.

Harmon, D. 1995. The status of the world's languages as reported in *Ethnologue. Southwest Journal of Linguistics* 14:1–33.

Koch, E., and S. Maslamoney. 1997. Words that click and rustle softly like the wind. *Mail and Guardian,* September 12–18, 1997:28–29.

Krauss, M. 1992. The world's languages in crisis. *Language* 68:4–10.

Maffi, L., T. Skutnabb-Kangas, and J. Andrianarivo. 1999. Linguistic diversity. In *Cultural and Spiritual Values of Biodiversity,* ed. D.A. Posey. Pp. 21–57. London and Nairobi: Intermediate Technology Publications and UNEP.

Rothenberger, A., comp. 1997. *Bibliography on the OSCE High Commissioner on National Minorities: Documents, Speeches and Related Publications.* The Hague: Foundation on Inter-Ethnic Relations.

Skutnabb-Kangas, T. 1984. *Bilingualism or Not: The Education of Minorities.* Clevedon: Multilingual Matters.

Skutnabb-Kangas, T. 1988. Multilingualism and the education of minority children. In *Minority Education: From Shame to Struggle,* ed. T. Skutnabb-Kangas and J. Cummins. Pp. 9–44. Clevedon: Multilingual Matters.

Skutnabb-Kangas, T. 1990. *Language, Literacy, and Minorities.* London: Minority Rights Group.

Skutnabb-Kangas, T. 1996a. Educational language choice—Multilingual diversity or monolingual reductionism? In *Contrastive Sociolinguistics,* Part III, *Language Planning and Language Politics,* ed. M. Hellinger and U. Ammon. Pp. 175–204. Berlin and New York: Mouton de Gruyter.

Skutnabb-Kangas, T. 1996b. Promotion of linguistic tolerance and development. In *Vers un agenda linguistique: Regard futuriste sur les Nations Unies / Towards a Language Agenda: Futurist Outlook on the United Nations,* ed. S. Léger. Pp. 579–629. Ottawa: Canadian Centre for Linguistic Rights, University of Ottawa.

Skutnabb-Kangas, T. 1998a. Human rights and language wrongs—A future for diversity. In *Language Rights,* ed. P. Benson, P. Grundy, and T. Skutnabb-Kangas. Special issue, *Language Sciences* 20(1):5–27.

Skutnabb-Kangas, T. 1998b. Human rights and language policy in education. In *The Encyclopedia of Language and Education,* vol. 1: *Language Policy and Political Issues in Education,* ed. R. Wodak and D. Corson. Pp. 55–65. Dordrecht: Kluwer Academic.

Skutnabb-Kangas, T. 2000. *Linguistic Genocide in Education—or Worldwide Diversity and Human Rights?* Mahwah, N.J.: Lawrence Erlbaum.

Skutnabb-Kangas, T., and S. Bucak. 1994. Killing a mother tongue—How the Kurds are deprived of linguistic human rights. In *Linguistic Human Rights: Inequality or Justice in Language Policy,* ed. T. Skutnabb-Kangas and R. Phillipson, in collaboration with M. Rannut. Pp. 347–370. Berlin and New York: Mouton de Gruyter.

Smolicz, J.J. 1979. *Culture and Education in a Plural Society.* Canberra: Curriculum Development Centre.

Smolicz, J.J. 1986. National language policy in the Philippines. In *Language and Education in Multilingual Settings,* ed. B. Spolsky. Pp. 96–116. Clevedon and Philadelphia: Multilingual Matters.

25

LANGUAGE, KNOWLEDGE, AND INDIGENOUS HERITAGE RIGHTS

Luisa Maffi

The existence of an "inextricable link" binding together linguistic, cultural, and biological diversity is strongly implied in indigenous peoples' views (cf. Posey 1999, this volume; see Appendix 1). For indigenous peoples, maintaining or restoring the integrity of their cultures, languages, and environments represents one interrelated goal. This holistic approach is increasingly evident in grassroots efforts around the world, as well as in the work of concerned scholars and advocates (Greaves 1994; Skutnabb-Kangas and Phillipson 1994; Brush and Stabinsky 1996; Posey 1996, this volume; Posey and Dutfield 1996; Daes 1997; Maffi 1998a; Maffi, Skutnabb-Kangas, and Andrianarivo 1999; Skutnabb-Kangas 2000, this volume; Brush this volume).

Over the past two decades, windows of opportunity have opened at the international level for the integrated protection of indigenous peoples' rights. This has resulted in particular from the passing, in 1989, of the International Labor Organization (ILO) Convention on Indigenous and Tribal Peoples, no. 169, and from the activities, since 1982, of the UN Centre for Human Rights' Working Group on Indigenous Populations (WGIP), leading to the elaboration of the 1994 Draft UN Declaration on the Rights of Indigenous Peoples (currently under examination by the Sub-Commission on the Prevention of Discrimination and the Protection of Minorities). Another key process is the implementation of the 1992 Convention on Biological Diversity (CBD)—especially its Articles 8j, 10c, 18.4, that are concerned with the respect, preservation and promotion of "knowledge, innovations and practices of indigenous and local communities embodying traditional lifestyles relevant for the conservation and sustainable use of biological diversity" (Article 8j), as well as with establishing appropriate agreements for the development and utilization of such "knowledge, innovations, and practices" and the equitable sharing of benefits deriving from this utilization (see Posey 1996, this volume; Posey and Dutfield 1996). In addition, UNESCO (the UN Educational, Scientific, and Cultural Organization), which has long been concerned with the protection and preservation of "world" cultural heritage (i.e., cultural heritage as the "patrimony of humanity"), has begun to move toward consideration of cultural heritage issues as they relate to cultural rights, including those of indigenous peoples (see, e.g., Nie 1998; UNESCO 1998; WCCD 1998). Since 1998, activities relevant to indigenous peoples' rights have also been undertaken by the World Intellectual Property Organization (WIPO), the specialized UN agency in charge of promoting the development, revision, harmonization, and application of international norms and standards on intellectual property rights (IPR), as well as of administering a number of IPR-related treaties. WIPO created a Global Intellectual Property Issues Division with the task, among others, of identifying the "needs and expectations" of potential "new beneficiaries" of IPR. "Indigenous peoples, local communities and holders of traditional knowledge" were identified as the first group of such potential new beneficiaries of the existing IPR system. WIPO was also to contemplate the possible evolution of the IPR system to meet indigenous peoples' needs for the protection of their traditional knowledge (Castelo 1998).[1]

It is interesting to consider this emerging convergence of different bodies of international law (human rights, labor rights, environmental protection, cultural heritage protection, IPR) around issues concerning the rights of indigenous peoples, in light of the conviction repeatedly expressed by indigenous peoples and others that, ultimately, a sui generis system will have to be put in place for the protection of indigenous peoples' rights. From that point of view, one might suggest that such a system will have to be sui generis in more ways than one. On the one hand, it will have to be newly developed—perhaps drawing in part on existing international legal instruments but by and large created anew to meet indigenous peoples' needs. On the other hand, it will require an unprecedented level of collaboration among a variety of UN bodies, such as WGIP, ILO, the CBD Secretariat, UNESCO, and WIPO, and between these and indigenous peoples—and an equally unprecedented form of "cross-fertilization" and "hybridization" among the distinct legal frameworks within which each of these different bodies, as well as indigenous peoples, operate. In such a context and through such processes, a holistic approach to protecting indigenous peoples' heritage might begin to emerge.

It is also of interest to observe, however, that while this international legal debate has been strongly concerned with the protection of indigenous knowledge—particularly traditional ecological knowledge—and traditional resources, as well as folklore and other forms of traditional culture, one key element has been conspicuously absent: explicit recognition of the *role of language*—specifically, in this case, indigenous and other local languages—*in creating, encoding, sustaining, and transmitting most of the cultural knowledge and patterns of behavior* that the related international instruments are intended to protect. This chapter argues that indigenous peoples' languages should be considered along with oral traditions and traditional knowledge as integral parts of indigenous peoples' heritage and that an integrated framework for the protection of these interrelated aspects of indigenous peoples' heritage should be sought. The chapter raises questions about what is implied in the process, with special reference to the application of the IPR framework to indigenous heritage, and offers some suggestions on how such integrated protection might be achieved.

It should be noted that the expression "indigenous heritage" used in this chapter builds on the terminology set forth by Erica-Irene Daes, chair of WGIP. Daes refers to the whole of "indigenous property" as "cultural heritage" (Daes 1997), including not only folklore, crafts, and indigenous knowledge but also biodiversity (Daes 1998). "Indigenous heritage" is understood here to include language as well. And because of the common use of the term "heritage" as equivalent to "legacy from the past," it is important to point out that, in the present context, the term should instead be understood as "living heritage" (cf. Posey this volume): creative, dynamic heritage that, while coming from the past, contains the seeds of future innovations. Or, as the Indigenous Peoples' Earth Charter (Kari-Oca Declaration of 1992) states in its Preamble: "We the indigenous peoples walk to the future in the footprints of our ancestors."

INDIGENOUS VIEWS OF PROPERTY RIGHTS

Many aspects of indigenous peoples' views of ownership and property rights have already been discussed in the literature (e.g., Greaves 1994; Brush and Stabinsky 1996; Posey 1996, this volume; Posey and Dutfield 1996; Daes 1997, 1998; Nieć 1998; Brush this volume). It may be useful, however, to summarize them here for the purposes of the following discussion (see Maffi 1998b for a more extensive overview). Since a great diversity of property-related indigenous concepts and institutions exists around the world, their features should not be taken as categorical, but rather as tentative generalizations.[2]

Among indigenous peoples, property, or better ownership, is mainly if not invariably communal—where "communal" may refer to collective entities of various sorts, from a whole community or tribe to a lineage or other kin group, to gender or age groups, professional guilds, and the like. In the indigenous framework, even those cases in which ownership may be individual involve both rights and obligations to the collectivity, its ancestors and descendants, as well as to the land and to higher spiritual powers. Although in numerous instances there may be known individual creators of given scientific or technological innovations, literary or artistic works or styles, creation is generally communal rather than individual, with individuals contributing to a collective creative process in close interaction with other group members, so that the creator is generally not identifiable (cf. Brush this volume). Even when there are known individual creators and/or owners, however, it is rare that individuals can consider themselves entirely free to dispose of their creation and/or property at their own will; here, too, collective obligations generally apply. While in numerous instances an indigenous scientific or technological idea may be identifiable as new and a literary or artistic work as original, perhaps in a majority of cases such scientific or technological innovations and literary or artistic creations are so much a part of a continuous collective creative tradition, that the moment and locus of emergence of the novel idea or original work cannot be identified, and the creation imperceptibly goes on to enrich, and be enriched by, the collective tradition.

Indigenous heritage is normally considered inalienable—inherent in the very nature and identity of the collectivity. Even when property is deemed alienable, collective obligations commonly apply. In traditional trade situations, people do not separate themselves from alienable property purely for material gain but in the context of complex interactions that are both economic and social in nature. Ownership rights are normally considered to be enjoyed in perpetuity, being handed down from one generation to the next. In some cases, and especially as concerns land and the local environment, ownership may even not be thought to reside in the human collectivity itself but in the Creator or other spiritual entities (cf. Posey 1999). In such instances, indigenous peoples may think of themselves not so much as owners but as guardians and stewards.

As for the source of innovation, in the indigenous framework it may or may not be considered to be the human mind: it may be thought to be the human spirit, or some other spiritual force; the creation itself may thus be seen as spiritual rather than intellectual in nature. Heritage may in some instances be considered sacred, such as when it is related to religious or other spiritual beliefs (e.g., sacred sites, ritual objects, ceremonies, etc.). Under the same circumstances, or when pertaining to special categories of people (such as shamans, healers, or religious leaders), traditional knowledge may be considered secret. Even common knowledge, however, does not fit well with Western notions of a "free-for-all" public domain, because of the existence of networks of mutual and communal obligations.

Furthermore, as is also the case with indigenous views of biodiversity (Posey 1999, this volume), there is a tendency among indigenous peoples toward a holistic, non-individualizing approach to the cultural as well as the natural world: a tendency to think not just in terms of parts or components but in terms of a whole and of the relationships among the elements of the whole—in other words, to think ecologically in both nature and culture. Indeed, the very distinction between "culture" and "nature" appears to be of little significance to indigenous peoples the world over. Thus, indigenous heritage has a dual nature, both cultural and natural (see Daes's definition of indigenous heritage above), as well as an inherent spiritual dimension (Posey 1999).

Concepts such as "indigenous peoples' heritage rights" (Daes 1997) or "traditional resource rights" (Posey 1996; Posey and Dutfield 1996) as bundles of rights—where "rights" should be understood to involve "traditional obligations" as well—are more suited to characteristic indigenous notions of ownership (cf. Nieć 1998). In this connection, Daes (1997:4) has suggested that not to take into account the specific features of indigenous concepts of ownership when considering the protection of indigenous peoples' heritage "would have the same effect on their identities, as the individualization of land ownership, in many countries, has had on their territories—that is, fragmentation into pieces, and the sale of the pieces, until nothing remains."

At the same time, there is little doubt that for indigenous peoples to engage in explicit discussion of the protection of their heritage rights—both among themselves and with international agencies—represents and brings about profound change, a new kind of development, in their ways of life. But then this process clearly is, in turn, the consequence of sweeping historical changes that have long been, and increasingly are, affecting indigenous communities around the world through contact with the West—from the ensuing sociocultural, political, and economic pressures exerted on their societies to the ongoing encroachment on, and despoiling of, their heritage. On the one hand, if a framework is not established for the protection of indigenous peoples' heritage, abuse can continue unchecked. On the other, the very establishment of such a framework, necessitated by exogenous circumstances, implies and carries with it exogenous change.

This raises complex issues for indigenous peoples, ones that almost assuredly will not have a single, universal answer for all groups everywhere and under all circumstances. It also raises difficult questions vis-à-vis the development of mechanisms that will tackle violations of their heritage rights both effectively and in a culturally appropriate manner, by limiting as much as possible any adverse consequences for the integrity of that heritage that may derive from the application of an exogenous framework. Some thinkers have even argued that the whole debate should be shifted away from this legal framework toward an ethic of cooperation (Brush this volume). Nevertheless, if international standards are to be established for the protection of indigenous heritage, they should reflect the general principles emerging from indigenous views, while incorporating the flexibility that will allow for appropriate protection in each specific case.

PROTECTION OF INDIGENOUS KNOWLEDGE, ORAL TRADITIONS, AND LANGUAGES

With these quandaries and caveats in mind, we can consider the specific issues of the links between indigenous knowledge, oral traditions, and languages and of a holistic framework for the protection of these interrelated aspects of indigenous peoples' heritage. Let us begin with an examination of the current state of protection of indigenous languages and linguistic human rights.

International Organizations and Language Issues

ILO Convention 169 and the Draft UN Declaration on the Rights of Indigenous Peoples do refer to indigenous languages as a part of the indigenous heritage that bears protecting, in the context of affirming indigenous peoples' right to continue to use and develop their own languages and to receive education in the mother tongue (see Skutnabb-Kangas 2000, this volume). When it comes to protection of their knowledge and folklore, however, mention of language disappears, obscuring the fundamentally language-based nature of knowledge and folklore.

Nor has there been, so far, a sustained effort by any UN body to give exhaustive treatment to the topic of indigenous languages and linguistic human rights, in spite of the fact that the fundamental importance of language rights for indigenous peoples was already affirmed as far back as 1981 at a UNESCO conference on ethnocide held in San José, Costa Rica, and that the Declaration of San José then advocated drafting a declaration of linguistic rights (Zinsser 1994:46). A Draft Universal Declaration of Linguistic Rights, a nongovernmental initiative, was handed over to UNESCO in 1996, as the first—if still imperfect—attempt at a comprehensive universal codification of linguistic human rights, both individual and collective (see Skutnabb-Kangas 2000, this volume). But this draft was not considered favorably by a majority of UNESCO member states, and so far there has been no official follow-up on it, although UNESCO has issued several statements to the effect that linguistic rights are high on its agenda, while renewed NGO efforts are underway to promote the declaration.

There are signs that this situation is beginning to change. At its sixteenth annual session (1998), WGIP had as its principal theme Indigenous Peoples: Education and Language, and the Working Group heard many statements by indigenous representatives stressing the central importance of their languages to them—not only to their education but indeed to their own identity and livelihood, to the continued development of their knowledge systems and cultural traditions, and to their relationship to land. Many indigenous representatives stated that, in their peoples' worldviews, language and land are intrinsically related.[3]

As for UNESCO, under whose international mandate language issues fall, until recently it devoted attention to language endangerment mostly through its long-time affiliated NGO, the International Council for Philosophy and Humanistic Studies, a coalition of learned societies and national academies. The council sponsored the publication of the *Atlas of the World's Languages in Danger of Disappearing* (Wurm 1996) and the compilation of the Red Books on Endangered Languages, modeled on the Red Books on Endangered Species. It has also been offering small grants for the study and documentation of endangered languages. This work has been carried out largely at an academic level, without direct links to indigenous peoples and issues of indigenous peoples' rights. UNESCO's Division of Languages, created in 1998 within the Education Sector, may be moving toward a more proactive role vis-à-vis indigenous and minority languages—in education as well as other contexts—and vis-à-vis linguistic diversity and linguistic rights generally. In turn, UNESCO's Culture Sector has on its agenda the protection of cultural rights, including linguistic rights, as human rights. In its 1995 report *Our Creative Diversity,* UNESCO's World Commission on Culture and Development (WCCD) makes a significant point: that cultural rights are *collective* rights, rights that refer to the freedom of "a group of people to follow or adopt a way of life of their choice" (WCCD 1998:25), a freedom that, "by protecting alternative ways of living encourages creativity, experimentation and diversity, the very essentials of human development" (WCCD 1998:26; see also Nieć 1998).

Matters concerning indigenous languages are also gaining some recognition in international instruments and processes relevant to the protection of biodiversity. The Code of Ethics of the International Society of Ethnobiology[4] affirms that "culture and language are intrinsically connected to land and territory, and cultural and linguistic diversity are inextricably linked to biological diversity. Therefore, the right of Indigenous Peoples to the preservation and continued development of their cultures and languages and to the control of their lands, territories and traditional resources are key to the perpetuation of all forms of diversity on Earth." Similarly, the Final Statement of the 1998 forum Biodiversity: Treasures in the World's Forests,[5] which was handed over to the Chair of the Subsidiary Body for Scientific, Technical, and Technological Advice to the CBD, affirmed that "Protecting forests and biodiversity requires protecting the cultural *and linguistic* diversity of indigenous peoples" (emphasis added). The connections between biodiversity and linguistic and cultural diversity have also been brought to the attention of conservation organizations such as the World Conservation Union (IUCN), World Wide Fund for Nature (WWF), and UN Environment Programme (e.g., McNeely 1997; *Nation* 1999; Posey 1999; Maffi, Oviedo, and Larsen 2000).

The Urgent Need for Action

These are all positive signs. Yet they are still vastly inadequate to face the urgent task of supporting indigenous peoples' efforts to maintain, revitalize, and continue to develop their languages in the face of constant pressures toward acculturation and assimilation—and, in many cases, of linguistic genocide proper, as defined in Article 3.1. of the 1948 UN International Convention for the Prevention and Punishment of the Crime of Genocide (see Skutnabb-Kangas this volume).

An appreciation of the nature, scope, and full import of the impending language extinction crisis (Krauss 1992)—that is, an appreciation of the dramatic loss of both existing knowledge and creative potential that the loss of indigenous languages implies for their speakers and humanity at large—has not yet reached the general public the way biodiversity loss has, although the two phenomena are intimately related in their causes and consequences (Maffi 1998a; Maffi, Skutnabb-Kangas, and Andrianarivo 1999; also see the introduction to this volume). By and large, both the general public and governments worldwide have yet to view cultural and linguistic diversity not merely as the "spice of life"—if not actually as an inconvenience or even a threat—but as an inestimable treasure and asset, just as biodiversity is. Lack of awareness or limited understanding of the language extinction crisis and its implications are also responsible for the fact that, as yet, nowhere in the world has there been an adequate mobilization of funds to help stem the loss of linguistic diversity, whether in governments, academic institutions, NGOs, international organizations, or the public and private funding sectors.

Language as Heritage

A sea change is needed in this connection, and perhaps the best chances to see it happen may reside in the currently emerging confluence between processes concerned with the development of basic human rights instruments for indigenous peoples and processes aiming at the protection of indigenous peoples' cultural heritage. On the one hand, the right to learn, use, and fully develop one's own language or languages in all contexts is undoubtedly a basic human right (as recognized in the Draft UN Declaration on the Rights of Indigenous Peoples, Article 14). On the other hand, if it is legitimate to speak of traditional knowledge as a part of an indigenous people's heritage, then it is also unquestionably legitimate to speak of languages—as the repositories of that knowledge—in those same terms. UNESCO has actually used the word *heritage* in relation to language: for example, in the WCCD report *Our Creative Diversity,* in the context of highlighting linguistic diversity as a "precious asset of humanity" and of warning that "the disappearance of any language means an impoverishment of the reservoir of knowledge and tools for intra-cultural and inter-cultural communication" (WCCD 1998:179). Yet, when it comes to the preservation of this heritage, the report is mostly concerned with the "preparation of descriptions and adequate documents such as grammars, lexicons, texts, and recordings" (WCCD 1998:181), that is, with the fixing of languages for posterity. It does not deal with what measures should be taken to help speakers of threatened languages keep their languages alive as part of their own (as well as the world's) *living* heritage. (For a more proactive treatment of this topic by authors inspired by the WCCD report, see Nieć 1998.)

Viewing language as heritage clearly lends support to the inclusion of indigenous languages side by side with indigenous knowledge and oral traditions in the context of the international processes concerned with the protection of indigenous peoples' rights. Furthermore, if the concept of the "inextricable link" between linguistic, cultural, and biological diversity—between language, knowledge, and land—is fully understood and embraced, then it can be argued that to include language in the evolution of these international processes and instruments is not only something we *can* do, it is something we *need* and *ought to* do.

Language, Knowledge, and Oral Tradition

Admittedly, these issues are highly complex, but then not much more so for language than for traditional knowledge or oral traditions. Although each of these aspects of human culture has its own specific characteristics and thus requirements for its protection, they also share many features. All three form part of a people's intangible heritage, comprising, according to Prott (1998a:223), "creations of the mind" as well as "patterns of behaviour and knowledge" and "information (know-how)"—all of which, arguably, both include and presuppose language. It is essentially impossible to envisage the perpetuation and continued development of traditional knowledge and oral traditions (and thus of an essential source of our "creative diversity") without considering the need to foster local languages. It is also of interest to note, as Prott reports, that a recent study of UNESCO's use of the concepts of "heritage of mankind," "world heritage," and "common heritage" in international instruments found that "these concepts are by no means precise, although in course of development" (Prott 1998a:228). One might submit that these concepts can never become precise unless and until their relationship with, and implications for, the concept of *local* heritage are made clear. Prott seems to agree, at least insofar as she proposes that perhaps "not only the substance, but also the methodology, of norm-creation at the international level designed to protect the cultural heritage must be oriented to the wisdom of normative systems the world over" (Prott 1998a:233)—which systems, of course, also include those of the world's indigenous peoples.

Another major issue that is common to indigenous knowledge, oral traditions, and languages alike is that, to the extent that they are protectable by existing international IPR instruments, such as copyright, they are protectable as individual elements (e.g., "expressions" of folklore) or as compilations of such elements, not as whole traditions, and as products rather than processes. There are several problems with this state of affairs.

First and foremost, the individualizing tendency implied in breaking down heritage into elements is counter to indigenous peoples' holistic approach and hampers the maintenance of the relationships among these and other aspects of their social and cultural life, as well as the relationship of all these aspects with the land. One crucial question that needs to be addressed here is: How can these aspects of indigenous heritage be protected as *bodies*—bodies of knowledge, bodies of folklore, bodies of language—and as *living,* constantly developing bodies, not as dead bodies from the past? Another question implied in the previous one is: How do we protect knowledge, oral traditions, and languages as *processes* (in the sense of "process" as "continuously unfolding change")?

Concerning biodiversity, it has been convincingly argued that what needs to be preserved, and protected, is not species as such, but the processes of speciation that bring species into existence (which, of course, in turn presuppose species in order to unfold)—and that to protect these processes what needs to be preserved are ecological relations. This is what ensures the continuity and diversity of biological life on Earth. Exactly the same argument has been made for cultural and linguistic diversity (Harmon 1996; Mühlhäusler 1996). It is essential to note that whatever amount of protection IPR systems may provide for indigenous heritage, it is protection against misappropriation, abuse, and misuse of this heritage—most definitely a just cause but not one that will per se protect heritage from *erosion.* Protection from erosion must then be the ultimate goal, one that requires mechanisms that will acknowledge and foster the integrity and dynamism of this heritage.

Second, insofar as traditional knowledge, oral traditions, and languages are documented and compiled, what kind of IPR protection can these compilations get? Current copyright laws provide that compilations of data (databases) are protected as long as they "by reason of the *selection or arrangement of their contents* constitute intellectual creations" and that "this protection *does not extend to the data or the material itself*" (Article 5 of the WIPO Copyright Treaty [WCT] of 1996; emphasis added). In other words, "copyright law protects only the *form of expression* of ideas, *not the ideas themselves*" (WIPO 1998b; emphasis added). Ideas as such can only be protected by IPR if they are inventions, under patent law, which requires that the invention be demonstrably new and nonobvious, as well as industrially applicable, and that the inventor be identified (cf. WIPO 1998a)—part or all of which, as we have seen, may not square with the process of formation and utilization of ideas in an indigenous context. Therefore, compiling indigenous knowledge, oral traditions, or languages under copyright law does not protect the *substance* of the compiled knowledge, oral tradition, or language. It only protects the *form of the compilation,* while putting the substance in the public domain, with nothing to prevent use of the substance by others.

The latter remark might raise objections. For is it not the free flow of ideas that is the very basis of the construction of knowledge and of the development of science and the arts? This is admittedly a very delicate issue, and one whose intricacies cannot be fully tackled here. It is possible, however, to point to at least some of its key aspects. While there is no doubt that in every society on Earth ideas have a way of flowing freely from one person to the next—which in every society is the source of all manner of innovation—this is not to say that each society will have developed the same concept of a "free-for-all" public domain of the kind that is largely taken for granted in Western societies.[6] In many, perhaps most indigenous societies, the notion that someone may feel free to take from a public-domain pool of ideas without feeling much of an obligation toward the originator of the idea—let alone the need to reciprocate—may seem as utterly foreign as the alternative concept may appear to people steeped in the Western public domain tradition.

As far as principles go, the fact that in the case of indigenous peoples the originator of an idea may be a collectivity rather than an individual does not change the picture. On the contrary, it explains how there may arise among these peoples an expectation of acknowledgment and reciprocation to the group as a whole. Furthermore, if we consider that the collectivities in question are ones that, after being despoiled of their freedoms and their lands, now see themselves increasingly despoiled of their identities and resources (as contained in their natural and cultural heritage), it should become understandable that they may be wary of the notion of "free-for-all." Or, as Lars-Anders Baer, a Sámi, puts it: "While some of these issues are considered 'secondary' rights for most sectors of civil society, for Indigenous Peoples and other traditional societies, the survival of their cultures, heritage, languages and environment determine the survival of their very existence and identity, and are as 'primary' as the right to life" (Baer 1998:12).

In the international processes in which they are increasingly making their voices heard, indigenous peoples have made it clear that they are more than willing to share their heritage with the rest of the world, subject to the principle of self-determination and thus to their right to decide whether, how, when, with whom, and what to share, and on condition that there be appropriate recognition and reciprocation of their contribution, whether in the form of benefit sharing or otherwise. This requirement might sound akin to the exclusive rights granted by patents, or to the features of trade secrets, and therefore might be thought of as related to the application of an IPR framework. In some cases, indigenous peoples may indeed seek to resort (and have actually resorted) to patents or trade secrets to protect their heritage. But the rights discussed above stem from and should be interpreted within the framework of fundamental (civil, political, social, cultural, economic) human rights in which the principle of self-determination is couched,[7] as well as of the emerging sui generis framework for indigenous peoples' rights. It makes a profound difference whether the community itself (or a designated subset or individual member of the community) is the compiler or co-compiler of any such database, or the database is the exclusive work of an external researcher. Indigenous peoples are advocating the fundamental right to retain control over these processes.

It has also been suggested that researchers working with indigenous peoples may need to consider extending the application of a principle originally developed to address issues concerning potential environmental and public health hazards: the Precautionary Principle (Bannister and Barrett forthcoming). This principle states that, where there are threats of serious or irreversible damage, lack of full scientific certainty should not be used as a reason for postponing measures to prevent such damage. The proposed extension suggests that, if there is reason to anticipate that publication of given data may result in harm for the people who were the source of the data (where "harm" should be defined by these people themselves),[8] either the research should not be carried out or its results should not be disseminated through the usual scientific channels (Bannister and Barrett forthcoming).

A third fundamental issue bearing on the protection of indigenous heritage by means of IPR is what happens to aspects of a people's knowledge, oral traditions, or language once they are compiled—that is, what happens in addition to how these aspects of heritage may be used by others. What are the consequences of compilation for the people themselves? Many scholars concerned with the preservation of indigenous heritage see compilation not only as a necessity in order to fix the heritage in tangible form lest it disappear, but also as a way of legitimating and building pride in that heritage among its owners. There is no doubt that compilation can serve all these purposes, and indigenous peoples themselves often see it this way, and devote themselves to it as well.

Not everyone among them agrees, however, for a variety of valid reasons. Some are concerned that, by compiling their heritage or by letting it be compiled, indigenous peoples may actually make that heritage more vulnerable to abuse, in part because protection of a compilation under copyright does not ensure protection of the ideas. Others, although they may not be opposed to compilation in general, worry that aspects of heritage may be compiled that should not be, such as sacred, secret, or otherwise restricted aspects of heritage, which might cause harm to parts of the community if divulged. Others are concerned that compilation may shift authority about their people's heritage from its traditional living locus in community elders and other experts to its new, impersonal, exogenous locus in a book, audiovisual recording, or multimedia database. Moreover, they fear that heritage not so compiled may become delegitimized, thus contributing to the breakdown of traditional social structures and means of heritage transmission.

Furthermore, it is worth noting that, while the Provisions of the Berne Convention for the Protection of Literary and Artistic Works (1971), referred to in the WCT, grant protection to every literary, scientific, and artistic production "whatever may be the mode or form of its expression" (Article 2.1.), the provisions then leave it up to member states to legislate on whether "works . . . shall not be protected unless they have been fixed in some material form" (Article 2.2.). Although one might expect that a compilation of intangible heritage will almost inevitably be something fixed in tangible form, this clause in the provisions allows for the possibility that intangible heritage that is not thus fixed may get no protection at all. And we should keep in mind that, even where this restriction may not apply, only the form of expression, not the substance, of the heritage is protected by copyright laws.

The Right to Orality

The complexity of these matters is such that only case-by-case examination—within the framework of an evolved, hybrid, integrated, flexible system for the protection of indigenous peoples' rights—may lead to appropriate solutions. When considering these points, however, it becomes apparent that one fundamental aspect of traditional cultures is being neglected in the international debates over indigenous peoples' rights, that is, orality. The shift from the oral to the written, from the intangible to the tangible, has already had and continues to have a profound impact on indigenous peoples in terms of its consequences on the nature of their cultural traditions and the integrity of their relationships with the land. For instance, in relation to the indigenous languages of Colombia, Seifart (1998:9) states: "it has become clear that linguistic analysis, alphabetization and finally bilingual education entail radical interference with originally illiterate indigenous cultures. The alphabetization of the language and the recording of traditional texts change the cultural dynamics of the oral tradition: from the moment these texts are recorded they lose their variable character. Therefore many cultures decide to take this step only after much deliberation."

As long as a tradition of orality still exists in indigenous societies, it bears protecting in and of itself, as a fundamental part of protecting the world's cultural diversity—and, through the intimate link between oral language and the transmission of knowledge (including traditional ecological knowledge), as a fundamental part of protecting biological diversity too. Among the rights to be accorded protection for indigenous peoples, therefore, *positive* recognition should be given to the *right to orality*—the right for indigenous peoples to maintain, transmit, develop their heritage, or parts of it, orally if they so wish, without any form of compilation or other fixation, and still have it protected. To leave it undetermined whether (thus making it highly likely that) heritage should be fixed in one form or another in order to receive protection—limited as this protection may currently be in regard to substance—amounts to an exogenous imposition on traditional cultures. It also amounts to a remarkable paradox, according to which, by and large, intangible, dynamic heritage can be protected only by being made into what it is not: that is, tangible and fixed.

To advocate the right to orality is most assuredly not to question the right of indigenous peoples to literacy, formal education, and each and every nonindigenous means and media of communication they may wish to have access to. To be sure, this is as fundamental a right as the right to orality being proposed here. Nor is it to suggest that indigenous peoples should never compile their heritage or have it compiled. Compilation presents its problems, as discussed above, but it can be and is a powerful tool, if built and used properly, and if it can be accorded adequate protection. The point is rather to suggest that the pervasiveness of literacy, writing, and all other forms of tangible fixation of ideas in the Western world should not blind us to the possibility that the very tissue that holds indigenous cultures together and the thread of their intimate symbiosis with the environment may be the *spoken* word. Listening to a traditional storyteller may be enough to convince one of this.[9] Yet, as the international processes concerned with the protection of indigenous peoples' rights unfold, there is a risk that the Western world's bias toward the written word and tangible expressions of reality will lead to overlooking and neglecting, and ultimately contribute to effacing, this fundamental, perhaps constitutive aspect of traditional cultures.

One can thus argue that, if what we care about is how to protect the integrity and continuity of indigenous peoples' heritage, then we should also put our minds to the protection of the right to orality—which, furthermore, goes back to the earlier question of how to protect heritage as "body" and "process." If what we care about is the perpetuation and continued development of indigenous heritage, we should be asking ourselves whether and to what extent fixation may help or hinder the process, may favor or slow indigenous innovation. In this connection, it is noteworthy that even in strongly patent-oriented countries such as the United States concern is beginning to be expressed about the possibility that patenting may be going overboard, stifling rather than promoting productive innovation (Downes 1997:9). More generally, questions about the actual pros and cons of global market forces vis-à-vis fostering creative development are now emerging (Prott 1998a:229).

Orality and Neighboring Rights

If nothing else, to think of orality should give one pause. But beyond that, it might be argued that the right to orality should be recognized as a basic right, even in education, as a part of the informal ways of teaching and learning—the right to whose perpetuation indigenous peoples advocate, while also affirming their right to formal education (e.g., Kari-Oca Declaration, Article 25; see Fettes 1999). It remains to be seen which legal framework might best protect this right or whether it is indeed appropriate to think of giving it protection in legal form. In this connection, it may be interesting to consider whether such protection might, at least in part, be offered through an extension and modification of the copyright-related concept of "neighboring rights."

Neighboring rights were established to protect, among others, the category of performers, in recognition of performers' creative contribution in bringing to life certain works of art and literature (music, dance, drama, etc.). The proposal has been made that these rights might thus be applied to "the largely unwritten and unrecorded cultural expression . . . generally known as folklore . . . since it is often through the intervention of performers that they are communicated to the public"

(WIPO 1998b:14). If appropriately informed by an anthropological understanding of how oral tradition works, such an application might perhaps begin to give recognition to orality in folklore and to get us beyond the obstacle that, in order to be protected, folklore should be compiled and fixed in tangible form. This application might also begin to give recognition and protection to indigenous performers as custodians and developers of their oral traditions. Anthropological understanding would require recognizing that in this context "performers" may have to be seen at once as the "creators," "renovators," and "innovators," not just as the "re-enactors," of these traditions, and that in many cases "performer" may refer not to a single individual, but instead to a whole community or subset of a community (e.g., elders, women, specialized guilds, etc.). It would also require acknowledging that, in both origination and ownership of the oral tradition, communal rights (and obligations) would in most cases have to be recognized.

In addition, it would be worthwhile to explore the possibility of extending the notion of neighboring rights to embrace other aspects of indigenous heritage, such as knowledge and language. Could indigenous peoples be recognized as the collective "performers" (as well as owners) of their own cultural traditions, including knowledge and language—again in the anthropological sense of ongoing collective creation and recreation of culture through a people's daily activities and through the interactions of community members with other people and with the local environment? This may sound far-fetched, yet perhaps less so if we remind ourselves what was included in Prott's definition of "intangible cultural heritage" mentioned above: "creations of the mind," "patterns of behaviour and knowledge," and "information" (which, as previously noted, at one and the same time include and presuppose language). It would be anthropologically accurate to argue that all aspects of this intangible cultural heritage emerge and develop through the daily performance of a people's social and cultural life. If this anthropological sense of "performance" and "performer" could find its way into the legal thinking related to neighboring rights, perhaps we might begin to see the emergence of a new framework for the protection not only of folklore but of knowledge and language as well.

Clearly, some careful rethinking of the concept of neighboring rights would be necessary if its applicability to this domain were under consideration. And the question of how to protect not just form of expression but also substance would have to be addressed.

The Need for a General Framework

Other possible avenues for the protection of indigenous knowledge and folklore have been suggested in the literature (including conceptualizing aspects of indigenous knowledge in terms of scientific discoveries, patentable technology, know-how, "prior art," or trade secrets, applying "domain public payant" provisions to the for-profit use of expressions of folklore, etc.). These and other proposed solutions cannot be discussed here except to note once again that in most such cases the traditional knowledge and oral traditions involved imply and depend on the use of indigenous languages. (For details, see Greaves 1994; Brush and Stabinsky 1996; Posey and Dutfield 1996; Daes 1997; Downes 1997; Posey 1999, this volume; Brush this volume.) All of these existing mechanisms, however, will require extension and modification if they are to be usefully employed for the protection of indigenous knowledge. Furthermore, it may be argued that, to the extent that they offer protection to indigenous peoples, in and of themselves they do so in a piecemeal and potentially inconsistent manner. What is crucially needed is a general framework within which specific provisions can be coherently coordinated. That general framework is still missing, although as previously suggested it may be in the process of emerging at the international level.

The proposal made above concerning the possible evolution of neighboring rights is one attempt to contribute to the elaboration of such a broader framework, at least as concerns the *intangible* aspects of indigenous heritage, such as knowledge, oral traditions, and languages. But what about the *tangible* aspects— from indigenous crafts to biodiversity? While no attempt can be made here to tackle this other major issue, it is at least possible to argue that crafts, biodiversity, and all other aspects of the tangible indigenous heritage are so inextricably bound to indigenous knowledge and to the creation and transmission of that knowledge through language that a common framework for the protection of all aspects of both tangible and intangible indigenous heritage should be found. That holistic framework will have to form the basis of a new sui generis system for the protection of indigenous peoples' rights.

TOWARD MORALLY DEFINED RIGHTS (AND DUTIES)?

It seems likely that much of that common framework will come from human rights. It might also derive from an in-depth reconsideration and reformulation of IPR concepts, as previously discussed in relation to neighboring rights. In addition, perhaps the IPR system might contribute to rethinking, developing and strengthening the concept of "moral rights"—currently limited to an author's right to claim authorship of his or her work and to be protected against "any distortion, mutilation, or other modification of, or other derogatory action in relation to, the work which would be prejudicial to the author's honor or reputation", the latter known as "right of integrity" (WIPO 1998b:8). Unlike economic rights, moral rights can never be transferred, that is, they are held by the author in perpetuity. They are also recognized in international human rights law, beginning with the Universal Declaration of Human Rights (Article 27.2). Much elaboration and evolution would be necessary if this concept were to be adapted to the needs of indigenous peoples, given the specific characteristics of their heritage, but this would be a worthwhile exercise. In particular, it would be interesting to consider whether and how the concept of "right to integrity" might be extended to refer to the right to integrity of an indigenous people's heritage as a whole—as both body and process.

Some have argued that "a focus on morally defined rights [for the protection of indigenous peoples" heritage] will not be successful, because it is too difficult to build arguments to bridge the wide gap between general human rights principles, and the specific details of existing or proposed IPR systems" (Downes 1997:7). Perhaps this skepticism is unjustified. There are signs that international processes may be opening up to the prospect of morally defined indigenous heritage rights, according to a global ethics built out of respect for cultural diversity and distinct world views. On the issue of how best to protect traditional knowledge, Prott (1998a) argues that, while the discussion has been mostly couched within the Western discourse on "rights," it is worth asking ourselves whether thinking in terms of "rights" may actually be our best or only option. She observes: "Many traditional communities think in terms of the *obligations* of the human community to earth and the other species on it. Perhaps the idea may be the beginning of a new normative framework for the preservation of heritage" (Prott 1998a:233; emphasis added). As a concrete example, Prott (1998b) points to the African Charter on Human and Peoples' Rights (1986),[10] whose first chapter is formulated in terms of "rights" and whose second is formulated in terms of "duties."

In a declaration issued in 1998,[11] the Scientific, Technical and Research Commission of the Organization of African Unity in charge of drafting its model Legislation on Access to Biological Resources and Community Rights for African States affirmed: "Natural resources and indigenous knowledge and technologies are a legacy that humanity owes to local communities." The expression "legacy that humanity owes to local communities" may be interpreted in two ways. One is that this legacy (of which indigenous languages should be understood to be a part) has come and continues to come to all of humanity from local communities (including indigenous communities). The other is that humanity owes it to local communities to support their efforts for the perpetuation and continued development of this invaluable legacy. There is little doubt that these issues pose extraordinary challenges for international law. Yet they are challenges of our times and in urgent need of being met. Innovative legal thinkers, drawing from both Western and indigenous legal traditions, can accomplish this if they seriously put their minds to it—and they have a chance to make history for all of humanity in the twenty-first century with the solutions they will devise.

NOTES

1. One of the Global Intellectual Property Issues Division's first initiatives in this context was the Roundtable on Intellectual Property and Indigenous Peoples in Geneva, Switzerland, 23–24 July 1998), WIPO's first meeting ever held with indigenous peoples and supporting nongovernmental organizations (NGOs). I was one of the invited participants in the Roundtable, representing the NGO Terralingua: Partnerships for Linguistic and Biological Diversity. The present chapter draws in part on a paper I submitted to WIPO as a follow-up to the Roundtable (Maffi 1998b). Thanks are due to David Harmon and Darrell Posey for commenting on a draft of that paper, and to Richard Owens and Shakeel Bhatti of WIPO for comments on an earlier version of this chapter. I have greatly benefited from these readers' suggestions, while retaining full responsibility for my statements herein. It should be noted that in 1999 WIPO held a second such roundtable—this time, however, titled Roundtable on Intellectual Property and Traditional Knowledge, indicating a move away from the politically charged notion of "indigenous peoples" toward the less contentious one of "holders of traditional knowledge" as IPR beneficiaries. It appears that some member states of WIPO want to limit the extent to which the special needs of indigenous peoples for the protection of their traditional knowledge will be taken into account by means of an evolution of IPR systems. Rather, they aim to restrict the scope of such protection to a purely economic context, especially in relation to issues relevant to the World Trade Organization (WTO) and the Trade-Related Aspects of Intellectual Property Rights agreement (TRIPS).

2. For some basic notions about the existing IPR system, in its two main guises of industrial property and copyright, see WIPO (1998a, b).

3. These remarks are based on my participation in the proceedings of the Sixteenth WGIP, in Geneva, 27–31 July 1998) as Terralingua representative. Terralingua presented an official submission on linguistic human rights in education (UN document E/CN.4/Sub.2/AC.4/1998/2), whose main points I summarized in a statement delivered during the hearings on the "education and language" item on the WGIP agenda.

4. As approved by the International Society Ethnobiology membership at the Sixth International Congress of Ethnobiology, Whakatane, Aotearoa/New Zealand, 23–28 November 1998.

5. Held in Schneverdingen, Germany, 3–7 July 1998, as a part of the Forests in Focus process leading up to the World Expo 2000 in Hannover. As an invited participant in the forum, I had the task of discussing the connections between linguistic and biological diversity. Some of these points were reflected in the Final Statement.

6. According to this concept, the "public domain" is essentially the space containing those ideas and information that are not otherwise protected by legal mechanisms such as copyright, patents, trade secrets, and the like (the latter defining the "private domain"). In the public domain, access to ideas and information is seen as free for all, not subject to other obligations (perhaps beyond the appropriate form of citation or acknowledgment). In traditional societies there exist customary laws and practices

that protect knowledge and thus configure a private domain distinct from the public domain. At the same time, in such societies even the public domain may not be conceived of as a space devoid of mutual social and moral obligations as the Western notion of "free-for-all" suggests. (I am grateful to Graham Dutfield, personal communication July 1999, for help in clarifying these concepts. See Dutfield 2000.)

7. As laid out in the UN International Covenant on Civil and Political Rights and International Covenant on Economic, Social, and Cultural Rights, both of 1966. See Posey this volume. For a model treatment of these issues as concerns ethnobiological knowledge, see the International Society of Ethnobiology's Code of Ethics in Appendix 1 of this volume.

8. Leaving it up to the people who are the source of the data to define possible "harm" does not necessarily imply any easy resolution of the matter. Even within small-scale communities people may disagree with one another on such issues. The point is that such negotiations with the "source" people should be conducted and all good-will efforts be made to find agreeable solutions and comply with local people's requests.

9. For a masterful analysis of orality among indigenous peoples, see Abram 1997. See also Sheridan 1991.

10. Concluded at Banjul in 1981, adopted in Nairobi the same year, and in force since 1986.

11. Following a meeting held in Addis Ababa, Ethiopia on 20–23 March 1998.

REFERENCES

Abram, D. 1997. *The Spell of the Sensuous.* New York: Vintage Books.

Baer, L.-A. 1998. Indigenous Peoples' cultural rights in the context of the Stockholm Conference. Report on the UNESCO Intergovernmental Conference on Cultural Policies for Development, Stockholm, 30 March–2 April 1998.

Bannister, K., and K. Barrett. Forthcoming. Weighing the proverbial "ounce of prevention" versus the "pound of cure" in a biocultural context: A role for the Precautionary Principle in ethnobotanical research. In *Ethnobotany and Conservation of Biocultural Diversity,* ed. L. Maffi and T. Carlson. Submitted to Advances in Economic Botany. The Bronx: New York Botanical Garden Press.

Brush, S., and D. Stabinsky, eds. 1996. *Valuing Local Knowledge: Indigenous People and Intellectual Property Rights.* Washington, D.C.: Island Press.

Castelo, R. 1998. Opening address, Roundtable on Intellectual Property and Indigenous Peoples, World Intellectual Property Organization, Geneva, Switzerland, 23–24 July 1998.

Daes, E.-I. 1997. *Protection of the Heritage of Indigenous People.* Study Series 10, Human Rights. New York and Geneva: United Nations.

Daes, E.-I. 1998. Opening address, Roundtable on Intellectual Property and Indigenous Peoples, World Intellectual Property Organization, Geneva, Switzerland, 23–24 July 1998.

Downes, D. 1997. Using intellectual property as a tool to protect traditional knowledge: Recommendations for next steps. CIEL discussion paper prepared for the CBD Workshop on Traditional Knowledge, Madrid, November 1997.

Dutfield, G. 2000. The public and private domains: Intellectual property rights in traditional knowledge. *Science Communication* 21(3):274–295.

Fettes, M. 1999. Indigenous education and the ecology of community. In *Indigenous Community-Based Education*, ed. by S. May. Pp. 20–41. Clevedon: Multilingual Matters.

Greaves, T., ed. 1994. *Intellectual Property Rights for Indigenous Peoples: A Sourcebook.* Oklahoma City: Society for Applied Anthropology.

Harmon, D. 1996. Losing species, losing languages: Connections between biological and linguistic diversity. *Southwest Journal of Linguistics* 15:89–108.

Krauss, M. 1992. The world's languages in crisis. *Language* 68(1):4–10.

Maffi, L. 1998a. Language: A resource for nature. In *Nature and Resources, the UNESCO Journal on the Environment and Natural Resources Research* 34(4):12–21.

Maffi, L. 1998b. Indigenous languages and knowledge and intellectual property rights. Follow-up document on the Roundtable on Intellectual Property and Indigenous Peoples, World Intellectual Property Organization, Geneva, Switzerland, 23–24 July 1998.

Maffi, L., G. Oviedo, and P.L. Larsen. 2000. *Indigenous and Traditional Peoples of the World and Ecoregion Conservation: An Integrated Approach to Conserving the World's Biological and Cultural Diversity.* A WWF Report. Gland, Switzerland: WWF International.

Maffi, L., T. Skutnabb-Kangas, and J. Andrianarivo. 1999. Linguistic diversity. In *Cultural and Spiritual Values of Biodiversity*, ed. by D.A. Posey. Pp. 21–57. London and Nairobi: Intermediate Technology Publications and UN Environment Programme.

McNeely, J.A. 1997. Interaction between biological and cultural diversity. In *Indigenous Peoples, Environment, and Development*, ed. by S. Buchi et al. Pp. 173–196. IWGIA Document 85.

Mühlhäusler, P. 1996. *Linguistic Ecology: Language Change and Linguistic Imperialism in the Pacific Rim.* London: Routledge.

Nation. 1999. Environmental destruction a threat to languages: UN Environment Programme. 7 September 1999. Distributed via Africa News Online by Africa News Service.

Nieć, H. (ed.) 1998. *Cultural Rights and Wrongs.* Paris and Leicester: UNESCO Publishing and Institute of Art and Law.

Posey, D.A. 1996. *Traditional Resource Rights: International Instruments for Protection and Compensation for Indigenous Peoples and Local Communities.* Cambridge, UK: IUCN Publications Services.

Posey, D.A., ed. 1999. *Cultural and Spiritual Values of Biodiversity.* London and Nairobi: Intermediate Technology Publications and UN Environment Programme.

Posey, D.A., and G. Dutfield. 1996. *Beyond Intellectual Property: Toward Traditional Resource Rights for Indigenous Peoples and Local Communities.* Ottawa: International Development Research Centre.

Prott, L.V. 1998a. International standards for cultural heritage. In *World Culture Report: Culture, Creativity, and Markets.* Pp. 222–236. Paris: UNESCO Publishing.

Prott, L. 1998b. Understanding one another on cultural rights. In *Cultural Rights and Wrongs*, ed. by H. Nie. Pp. 161–175. Paris and Leicester: UNESCO Publishing and Institute of Art and Law.

Seifart, F. 1998. Situation of the indigenous languages of Colombia, especially Chimila. *Ogmios* 9:8–10.

Sheridan, J. 1991. The silence before drowning in alphabet soup. *Canadian Journal of Native Education* 18(1):23–31.

Skutnabb-Kangas, T. 2000. *Linguistic Genocide in Education—Or Worldwide Diversity And Human Rights?* Mahwah, N.J.: Lawrence Erlbaum.

Skutnabb-Kangas, T., and R. Phillipson (eds.). In collaboration with Mart Rannut. 1994. *Linguistic Human Rights: Overcoming Linguistic Discrimination.* Berlin and New York: Mouton de Gruyter.

UNESCO. 1998. *World Culture Report: Culture, Creativity, and Markets.* Paris: UNESCO Publishing.

WCCD. 1998 [1995]. *Our Creative Diversity.* Report of the World Commission on Culture and Development. Paperback edition. Paris and Leicester, UK: UNESCO Publishing and Oxford and IBH Publishing.

WIPO. 1998a. Main aspects of industrial property. Document prepared for the Roundtable on Intellectual Property and Indigenous Peoples, World Intellectual Property Organization, Geneva, Switzerland, 23–24 July 1998.

WIPO. 1998b. Basic notions of copyright and neighboring rights. Document prepared for the Roundtable on Intellectual Property and Indigenous Peoples, World Intellectual Property Organization, Geneva, Switzerland, 23–24 July 1998.

Wurm, S.A. 1996. *Atlas of the World's Languages in Danger of Disappearing.* Paris and Canberra: UNESCO Publishing and Pacific Linguistics.

Zinsser, J.P. 1994 *A New Partnership: Indigenous Peoples and the United Nations System.* Educational Studies and Documents 62. Paris: UNESCO Publishing.

26

A TAPE DOCUMENTATION PROJECT
FOR NATIVE BRAZILIAN LANGUAGES

Denny Moore

A project for the documentation of Brazilian indigenous languages by means of audio and video recordings is being organized by the Linguistics Division of the Museu Paraense Emílio Goeldi, a Brazilian federal research institute located in Belém, Pará, Brazil.[1] The first phase, equipment selection and purchase, is largely complete, and the second phase, testing of equipment and procedures, is under way.

Although the recording phase is not yet operational, it is perhaps worthwhile describing the goals of the project and the general and specific considerations and strategies that underlie it, since the lack of documentation it addresses is common to many other regions of the world. As far as we know, no project of similar form and scope has been previously attempted. The basic idea is to use technological advances in low-cost, high-quality portable recording equipment to preserve at least some knowledge of endangered linguistic diversity, for science and for the speakers themselves and their descendants.

THE STATE OF THE DOCUMENTATION OF BRAZILIAN
INDIGENOUS LANGUAGES

There are approximately 180 known indigenous languages in Brazil, although new ones are continually being discovered and there are still some indigenous groups without contact with national society. Moore and Storto (1992) estimate that less than 10 percent of the indigenous languages of Brazil have complete descriptions of good scientific quality and less than half have received some published description, usually of segmental phonology or details of grammar.

There are approximately 150,000 speakers of indigenous languages in Brazil. A few languages have numerous speakers; Makuxi, for example, has 15,000, but many have far fewer. There is one speaker of Umotina, one of Máku, and two of Xipaya. Moore and Storto (1992) calculate the situation of the twenty-five native languages of the state of Rondônia as:

- 10 percent are no longer in use;
- 30 percent have a low number of speakers and lack young speakers;
- 25 percent have either a low number of speakers or a lack of young speakers; and
- 35 percent have numerous speakers, including young ones.

The situation of languages in this state is probably somewhat worse than average in Brazil, since the indigenous groups in the south of the state have been in contact with national society for a number of decades. But all regions have languages in precarious circumstances.

Even when languages have been studied, the analysis is by no means always accurate. For example, both the Gavião and the Surui languages of Rondônia were thought for years not to be tonal, in spite of the presence of numerous minimal pairs for tone and the existence of whistled speech. The famous "long verbs" of the Karajá language were recently shown actually to be sequences of verbs, particles, and auxiliaries (Ribeiro 1996). The questionable state of existing analyses is another reason for wanting to record actual samples of languages, even if they have been described previously.

Indigenous groups who no longer have any active speakers of their native language, such as the Mura of the state of Amazonas, are eager to find any sample or record of it. Unfortunately, the chances of this are usually not good.

In spite of the desirability of tape documentation of indigenous languages in Brazil, there are very few individuals or institutions prepared to carry out such work. Even linguists who have worked for twenty years on a language frequently have little recorded material of quality. Many linguists still record using the internal omnidirectional microphone of monophonic cassette recorders without noise reduction. Import duties on electronic equipment make the acquisition and maintenance of semiprofessional or professional recording equipment difficult and expensive. There is little tradition of data collection for diachronic or areal studies—perspectives which are valuable for planning documentation content. A certain knowledge of the natural environment and a methodology and stimulus materials for eliciting names of plant and animal species is also desirable for documentation but is often lacking because linguistics is usually housed in departments of letters in Brazil. A passive tape archive, that is, an archive that accepts miscellaneous donations of taped materials, would be of little value, given this context.

One way to improve this situation is for one institution to develop a specialized capacity for tape documentation, meet the technical challenges that this involves, and then conduct an active taping campaign, in cooperation with other institutions. It was with this intention that the Linguistics Division of the Museu Goeldi began development of its documentation program.

OVERVIEW OF THE PROJECT

There will be two types of documentation taping: standardized and nonstandardized. The preliminary goal of the standardized documentation will be ten hours of taped material from each of the 180 languages and approximately twenty major dialects—a total of 2000 hours. The material will be taped in both digital audio (DAT) and in video (Hi8). The original plan was to focus on the DAT recording, with 10 percent of the material also recorded simultaneously in video to see the visual cues of speech production (lip rounding, interdental articulation, etc.), as a sort of calibration of the audio recording. However, most people viewing the video test recordings felt that the video version was so useful that it would be better, for the relatively small extra cost, to document all the standardized material in video as well as audio. The goal of ten hours is arbitrary and may be adjusted after more practical experience is gained. The idea is to tape sufficiently to get a substantial amount of useful information, but not to tape so much that the operation becomes too cumbersome and time consuming and threatens wide coverage of a large number of languages.

The standardized taping will follow a protocol. This will, in general, be administered in Portuguese, since it is not possible to learn enough of the indigenous languages in a short time to use them for this purpose. Most of the data collected will be linguistic, but questions will be included about the speakers, their community, their culture, and the history and current situation of their group. Since the taping will follow a protocol, it does not need to be done by a professional linguist. Students or other trained workers can carry out the recording. Tapes can be inspected for quality control. For administrative, logistical, and time considerations, it is not possible to conduct the recording of so many indigenous languages in the villages of the native speakers. Fortunately, there are regional centers of the FUNAI (Brazil's National Indian Foundation) where Indians go with a certain frequency for health, economic, or administrative purposes. It is generally possible to locate native speakers near these centers. These may not necessarily be the most apt members of the community to act as linguistic informants, but taping is less demanding than analytical work, which requires producing repetitions, formulating grammaticality judgments, and having a certain grasp of analytic objectives. Still, it is important to choose those speakers who are more knowledgeable, and this requires experience. Speakers who are essentially monolingual are problematical, since in these cases it is difficult to follow a protocol in Portuguese. The general strategy will be to make a once-over pass of each region, collecting what is possible, and then, when practicable, to return for what was missed.

Originals of tapes will be catalogued and stored away from humidity and magnetic fields. Duplicates will be stored in a separate location. Inevitably, there will be some deterioration of the tapes. One solution being contemplated is digitalizing the contents of the tapes and storing them as computer files. This becomes increasingly attractive as mass storage capacity becomes ever cheaper and compression algorithms for video become more effective. (The large amount of information on video tapes means that their digitalized form occupies a lot of computer storage space.) It may be possible to index the taped data when it is in the form of computer files, so that specific information could be recovered easily by instructions to the computer. Laser storage media, such as compact disks (CDs) or their larger capacity (over four gigabytes) successors, the digital video disks (DVDs), are more resistant to deterioration.

An obvious problem is the transcription and utilization of the massive amount of taped data to be collected. Aside from the native speech community itself, the principal users of the taped data would probably be researchers (1) interested in other languages genetically related to the one they study, (2) studying a particular area or region, (3) planning to work intensively with the language in question, or (4) interested in a particular phenomenon (for example, tone or nasalization spread) known or suspected to exist in a certain language. If each person using a tape adds value to the collection by returning a transcription of it, then the taped material will be transcribed over time as it is consulted, even if no special projects for transcription and analysis are organized, although that, too, is possible.

Languages in serious danger of extinction will be given priority for documentation. Aside from the basic ten-hour protocol, a second, extended protocol will be taped in these cases.

Less will be said here about nonstandardized taping, which will be carried out in the villages, in cooperation with the native community. Many groups would like to document their culture but lack the equipment, technical know-how, and safe storage conditions. More and more communities have some access to cassette players and VCRs, so copies in the form of audio cassettes or VHS video tapes can be returned to the source community. Some indigenous groups, such as the Kayapó, already shoot and edit video. If a topic to be documented is discussed within the community for several days (a sort of mini–research project) and then taping is planned to document it, the result will be richer and more accurate. Topics that have been suggested by several individual Indians are music, narratives, oral history, and traditional knowledge of the environment. The process of researching the topic and recording the tapes confers prestige on traditional culture, especially if its beauty and intelligence can be captured on a well-executed tape.

Certain practices (for example, luring game animals by imitating their calls) are easily filmed in video but difficult to document precisely in words.

The use of video described above is different from the two uses more familiar to linguists: video as a stimulus for texts (which are then studied to see how the speaker analyzed a stretch of video and organized his description of it in his language) and video as a means of studying discourse in natural context.

THE UTILITY AND THE LIMITATIONS OF TAPE DOCUMENTATION

Documentation recordings cannot, of course, replace real fieldwork, in which analysis and hypothesis testing accompany data collection and direct it. Taped data is limited, and conclusions that can be drawn from it are correspondingly limited. The data is only as good as the planning that went into its collection. Inevitably, errors will be taped unknowingly. Furthermore, the time required to transcribe and analyze ten hours of taped data is enormous.

On the other hand, documentation tapes, recorded in the manner planned here, have a number of unique advantages. One is that they permit gathering a great amount of data in a relatively short amount of time, allowing for extensive coverage of a large number of languages. This sort of data collection is doable. While it must be planned carefully by advanced researchers, the actual taping can be done by less trained personnel. The data being taped is original; it is not simply data previously gathered being transferred from point A to point B. Since it is a recorded sample of the language, it permits verification in the sense that anyone can work with a tape copy and come to his or her own conclusions, in a manner that is not true of traditional word lists. Instrumental analysis, such as computer-generated sound spectrograms or measures of pitch and intensity, can be applied to a tape but not to a word list.

An important reason for standardizing the elicitation protocol is that comparable data from all the languages is much more valuable. For example, if a traditional word list for one language contains 200 items and a word list for a sister language, collected by someone else, contains 150 items, there may only be, say, fifty items contained in both lists, severely limiting comparability, which is essential for finding cognate terms and borrowings. Standardization also helps in evaluating the reliability of the data. Items that correspond closely to their word list equivalents in related languages and dialects are unlikely to be incorrect. On the other hand, those that correspond to something very different (say, "stone" instead of "house") in related languages and dialects would be suspect.

Documentation tapes can facilitate future fieldwork by providing a sample of the language that can be studied ahead of time to identify phenomena of theoretical interest. Data about the community, such as village size, location, and access, can facilitate planning. Documentation tapes serve as a permanent record for the speech community and have an intuitive accessibility to the speakers that written analysis does not have. For most of the indigenous languages of Brazil, the documentation tapes made by this project may well be the only ones of quality which will ever be done.

CONTENTS OF THE STANDARDIZED DOCUMENTATION

The elicitation protocol is still being planned. Preliminary versions will be tested in pilot tapings and then revised. The relative proportions of the various types of data being collected will be established after pilot tapings, in order to determine the time required for each type and the degree of coverage desired. The kinds of data to be gathered include lexical items, phonetic, phonological, morphological, and syntactic data, and spontaneous texts.

Lexical Items

Lexical items will be grouped semantically for ease of elicitation and for clarification of meaning. A preliminary idea is to include many of the items of Kaufman's (1973) list of words shown empirically to be stable over time in a number of world areas.[2] Stable items increase the number of cognates found among genetically related languages.

A number of cultural words will also be included—although they are usually less stable—since they are essential both for reconstructing protoculture and for tracing diffusion. For example, according to Rodrigues and Dietrich (1997:274) the word for "hammock" can be reconstructed in Proto-Tupí, indicating considerable antiquity for this element of tropical forest culture. Most of the indigenous languages in southern Rondônia, Brazil, regardless of genetic affiliation, have a word for "maize" that is phonetically similar to atiti, presumably because this domesticate diffused through the region, along with its name, from some as yet unidentified source.

Words for objects of the natural environment will be included, in part because of the inferences they permit about the past. In the case of Proto-Indo-European, for example, the fact that one can reconstruct words such as "snow" and "salmon" indicates that speakers of the protolanguage inhabited temperate regions, since they were acquainted with items characteristic of such climate. Newcomers to a geographical region frequently adopt names for plant and animal species from the people already living there, as in the case of European settlers in the Americas. For comparative purposes, words for native items should preferably be stable. Balée and Moore (1991) found that monomorphemic plant names, usually designating species associated with human management, are more stable than descriptive, metaphorical names, which are more likely to designate wild species. Words for species that have (or had, since we are dealing with prehistory) restricted distributions are useful for inferences about the past. For example, if it is possible to reconstruct the word for "rat" in a protolanguage, this tells us little about the environment of the speakers, since rats are ubiquitous. Reconstruction of the word for "dolphin," however, would indicate that speakers of the protolanguage lived near large rivers or the sea but not on the high semiarid *planalto* region of central Brazil.

Given the geographical diversity of a country as large as Brazil, the task of selecting an optimal set of nature terms for elicitation is a matter that requires some study. Another problem is identifying the exact referent of an indigenous word for a plant or animal. Portuguese terms vary from one region to another and may not reflect the distinctions recognized by speakers of an indigenous language. Presumably, photographs or other stimulus material will be necessary for identification. As part of the testing of the elicitation protocol, the stimulus materials will be checked to see if they produce consistent responses from members of the same speech community.

Phonetic Data

The lexical items and the simple syntactic constructions will provide a sample of the language for phonetic studies, though it will be difficult to determine contrasts. A unidirectional lavalier microphone will be used to minimize extraneous sounds. Individual words will be pronounced twice, to help determine the range of variation of sounds. When a language is known to be tonal, it would be useful to record whistled versions of words and simple constructions also, though it takes experience to elicit whistled speech.

Phonological Data

The individual words and morphological processes recorded will give some basis for a preliminary analysis of the phonological inventory and morphophonological alternations. If the person eliciting occasionally repeats the informant's speech, the informant's corrections give additional clues to his phonological intuitions and perceptions.

Morphological Data

There are obvious processes of inflection that can be elicited, such as person marking on various word classes (nouns, intransitive verbs, etc.). At least the more common processes of derivation can be found by eliciting related words, such as, in English, "hot" (adjective), "heat" (intransitive verb), "heat" (transitive verb), and "heat" (noun). Compounding is more difficult to elicit but will appear in lexical items if it exists. Singular and plural verbs and adjectives are fairly common in native Brazilian languages.

Syntactic Data

The general strategy for identifying the component words and morphemes of elicited syntactic constructions is to build them out of words already elicited as lexical items and to use controlled variation to correlate meaning and form. Basic topics to be investigated will include sentence types, clauses with one, two, or three arguments, changes in sentence structure due to pragmatic processes (for example, questions, focus, or imperatives), embedded clauses or their equivalents, composition of noun phrases and verb phrases (if these exist), and the behavior of noun phrase types (human, animate, etc.).

Spontaneous Texts

Texts provide the best sample of the grammar of a language and control for the simplifications and distortion of bilingual elicitation, but it is difficult to identify the component words from a tape without the assistance of a speaker, though passive speakers and speakers of closely related languages may be able to follow a text word by word. Texts will be immediately followed by a free translation into Portuguese (or perhaps into another language if Portuguese is not possible). A variety of texts provides a better sample of the language. Oral history, traditional narrative, or other texts that contain useful information about culture are preferable.

NONLINGUISTIC QUERIES

Brief nonlinguistic inquiries provide an opportunity to obtain some basic information and also some vocabulary items. Questions about the speaker include name, age, languages spoken, ethnic identity, place of birth, and place of residence. Questions about an informant's group include number, location and size of villages, names of the group, and existence of any internal subgroups. Questions that facilitate field visits in the future and give some idea of the situation of the group include those dealing with access, presence of missionaries or government officials, land invasions and relations with neighboring national communities, local market economy and preferred trade goods, schools, health care, relations between villages, observance (or not) of traditional festivals, and shamanism. Questions having to do with oral history and prehistory include, for example, "Before your people arrived where they are today, where did they live, long ago?" "Which groups were neighbors of yours then?" "Did you fight with them?" "Marry them?" Some basic aspects of culture can be briefly investigated, and native terms for culture elements noted. These aspects include subsistence activities, economic relations, cosmology and ritual calendar, and kinship and social organization.

EQUIPMENT INFORMATION FOR DOCUMENTATION TAPING

Some of the information gathered in the process of equipment selection for the Museu Goeldi documentation taping project will be summarized below, in the hope that it will be helpful to others.[3] Equipment is rapidly evolving and any information is soon out of date, but general principles and approximate price ranges (given here in U.S. dollars) are useful. In addition, three catalogues (B&H Photo-Video 1994, Schaeffer 1994, Sony Electronics Co. 1994) are listed in the References, with addresses for contact in brackets. Prices cited in the text below are rounded discount prices from 1996. One manufacturer, Sony, is overrepresented because of the formats and price ranges selected, the need for compatibility, and the availability of maintenance in Brazil.

Audio Equipment

There are two types of audio recorders and tapes: *analog* and *digital.* Analog devices record sound by varying the degree of magnetization of metal particles in the tape. Digital devices represent sound as binary numbers, like a computer file on a diskette. Analog recordings, such as ordinary cassette tapes, are cheaper but have a certain level of inherent noise and each successive copy increases this noise, though there are systems of noise reduction (for example, Dolby A, B, C, or SR, as well as dbx) that reduce the noise (but also, according to some purists, alter the signal slightly). Digital recordings have negligible inherent noise, and, as in the case of diskette files, copies can be made of copies indefinitely. Consumer DAT (digital audio tape) recorders sold in the United States permit copies to be made of an original, but copies cannot be made of copies, so that pirating will at least be kept to a lower level.

DAT cassettes are smaller than analog cassettes and have time code written on them as they are recorded, so that any point on the tape can be specified and located. A 120-minute DAT cassette costs about $9. Whereas DAT equipment is already third generation (at least in Japan), minidisks are a newer rival digital medium almost equal in sound quality. Minidisk recorders use digital data compression and magneto-optical technology. They cost slightly more than consumer-level DAT recorders. Because the technology is more established, the Museu Goeldi project selected DAT recorders: consumer-level Walkman models (Sony TCD-D8, $650) and portable professional recorders (Sony TCD-D10 PRO II, $3,000). The consumer model operates on four AA batteries, but the professional model consumes about ten watts, making it wiser, for field use, to use it with its optional DC adapter and a rechargeable 12-volt NiCad (nickel-cadmium) battery.

Three measures that are frequently used for audio equipment are *frequency range, dynamic range,* and *signal-to-noise ratio.* We will only discuss the first. Young people can hear sounds as low as 20 Hz and as high as 20,000 Hz, though the average person hears up to 16,000 Hz (Talbot-Smith 1997:41). Human hearing is most sensitive at about 3000 Hz. To qualify as "high fidelity," a device must capture high frequencies of at least 12,000 Hz. A professional portable analog recorder captures frequencies of approximately 40–15,000 Hz with iron tape, 40–16,000 Hz with chrome tape, and 40–17,000 Hz with metal tape. DAT equipment typically captures 20–20,000 Hz.

Microphone capacity should match that of the recorder. More expensive microphones ($300–500) may reach 20–20,000 Hz. A good consumer-level stereo microphone for use with analog equipment, costing $80, typically captures 100–15,000 Hz. *Omnidirectional* microphones pick up sounds from all directions. *Unidirectional* or *cardioid* microphones minimize lateral sound pickup, which is better for recording speech in noisy environments. *Shotgun* microphones are still more unidirectional. *Stereophonic* microphones are essentially two unidirectional microphones that simulate the directionality of human hearing. Professional-level microphones and other audio equipment are usually connected by *balanced* cables, which contain three wires (though monophonic), one of which acts as shielding. Consumer-level equipment typically uses *unbalanced* cables and small plugs.

A variety of microphones were selected for the documentation project. Two worth mentioning specifically are a cardioid lavalier (small clip-on) microphone (Sennheiser MKE 40–60, frequency range 40–20,000 Hz, $250) with its phantom power supply (Sennheiser K-6, $190, usable with other microphones of the same series) for clear, sensitive recording at a short distance from the speaker's mouth, and a short shotgun camcorder microphone (Sennheiser MKE-300, 150–17,000 Hz, $180).

Video Equipment

The two older video tape formats, used only for consumer-level recording, are *VHS* and *8mm*. Both have about 230 *lines of horizontal resolution*. High-fidelity VHS has better sound quality, but the 8mm format is much more compact. The semi-professional formats that grew out of these two rivals (with which they are backwardly compatible) are *S-VHS* and *Hi8*, both of which have about 400 lines of resolution. (By comparison, commercial television has 330 lines.) The S-VHS format has superior sound, 20–20,000 Hz. The Hi8 tape records audio at 30–15,000 Hz, interleaved on the tape with the video signal, or at 20–15,000 Hz on a separate PCM (pulse code modulator) digital stereo track (used only by larger cameras, not camcorders, but available for dubbing on VCRs or editors).

Given the frequent necessity for foot and canoe transportation in the Brazilian Amazon region, the documentation project chose the compact Hi8 format, though S-VHS is an excellent alternative. The camcorders acquired (Sony CCD-TR-3000, $1,450 each) were much less expensive than professional Hi8 cameras (easily $7,000), weigh less than a kilo, and offer *image stabilization* (neutralizes small shakes and tremors), *time code* (for easy and exact editing), 16:1 *optical zoom,* and 32:1 *digital zoom.* External microphones plug into the camcorder and increase the audio performance. A 120-minute Hi8 tape costs about $9 (metal particle type) or $15 (evaporated metal type). Tapes of 90 minutes have less *dropout*—magnetized particles falling off (resulting in white spots in the image). Dropout is a greater problem on compact tape formats. A semiprofessional Hi8 *editor* (Sony EVO-9720, $5,700) attached to a *title generator* (Videonics Video Titlemaker 2000, $450) and to two *production monitors* (Sony PVM14N1U / N2U, $630 each) is sufficient equipment to produce well-edited, television-ready documentation videos. *Nonlinear* editing is possible on a personal computer and has certain advantages and disadvantages. Tripods and illumination equipment are important for image quality. Ambulatory stabilization suspension devices (Steadycam Jr., $450) allow filming stable images while walking or even running. Where low-energy consumption is essential, daylight fluorescent lamps or compact fluorescents with a *color temperature* of about 3000° K (more toward yellow) offer lighting with good color rendering.

National video formats are different from tape formats. The United States and Japan use the NTSC format, Brazil has its own, PAL-M, and Europe is PAL or SECAM. Within the same national format one tape format can be dubbed into another. For example, a Hi8 signal in NTSC can be transferred to an NTSC VHS tape (at a lower resolution) in an NTSC VCR. But the same NTSC signal could not be recorded onto a PAL VHS tape in a PAL VCR, unless it were first converted, a somewhat complicated process.

Digital video cameras are now available. Lightweight models (less than two kilograms) can film 500 lines of resolution, have image stabilization, and record PCM digital stereo sound. These are more expensive than analog camcorders but less expensive than analog professional cameras. The tapes are expensive and last thirty minutes.[4]

Portable Energy

Recall that there are two common types of electricity: *alternating* current (AC, the kind from a wall plug) and *direct* current (DC, the kind found in all batteries and produced by solar panels). In a DC circuit, one wire is plus and one wire is negative. In an AC circuit of 110 volts, one wire alternates between plus and minus and the other is neutral. In general one cannot switch the two DC wires (change *polarity*) as one generally can with the two AC wires. *Voltage* is like the pressure of a flow of water; *amperes* are like the quantity of water flowing, and *watts* are volts multiplied by amperes—the total force of the flow. The work done is measured in *watt-hours*—the number of watts over a period of time. Battery storage capacity is usually given in *ampere-hours.* Multiply amp-hours by the voltage of the battery to find the capacity in watt-hours. Electric current flows from higher voltage to lower voltage, so that solar panels, for example, must produce a voltage higher than that of the battery they are charging and often have *diodes* to keep energy from flowing backward at night.

These facts give a sufficient basis for measuring electricity and planning mini-systems to supply energy to documentation equipment. For example, the Linguistics Division at the Goeldi uses a flexible solar panel (Uni-Solar MBC-262, $169), which is nominally eleven watts but actually produces over twelve watts (since flexible panels do well in heat, unlike others), to charge a one-kilogram rechargeable NiCad battery (NRG Compak, $160, sold for video use). The battery holds about sixty watt-hours of energy, of which fifty are usable (leaving the other ten to avoid deep discharge), and the usable amount can be supplied by the panel in a bit more than four hours of full sun. This fifty watt-hours could run a ten-watt camcorder for five hours, if charging efficiency were perfect, or a sixteen-watt fluorescent lamp for just over three hours. To use both the camcorder and the lamp it is necessary to have two solar panels and two batteries. Laptop energy consumption varies from under seven to over twenty watts, depending on the machine.

NiCad batteries are lighter and last for many years, but they are more expensive and need to be charged rather slowly unless a special quick-charge device is used. *Lead-acid* batteries are heavy but cheap and available. *Deep cycle* lead batteries have thicker plates and can be repeatedly discharged to a deeper level without harm. *Transformers* change AC current to DC current. *Inverters* (Radio Shack models are $70 and $100) change 12-volt DC electricity to 110-volt AC current, wasting perhaps five watts in the process. DC *voltage adapters* change the voltage level of DC current but do not convert it to AC.

Protection of Electronic Equipment in the Field

Airtight rigid cases made of metal or fiberglass, with foam rubber internal cushioning, offer protection in difficult field conditions (Pelikan Case, medium-large, $120). Reheatable silica gel canisters ($5) dehumidify the interior of a closed space of less than a cubic meter. These are available from suppliers of conservation materials, whose catalogues can be found in the conservation and repair sector of any library.

NOTES

1. Michael Silverstein kindly offered excellent advice on the linguistic content, though most of it does not appear in the brief summary presented here. Aryon Rodrigues and Ken Hale are consultants to the project. None of these three necessarily agree with the views presented here.
2. This approach was followed by Berlin and Kaufman in their South American Indian Languages Documentation Project.

3. The Centers of Excellence program of the G-7 for the Museu Goeldi and also IRD (Institut de Recherche pour le Développement, formerly ORSTOM) generously supported equipment acquisition.

4. As of August 1999, the cost of newer technology has been reduced considerably since the above was written. Minidisk recorders can be purchased for less than $300. Good consumer Hi8 camcorders are available for less than $800. Consumer Hi8 cassettes lasting 120 minutes now cost $5 (metal particle) or $9 (evaporated metal). Digital video is rapidly becoming popular, especially in the miniature format (mini-DV). A 60-minute mini-DV cassette costs $9, and a good consumer mini-DV camcorder costs about $1600. The digital format offers up to 500 horizontal lines of resolution with almost no noise, as well as CD quality PCM audio. An important advantage of this format is that successive generations of copies are essentially identical, unlike analog video formats, which lose quality noticeably with each generation. Digital video can be transferred to a microcomputer as a digital file via the IEEE 1394 FireWire connection with no loss. Sony recently introduced the Digital8 format, which records digital video and audio on 8mm and Hi8 tapes, with performance almost equal to mini-DV, while costing less ($1000 for a good consumer camcorder). But one hour of video requires two hours of tape, since the speed is increased so as to hold more information. These devices can read (but not write) 8mm and Hi8 tapes also.

REFERENCES

B&H Photo-Video 1994. *The Professional Video Sourcebook.* 2d ed. [B&H Photo-Video, 420 Ninth Avenue, New York, NY 10011, USA; <www.bhphotovideo.com>. (Knowledgeable and honest, with extremely competitive pricing on professional and semiprofessional video and audio.)]

Balée, W.L., and D. Moore. 1991. Similarity and variation in plant names in five Tupi-Guarani languages (Eastern Amazonia). *Bulletin of the Florida Museum of Natural History, Biological Sciences* 35(4):209–262.

Kaufman, Terrence. 1973. Kaufman's basic concept list on historical principles. Department of Linguistics, University of Pittsburgh.

Moore, D., and L. Storto. 1992. Linguística indígena no Brasil. Manuscript. Museu Goeldi, Belém, Brazil.

Ribeiro, E.R. 1996. Morfología do verbo Karajá. Master's thesis, Universidade Federal de Goiás.

Rodrigues, A.D., and W. Dietrich. 1997. On the linguistic relationship between Mawé and Tupí-Guaraní. *Diachronica* 14(2):265–304.

Schaeffer, J., ed. 1994. *Real Goods Solar Living Sourcebook: The Complete Guide to Renewable Energy Technologies and Sustainable Living.* 8th ed. White River Junction, Vt.: Chelsea Green Publishing Co. [Real Goods Trading Corporation, 966 Mazzoni Street, Ukiah, CA 95482-3471, USA; <realgood@well.sf.ca.us>]

Sony Electronics Co. 1994. *Sony Style.* [Sony Plaza, 550 Madison Avenue, New York, NY 10022, USA; (212) 833-8800. (Sony's glossy consumer catalogue)].

Talbot-Smith, M. 1997. *Sound Assistance.* Oxford: Focal Press.

27

THE ROLE OF THE GLOBAL NETWORK OF INDIGENOUS KNOWLEDGE RESOURCE CENTERS IN THE CONSERVATION OF CULTURAL AND BIOLOGICAL DIVERSITY

D. Michael Warren

The terms *indigenous technical knowledge* and *indigenous knowledge* (IK) were first used in publications in 1979 and 1980 (Chambers 1979; Brokensha, Warren, and Werner 1980), but it was the influence of the United Nations Conference on Environment and Development held in June 1992 in Rio de Janeiro that provided a global awareness of the complementary nature of biodiversity and the indigenous knowledge about these natural resources and their uses by human communities. IK refers to the knowledge generated by communities and ethnic groups that usually pass the knowledge from one generation to the next through oral transmission; it is focused on the microenvironment in which it is generated. But from Berlin's comparative study (1992), it appears that there are more commonalities than had been anticipated between IK systems and their counterparts within the global knowledge system generated by the world's network of universities and research laboratories. This has important implications for understanding the universal nature of knowledge systems and how they are generated.

Editor's note: Permission to posthumously include this chapter in the book was graciously granted by Medina Warren, Mike Warren's daughter, on behalf of Mike's wife Mary Salawuh Oyelade Warren, through the kind offices of Dr. Norma Wolff, Mike's colleague in the Anthropology Department at Iowa State University. Dr. Wolff also kindly provided a copy of Mike's diagram of the IK cycle, reproduced here as figure 27.1. Permission for reproduction of this diagram, which Mike used extensively in his writing and teaching, was also obtained from Medina Warren. The minor editing that the chapter draft needed was performed by the editor.

The Agenda 21 documents from the Rio conference refer to IK and the need for its preservation numerous times. One reference states that "governments . . . with the cooperation of intergovernmental organizations . . . should . . . take action to respect, record, protect and promote the wider application of the knowledge, innovations and practices of indigenous and local communities . . . for the conservation of biological diversity and the sustainable use of biological resources" (CGIAR 1993:8). The Global Biodiversity Strategy included as one of its ten principles for conserving biodiversity the principle that "Cultural diversity is closely linked to biodiversity. Humanity's collective knowledge of biodiversity and its use and management rests in cultural diversity; conversely conserving biodiversity often helps strengthen cultural integrity and values" (World Resources Institute, World Conservation Union, and United Nations Environment Programme 1992:21).

The U.S. National Research Council stated that "development agencies should place greater emphasis on, and assume a stronger role in, systematizing the local knowledge base—indigenous knowledge, 'gray literature,' anecdotal information. A vast heritage of knowledge about species, ecosystems, and their use exists, but it does not appear in the world literature, being either insufficiently 'scientific' or not 'developmental'" (National Research Council 1992a:10). "If indigenous knowledge has not been documented and compiled, doing so should be a research priority of the highest order. Indigenous knowledge is being lost at an unprecedented rate, and its preservation, preferably in data base form, must take place as quickly as possible" (National Research Council 1992a:45).

This chapter describes the active role that humans and their communities play at the global level with regard to biodiversity and its use by humans, and the role of the growing global network of IK resource centers in recording this important last frontier of human knowledge.

IK AS CULTURAL CAPITAL

Much research has been conducted on the role of colonialism and racism in shaping attitudes and stereotypes that have devalued the contributions to global knowledge by non-Euro-American communities. Only recently have scholars and international development agencies begun to recognize IK as an invaluable resource now regarded as cultural capital (Berkes and Folke 1992; Hyndman 1994), that is, as important for preservation as biological capital. In recent years, many organizations have recognized the importance of this overlooked resource, including the Consultative Group on International Agricultural Research (CGIAR 1993), the International Labor Organization (Warren 1997a), the United Nations Environmental Programme (Dowdeswell 1993), the Food and Agricultural Organization of the United Nations (Herbert 1993; Saouma 1993; den Biggelaar and Hart 1996), the International Board for Plant Genetic Resources (1993), the International Plant Genetic Resources Institute (Guarino 1995; Eyzaguirre and Iwanaga 1996), the United Nations Development Programme (Rural Advancement Fund International 1994), the U.K. Department of International Development (Sillitoe 1998), the World Bank (Warren 1991b, 1995a; Davis 1993), the International Development Research Centre (1993), the International Center for Living Aquatic Resource Management (Pauly, Palomares, and Froese 1993), the U.S. National Research Council (1991, 1992a, b, 1993), and UNESCO (1994a, b). Now regarded as intellectual property, the issue of the rights of communities to their own knowledge is being actively discussed and debated (Brush and Stabinsky 1996; Posey and Dutfield 1996).

Of major concern is the rapid loss of the knowledge of many communities as universal formal education is enforced with a curriculum that usually ignores the contributions of local communities to global knowledge. The loss of knowledge is linked indelibly to language extinction since language is the major mechanism for preserving and transmitting a community's knowledge from one generation to another (Hunter 1994).

The growing array of recorded knowledge systems related to biological materials include domesticated and nondomesticated plants and animals. A survey of this literature finds case studies for plants, forestry, fisheries, and animals (plants: Altieri and Merrick 1988; Johannes 1989; Juma 1989; Warren 1989, 1995a, b; Oldfield and Alcorn 1991; Berlin 1992; National Research Council 1992a, b, 1993; Gadgil, Berkes, and Folke 1993; Inglis 1993; Morrison, Geraghty, and Crowl 1994; Rajasekaran and Warren 1994; Abbink 1995; Martin 1995; Warren, Slikkerveer, and Brokensha 1995; Aregbeyen 1996; Ferguson and Mkandawire 1996; Setyawati 1996; Adams and Slikkerveer 1997; Innis 1997; Warren and Pinkston 1997; forestry: Posey 1985; Mathias-Mundy et al. 1992; Castro 1995; Ranasinghe 1995; Warren, Slikkerveer, and Brokensha 1995; Hanyani-Mlambo and Hebinck 1996; fisheries: Pinkerton 1989; Hviding and Baines 1992; Pauly, Palomares, and Froese 1993; Ruddle 1994; Warren, Slikkerveer, and Brokensha 1995; and animals: Gunn, Arlooktoo, and Kaomayok 1988; Kohler-Rollefson 1993; Mathias-Mundy and McCorkle 1993; Slaybaugh-Mitchell 1995; McCorkle, Mathias, and Schillhorn van Veen 1996).

It is difficult to discuss biological resources without referring to indigenous approaches to natural resource management of soils and water. There are also numerous resources now available in these areas (Hecht and Posey 1989; Carney 1991; Rajasekaran, Warren, and Babu 1991; Pawluk, Sandor, and Tabor 1992; Quintana 1992; Warren 1992; Davis 1993; Reij 1993; DeWalt 1994; Dialla 1994; Rajasekaran and Warren 1995; Roach 1997).

INDIGENOUS EXPERIMENTATION AND BIODIVERSITY

All communities have individuals who are regarded as particularly creative in terms of active experimentation and innovation. Often this creativity extends into the realm of biodiversity. Studies among the Yoruba of Nigeria by Warren and Pinkston (1997) and among the Kayapó of Brazil by Posey (1985) demonstrate how biodiversity is actively cultivated, leading to higher indices of biodiversity. We now understand that farmers actively conduct formal on-farm and backyard breeding with similar objectives to those of crop breeders currently working with the new array of biotechnology methodologies (Warren 1996; Bunders, Haverkort, and Hiemstra 1997). There is a rich literature on indigenous approaches to experimentation and innovation (for example, Chambers 1983, 1997; Richards 1986; Ashby, Quiros, and Rivers 1989; Chambers, Pacey, and Thrupp 1989; Juma 1989; Warren 1989, 1991a, 1994, 1996, 1997b; Gamser, Appleton, and Carter 1990; Gupta 1990; den Biggelaar 1991; Haverkort, van der Kamp, and Waters-Bayer 1991; Pretty 1991, 1995; Hiemstra 1992; Atte 1992; Moock and Rhoades 1992; Reijntjes, Haverkort, and Waters-Bayer 1992; Thurston 1992, 1997; de Boef, Amanor, and Wellard 1993; Warren and Rajasekaran 1993, 1994; Haverkort and Millar 1994; McCorkle 1994; Prain and Bagalanon 1994; Rajasekaran 1994; Rural Advancement Fund International 1994; Scoones and Thompson 1994; Ashby et al. 1996; Berg 1996; den Biggelaar and Hart 1996; McCorkle, Mathias, and T.W. Schillhorn van Veen 1996; Selener, Purdy, and Zapata 1996; Innis 1997; Sumberg and Okali 1997; Van Veldhuizen et al. 1997; Warren and Pinkston 1997; Prain, Fujisaka, and Warren 1999; Warren, Slikkerveer, and Brokensha 1995).

GENDERED KNOWLEDGE AND BIODIVERSITY

The interest in IK resulted in the discovery that such knowledge is often variable depending on a person's gender, age, and occupation. New case studies indicate clearly the important role of women in many communities in preserving and extending knowledge about the biological resources in the area (including Carney 1991; Appleton and Hill 1994; Mishra 1994; Moreno-Black, Somnasang, and Thamthawan 1994; Quiroz 1994; Ulluwishewa 1994; Nazarea-Sandoval 1995; Systemwide . . . 1997; and Zweifel 1997).

THE IK CYCLE AND THE ROLE OF IK IN EDUCATION

In the past decade the generation of IK in dynamic ways within any given community has been conceived as a cycle (see fig. 27.1). IK provides the starting point in a continual process with the IK serving as the basis for both individual and community decision making. Often the decision making is carried out within indigenous organizations that have development functions such as providing forums for the identification, discussion, and prioritization of community problems as well as the search for solutions to them (Blunt and Warren 1996). The search often involves a variety of indigenous approaches to creativity that include experimentation and innovation, the results of which are evaluated, often through indigenous modes of communication, and new knowledge found to be useful is incorporated into the IK. Case studies of all of these functions within the IK cycle are presented in Warren, Slikkerveer, and Brokensha (1995).

Because IK has not been formally recorded, it has not been convenient to add it into the educational curricula for primary, secondary, and tertiary educational institutions. This has led many persons to believe that the only knowledge worthy of the label comes from the Euro-American scientific tradition. By taking interesting case studies and developing them into teaching modules, one quickly learns that students react very positively to them. There is a growing global network of individuals and institutions engaged in changing educational curricula to provide the visibility and value to IK that it deserves (for example, Kothari 1995; Kroma 1995; Warren, Egunjobi, and Wahab 1996; Kreisler and Semali 1997; Semali 1997).

ROLES OF IK RESOURCE CENTERS IN FOSTERING BIOCULTURAL DIVERSITY

The first two IK resource centers, the Center for Indigenous Knowledge for Agriculture and Rural Development (CIKARD) and the Leiden Ethnosystems and Development (LEAD) Programme, were established in 1987. By the end of 1997 there were 33 formally established centers, four universities with formal IK study groups, with another 20 centers in the process of being organized. These centers now exist on every inhabited continent. In September 1992 Canada's International Development Research Centre funded the International Symposium on Indigenous Knowledge and Sustainable Development conducted at the Regional Program for the Promotion of Indigenous Knowledge in Asia (REPPIKA) based at the International Institute of Rural Reconstruction in Silang, the Philippines (Flavier, De Jesus, and Navarro 1995). This was the first opportunity for directors of existing centers as well as individuals working to establish centers to meet personally and think through the role of the centers at the global level (CIRAN 1993a, b).

In half a decade the results of the 1992 meeting have been impressive. Manuals for recording IK have been produced by REPPIKA (IIRR 1996) and the Centre for Traditional Knowledge (1997). Directors and research associates at other centers have developed additional guidelines to facilitate the recording and archiving of IK, for example, REPPIKA (Mathias 1996a, b), the Kenya Resource Centre for Indigenous Knowledge (KENRIK) (Maundu 1995), CIKARD (Warren and McKiernan 1995), and the African Resource Centre for Indigenous Knowledge (ARCIK) (Phillips and Titilola 1995).

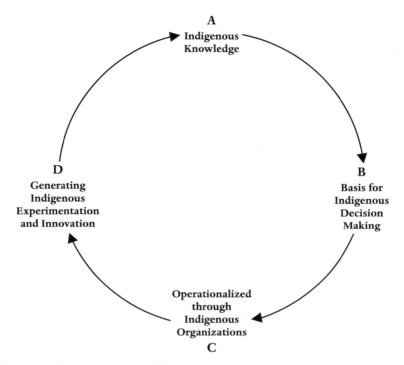

Figure 27.1. The indigenous knowledge cycle.

National and international policy related to the role of IK in sustainable approaches to natural resource management and development have been discussed by various centers including CIKARD (Rajasekaran, Warren, and Babu 1991) and the Venezuela Resource Center for Indigenous Knowledge (VERCIK) (Quiroz 1996).

The global networking has been facilitated through the publication of the *Indigenous Knowledge and Development Monitor* by the Centre for International Research and Advisory Networks (CIRAN), published thrice yearly since 1993 and available in hardcopy as well as on the Internet at http://www.nuffic.nl/ciran/ikdm (von Liebenstein, Slikkerveer, and Warren 1995; Warren, von Liebenstein, and Slikkerveer 1993). The *Monitor* now links individuals in more than 130 countries.

Databases of recorded IK have been established at CIKARD (Warren and McKiernan 1995) and LEAD (Slikkerveer 1995). CIKARD now provides a very fast keyword search engine that leads the searcher to citations and abstracts of more than 5000 documents housed at the CIKARD Library (http://www.iitap.iastate. edu/cikard/cikard.html), as well as IK teaching modules, and full texts of French and Spanish translations of key documents. International Center for Living

Aquatic Resource Management has established a global database for IK related to fish (Pauly, Palomares, and Froese 1993).

Several centers and study groups have organized national workshops on the role of IK in sustainable development. Proceedings of several workshops have now been published, for example, the South African Resource Centre for Indigenous Knowledge (SARCIK) (Normann, Snyman, and M. Cohen 1996), the Sri Lanka Resource Centre for Indigenous Knowledge (SLARCIK) (Ulluwishewa and Ranasinghe 1996), Obafemi Awolowo University Indigenous Knowledge Study Group (Warren 1996), and the University of Ibadan Indigenous Knowledge Study Group (Warren, Egunjobi, and Wahab 1996). Other institutes in the process of establishing centers have also carried out workshops, for example, the Center for Integrated Agricultural Development (1994) in Beijing. These conferences and workshops provide opportunities to involve policymakers from government as well as representatives of international donor agencies. They also result in excellent publicity through newspaper, television, and radio coverage.

CIKARD and the Interinstitutional Consortium for Indigenous Knowledge (ICIK), as well as the Indigenous Knowledge Study Group at the University of Ibadan have been instrumental in establishing an agenda for incorporating IK case studies into educational curricula through a growing global network available on the CIKARD home page (Kroma 1995; Warren, Egunjobi, and Wahab 1996; Kreisler and Semali 1997; Semali 1997).

The African Resource Centre for Indigenous Knowledge (ARCIK) has carried out a cost-benefit analysis of using IK in development projects (Titilola 1990). Some centers have provided guidelines for other centers to emulate, for example, ARCIK (Phillips and Titilola 1995), the Philippines Resource Center for Sustainable Development and Indigenous Knowledge (PHIRCSDIK) (Serrano, Labios, and Tung 1993), LEAD (Slikkerveer and Dechering 1995), SLARCIK (Ulluwishewa 1993), CIRAN (von Liebenstein, Slikkerveer, and Warren 1995), CIKARD (Warren and McKiernan 1995), and REPPIKA (Flavier, De Jesus, and Navarro 1995).

In order to assure easier access to new case studies, three publication series have been established. Bibliographies in Technology and Social Change and Studies in Technology and Social Change are published at CIKARD, while the IT Studies in Indigenous Knowledge and Development book series is published in London by Intermediate Technology Publications.

Guidelines and recommendations for recording IK, the development of methods and manuals for recording IK, the archiving and sharing of IK, the utilization of IK by local groups, extension workers, educators, researchers and policymakers, research on IK, policy issues, and an action plan for the IK centers are all available (CIRAN 1993a, b).

CONCLUSIONS

The relationship between the viability of a language and the knowledge that has been created, preserved, and maintained through that language is inextricable. The recording of IK systems has clearly indicated that they are dynamic, reflecting community reactions to changing sets of circumstances. Many of them are complex and sophisticated. Some of them incorporate exciting discoveries that result from systematic experimentation, results that are being shared with other communities in various parts of the world struggling with similar problems. The use of both the neem tree and vetiver grass and the knowledge generated about them in South Asia has recently been spread globally (National Research Council 1992b, 1993). Communities that live in close contact with the natural environment have extensive knowledge of their natural resources including the biological realm and the soil and water that nurture them. They are the true managers of in situ conservation of biodiversity and the knowledge of their biological realm. With a growing number of committed persons and established IK resource centers around the globe, it is anticipated that the stores of knowledge of communities worldwide will become recorded so they can be recognized as contributions to global knowledge.

REFERENCES

Abbink, J. 1995. Medicinal and ritual plants of the Ethiopian Southwest: An account of recent research. *Indigenous Knowledge and Development Monitor* 3(2):6–8.

Adams, W.M., and L.J. Slikkerveer, eds. 1997. *Indigenous Knowledge and Change in African Agriculture.* Studies in Technology and Social Change no. 26. Ames: CIKARD, Iowa State University.

Altieri, M.A., and L.C. Merrick. 1988. Agroecology and in situ conservation of native crop diversity in the Third World. In *Biodiversity,* ed. E.O. Wilson and F.M. Peter. Pp. 15–23. Washington, D.C.: National Academy Press.

Appleton, H.E., and C.L.M. Hill. 1994. Gender and indigenous knowledge in various organizations. *Indigenous Knowledge and Development Monitor* 2(3):8–11.

Aregbeyen, J.B.O. 1996. Traditional herbal medicine for sustainable PHC. *Indigenous Knowledge and Development Monitor* 4(2):14–15.

Ashby, J.A., C.A. Quiros, and Y.M. Rivers. 1989. Farmer participation in technology development: Work with crop varieties. In *Farmer First: Farmer Innovation and Agricultural Research,* ed. R. Chambers, A. Pacey, and L.A. Thrupp. Pp. 115–122. London: Intermediate Technology Publications.

Ashby, J.A., et al. 1996. Innovation in the organization of participatory plant breeding. In *Participatory Plant Breeding,* ed. P. Eyzaguirre and M. Iwanaga. Pp. 77–97. Rome: International Plant Genetic Resources Institute.

Atte, O.D. 1992. *Indigenous Local Knowledge as a Key to Local Level Development: Possibilities, Constraints, and Planning Issues.* Studies in Technology and Social Change no. 20. Ames: CIKARD, Iowa State University.

Berg, T. 1996. The compatibility of grassroots breeding and modern farming. In *Participatory Plant Breeding,* ed. P. Eyzaguirre and M. Iwanaga. Pp. 31–36. Rome: International Plant Genetic Resources Institute.

Berkes, F., and C. Folke. 1992. A systems perspective on the interrelations between natural, human-made, and cultural capital. *Ecological Economics* 5:1–8.

Berlin, B. 1992. *Ethnobiological Classification: Principles of Categorization of Plants and Animals in Traditional Societies.* Princeton: Princeton University Press.

Blunt, P., and D.M. Warren, eds. 1996. *Indigenous Organizations and Development.* London: Intermediate Technology Publications.

Brokensha, D.W., D.M. Warren, and O. Werner, eds. 1980. *Indigenous Knowledge Systems and Development.* Lanham, Md.: University Press of America.

Brush, S.B., and D. Stabinsky, eds. 1996. *Valuing Local Knowledge: Indigenous People and Intellectual Property Rights.* Washington, D.C.: Island Press.

Bunders, J., B. Haverkort, and W. Hiemstra, eds. 1997. *Biotechnology: Building on Farmers' Knowledge.* Basingstoke, UK: Macmillan Education.

Carney, J. 1991. Indigenous soil and water management in Senegambian rice farming systems. *Agriculture and Human Values* 8(1/2):37–48.

Castro, P. 1995. *Facing Kirinyaga: A Social History of Forest Commons in Southern Mount Kenya.* London: Intermediate Technology Publications.

Center for Integrated Agricultural Development. 1994. *Indigenous Knowledge Systems and Rural Development in China: Proceedings of the Workshop.* Beijing: Beijing Agricultural University.

Centre for Traditional Knowledge. 1997. *Guidelines for Environmental Assessments and Traditional Knowledge.* Ottawa: Centre for Traditional Knowledge.

CGIAR 1993. Indigenous knowledge. In *People and Plants: The Development Agenda.* P. 8. Rome: Consultative Group on International Agricultural Research.

Chambers, R. 1979. *Rural Development: Whose Knowledge Counts?* Special issue of *IDS Bulletin,* Institute of Development Studies, University of Sussex, 10:2.

Chambers, R. 1983. *Rural Development: Putting the Last First.* London: Longman.

Chambers, R. 1997. *Whose Reality Counts? Putting the First Last.* London: Intermediate Technology Publications.

Chambers, R., A. Pacey, and L.A. Thrupp, eds. 1989. *Farmer First: Farmer Innovation and Agricultural Research.* London: Intermediate Technology Publications.

CIRAN. 1993a. Background to the International Symposium on Indigenous Knowledge and Sustainable Development. *Indigenous Knowledge and Development Monitor* 1(2):2–5.

CIRAN. 1993b. Recommendations and action plan. *Indigenous Knowledge and Development Monitor* 1(2):24–29.

Davis, S.H., ed. 1993. *Indigenous Views of Land and the Environment.* World Bank Discussion Papers no. 188. Washington, D.C.: World Bank.

De Boef, W., K. Amanor, and K. Wellard. 1993. *Cultivating Knowledge: Genetic Diversity, Farmer Experimentation, and Crop Research.* London: Intermediate Technology Publications.

DeWalt, B. 1994. Using indigenous knowledge to improve agriculture and natural resource management. *Human Organization* 53(2):123–131.

Den Biggelaar, C. 1991. Farming systems development: Synthesizing indigenous and scientific knowledge systems. *Agriculture and Human Values* 8(1/2):25–36.

Den Biggelaar, C., and N. Hart. 1996. *Farmer Experimentation and Innovation: A Case Study of Knowledge Generation Processes in Agroforestry Systems in Rwanda*. Rome: FAO.

Dialla, B.E. 1994. The adoption of soil conservation practices in Burkina Faso. *Indigenous Knowledge and Development Monitor* 2(1):10–12.

Dowdeswell, E. 1993. Walking in two worlds. Address presented at the InterAmerican Indigenous People's Conference, Vancouver, 18 September 1993.

Eyzaguirre, P., and M. Iwanaga, eds. 1996. *Participatory Plant Breeding*. Proceedings of a workshop on Participatory Plant Breeding, 26–29 July 1995, Wageningen, The Netherlands. Rome: International Plant Genetic Resources Institute.

Ferguson, A.E., and R.M. Mkandawire. 1996. A crop diversity improvement strategy. *Indigenous Knowledge and Development Monitor* 4(1):6–7.

Flavier, J.M., A. De Jesus, and C.S. Navarro. 1995. The Regional Program for the Promotion of Indigenous Knowledge in Asia (REPPIKA). In *The Cultural Dimension of Development: Indigenous Knowledge Systems*, ed. D.M. Warren, L.J. Slikkerveer, and D. Brokensha. Pp. 479–487. London: Intermediate Technology Publications.

Gadgil, M., F. Berkes, and C. Folke. 1993. Indigenous knowledge for biodiversity conservation. *Ambio* 22(2/3):151–156.

Gamser, M.S., H. Appleton, and N. Carter, eds. 1990. *Tinker, Tiller, Technical Change*. London: Intermediate Technology Publications.

Guarino, L. 1995. Secondary sources on cultures and indigenous knowledge systems. In *Collecting Plant Genetic Diversity: Technical Guidelines*, ed. L. Guarino, V. Ramanatha Rao, and R. Reid. Pp. 195–228. Wallingford, UK: CAB International on behalf of the International Plant Genetic Resources Institute in association with the FAO, IUCN, and UNEP.

Gunn, A., G. Arlooktoo, and D. Kaomayok. 1988. The contribution of the ecological knowledge of Inuit to wildlife management in the Northwest Territories. In *Traditional Knowledge and Renewable Resource Management*, ed. M.M.R. Freeman and L.N. Carbyn. Pp. 22–30. Edmonton: Boreal Institute for Northern Studies.

Gupta, A. 1990. *Honey Bee*. [A quarterly journal devoted to indigenous innovations]. Ahmedabad, India: Indian Institute of Management.

Hanyani-Mlambo, B.T., and P. Hebinck. 1996. Formal and informal knowledge networks in conservation forestry in Zimbabwe. *Indigenous Knowledge and Development Monitor* 4(3):3–6.

Haverkort, B., and D. Millar. 1994. Constructing diversity: The active role of rural people in maintaining and enhancing biodiversity. *Etnoecológica* 2(3):51–64.

Haverkort, B., J. van der Kamp, and A. Waters-Bayer, eds. 1991. *Joining Farmers' Experiments*. London: Intermediate Technology Publications.

Hecht, S.B., and D.A. Posey. 1989. Preliminary results on soil management techniques of the Kayapo Indians. *Advances in Economic Botany* 7:174–188.

Herbert, J. 1993. A mail-order catalog of indigenous knowledge. *Ceres: The FAO Review* 25(5):33–37.

Hiemstra, W., with C. Reijntjes, and E. van der Werf. 1992. *Let Farmers Judge: Experiences in Assessing Agriculture Innovations*. London: Intermediate Technology Publications.

Hunter, P.R. 1994. *Language Extinction and the Status of North American Indian Languages.* Studies in Technology and Social Change no. 23. Ames: CIKARD, Iowa State University.

Hviding, E., and G.B.K. Baines. 1992. *Fisheries Management in the Pacific: Tradition and the Challenges of Development in Marovo, Solomon Islands.* Discussion Paper no. 32. Geneva: United Nations Research Institute for Social Development.

Hyndman, D. 1994. Conservation through self-determination: Promoting the interdependence of cultural and biological diversity. *Human Organization* 53(3):296–302.

IIRR. 1996. *Recording and Using Indigenous Knowledge: A Manual.* Silang, Cavite, Philippines: REPPIKA, International Institute of Rural Reconstruction.

Inglis, J.T., ed. 1993. *Traditional Ecological Knowledge: Concepts and Cases.* Ottawa: International Program on Traditional Ecological Knowledge and International Development Research Centre.

Innis, D.Q. 1997. *Intercropping and the Scientific Basis of Traditional Agriculture.* London: Intermediate Technology Publications.

International Board for Plant Genetic Resources. 1993. Rural development and local knowledge: The case of rice in Sierra Leone. *Geneflow* 1993:12–13.

International Development Research Centre. 1993. Special issue on Indigenous and Traditional Knowledge. *IDRC Reports* 21(1).

Johannes, R.E., ed. 1989. *Traditional Ecological Knowledge: A Collection of Essays.* Gland, Switzerland: International Union for the Conservation of Nature.

Juma, C. 1989. *Biological Diversity and Innovation: Conserving and Utilizing Genetic Resources in Kenya.* Nairobi: African Centre for Technology Studies.

Kohler-Rollefson, I. 1993. Traditional pastoralists as guardians of biological diversity. *Indigenous Knowledge and Development Monitor* 1(3):14–16.

Kothari, B. 1995. From oral to written: The documentation of knowledge in Ecuador. *Indigenous Knowledge and Development Monitor* 3(2):9–12.

Kreisler, A., and L. Semali. 1997. Towards indigenous literacy: Science teachers learn to use IK resources. *Indigenous Knowledge and Development Monitor* 5(1):13–15.

Kroma, S. 1995. Popularizing science education in developing countries through indigenous knowledge. *Indigenous Knowledge and Development Monitor* 3(3):13–15.

Martin, G.J. 1995. *Ethnobotany: A Methods Manual.* London: Chapman and Hall.

Mathias, E. 1996a. Framework for enhancing the use of indigenous knowledge. *Indigenous Knowledge and Development Monitor* 3(2):17–18.

Mathias, E. 1996b. How can ethnoveterinary medicine be used in field projects? *Indigenous Knowledge and Development Monitor* 4(2):6–7.

Mathias-Mundy, E., and C.M. McCorkle. 1993. *Ethnoveterinary Medicine: An Annotated Bibliography.* Bibliographies in Technology and Social Change no. 6. Ames: CIKARD, Iowa State University.

Mathias-Mundy, E., et al. 1992. *Indigenous Technical Knowledge of Private Tree Management: A Bibliographic Report.* Bibliographies in Technology and Social Change no. 7. Ames: CIKARD, Iowa State University.

Maundu, P. 1995. Methodology for collecting and sharing indigenous knowledge: A case study. *Indigenous Knowledge and Development Monitor* 3(2):3–5.

McCorkle, C.M. 1994. *Farmer Innovation in Niger.* Studies in Technology and Social Change no. 21. Ames, Iowa: CIKARD, Iowa State University.

McCorkle, C.M., E. Mathias, and T.W. Schillhorn van Veen, eds. 1996. *Ethnoveterinary Research and Development*. London: Intermediate Technology Publications.

Mishra, S. 1994. Women's indigenous knowledge of forest management in Orissa (India). *Indigenous Knowledge and Development Monitor* 2(3):3–5.

Moock, J.L., and R.E. Rhoades, eds. 1992. *Diversity, Farmer Knowledge, and Sustainability*. Ithaca: Cornell University Press.

Moreno-Black, G., P. Somnasang, and S. Thamthawan. 1994. Women in northeastern Thailand: Preservers of botanical diversity. *Indigenous Knowledge and Development Monitor* 2(3):24.

Morrison, J., P. Geraghty, and L. Crowl, eds. 1994. *Science of Pacific Island Peoples*. 4 vols. Suva, Fiji: Institute of Pacific Studies, The University of the South Pacific.

National Research Council. 1991. *Toward Sustainability: A Plan for Collaborative Research on Agriculture and Natural Resource Management*. Washington, D.C.: National Academy Press.

National Research Council. 1992a. *Conserving Biodiversity: A Research Agenda for Development Agencies*. Washington, D.C.: National Academy Press.

National Research Council. 1992b. *Neem: A Tree for Solving Global Problems*. Washington, D.C.: National Academy Press.

National Research Council. 1993. *Vetiver Grass: A Thin Green Line against Erosion*. Washington, D.C.: National Academy Press.

Nazarea-Sandoval, V.D. 1995. Indigenous decision-making in agriculture: A reflection of gender and socioeconomic status in the Philippines. In *The Cultural Deminsion of Development: Indigenous Knowledge Systems,* ed. D.M. Warren, L.J. Slikkerveer, and D. Brokensha. Pp. 155–173. London: Intermediate Technology Publications.

Normann, H., I. Snyman, and M. Cohen, eds. 1996. *Indigenous Knowledge and Its Uses in Southern Africa*. Pretoria: Human Sciences Research Council.

Oldfield, M.L., and J.B. Alcorn, eds. 1991. *Biodiversity: Culture, Conservation, and Ecodevelopment*. Boulder: Westview Press.

Pauly, D., M.L.D. Palomares, and R. Froese. 1993. Some prose on a database of indigenous knowledge on fish. *Indigenous Knowledge and Development Monitor* 1(1):26–27.

Pawluk, R.R., J.A. Sandor, and J.A. Tabor. 1992. The role of indigenous soil knowledge in agricultural development. *Journal of Soil and Water Conservation* 47(4):298–302.

Phillips, A.O., and S O. Titilola. 1995. Sustainable development and indigenous knowledge systems in Nigeria: The role of the Nigerian Institute of Social and Economic Research (NISER). In *The Cultural Dimension of Development: Indigenous Knowledge Systems,* ed. D.M. Warren, L.J. Slikkerveer, and D. Brokensha. Pp. 475–478. London: Intermediate Technology Publications.

Pinkerton, E., ed. 1989. *Co-operative Management of Local Fisheries: New Directions for Improved Management and Community Development*. Vancouver: University of British Columbia Press.

Posey, D.A. 1985. Management of tropical forest ecosystems: The case of the Kayapó Indians of the Brazilian Amazon. *Agroforestry Systems* 3(2):139–158.

Posey, D.A., and G. Dutfield. 1996. *Beyond Intellectual Property: Toward Traditional Resource Rights for Indigenous Peoples and Local Communities*. Ottawa: IDRC Books.

Prain, G., and C.P. Bagalanon, eds. 1994. *Local Knowledge, Global Science, and Plant Genetic Resources: Towards a Partnership*. Los Baños: UPWARD.

Prain, G., S. Fujisaka, and D.M. Warren, eds. 1999. *Biological and Cultural Diversity: The Role of Indigenous Agricultural Experimentation in Development.* London: Intermediate Technology Publications.

Pretty, J.N. 1991. Farmers' extension practice and technology adaptation: Agricultural revolution in seventeenth–nineteenth century Britain. *Agriculture and Human Values* 8(1/2):132–148.

Pretty, J.N. 1995. *Regenerating Agriculture: Policies and Practice for Sustainability and Self-Reliance.* London: Earthscan.

Quintana, J. 1992. American Indian systems for natural resource management. *Akwe:kon Journal* 9(2):92–97.

Quiroz, C. 1994. Biodiversity, indigenous knowledge, gender, and intellectual property rights. *Indigenous Knowledge and Development Monitor* 2(3):12–15.

Quiroz, C. 1996. Local knowledge systems contribute to sustainable development. *Indigenous Knowledge and Development Monitor* 4(1):3–5.

Rajasekaran, B. 1994. *A Framework for Incorporating Indigenous Knowledge Systems into Agricultural Research, Extension, and NGOs for Sustainable Agricultural Development.* Studies in Technology and Social Change no. 22. Ames: CIKARD, Iowa State University.

Rajasekaran, B., and D.M. Warren. 1994. IK for socioeconomic development and biodiversity conservation: The Kolli Hills. *Indigenous Knowledge and Development Monitor* 2(2):13–17.

Rajasekaran, B., and D.M. Warren. 1995. Role of indigenous soil health care practices in improving soil fertility: Evidence from South India. *Journal of Soil and Water Conservation* 50(2):146–149.

Rajasekaran, B., D.M. Warren, and S.C. Babu. 1991. Indigenous natural-resource management systems for sustainable agricultural development: A global perspective. *Journal of International Development* 3(4):387–401.

Ranasinghe, H. 1995. Traditional tree-crop practices in Sri Lanka. *Indigenous Knowledge and Development Monitor* 3(3):7–9.

Reij, C. 1993. Improving indigenous soil and water conservation techniques: Does it work? *Indigenous Knowledge and Development Monitor* 1(1):11–13.

Reijntjes, C., B. Haverkort, and A. Waters-Bayer. 1992. *Farming for the Future: An Introduction to Low-External-Input and Sustainable Agriculture.* Leusden: ILEIA.

Richards, P. 1986. *Coping with Hunger: Hazard and Experiment in an African Rice-farming System.* London: Allen & Unwin.

Roach, S.A. 1997. *Land Degradation and Indigenous Knowledge in a Swazi Community.* M.A. thesis. Ames: Department of Anthropology, Iowa State University.

Ruddle, K. 1994. *A Guide to the Literature on Traditional Community-Based Fishery Management in the Asia-Pacific Tropics.* FAO Fisheries Circular no. 869. Rome: FAO.

Rural Advancement Fund International. 1994. *Conserving Indigenous Knowledge: Integrating Two Systems of Innovation.* New York: United Nations Development Programme.

Saouma, E. 1993. Indigenous knowledge and biodiversity. In *Harvesting Nature's Diversity.* Pp. 4–6. Rome: FAO.

Scoones, I., and J. Thompson, eds. 1994. *Beyond Farmer First: Rural People's Knowledge, Agricultural Research, and Extension Practice.* London: Intermediate Technology Publications.

Selener, D., with C. Purdy and G. Zapata. 1996. *Documenting, Evaluating, and Learning from Our Development Projects: A Participatory Systematization Workbook.* New York: International Institute of Rural Reconstruction.

Semali, L. 1997. Cultural identity in African context: Indigenous education and curriculum in East Africa. *Folklore Forum* 28:3–27.

Serrano, R.C., R.V. Labios, and L. Tung. 1993. Establishing a national IK resource centre: The Case of PHIRCSDIK. *Indigenous Knowledge and Development Monitor* 1(1):5–6.

Setyawati, I. 1996. Environmental variability, IK, and the use of rice varieties. *Indigenous Knowledge and Development Monitor* 4(2):11–13.

Sillitoe, P. 1998. The development of indigenous knowledge: A new applied anthropology. *Current Anthropology* 39 (2):223–252.

Slaybaugh-Mitchell, T.L. 1995. *Indigenous Livestock Production and Husbandry: An Annotated Bibliography.* Bibliographies in Technology and Social Change no. 8. Ames: CIKARD, Iowa State University.

Slikkerveer, L.J. 1995. INDAKS: A bibliography and database on indigenous agricultural knowledge systems and sustainable development in the tropics. In *The Cultural Dimension of Development: Indigenous Knowledge Systems,* ed. D.M. Warren, L.J. Slikkerveer, and D. Brokensha. Pp. 512–516. London: Intermediate Technology Publications.

Slikkerveer, L.J., and W.H.J.C. Dechering. 1995. LEAD: The Leiden Ethnosystems and Development Programme. In *The Cultural Dimension of Development: Indigenous Knowledge Systems,* ed. D.M. Warren, L.J. Slikkerveer, and D. Brokensha. Pp. 435–440. London: Intermediate Technology Publications.

Sumberg, J., and C. Okali. 1997. *Farmers' Experiments: Creating Local Knowledge.* Boulder: Lynne Rienner Publishers.

Systemwide Programme on Participatory Research and Gender Analysis. 1997. *A Global Programme on Participatory Research and Gender Analysis for Technology Development and Organisational Innovation.* AgREN Network Paper no. 72. London: Agricultural Research and Extension Network, UK Overseas Development Administration (ODA).

Thurston, H.D. 1992. *Sustainable Practices for Plant Disease Management in Traditional Farming Systems.* Boulder: Westview Press.

Thurston, H.D. 1997. *Slash/Mulch Systems: Sustainable Methods for Tropical Agriculture.* Boulder/London: Westview Press/Intermediate Technology Publications.

Titilola, S.O. 1990. *The Economics of Incorporating Indigenous Knowledge Systems into Agricultural Development: A Model and Analytical Framework.* Studies in Technology and Social Change no. 17. Ames: CIKARD, Iowa State University.

Ulluwishewa, R. 1993. Indigenous knowledge, national IK resource centres, and sustainable development. *Indigenous Knowledge and Development Monitor* 1(3):11–13.

Ulluwishewa, R. 1994. Women's indigenous knowledge of water management in Sri Lanka. *Indigenous Knowledge and Development Monitor* 2(3):17–19.

Ulluwishewa, R., and H. Ranasinghe, eds. 1996. *Indigenous Knowledge and Sustainable Development.* Proceedings of the First National Symposium on Indigenous Knowledge and Sustainable Development, Colombo, March 19–20, 1994. Nugegoda: SLARCIK.

UNESCO. 1994a. Special issue, Traditional Knowledge in Tropical Environments. *Nature and Resources* 30(1).

UNESCO. 1994b. Special issue, Traditional Knowledge into the Twenty-First Century. *Nature and Resources* 30(2).

Van Veldhuizen, L., et al., eds. 1997. *Farmers' Experimentation in Practice: Lessons from the Field.* London: Intermediate Technology Publications.

Von Liebenstein, G., L.J. Slikkerveer, and D.M. Warren. 1995. CIRAN: Networking for indigenous knowledge. In *The Cultural Dimension of Development: Indigenous Knowledge Systems,* ed. D.M. Warren, L.J. Slikkerveer, and D. Brokensha. Pp. 441–444. London: Intermediate Technology Publications.

Warren, D.M. 1989. Linking scientific and indigenous agricultural systems. In *The Transformation of International Agricultural Research and Development,* ed. J.L. Compton. Pp. 153–170. Boulder: Lynne Rienner Publishers.

Warren, D.M. 1991a. The role of indigenous knowledge in facilitating a participatory approach to agricultural extension. In *Proceedings of the International Workshop on Agricultural Knowledge Systems and the Role of Extension,* ed. H.J. Tillmann, H. Albrecht, M.A. Salas, M. Dhamotharah, and E. Gottschalk. Pp. 161–177. Stuttgart: University of Hohenheim.

Warren, D.M. 1991b. *Using Indigenous Knowledge in Agricultural Development.* World Bank Discussion Papers no. 127. Washington, D.C.: World Bank.

Warren, D.M. 1992. *A Preliminary Analysis of Indigenous Soil Classification and Management Systems in Four Ecozones of Nigeria.* Ibadan: African Resource Centre for Indigenous Knowledge/International Institute of Tropical Agriculture.

Warren, D.M. 1994. Indigenous agricultural knowledge, technology, and social change. In *Sustainable Agriculture in the American Midwest,* ed. G. McIsaac and W.R. Edwards. Pp. 35–53. Urbana: University of Illinois Press.

Warren, D.M. 1995a. Indigenous knowledge for agricultural development. Keynote speech given at the Workshop on Traditional and Modern Approaches to Natural Resource Management in Latin America, World Bank, April 25–26, 1995.

Warren, D.M. 1995b. Indigenous knowledge, biodiversity conservation, and development. In *Conservation of Biodiversity in Africa: Local Initiatives and Institutional Roles,* ed. L.A. Bennun, R.A. Aman, and S.A. Crafter. Pp. 93–108. Nairobi: Centre for Biodiversity, National Museums of Kenya.

Warren, D.M. 1996. The role of indigenous knowledge and biotechnology in sustainable agricultural development. Pp. 6–15. In *Indigenous Knowledge and Biotechnology.* Ile-Ife, Nigeria: Indigenous Knowledge Study Group, Obafemi Awolowo University.

Warren, D.M. 1997a. The incorporation of indigenous knowledge into project implementation for the development of indigenous peoples. Address given at the ILO-INDISCO Donor Consultation and Planning Workshop on Employment and Income-Generating for Indigenous and Tribal Peoples: Lessons Learned in Asia, New Delhi, 4–8 November 1997.

Warren, D.M. 1997b. The role of indigenous knowledge systems in facilitating sustainable approaches to development. Proceedings of the International Conference on Nature Knowledge, Istituto Veneto di Scienze, Lettere ed Arti, Venice, 4–6 December 1997.

Warren, D.M., and G. McKiernan. 1995. CIKARD: A global approach to documenting indigenous knowledge for development. In *The Cultural Dimension of Developent: Indigenous Knowledge Systems,* ed. D.M. Warren, L.J. Slikkerveer, and D. Brokensha. Pp. 426–434. London: Intermediate Technology Publications.

Warren, D.M., and J. Pinkston. 1997. Indigenous African resource management of a tropical rainforest ecosystem: A case study of the Yoruba of Ara, Nigeria. In *Linking Social and Ecological Systems,* ed. F. Berkes and C. Folke. Pp. 158–189. Cambridge: Cambridge University Press.

Warren, D.M., and B. Rajasekaran. 1993. Indigenous knowledge: Putting local knowledge to good use. *International Agricultural Development* 13(4):8–10.

Warren, D.M., and B. Rajasekaran. 1994. Using indigenous knowledge for sustainable dryland management: A global perspective. In *Social Aspects of Sustainable Dryland Management,* ed. D. Stiles. Pp. 193–209. New York: John Wiley.

Warren, D.M., L. Egunjobi, and B. Wahab, eds. 1996. *Indigenous Knowledge in Education.* Proceedings of a Regional Workshop on Integration of Indigenous Knowledge into Nigerian Education Curriculum. Ibadan: Indigenous Knowledge Study Group, University of Ibadan.

Warren, D.M., L.J. Slikkerveer, and D. Brokensha, eds. 1995. *The Cultural Dimension of Development: Indigenous Knowledge Systems.* London: Intermediate Technology Publications.

Warren, D.M., G.W. von Liebenstein, and L.J. Slikkerveer. 1993. Networking for indigenous knowledge. *Indigenous Knowledge and Development Monitor* 1(1):2–4.

World Resources Institute, World Conservation Union, and United Nations Environment Programme. 1992. *Global Biodiversity Strategy: Policy-makers' Guide.* Baltimore: WRI Publications.

Zweifel, H. 1997. Biodiversity and the appropriation of women's knowledge. *Indigenous Knowledge and Development Monitor* 5(1):7–9.

28

INDIGENOUS PEOPLES AND CONSERVATION

Misguided Myths in the Maya Tropical Forest

James D. Nations

During the mid-1970s, I spent three years doing anthropological fieldwork with an indigenous rainforest group called the Lacandon Maya, in Chiapas, Mexico. That experience led me to understand that the maintenance of indigenous traditions and the conservation of biological diversity frequently go hand in hand. Later fieldwork as a conservationist in Central and South America reinforced this understanding, and I came to support two of the unspoken tenets of work with indigenous communities and environmental conservation.

The first of these tenets is that, where there are indigenous people, there are tropical forests, and where there are tropical forests, there are indigenous people. Because tropical forests are the world's foremost repositories of biological diversity, it follows that a positive correlation exists between indigenous people and conservation. Geographer Bernard Nietschmann refined this concept into "the rule of indigenous environments," which states that, "Where there are indigenous peoples with a homeland, there are still biologically-rich environments" (Nietschmann 1992:3).

The second tenet of conservation that I came to support is this: Twenty-five years from now, most of the tropical forests that remain on earth—and thus, the largest repositories of biological diversity—will be those under the control of indigenous people. Indigenous communities have more than an economic attachment to the land and forest. They have spiritual, ancestral, and linguistic ties to the land that are rarely shared by the loggers, ranchers, and slash-and-burn colonists who also seek to utilize tropical forests.

While it took me only a few years of anthropological fieldwork to learn to support these two beliefs of contemporary conservation, it took me only a few years as a conservationist to begin to question them. In the cold, hard light of the early twenty-first century, these two beliefs are complicated by the only statement that a social scientist can accurately make about any element of human behavior: It depends.

This law of social science holds true in the case of indigenous people and biodiversity conservation, as well. It is not always true that where there are indigenous people there is also tropical forest, nor is it always true that where there is tropical forest, one will find indigenous people. In at least one region of the Latin American tropics, the Maya tropical forest of Mexico, Guatemala, and Belize, these beliefs about cultural diversity and conservation potentially threaten, rather than promote, the maintenance of the region's biological heritage.

THE LOWLAND MAYA

The Maya tropical forest is a single rainforest ecosystem divided by history into three independent countries. Crisscrossed by tributaries of the Usumacinta River, the forest stretches from the Selva Lacandona of eastern Chiapas, Mexico, across the Department of the Petén in northern Guatemala, and through much of the nation of Belize.

This lowland tropical forest was the heartland of Classic Maya civilization, and for more than a thousand years, millions of Maya people used the region's natural resources as the biological foundation for the most advanced New World civilization of its time. When Classic Maya civilization disintegrated around 900 a.d., hundreds of thousands of Maya people continued to live in the region, although only in a few isolated pockets did they maintain the written records and mathematical expertise the Maya had developed over the centuries.

When Spaniards invaded the Maya world in the early sixteenth century, as much as 90 percent of the Maya population died from introduced diseases. That any cultural traditions survived at all is a miracle. Yet, some of the Maya peoples who held on and who continue to live in the Maya tropical forest today—especially the Lacandon, Itzaj, and Mopan—maintain an amazing array of traditional knowledge about the forest, its wildlife, and its ecology. These groups conserve this knowledge in their original lowland Maya languages, each of which preserves the cultural memory of centuries of biological trial and error.

INDIGENOUS (AND OTHER) THREATS TO BIODIVERSITY

Despite the efforts of these lowland Maya groups to protect their cultural and biological heritage, the invasion of their territory by thousands of highland Maya people, as well as loggers, cattlemen, and Spanish-speaking colonists, threatens to eliminate the knowledge and resources they have kept alive. This reality belies the statement that, where there are indigenous people there are tropical forests, and vice versa.

The first tenet of indigenous people and biodiversity is countered by the reality that tens of thousands of indigenous people live today in areas of the Maya forest where biological diversity has been seriously depleted. The lowland tropical forests of the northern Selva Lacandona and of the Ocosingo Valley of Chiapas, for example, have been almost entirely eradicated. Throughout much of the rest of eastern Chiapas, Tzeltal, Tzotzil, and Tojolabal Maya families are clearing and burning the remaining forest at breakneck speed to grow corn, chiles, and beef cattle (Nations 1994).

Also bending under the weight of reality in the Maya tropical forest is the second tenet I once held, that the only forests to survive in the future will be those controlled by indigenous communities. The largest natural areas in the Maya tropical forest today are found in national parks and biosphere reserves, and one of the major threats to their survival is the expanding agricultural frontier, pioneered in most cases by indigenous families.

The biggest block of forest in the region, the 1.6 million hectare Maya Biosphere Reserve of northern Guatemala, is threatened by the expansion of oil roads, cattlemen, and Q'eqchi' Maya farmers moving in from the southern Petén, an area they originally migrated into from the mountainous area of Alta Verapaz, Guatemala (see Atran this volume). In a strange reversal of the original belief, the communities working to protect the Maya Biosphere Reserve are rural, Spanish-speaking Ladinos (mestizos). As centuries-long harvesters of chicle, allspice, and the *Chamaedorea* floral palms called *xate,* these 6000 Spanish-speakers are watching their economic and spiritual ties to the forest go up in the smoke of agricultural fires set by indigenous families and cattlemen. In all but a very few cases, Maya indigenous communities of the Maya tropical forest have become a threat to biodiversity rather than guardians of its future.

UNDERLYING CAUSES

It is crucial to point to the underlying element that explains this situation. All but a few of the Maya groups living in the Maya tropical forest today represent displaced communities that are expanding out of their original homelands in the highlands in reaction to economic factors and population growth. The acquisition of their original homelands by outsiders during the past five hundred years—in some cases during the past hundred years—set the stage for this out-migration. Hence, the qualifying phrase in Nietschmann's rule of indigenous environments: "Where there are indigenous peoples *with a homeland,* there are still biologically-rich environments" (emphasis added).

In the specific case of northern Guatemala, thousands of Q'eqchi' Maya families are migrating into the remaining tropical forest of the Maya Biosphere Reserve. Traditional agricultural techniques that served the Q'eqchi' in their homeland of Alta Verapaz evolve into destructive patterns of deforestation in the lowland forest. The Q'eqchi's production of cattle, chiles, and corn as cash crops degrades the forest land they clear and continually pushes families to expand into new forest areas, aiming them toward a future that serves neither them nor the sustainable use of biological resources.

Anyone determined to find a villain in this process would have to look back a hundred years to the social and political movements that established Alta Verapaz as one of Guatemala's premier coffee-producing areas. During the last quarter of the nineteenth century, the central governments of Guatemala declared almost a half million hectares of Q'eqchi' homeland as "empty lands," thus allowing foreign, mostly German, entrepreneurs to create a coffee export empire based on Q'eqchi' land and labor (Wilson 1995:35). By the end of the 1870s, "there was no untitled land left in Alta Verapaz," and by 1885, two-thirds of the commerce in the department was in the hands of German coffee magnates (Wilson 1995:36).

Anthropologist Richard Wilson cites an 1867 letter from the inhabitants of a Q'eqchi' community to the president of Guatemala: "After having had our houses and farms, which are the fruit of our labor, taken away from us . . . the Commissioner of Panzos has forced us to plant coffee in the mountains where we grow corn. This appears to be nothing more than an attempt to exterminate us" (Wilson 1995:36). It is unlikely, however, that extermination was the goal of the coffee producers—they depended for success on Q'eqchi' labor on their *fincas*—and the Q'eqchi' population continued to grow during the nineteenth century and into the twentieth.

As Wilson indicates, citing anthropologist Richard Adams (1965), the pressure for food-production land created by expanding coffee plantations was ameliorated somewhat, "by migration to the vast areas of virgin forest in the north of the department and in the Petén and Belize" (Wilson 1995:38). This expansion into virgin territory continues today, fueled more by the Q'eqchi's own export production and population growth than by continuing expulsion from their home territory. The population growth results from the high value placed on child labor in an agrarian economy, improved access to antibiotics, the lack of access to contraceptives, and old-fashioned *machismo* that denigrates the right of women to make their own decisions about the number of children they bear.

THE PROBLEM OF POPULATION GROWTH

Population growth rates in the Maya tropical forest, especially among indigenous families, are among the highest in the world (CARE and CI 1995). The combination of natural increase and in-migration gives the lowland forests of the Maya region some of the fastest growing human populations on earth. In the Selva Lacandona of eastern Chiapas, the current population of 450,000 people is growing at the rate of 7 percent a year, a rate that, unchanged, will double the population within ten years. Fifty-two percent of the population of the Selva Lacandona is under the age of fifteen years and has not yet reached reproductive age (World Bank 1994).

In the neighboring Guatemalan Petén, the population of 400,000 is growing at the rate of 10 percent a year, due one-third to natural increase and two-thirds to in-migration. According to Petén conservation workers, 60 percent of the immigrants are highland Maya, mostly Q'eqchi'. The average number of children per woman in the Petén is 7.1, one of the highest rates in Latin America (Forth and Grandia 1999:89).

As Q'eqchi' Maya families expand northward into the Maya Biosphere Reserve and across the international border into Belize, they are moving into protected areas designed to conserve the biological diversity of the Maya tropical forest. Some of these families openly state that they have a historical right to occupy these lands. Challenged on his family's presence in the Laguna del Tigre National Park, inside Guatemala's Maya Biosphere Reserve, one Q'eqchi' leader in the newly formed community of Paso Caballos responded, "This is the Maya Biosphere Reserve, we're Maya, what's the problem?" (Soza, personal communication 1997).

THE PAN-MAYAN MOVEMENT

The Q'eqchi' leader's response is an extreme expression of a social movement newly emerging in the Maya region. Known as Pan-Mayanism, or *el Movimiento Maya* in Spanish, it is, according to linguist Charles Andrew Hofling, "one of the largest and most powerful revitalization movements in the world today" (1996:108). The Movimiento Maya gained momentum with the awarding of the 1992 Nobel Peace Prize to Rigoberta Menchú, a highland Guatemala Maya woman, in recognition of her struggle for indigenous rights during Guatemala's civil war. The Maya soldiers of the 1994 Zapatista rebellion in Chiapas attracted even more international attention to the movement.

In its original intent, Pan-Mayanism was welcomed by most observers as a reassertion by Maya peoples of rights and traditions denied them by military governments, Ladino politicians, and Spanish-speaking landowners during the last five hundred years. As Victor Montejo, of University of California-Davis, has stated: "To call ourselves Maya is to reject the colonial world of being *Indios*." (remarks as a discussant in the panel session Bordering on the Essence: Pan-Mayan Projects in Comparative Perspective at the 1997 meeting of the American Anthropological Association).

In its extremist expression, however, Pan-Mayanism takes on the mantle of deforestation and land speculation in the name of a proud social movement. Highland Maya communities are expanding into the lowland forest in Chiapas, the Guatemalan Petén, and southern Belize, carrying the banner of Pan-Mayanism and a questionable historical claim that, because the Classic Maya once occupied all this territory, it should be open to any Maya peoples today. In reality, during the Classic, Post-Classic, Conquest, and Colonial eras, the Maya tropical forest was occupied by Yucatec and Cholan (Chol, Cholti, and Chorti) speakers, who included the ancestors of today's lowland rainforest Maya: the Lacandon, Itzaj, and Mopan.

Combined with incongruous farming practices and rapid population growth, the geographical expression of Pan-Mayanism portends an environmental and social disaster. It is tempting to justify highland Maya invasions of lowland forests as a case of oppressed communities rightfully taking the land that has been denied them, but it is the environment, and subsequently indigenous communities themselves, that suffer the consequences.

In lowland Chiapas, for example, the expansion of highland Maya families into the traditional territory of lowland Maya peoples is already creating land disputes between indigenous Maya groups who normally would be allies. The movement of Tzeltal families from the Ocosingo Valley into the indigenous reserve of the Comunidad Lacandona is pitting thousands of invading Tzeltal Maya farmers against the 500 Lacandon Maya, 3000 Chol Maya, and 7000 Tzeltal Maya who have legal title to 614,000 hectares of lowland forest land in and around the Montes Azules Biosphere Reserve.

During the 1994 Zapatista rebellion, the environment of lawlessness created by the military stalemate between rebels and Mexican army troops led to a half dozen invasions of the Comunidad Lacandona by their neighboring Maya cousins, sometimes pitting Tzeltales against other Tzeltales. When the Zapatistas initially stated in press releases that their goal was to secure land for the indigenous families of the Selva Lacandona, the elected representatives of the Comunidad Lacandona—one each from the Lacandon, Chol, and Tzeltal settlements that comprise it—immediately issued an open letter denying involvement in the rebellion and seeking government support for protection of their land.

INDIGENOUS PEOPLES AND CONSERVATION

Market forces, land degradation, and population growth are pushing highland Maya families into lowland forests throughout the Maya region, turning upside down the rule that indigenous people and conservation are natural allies. The three clearest cases of indigenous communities who can be called conservationists in the Maya tropical forest are the Lacandones in Chiapas, the Itzaj in the Guatemalan Petén, and the Mopan in southern Belize. In all three instances, the groups are small in number, are still living on what remains of their traditional homeland, and are attempting to defend their forested territory against loggers, cattle ranchers, and invading Maya cousins alike.

The situation in the Maya tropical forest illustrates the politically correct but shortsighted view that Kent Redford of the Wildlife Conservation Society has called "the myth of the ecologically noble savage," the idea that indigenous communities can automatically be equated with conservation (Redford 1991). As the case of the Maya forest illustrates, the priority of many indigenous peoples is not conservation, but land and rights to resources.

That reality should lead us, as anthropologists and conservationists, to a redefined view of indigenous communities in the Latin American tropics. This view was best expressed by University of Idaho anthropologist Tony Stocks, and can be paraphrased as follows: Biological diversity may be conserved in traditional indigenous communities with secure land tenure, low population density, and low involvement in the cash economy, but in most other cases, increasing cash needs and high population growth are changing indigenous peoples' relationship with the ecosystems they inhabit (Stocks 1996:23).

This new perspective does not mean that the positive dialogue between conservationists and indigenous peoples has run its course. On the contrary, indigenous peoples can still be one of the best hopes for maintaining large landscapes in a more or less natural state (Stearman 1996:264). But to benefit from this hope, conservationists must address the priority issues of indigenous peoples: land, resource rights, education, and health care.

In the Maya tropical forest, there are very specific actions we can take to help achieve these goals:

1. We can assist indigenous communities in securing legal title to their original homelands so they can develop sustainable land-use practices on territory they own and control. We must recognize, however, that having a homeland does not automatically end the destruction of biodiversity, if human and livestock populations continue to expand at rapid rates.

2. We can help indigenous communities adapt their agricultural systems to the reality of the twenty-first century. The reality is that we will watch the world's population expand over the next forty years to probably 8 to 10 billion people (Cohen 1995). Even if per capita consumption remains constant, this population growth will require us to almost double world food production, and to do so in environmentally sustainable ways.

These increases in food production will have to come from land already under production if the biological diversity of tropical forests is to survive. Tropical forests and tropical forest peoples will not endure if we feed the world's expanding population by increasing the amount of land under cultivation. Traditional indigenous agriculture, with its centuries of trial and error and human-guided crop diversity, could be one of the keys to increasing yields on existing agricultural lands.

3. In the Maya tropical forest, we have to move beyond the romantic notion of the Maya as *Hombres de Maiz*, Men of Corn. For example, "Today corn farming remains not just a way to feed the family, but an affirmation of Mayan identity" (Simon 1997:85). The misguided idea that every Maya family has a genetic right to a plot of forest and a bag of corn seed runs counter to that family's survival and to ours as biological beings. Deforestation rates and population growth have relegated this vision to history. Even more, in Mexico, the economic viability of the Men of Corn died with the arrival of the North American Free Trade Agreement, NAFTA, which—like it or not—established a yearly, duty-free import quota of 2.5 million metric tons of corn into Mexico each year and a total phase-out of tariffs on corn imports over that amount during the next 15 years (Hufbauer and Schott 1993:47–57). The political reality is that Mexico will now be importing corn from Iowa and selling vegetables and petroleum in exchange.

4. We can work to defuse land conflicts between indigenous communities by financing and providing logistics for dialogue between the groups in their native languages. Once we assist the groups in coming together to talk, we—as outsiders—can stay out of the way, or better yet, serve coffee.

5. We can support projects that salvage and disseminate indigenous conservation traditions. In order to do so, we must assist indigenous groups in studying these environmental traditions themselves. There is a great deal to be gained by paying interested young people a salary to study under their tribal elders and shamans.

We also need a new generation of professionally trained fieldworkers who are willing to learn indigenous languages, live in the field, and train their indigenous colleagues to record the centuries-old information that is evaporating with the death of traditional leaders and the acculturation of those who survive. Training graduate students to produce fusty, theoretical treatises on this year's fad in social science is a waste of time and a waste of minds. Fashions in science come and go, but accurate field data last forever. Aldo Leopold's warning about natural ecosystems holds true for traditional knowledge as well: "The first rule of intelligent tinkering is to save all the pieces" (Leopold 1989).

It is also imperative that we work with nonindigenous traditional forest dwellers, who sometimes display the type of conservation ethic frequently associated with indigenous communities. One of the best examples is that of the Spanish-speaking chicle and allspice harvesters of the Guatemalan Petén. These groups can be conservationists for the same reasons that indigenous people can be: they want continued access to land and sustainable natural resources for their own economic survival.

6. We need to work on the challenge of population growth, in the Maya tropical forest and throughout the world, especially in our own overconsuming societies. The best way to do this in traditional communities is to improve the status of women and educate young girls. United Nations studies indicate that improv-

ing the economic status of women increases their freedom to make their own decisions within the family and within the community, and when women in the developing world are allowed to make their own decisions about how many children they produce, one-third fewer children are born (Cain 1984; World Bank 1985; Jacobson 1993).

We can help women improve their status within the family and within the community by helping them create microenterprises and by financing the education of young girls. This education can be bilingual and it can—and should—include education in cultural traditions, especially traditions that promote the sustainable use of land and natural resources. These conservation traditions, expressed in native languages, are what Hazel Henderson called "the cultural DNA" that can help us create sustainable economies in healthy ecosystems on this, the only planet we have (Gell-Mann 1994:292).

7. Finally, as anthropologists and conservationists, we must slough off misguided visions of biological conservation and indigenous movements and focus on the environmental reality that confronts all human beings as we stumble into the new millennium.

REFERENCES

Adams, R. 1965. *Migraciones internas en Guatemala: Expansión agraria de los indígenas Kekchíes hacia el Petén*. Guatemala: Seminario de Integración Social Guatemalteca.

Cain, M. 1984. *Women's Status and Fertility in Developing Countries: Son Preference and Economic Security*. World Bank Staff Working Paper no. 682. Washington, D.C.: World Bank.

CARE and CI. 1995. *Análisis de los impactos ambientales actuales y potenciales del proceso de reintegración de los retornados a Guatemala y recomendaciones para su mitigación: Informe final*. Guatemala: CARE, Conservation International.

Cohen, J. 1995. *How Many People Can the Earth Support?* New York: W.W. Norton.

Gell-Mann, M. 1994. *The Quark and the Jaguar*. New York: W.H. Freeman.

Forth, M., and L. Grandia. 1999. Population and environment in the Petén, Guatemala. In *Thirteen Ways of Looking at a Tropical Forest: Guatemala's Maya Biosphere Reserve*, ed. J.D. Nations. Pp.85–91. Washington, D.C.: Conservation International.

Hofling, C.A. 1996. Indigenous linguistic revitalization and outsider interaction: The Itzaj Maya case. *Human Organization* 55(1):108–116.

Hufbauer, G.C., and J.J. Schott. 1993. *NAFTA: An Assessment*. Rev. ed. Washington, D.C.: Institute for International Economics.

Jacobson, J.L. 1993. Closing the gender gap in development. In *The State of the World 1993*. Pp. 61–79. Washington, D.C.: Worldwatch Institute.

Leopold, A. 1989. *A Sand County Almanac: Sketches Here and There*. New York: Oxford University Press.

Nations, J.D. 1994. The ecology of the Zapatista revolt. *Cultural Survival Quarterly* 18(1): 31–33.

Nietschmann, B.Q. 1992. *The Interdependence of Biological and Cultural Diversity.* Occasional Paper no. 21. Center for World Indigenous Studies, December 1992.

Redford, K.H. 1991. The ecologically noble savage. *Cultural Survival Quarterly* 13(1):46–48.

Simon, J. 1997. The long road home. *Native Peoples* Fall 1997:84–89.

Stearman, A. M. 1996. On common ground: The Nature Conservancy and traditional peoples: The Rio Chagres, Panama, workshop. In *Traditional Peoples and Biodiversity Conservation in Large Tropical Landscapes,* ed. K.H. Redford and J.A. Mansour. Pp. 237–250. America Verde Publications. Arlington, Va.: Nature Conservancy, Latin America and Caribbean Division.

Stocks, A. 1996. The BOSAWAS Natural Reserve and the Mayangna of Nicaragua. In *Traditional Peoples and Biodiversity Conservation in Large Tropical Landscapes,* ed. K.H. Redford and J.A. Mansour. Pp. 1–32. America Verde Publications. Arlington, Va.: Nature Conservancy, Latin America and Caribbean Division.

Wilson, R. 1995. *Maya Resurgence in Guatemala: Q'eqchi' Experiences.* Norman: University of Oklahoma Press.

World Bank. 1985. Slowing population growth. In *Population Change and Economic Development.* Pp. 66–86. New York: World Bank and Oxford University Press.

World Bank. 1994. Interviews with World Bank personnel, Washington, D.C., conducted by Felipe Villagran, consultant on population in the Selva Lacandona. Notes on file with Conservation International, Washington, D.C.

29

BIOCULTURAL DIVERSITY AND LOCAL POWER IN MEXICO

Challenging Globalization

Victor M. Toledo

Most people experience contemporary globalization as threatening local and microregional self-reliance, diminishing participatory democracy, and destroying diversity of cultures, while devouring the last remnants of natural resources and wilderness. In the rural areas, this perverse process is being made possible in all those spaces where individuals and communities are unable to resist the three main mechanisms of globalization: dependency, specialization, and centralization of power.

Challenging these forces, a myriad of rural communities in countries as different as Finland, Japan, India, Australia, Peru, and Mexico have initiated, in the last decades, successful experiences of local organization and control over the global forces (Imhoff 1996; Toledo 1996a, 1997a; Apffel-Marglin 1997; Pietilä 1997). Diversity, self-sufficiency, grassroots democracy, equity, and decentralization of power are basic principles guiding the actions of these local movements. As a whole, these principles represent new paradigms in the construction of an alternative modernity; that is, the creation of a sustainable, ecologically inspired, post-neoliberal society.

By reviewing the successful experiences of hundreds of indigenous communities in Mexico, a megadiversity country in both biological and cultural terms, I will illustrate that the conservation and wise utilization of biodiversity among indigenous communities is a key mechanism of cultural preservation and that the maintenance of this double diversity (biological and cultural) is only possible through the reinforcement and/or maintenance of local power. Thus, in rural areas, biocultural diversity and local empowerment are but the two faces of the same challenging story against perverse globalization.

BIOCULTURAL DIVERSITY OF MEXICO

Mexico is a privileged country, from a biological as well as a cultural point of view. Its territory is home to a large number of species of plants, animals, fungi, and other organisms as well as to a large variety of cultures that speak different languages. As a result, the country has been ranked ninth among the top nations in terms of biocultural wealth (Ryan 1993) and is considered one of the main diversification centers in the world.

Although much speculation has arisen about this underused wealth, the majority of researchers agree that the wide biological and cultural diversity of Mexico is, in turn, the product of another equally important feature, namely, ecological heterogeneity. Ecological diversity in Mexico is the combined result of its topographic and climatic variations, which create a rich mosaic of environmental and microenvironmental conditions. The prominence of mountain ranges throughout the territory and the existence of thirty peaks of over 3000 m above sea level produce an unusual diversity of natural landscapes.

Mexico also has desert regions where it does not rain for many years (e.g., the Baja California Peninsula) and rain forests where the annual precipitation exceeds 5000 mm (e.g., the Tabasco-Chiapas border area). Furthermore, some specialists assert that the Mexican territory includes practically all major types of natural environments known on the planet and that only India and Peru have a comparable environmental diversity. As a result, it is possible to distinguish six main ecological regions in the Mexican territory plus inland waters and the coastal zone (Toledo and Ordoñez 1993).

Diversity of environments and wealth of natural resources were undoubtedly decisive factors in making Mexico the birthplace of major early civilizations of Mesoamerica. As a result contemporary rural Mexico is still, after many centuries of social change, an agrarian space dominated by indigenous or Indian-descended peasants. In fact, according to the 1991 Seventh Agricultural, Forestry, and Peasant National Census of Mexico, nearly 3 million peasant productive units owned half of the Mexican territory (103 million hectares), principally forested areas (70 percent of national total) and agricultural lands (80 percent).

These peasant units (including artisanal fishers) are embodied in two forms of community-based corporate ownership currently recognized and supported by law: *ejidos* and *comunidades*. The *ejido* is a creation of the Mexican Revolution of 1910–17 that enables groups of people to petition for access to resources to which they have no prior claim, whereas the *comunidad* is a preexisting corporate entity

Table 29.1 Land Ownership in Rural Mexico

Ownership	Holdings	Percent	Owned Area[a]	Percent
Community-based	3,040,495	66.3	103.3	59.0
Private	1,410,742	30.8	71.7	40.9
Mixed	133,912	2.9	0.13	0.1
Total	4,585,149	100.0	175.1	100.0

Source: Seventh Agricultural, Forestry, and Peasant National Census of Mexico, 1991.

[a]In millions of hectares.

whose rights are recognized if members can demonstrate prior, longstanding, community-based use of the land and waters. Constitutional recognition of community-based tenure has provided a protective shell for the functioning and evolution of resource management systems responsive to local ecological and biodiversity conditions (Alcorn and Toledo 1995). In summary, while entrepreneurial farms dominate primary activities such as cattle ranching, irrigated areas, and some profitable branches of rain-fed agriculture, peasant units are the main economic agents of forested regions and rain-fed agricultural lands (see table 29.1).

In addition to the mestizo, or Spanish-speaking, producers, there are among the Mexican peasantry 54 main indigenous peoples speaking 240 languages and dialects other than Spanish (Grimes 1988), who live in practically all the main natural habitats of the country from tropical lowlands and temperate mountains to coastlands and deserts or semidesert areas (see map 29.1). According to studies I have carried out with collaborators, most of the indigenous population is distributed through the tropical humid region (1.54 million people), the temperate subhumid area (1.79 million) and the tropical subhumid area (1.11 million) (see table 29.2). Thus, over 90 percent of the indigenous population lives in the forested area of the country (tropical and temperate forests), and less than 10 percent in the arid and semiarid regions.

THE STRONG RELATIONSHIP BETWEEN BIOLOGICAL AND CULTURAL DIVERSITY IN MEXICO

The profound relationship that has existed since time immemorial between nature and culture is a feature that can be witnessed in contemporary Mexico. Each species of plant, group of animals, and type of soil and landscape nearly always has a corresponding linguistic expression, a category of knowledge, a practical use, a religious meaning, or an individual or collective vitality, for the indigenous inhabitants of the country (Toledo 1996b).

Moreover, in the nation with the third richest biological diversity on earth

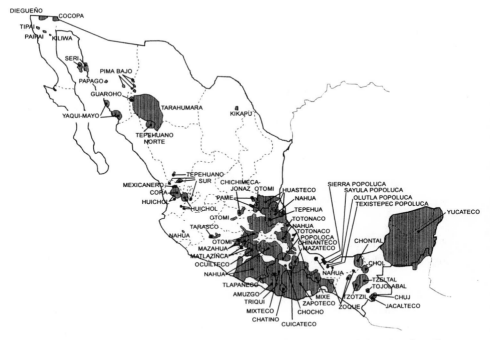

Map 29.1. Geographic distribution of the main indigenous peoples of Mexico. Adapted from E. Díaz-Couder 1987. Map of Indian Languages of Mexico and Central America. Department of Anthropology, University of California, Berkeley.

(Ramamoorthy et al. 1993), the geographical distribution of peasant communities coincides with most of the areas of extraordinarily high biodiversity. For instance, half of the almost 30,000 ejidos and indigenous communities are distributed in the ten biologically richest states of Mexico (see map 29.2), and nearly 80 percent of the territory of the protected areas established by the Mexican government is inhabited by peasants and indigenous peoples. As a consequence, the main biosphere reserves of Mexico are surrounded by or belong to indigenous communities, most of which claim active participation in the management of protected areas. This situation is especially notable in the south (Montes Azules, Chiapas; Calakmul, Campeche; Santa Marta, Veracruz), but it also exists in the central (Sierra of Manantlán, Jalisco, and Reserve of Monarch Butterfly in Michoacán) and northern (El Pinacate, Sonora) portions of the country.

In addition, the high overlap between indigenous peoples and regions with large numbers of species and endemics can be illustrated by the fact that 65 of the 155 priority regions recommended by a selected group of experts convened in 1996 by CONABIO (Comisión Nacional para el Conocimiento y Uso de la Biodiversidad), the governmental institution working on the conservation of Mexican

Table 29.2 Indigenous Population in Each
Ecological Zone of Mexico

Zone	Population	Percentage
Humid tropics	1,574,926	30.0
Subhumid tropics	1,115,948	21.2
Humid temperate	276,829	5.3
Subhumid temperate	1,795,850	34.3
Arid and semiarid	444,610	8.4
Coastal	40,613	0.8
Total	6,411,931	100.0

Source: Author's calculation of statistics from the Eleventh Population
Census of Mexico, 1990.

biodiversity, are indigenous homelands (see map 29.3). As a consequence, nearly
60 percent of the areas located below the Tropic of Cancer that are recommended
for protection overlap, partially or totally, with indigenous territories.

The coexistence of indigenous peoples and biodiversity reaches its maximum
expression in the southern state of Oaxaca. This portion of the Mexican territory
has been recognized by experts as the state with the highest numbers of species
and endemics of plants and animals (vertebrates) (Flores-Villela and Gerez 1988).
Moreover, according to Grimes (1988), Oaxaca is home to indigenous peoples
speaking 104 different languages and dialects; it is ranked first in terms of linguistic diversity. In Oaxaca, 1406 agrarian units (ejidos and indigenous communities)
are the owners of more than 7 million hectares, which represent almost two-
thirds of the state's territory.

BIODIVERSITY AND INDIGENOUS STRATEGIES
OF NATURAL RESOURCE MANAGEMENT

Since peasant production is based more on ecological exchanges than on economic exchanges, indigenous households tend to adopt survival mechanisms that
guarantee an uninterrupted flow of goods, materials, and energy from nature
(Toledo 1990). Thus, they carry out a nonspecialized production based on the
principles of diversity of landscapes, resources, species, genes, and as a consequence, of productive practices. This multiuse strategy, through which indigenous peasants maintain and reproduce their productive systems, constitutes an
ecologically valuable characteristic that tends to conserve natural resources and
that buffers peasant households against both market fluctuations and environmental changes and hazards.

It has been demonstrated that some natural disturbances can increase biodiver-

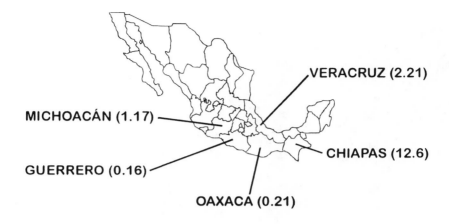

Map 29.2. Number of agrarian units (*ejidos* and indigenous communities) and owned area in million hectares in the 10 biologically richest states of Mexico. The number between parentheses indicates the percentage of the state's territory under some kind of protection by 1996. Sources: Flores-Villela and Gerez (1988); Seventh Agricultural, Forestry and Peasant National Census of Mexico (1991). Statistics from the Instituto Nacional de Ecología, SEMARNAP, Mexico.

sity if they increase habitat heterogeneity, reduce the influence of competitively dominant species, or create opportunities for new species to invade (Harris 1984; Brown and Brown 1992). The creation of landscape mosaics under the indigenous multiuse strategy in areas originally covered by only one natural community is a human-originated mechanism that theoretically tends to maintain and even increase biodiversity. In fact, several authors have stressed the importance of the models of low-intensity mosaic usage of the landscape (including polycultures) by indigenous peoples and other smallholder populations for biodiversity conserva-

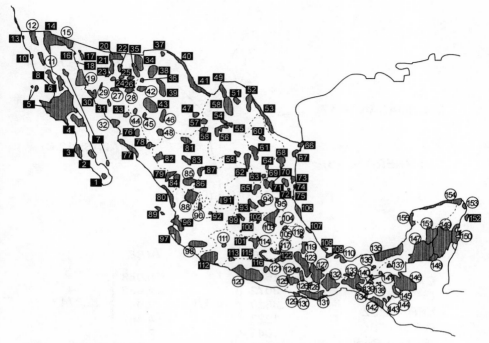

Map 29.3. Overlap of the 155 conservation priority areas of Mexico recommended by CONABIO (Comisión Nacional para el Conocimiento y Uso de la Biodiversidad) with territories of indigenous communities (numbers in circles).

tion (Altieri, Merrick, and Anderson 1987; Alcorn 1991, 1994; Gonzalez-Bernaldez 1991; Brown and Brown 1992; Toledo, Ortiz, and Medellín-Morales 1994).

In Mexico, the indigenous multiuse strategy has been documented by numerous studies (see, for instance, the papers included in Leff and Carabias 1993), especially in the humid tropics where this strategy is a historical synthesis resulting from the combination of elements of pre-Hispanic origin (such as maize fields, forest products, and fishing) with crops derived from European contact (coffee, sugar cane, citrics, or cattle raising) and even "modern" practices and inputs (such as the use of improved seeds or agrochemicals).

The indigenous strategies tend to include: (1) the Mesoamerican milpa system that generally constitutes a polyculture including up to 20–25 agricultural and forestry species; (2) the extraction of timber and nontimber products from the primary or mature forests and their secondary forests of different ages resulting from the succession process; (3) the manipulation of forests in different stages of anthropic disturbance, from which different products are obtained (mainly vanilla, coffee, and cocoa); (4) the management of home gardens, which are agroforestry systems located close or next to dwellings; (5) the extraction of products from available water bodies (rivers, lakes, and swamps); (6) the management of small-scale cattle-raising areas, which are generally pasture areas combined with legume species of trees and shrubs; and (7) management of forestry or agricultural plantations for cash crops such as sugarcane, sesame, citrics, or rubber.

SUSTAINABLE DEVELOPMENT AT COMMUNITY LEVEL:
A KEYSTONE CONCEPT

Since 1987, when the World Commission on Environment and Development published the Bruntland report, the academic, scientific, and policy-making communities have focused considerable attention on the application of the concept of sustainability to the primary or agricultural sector. This interest took on an international dimension through the governmental agreements derived from the Earth Summit held in Rio de Janeiro in 1992 (especially Agenda 21). But community development has been a main concern for academics for a long time, and recent contributions have emphasized two main social aspects: farmer participation and local empowerment.

Despite this interest, the new paradigm of sustainability has been adopted more as a technical or productive factor in order to generate an alternative or sustainable agriculture than from an holistically visualized perspective. For this reason, a broader and more integrative approach must focus on sustainable rural development and on the more specific topic of sustainable community development, which is the main concern in Third World countries.

It is possible to define sustainable community development as an endogenous mechanism that permits a community to take (or retake) control of the processes that affect it (Toledo 1997b). This definition derives from a general principle of political ecology that affirms that the fundamental reason why contemporary society and nature suffer generalized processes of exploitation and deterioration is the *loss of control* of human society over nature and itself. From this perspective, the history of humanity can be seen as a movement toward an ever greater loss of control over the processes that affect people and their surroundings, in stark contrast to the paradigm of progress. In other words, self-determination and local empowerment, conceived as a "taking of control," have to be the central objectives in all community development.

In fact, local self-reliance is visualized as the primary goal of sustainable development of rural communities, which have remained largely exploited by centralized urban powers throughout history (see a global review of this situation in Powelson and Stock 1987). It is possible to distinguish among six different types of process, which, in turn, are based on the application of nine practical and philosophical principles (see fig. 29.1).

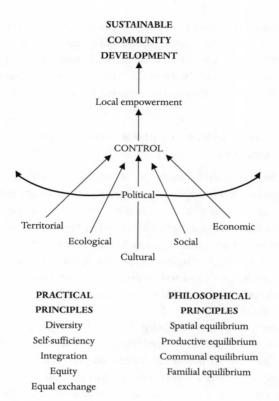

Figure 29.1. Principles and processes of sustainable development at the village community level.

The first action that a community should take is to establish control over its territory. This action implies the definition of the territory's boundaries and the legal or judicial recognition by national states of this territory as belonging to local communities and their members. The adequate or nondestructive use of natural resources (flora, fauna, land, water resources, etc.) that form part of a community's territory constitutes the second, ecological, form of taking control of a rural area. It succeeds through the design of, and a stake in, the execution of management plans for natural resources capable of standardizing and regulating agriculture, cattle-raising, forestry, and fishing. Such a management plan would imply the elaboration of ecological diagnostic techniques, some kind of inventory and, if possible, a geographical information system that would be able to evaluate the ecological resources of a community's territory.

Cultural control—that is, the community making decisions that safeguard its cultural values—includes language, customs, knowledge, beliefs, and lifestyles. The community must create mechanisms that guarantee cultural rescue and the consciousness of its members regarding the necessity of maintaining their main cultural traits (ethnic resistance). The improvement of the quality of life of the members of a community is a central task of all community development, and it is a part of establishing social control. This includes such aspects as nutrition, health, education, housing, sanitation, recreation, and information.

The regulation of economic exchanges that link a community with the rest of society, or with international markets, constitutes the taking of economic control. It implies confronting from a community perspective the external economic phenomena that affect a community's productive life, such as the politics of price-fixing (for the market or the state), macroeconomic politics, subsidies, loans, and the like. It implies attenuating the mechanisms that inhibit and constrain the productive sphere of a community.

Finally, the last dimension is the taking of political control. It implies a capacity for a community to organize itself socially and productively in such a way as to promulgate or ratify the norms, rules, and principles that govern the community's political life. The taking of political control would not succeed without the exercise of real community democracy. It implies decision-making based on the consensus of the members of a community, the rights and aspirations of individuals and families, and the defense of the community for all.

Each of these six dimensions of community development (territorial, ecological, cultural, social, economic, and political) cannot easily exist without the others. Put in another way, the reclaiming of control must be integrated and complete, with these six dimensions included together. For example, it is not possible to maintain and defend culture while the destruction of natural resources persists, which in turn tends to affect the quality of life. But the defense of culture and nature, the maintenance and improvement of the quality of life of the members of a community (the producers and their families), and the suppression of those economic injustices that perpetuate unequal exchange within society, will prove difficult tasks if a true political organization does not exist. The taking of political control (that is, community democracy) is, without doubt, the pivotal action on which the other dimensions of control depend.

These six processes that form true sustainable community development can only succeed to the extent that the members of an indigenous community acquire, increase, and consolidate a community consciousness. In the majority of Third World countries, rural communities find themselves permanently under siege by the destructive forces of "modernizing development," based on the destruction of nature and collective wealth, and the consecration of individual interests—forces that an industrial, technocratic, materialistic society increasingly imposes in all corners of the world. For this reason, initiatives of sustainable development must take into account the social situation of the communities to be considered. It is possible to find not only well-organized communities but also ones clearly experiencing social disruption or degradation. In all these cases, however, the community itself must elaborate, as a prerequisite, a plan for its development. This is the essential political and practical instrument of a community's struggle, for its resistance to current forces of disintegration and as the departure point for integrating its actions.

REVIEW OF EXPERIENCES

Observers all over the world have been paying particular attention to the actions, initiatives, and negotiations of the Zapatista rebels, to such an extent that the indigenous rebellion of Chiapas has been identified in media, academic, and political circles as the main social mobilization movement of rural Mexico (see, for instance, Collier and Lowery-Quaratello 1994; Barry 1995). The Zapatistas, however, are not the only actors waging a prominent battle in the countryside of Mexico. Paralleling the Zapatista revolt, an explosive but pacific social movement of ecological inspiration has been growing during the last decade in practically each main indigenous region of rural Mexico.

These invisible but successful actions have been undertaken by numerous peasant and indigenous communities and their regional and national organizations, and developed by themselves or with the help of nongovernmental organizations, scientific and technical institutions, some state agencies, and progressive international foundations (for a detailed analysis of this movement, see Toledo 1999, and other approaches in Nigh 1992, Toledo 1992, Carruthers 1997). All these actions are explicitly or implicitly based on the principles discussed in the previous section. In other words, these community-based movements of the Mexican countryside are searching for political objectives similar or close to sustainable development as defined previously.

It is estimated that more than 2000 rural communities, principally in the central and southern portions of Mexico, are involved in some kind of environmentally motivated action (Toledo 1999). These rural movements cover the following aspects: defense and sustainable management of temperate lakes in the west (Pátzcuaro, Zirahuén, Chapala), sustainable management of tropical forests in the southeast (principally Quintana Roo, Campeche, and Chiapas) and of mountain temperate forests (Oaxaca, Michoacán, Durango, and five other states), and organic production, based on agroecological principles, of vanilla (Chinantla, Oaxaca), honey (Colima), brown sugar (Las Huastecas), cocoa (Chiapas), and especially coffee (several states). Other indigenous movements include demands to protest a dam-building program (by the Consejo de Pueblos Nahuas del Alto Balsas) or a tourist megaproject (Tepoztlán, Morelos), as well as defenses of local or regional natural resources threatened by urban overuse or industrial contamination (oil pollution), such as fisheries of coastal lagoons (Tabasco, Campeche, and

Michoacán), water springs (in Guerrero and Mexico), or rivers (Morelos). Finally, there are also indigenous communities involved in conservation actions or microregional projects of ecotourism (Oaxaca, Yucatan, Sonora, Chihuahua).

The geographical, ecological and cultural importance of these experiences can be illustrated by the following facts. The forestry communities, which have been winning important political and organizational battles, are principally represented by the National Union of Forestry Community Organizations (UNOFOC). This union, with nearly 550 ejidos and comunidades covering a forested area of over 4 million hectares, produces 40 percent of the Mexican timber and includes communities belonging to 12 indigenous peoples in 10 states. The UNOFOC is demanding from the Mexican government changes in forest law, technical support, and economic incentives to realize sustainable management of the forests. Its strategy includes forestry management for timber and nontimber products, food self-sufficiency, conservation of biodiversity, soils, and water, and improvement of the quality of life of local people. These circumstances leave Mexico supporting the largest experiment with community-based forestry in the world (Bray 1995).

Within the context of worldwide coffee production, Mexico at present stands fourth in terms of volume, fifth in terms of harvested surface, and ninth in terms of yield performance. It is estimated that the number of coffee producers reaches approximately 200,000. In Mexico, 70 percent of coffee production is performed by social groups pertaining to rural communities (ejidos and indigenous communities). To a large extent, this community-based sector is made up of producers belonging to 28 indigenous peoples, among whom the Zapotecs, Mixtecs, Mixes, Totonacs, Nahuas, Huastecs, Tzeltales, Zoques, and Chatinos stand out (Moguel and Toledo 1996). Thus, in Mexico, small-scale growers dominate in terms of population and of land planted in coffee. They maintain multilayered, shaded coffee agroforests that contrast with the modern, agroindustrial, unshaded coffee plantations, which utilize chemical inputs and year-round labor.

An interesting consequence of these methods is that Mexico is the world's primary exporting country for organic coffee (accounting for one-fifth of the total volume) and that this production, carried out without using agrochemical additives and under a shade of native trees, is performed by indigenous producers of 450 communities of Oaxaca, Chiapas, Guerrero, and other states. In Oaxaca alone, the state with the highest biodiversity in the country, there are 35 indigenous organizations with about 18,000 coffee producers (Moguel 1991), one quarter of which are moving in an ecologically sound direction in order to export organic coffee.

More interesting yet is that the National Coordination of Coffee Organizations (CNOC), which is the most important organization of small coffee growers in Mexico, is adopting a development policy based on the new paradigm of sustainability. Through this policy, CNOC is promoting and/or reinforcing ecologically and socially sound coffee-growing systems among its roughly 75,000 affiliated producers.

During the last two decades, Mexican coastal ecosystems have been suffering increasing coastal pollution and accelerated destruction of estuarine and coastal habitats. These processes have been especially strong in the tropical waters of the south, where oil extraction, urban coastal development, and tourism megaprojects have contributed to the depletion of marine and coastal resources.

As a response, during the last few years an important social movement of indigenous or artisanal fishing villages has initiated a variety of political actions against oil interests and other industrial polluters in defense of its natural resources. Through the National Movement of Riverine Fishermen, these fishery communities have fought against oil pollution in different regions of Michoacán, Veracruz, Campeche, and especially Tabasco, where the Chontal Indians are the main agents of these actions. In addition, while negotiating with the Mexican government, they have initiated new development programs inspired by the principles of sustainability.

In summary, despite the fact that indigenous peoples of Mexico possess a very long tradition of political struggles, principally for human and land rights, one of the characteristics of the new actions is the shift away from exclusive focus on land tenure and human right issues. In fact, the recent political processes have been demonstrating that the best way to defend the human rights of indigenous peoples is by encouraging self-organization and the economic improvement of local communities and that this situation is only possible through correct, sustainable management of natural resources. Thus, all these new social movements are acutely aware that the defense of their culture and modes of life is linked to controlling and managing their lands and natural resources as well as to their socioeconomic relations with the markets and the entire society.

As pointed out by Nigh (1992), most of these new movements are utilizing the same collective organization based on traditional concepts of reciprocity, communal property, and voluntary work in order to create business corporations capable of providing quality products at competitive prices in the open market. This allows what Bonfil (1991) called *cultural control,* which is to say, the capacity of local communities and people to make their own decisions on different aspects of daily life. This is, ultimately, the main objective of any sustainable development initiative at the community level.

BIOCULTURAL DIVERSITY AS THE BASIS FOR AN ALTERNATIVE MODERNITY

According to linguists, during the twenty-first century, 3000 of the existing 6000 languages will perish and another 2400 will come near to extinction, leaving just 600 languages in the "safe" category of 100,000 speakers or more (Hale 1992). If the critical mass is 100,000 speakers, then in Mexico the majority of indigenous

Table 29.3 Size Classification of Indigenous Languages of Mexico by Number of Speakers

1,000,000 or more	Nahuatl
500,000 to 1,000,000	Maya
250,000 to 500,000	Mixtec, Zapotec, Tzeltal, Tzotzil, Otomí, Totonac
100,000 to 250,000	Chol, Chinantec, Huastec, Mazahua, Mazatec, Purépecha
50,000 to 100,000	Tarahumara, Tlapanec, Zoque
10,000 to 50,000	Amuzgo, Chatino, Chontal ofTabasco, Cora, Cuicatec, Huave, Huichol, Kanjobal, Mame, Mayo, Popoluca, Tepehua, Tepehuan, Tojolabal,Trique
Less than 10,000	All other (24)

Source: Eleventh Population Census of Mexico, 1990.

groups are in the endangered category. In fact, 12 ethnic groups have populations over 150,000, five groups have between 100,000 and 150,000, and twenty-nine are under the survival threshold of 100,000 (see table 29.3).

Although many factors have been invoked as causes of the depletion of cultural identity, "development" seems to be one of the main agents of this ethnic erosion. Where indigenous peoples do not face physical destruction and even maintain their territories and lands in a safe way, they may, nevertheless, face disintegration through practices promoted by "modernization." The vigorous criticism that political ecology has spawned in relation to rural production allows us to glimpse two alternative methods for rural development in the contemporary world: the "modern" (or technocratic) and the "postmodern" (based on the new paradigm of sustainability). The "modern" way assumes that the only valid knowledge for managing natural resources is the one generated at Western academic institutions (universities and research centers). This alternative tends to ignore the experiences of indigenous peoples that inhabit a specific region.

The "modern" way, based on the intensive use of fossil fuels, capital, machinery, and other inputs, is essentially ecologically destructive since it is unable to maintain production systems over long periods of time without the deterioration of natural resources. Mainly directed toward establishing specialized production units, the "modern" way promotes the depletion of biodiversity by molding natural resources and converting them in a "factory floor" of merchandise. Thus, rural modernization as a technocratic process tends to consist in the compulsive substitution of peasants' small-scale, solar-energized productive units based on biodiversity management with entrepreneurial or collective, large-scale, fossil-dependent holdings based on biodiversity suppression. Finally, under the pressure of these tendencies, both cultural diversity and biological diversity become endangered.

In contrast, a new strategy of sustainable development at the community scale can be converted into appropriate options for rural modernization based on the evolution—not the substitution—of indigenous community-based practices, on the adequate management of biodiversity, and on effective actions of local empowerment. In Mexico, indigenous communities are playing an active role in the search for alternative routes to modernization. These actions contrast with, and challenge, the recent agrarian policies promoted by the government in the Mexican countryside, which are the national-scale expression of a global process of integration. For instance, the changes to the Agrarian, Forestry, and Fisheries laws promoted by the Mexican government during 1992–93 are part of a political strategy in the midst of the opening of the Mexican economy to world trade and foreign investment. These reforms have altered key structural aspects of land tenure and redefined how agriculture, forestry, and fisheries fit into the national and international economies. Thus, the North American Free Trade Agreement (NAFTA) is leaving without protection both cultural diversity, represented by the peasant small-scale farmers, and biological diversity, embodied in the natural resources of Mexico (Toledo 1996c).

In Mexico, nature has played a decisive role for indigenous peoples, as an ally and a refuge, as a source of inspiration and as a catalyst for the productive and creative aspirations of thousands of communities. Therefore, in Mexico—where one hundred plant species including corn were domesticated, and where the ancient cultures are still present in three million peasant production units (ejidos and comunidades) that make use of the resources of the country ranked third in the world in biological richness—an approach based on sustainability is the most important path to an alternative rural modernization. This is perhaps the only, certainly the best, way to guarantee biocultural diversity.

REFERENCES

Alcorn, J.B. 1991. Ethics, economies and conservation. In *Biodiversity: Culture, Conservation, and Ecodevelopment*, ed. M. Oldfield and J.B. Alcorn. Pp. 317–349. Boulder, Colo.: Westview Press.

Alcorn, J.B. 1994. Noble savage or noble state? Northern myths and southern realities in biodiversity conservation. *Etnoecológica* 3:7–19.

Alcorn, J.B., and V.M. Toledo. 1995. The role of tenurial shells in ecological sustainability. In *Property Right in a Social and Ecological Context*, ed. S. Hanna and M. Munasinghe. Pp. 123–140. Washington, D.C.: World Bank.

Altieri, M., L. Merrick, and M.K. Anderson. 1987. Peasant agriculture and the conservation of crop and wild plant resources. *Conservation Biology* 1:49–53.

Apffel-Marglin, F. 1997. Counter-development in the Andes. *Ecologist* 27(6):221–224.

Barry, T. 1995. *Zapata's Revenge: Free Trade and the Farm Crisis in Mexico*. Boston, Mass.: South End Press.

Bonfil, G. 1991. *Pensar Nuestra Cultura*. México, D.F.: Alianza Editorial.

Bray, D. 1995. Peasant organizations and the "permanent reconstruction of nature": Grassroots sustainable development in rural Mexico. *Journal of Environment and Development* 7:67–77.

Brown, K.S., and G.G. Brown. 1992. Habitat alteration and species loss in Brazilian forests. In *Tropical Deforestation and Species Extinction*, ed. T.C. Whitmore and J.A. Sayer. Pp. 119–142. London: Chapman & Hall.

Carruthers, D.V. 1997. Agroecology in Mexico: Linking environmental and indigenous struggles. *Society and Natural Resources* 10:259–272.

Collier, G.A., and E. Lowery-Quaratiello. 1994. *Basta! Land and the Zapatista Rebellion in Chiapas*. Oakland, Calif.: Institute for Food and Development Policy.

Flores-Villela, O., and P. Gerez. 1988. *Conservación en México: Síntesis sobre vertebrados terrestres, vegetación, y uso del suelo*. Xalapa, Ver., Mexico: Instituto Nacional de Investigaciones sobre Recursos Bióticos and Conservación Internacional.

Gonzalez-Bernaldez, F. 1991. Diversidad biológica, gestión de ecosistemas, y nuevas políticas agrarias. In *Diversidad Biológica/Biological Diversity*, ed. F.F. Pineda et al. Pp. 23–32. Madrid: Fundación R. Areces.

Grimes, B., ed. 1988. *Ethnologue: Languages of the World*. Dallas: Summer Institute of Linguistics.

Hale, K. 1992. On endangered languages and the safeguarding of diversity. *Language* 68:1–2.

Harris, L.D. 1984. *The Fragmented Forest*. Chicago: University of Chicago Press.

Imhoff, D. 1996. Community supported agriculture. In *The Case against the Global Economy*, ed. J. Mander and E. Goldsmith. Pp. 425–433. San Francisco: Sierra Club Books.

Leff, E., and J. Carabias, eds. 1993. *Cultura y manejo sustentable de los recursos naturales*. 2 vols. México, D.F.: Centro de Investigaciones Interdisciplinarias en Humanidades, UNAM and Editorial Porrúa.

Moguel, J. 1991. La coordinadora estatal de productores de café en Oaxaca. In *Cafetaleros: La construcción de la autonomía*, ed. G. Egea and L. Hernández. Pp. 103–122. México, D.F.: CNOC and Servicios de Apoyo Local.

Moguel, P., and V.M. Toledo. 1996. El café en México: Ecología, cultura indígena, y sustentabilidad. *Ciencias* 43:40–51.

Nigh, R. 1992. La agricultura orgánica y el nuevo movimiento campesino en México. *Antropológicas* 3:39–50.

Pietilä, H. 1997. The villages in Finland refuse to die. *Ecologist* 27(5):178–181.

Powelson, J.P., and R. Stock. 1987. *The Peasant Betrayed: Agriculture and Land Reform in the Third World*. Boston, Mass.: Lincoln Institute of Land Policy.

Ramamoorthy, T.P., et al., eds. 1993. *Biological Diversity of Mexico: Origins and Distribution*. New York and Oxford: Oxford University Press.

Ryan, A.T. 1993. Supporting indigenous peoples. In *The State of the World 1993*. Washington, D.C.: Worldwatch Institute.

Toledo, V.M. 1990. The ecological rationality of peasant production. In *Agroecology and Small-Farm Development*, ed. M. Altieri and S. Hecht. Pp. 51–58. Boca Raton, Fla.: CRC Press.

Toledo, V.M. 1992. Toda la utopía: El nuevo movimiento ecológico de los indígenas y campesinos de México. In *Autonomía y Nuevos Sujetos Sociales en el Desarrollo Rural*, ed. J. Moguel et al. Pp. 33–54. México, D.F.: Siglo XXI Editores.

Toledo, V.M. 1996a. Los ejidos y las comunidades: Lugar de inicio del desarrollo sustentable en México. *Revista de la Universidad de Guadalajara* 6:28–34.

Toledo, V.M. 1996b. *Mexico: Diversity of Cultures*. México, D.F.: CEMEX and Sierra Madre.

Toledo, V.M. 1996c. The ecological consequences of the agrarian law of 1992. In *The Reform of Agrarian Reform in Mexico*, ed. L. Randall. Pp. 247–260. Armonk, N.Y.: M.E. Sharpe.

Toledo, V.M. 1997a. La utopía realizándose: El desarrollo sustentable de comunidades y ejidos. *Ojarasca* (nueva época) 4:3–9.

Toledo, V.M. 1997b. Sustainable development at the village community level: A Third World perspective. In *Environmental Sustainibility*, ed. F. Smith. Pp. 233–250. Boca Raton, Fla.: St. Lucie Press.

Toledo, V.M. 1999. El otro zapatismo: Luchas indígenas de inspiración ecológica en México. *Ecología Política* 18:11–22.

Toledo, V.M., and M. de Jesús Ordoñez. 1992. The biodiversity scenario of Mexico: A review of terrestrial habitats. In *The Biodiversity of Mexico: Origins and Distribution*, ed. T.P. Ramamoorthy et al. Pp. 81–101. New York and Oxford: Oxford University Press.

Toledo, V.M., B. Ortiz, and S. Medellín-Morales. 1994. Biodiversity islands in a sea of pasturelands: Indigenous management in the humid tropics of Mexico. *Etnoecológica* 3:37–50.

30

LANGUAGE, ETHNOBOTANICAL KNOWLEDGE, AND TROPICAL PUBLIC HEALTH

Thomas J. Carlson

Most tropical rural populations do not have access to modern pharmaceuticals for their health care needs. Traditional botanical medicine provides a low-cost or free form of medical care for tropical rural communities. The World Health Organization (WHO) has estimated that 80 percent of the world's population use botanical medicines for their primary health care (Farnsworth et al. 1985). The WHO recognizes the importance of traditional medicine and has established the Traditional Medicine Program as a mechanism to assess the efficacy and safety of traditional botanical medicines (Bannermann 1983; Farnsworth et. al. 1985; WHO 1991; Akerele 1992). This program helps the WHO understand how traditional medicine contributes to the public health needs of tropical countries.

UNDERSTANDING BOTANICAL MEDICINES

In tropical rural areas, medicinal plants grow locally and knowledge of how to use these plants as medicine is commonly present within the local communities and cultures. Traditional healers have their own knowledge of the taxonomy and ecology of medicinal plants (Berlin 1992; Berlin, Breedlove, and Raven 1966). They also know the natural history and signs and symptoms of different diseases and have botanical medicine preparations to treat many of these ailments (Maffi 1994; Carlson and King 1997). Physicians have contributed to understanding the knowledge of traditional healers through ethnomedical field research on medicinal plants (see Lozoya 1994; Carlson and King 1997). The presence within traditional medicine of detailed ethnopharmacological systems of medicinal plant preparation, extraction, dose size, dose interval, and mode of administration has been described (Elisabetsky and Setzer 1985; Johns 1990; Elisabetsky and Wannmacher 1993; Iwu 1993; Etkin 1996; Sousa Brito 1996; Zent 1996; Carlson and King 1997). Ethnobiological knowledge ensures that these tropical rural communities know how to collect, prepare, and administer affordable and locally available botanical medicines to treat many of their health problems.

There is a strong interrelationship between botanical resources, language, and ethnobiological knowledge of medicinal plants. Tropical countries contain the most biologically and culturally diverse traditional medicine systems in the world (Durning 1992). Reference has been made to loss of potential new medicines from loss of vascular plant species as ecosystems around the world are being destroyed (Abelson 1990). At even greater risk of being lost are the languages used to describe the ethnomedical knowledge of these medicinal plants (Krauss 1992). In many tropical rural cultures the ethnobotanical information is not written but is passed on orally from generation to generation in the local languages. When the local languages and cultures become endangered or extinct, the knowledge of how to use plants as medicines is lost or diminished. Therefore, it is essential to conserve the cultures and languages along with the biological ecosystems and species so that the knowledge of how to use these medicinal plants is maintained.

Guinea in West Africa is a salient example of these processes. In 1996 and 1997, I conducted ethnobotanical field research collaboration in Guinea with 31 different traditional healers (10 females, 21 males) representing 11 different communities, 3 major ecosystems, and 6 different languages (Carlson et al. in press). The 3 ecosystems were coastal, savanna, and tropical forest. The ethnolinguistic groups were Foula (= Pular), Maninke (= Malinke), Kpelewo (= Guerze), Sousou (= Susu), Toma, and Mano. While collaborating with the traditional healers from these different ethnolinguistic groups, it became apparent that they provided free, bartered, or inexpensive medical care for rural populations and many people in urban areas. These traditional healers served communities of people who did not have access to modern medicines and could not afford them if they were available. These healers' collective pharmacopoeia included hundreds of different medicinal plant species used to treat a broad spectrum of diseases. For example, these healers used more than 50 different species of plants to treat Type 2 diabetes mellitus in adults. Certain medicinal plant species were used by different healers from different ethnolinguistic groups in different regions of Guinea to treat the same disease. Even though French was the colonial language in Guinea, many of the healers did not speak French or any other European languages. Their local names and classification systems of diseases, medicinal plants, and botanical species were typically in the local indigenous language; most of these healers did not have names for these plants in French. Without the use of their local indigenous language, the knowledge on the medicinal plants would be greatly diminished.

ETHNOMEDICALLY VS. RANDOMLY SELECTED PLANTS

Plants have been collected and analyzed for medicinal qualities by scientific institutions and pharmaceutical companies. In recent decades, the typical approach by the larger drug companies has been to collect large numbers of plants at random and screen them for a variety of diseases without consideration of whether or not the plant was used traditionally as a medicine. This is in contrast to ethnomedically directed pharmacological evaluation of plants species used by traditional healers, which was a more typical strategy in the earlier part of the twentieth century. But today there are some small companies such as Shaman Pharmaceuticals in California (King and Carlson 1995) that focus their research on evaluating medicinal plants based on strong ethnomedical information. Ethnomedically directed pharmacological evaluation of plants has played a very important role in the development of plant-derived drugs now used in modern medicine. It has been estimated that 25 percent of the modern medical drug prescriptions (119 different chemical substances) written between 1959 and 1980 in the United States were pharmaceuticals derived from 90 different botanical species; 74 percent of these 119 pharmaceuticals were derived from medicinal plants used ethnomedically for the same therapeutic use (Farnsworth et al. 1985). Hence, approximately 20 percent of all pharmaceutical prescriptions written between 1959 and 1980 were pharmaceuticals derived from ethnobotanical leads. This highlights the value of integrating ethnolinguistic knowledge on the therapeutic qualities of locally used medicinal plants into the study of plant-derived compounds through modern medical scientific evaluation in humans.

In ethnomedically directed screening of plants for pharmacological activity, plant species are screened that are used by traditional healers to treat the disease(s) under evaluation. Traditional cultures around the world have access to hundreds of different plant species in the biological ecosystems in which they live. The total number of plants used as medicines is typically large; however, there is usually a relatively small subset of plants selected to treat each specific disease. A very important feature of prioritizing the therapeutic value of medicinal plants is evaluating both quantitative and qualitative features of the ethnomedical information (Carlson and King 1997). Studies have shown that taking into account the ethnolinguistic knowledge on the plants to guide biological screening results in identifying a high percentage of biologically active plants. Forty out of seventy plants (57 percent) used ethnomedically in tropical countries to treat Type 2 diabetes mellitus demonstrated activity in mice with diabetes (Carlson et al. 1997a).

Recent publications on antidiabetic activity in plants (Bierer et al. 1998; Luo et al. 1998a, b) describe several plant species used ethnomedically to treat Type 2 diabetes mellitus in adults that have demonstrated antidiabetic activity in rodents and in some cases humans.

The benefits of screening ethnomedical plant collections versus random plant collections is discussed by Lewis and Elvin-Lewis (1995). Several studies will be described below that have demonstrated that plants used ethnomedically to treat specific diseases show a significantly higher rate of pharmacological activity compared with plants collected randomly. These data refute the notion that the therapeutic use of these medicinal plants in traditional rural cultures is only due to the placebo effect and/or the cultural context. A study conducted in Brazil showed that 18 percent (4 of 22) of the plants used ethnomedically to treat malaria demonstrated in vitro antimalarial activity compared with .07 percent (2 of 273) of plants collected randomly (Carvalho and Krettli 1991). Plant extracts have been evaluated by the U.S. National Cancer Institute (NCI) in in vitro screens for anticancer and anti-HIV activity. Spjut and Perdue (1976) evaluated a variety of published studies on natural products screened against in vitro anticancer NCI assays and reported that plants collected randomly had an activity rate of 10.4 percent, whereas plants used in traditional medicine had significantly higher activity rates with 29.3 percent (plants used as vermifuges), 38.6 percent (plants used as fish poisons), and 52.2 percent (plants used as arrow or dart poisons). The NCI anti-HIV assays were used to evaluate the biological activity of "powerful plants" from a traditional healer in a Belize village compared with plants collected randomly from this country. When the plant extracts were prepared and tested, the tannins, which are present in some traditional botanical medicines, were not selectively removed. The first comparison showed that 25 percent (5 of 20) of the ethnobotanically selected plants were active in the in vitro anti-HIV NCI screen compared with 6 percent (1 of 18) of the randomly collected plants (Balick 1990). A larger group of plants subsequently evaluated in the same assay by the same author (Balick 1994) showed that 15 percent (11 of 73) of the plants presented by the traditional healer were active, compared with 1.6 percent (1 of 61) of the plants collected randomly. Research by Vlietnick and Vanden Berghe (1991) demonstrated that 25 percent of the ethnobotanical plant collections evaluated were active in in vitro antiviral assays compared with 5 percent of the plants collected randomly.

Research at Shaman Pharmaceuticals compared the frequency of isolating antiviral compounds from plants used ethnomedically to treat viral infections versus from natural products that were randomly collected (Carlson et al. 1997a). In these studies, the plants collected ethnobotanically were screened against three different antiviral assays (respiratory syncytial virus, influenza, and cytomegalovirus) while the randomly collected natural products were screened against an antiviral assay for herpes simplex virus at a different pharmaceutical company. The same virologist supervised the screening programs for each of the four different viruses and used comparable biological evaluation methods: 15,000 randomly collected natural products screened against herpes simplex virus yielded two active compounds with 0.013 percent isolation frequency; 231 plants used ethnomedically to treat viral infections screened against cytomegalovirus yielded five active compounds with 2.2 percent isolation frequency; 123 plants used ethnomedically to treat viral infections screened against influenza virus yielded two active compounds with 1.6 percent isolation frequency; 97 plants used ethnomedically to treat viral infections screened against respiratory syncytial virus yielded eight active compounds with 8.2 percent isolation frequency. The frequency of isolating pure compounds with antiviral activity from plants used ethnomedically to treat viral infections is respectively 123 times, 169 times, or 630 times higher depending respectively on the virus (influenza, cytomegalovirus, or respiratory syncytial virus) being tested when compared with the frequency of isolating anti–herpes simplex virus compounds from randomly collected natural products. While these studies compared screening of different viruses, the data still suggest that evaluating plants used ethnomedically to treat viral infections can enhance laboratory scientists' ability to isolate pure pharmacologically active antiviral compounds from plants. A diverse spectrum of traditional botanical medicines from tropical countries contain active antiviral compounds.

NONVIABILITY OF MODERN MEDICINES IN TROPICAL RURAL AREAS

Modern medicines are not viable in most tropical village settings. Factors that prevent sustainable use include high cost, perishability of medicines, or lack of medically trained personnel in the village to administer the medicines appropriately. Modern medicines are far too expensive for most tropical rural communities to afford. If medicines are donated to a village, seldom is there a steady supply of these medicines over a long period. Whether medicines are purchased by or donated to a community, there is often the problem that many modern medicines, including antibiotics, perish in hot humid environments where refrigeration is not available. When these medicines are ingested they may have a significantly diminished activity. Some medicines that are donated have expired and thus cannot be sold in Europe, Canada, or the United States (Berkmans et al. 1997).

Well-intentioned organizations that donate medicines may not adequately take into consideration the health care needs of the region. Many of the drugs donated are used to treat diseases that either are rare or not present in the communities where the medicines are donated (WHO 1992). These medicines are then often used inappropriately in tropical villages; often administered at the wrong dose size, dose interval, and length of treatment; typically not used for the disease(s) they are active against and instead taken for the wrong diseases. Examples of this are penicillin injections or oral antibacterial medicines given inappropriately for a broad spectrum of diseases such as headache, backache, and viral infections. Another potential major hazard is the inappropriate use of topical steroid creams and ointments. While these are useful for certain skin diseases, they can cause some infectious skin diseases to get worse. There are even topical steroid eye ointments available in tropical villages that are known to be used for any type of eye disease, although there is only a narrow range of noninfectious inflammatory eye disorders that these medicines benefit. Application of steroid ointment onto the eye of a person with herpes simplex infection of the cornea may result in a progression of the infection and even loss of vision. Another serious hazard of modern medicines is the use of needles to inject drugs. The needles may not be sterile and are commonly reused without appropriate sterilization, which contributes to the spread of HIV (Garrett 1994), hepatitis B, and other blood-borne infections.

In addition, the presence and consumption of modern medicines in the village setting acts to weaken the confidence and respect of the local people in their own locally available traditional medicine. Local people may develop the belief that their health care solutions are from pills in bottles or worse yet from injections.

CONTRIBUTIONS OF TRADITIONAL BOTANICAL MEDICINES TO TROPICAL RURAL PUBLIC HEALTH

Traditional botanical medicine plays a central role in providing health care in tropical countries. Many of these botanical medicines contain safe and effective pharmacologically active compounds. These locally available plant-based medicines are also typically free or affordable to all members of rural communities. Many of the medicinal plants used locally in tropical village settings are cultivated in gardens or harvested from local ecosystems in a sustainable manner. Since the medicinal plant information is present within the local community, its persistence reinforces the community's ethnolinguistic identity and can strengthen the value of conserving their biological, ecological, and linguistic diversity.

Traditional botanical medicines used in tropical villages provide treatments for a variety of common health problems including diarrhea, respiratory infections, and malaria. Rotavirus diarrhea and viral lower respiratory infections in infants and young children are the most common causes of death in the tropical world. Modern medicine does not have a pharmacological treatment for Rotavirus diarrhea or parainfluenza pneumonia. Respiratory syncytial virus pneumonia in infants is treated with ribavirin, a modern medical treatment that is very expensive and not available to tropical rural children. Traditional botanical medicines are commonly used to treat respiratory syncytial virus pneumonia in children and many different plant species have demonstrated in vitro and in vivo biological activity against this virus (Carlson et al. 1997a; Kernan et al. 1998; Chen et al. 1998).

Leaf tea from guava (*Psidium guajava* L.)is used widely in traditional medicine to treat diarrhea, and its antidiarrheal activity in animals has been demonstrated in the literature (Maikere-Faniyo et al. 1989; Lutterodt 1992). *Croton lechleri* (Muell.) Arg. is a common medicinal plant that produces red bark latex, which is used to treat diarrhea in South America. A compound was isolated from this latex that has demonstrated antidiarrheal activity in in vitro and rodent studies (Gabriel et al. 1999). A human double-blind placebo controlled study was conducted on a compound from this plant that showed a statistically significant reduction in diarrhea in AIDS patients with chronic diarrhea (Holodniy et al. 1999).

Artemisia annua L. is a medicinal plant used traditionally to treat malaria in Asia. Artemisinin, a compound derived from this plant, and its analogue, artemether, have demonstrated antimalarial activity in human studies (Hien et al. 1996; Van Hensbroek et al. 1996). A study conducted in Brazil demonstrated in vitro antimalarial activity in different medicinal plants (Carvalho and Krettli 1991).

Fungal skin infections are also commonly treated with traditional botanical medicines. Two different medicinal plants (*Chelonanthus alata* [Aubl.] Maas from Thailand and *Anthocleista djalonensis* A. Chev. from Nigeria) used ethnomedically to treat fungal skin infections were analyzed; the same compound, irlbacholine, was isolated from both plants and demonstrated antifungal activity (Bierer et al. 1995).

Traditional botanical treatment is also administered for psychiatric illnesses. *Raulvolfia vomitoria* Afzel. is a common West African medicinal plant used to treat psychosis. The ground-up root of this species taken orally demonstrated therapeutic antipsychotic effects in human patients in Nigeria (Obembe et al. 1994).

There are numerous examples of a plant species used to treat the same common disease in many different cultures, countries, and in some cases even different continents. Well-known examples include: *Psidium guajava* L. leaf tea used for diarrhea in Latin America (Caceres et al. 1993), Africa (Maikere-Faniyo et al. 1989), and Asia (Grosvenor et al. 1995); *Momordica charantia* L. decoction to treat adult onset diabetes mellitus in Latin America (Zamora-Martinez and Pola 1992), Middle East (Mossa 1985), and Asia (Lotlikar and Rajarama Rao 1966); and *Senna alata* (L.) Roxb. leaf sap to treat fungal skin infections in Latin America (Caceres et al. 1991), Africa (Crockett et al. 1992), and Asia (Ibrahim and Osman 1995).

ETHNOBIOLOGICAL KNOWLEDGE AND REFUGEE ADAPTATION TO NEW ENVIRONMENTS

Medicinal plants are very important for urban, forest, agrarian, and pastoralist communities. They can also be extremely valuable for refugee populations who have been forced to leave their communities and countries because of civil wars, political strife, or environmental disasters. These processes typically result in a tremendous amount of physical and psychological suffering. Publicity by the media and human rights groups may stimulate a mobilization of host government and international relief organizations to help provide food, medicine, and lodging for these displaced peoples. There is a tendency for these refugees to be depicted as helpless populations whose survival depends on outside organizations providing for their needs. A feature not adequately emphasized is how these displaced populations adapt traditional ecological and ethnobiological knowledge from their home region to the biological environments in their new region to help meet their needs for food, medicine, and shelter.

Many refugees are from villages and have extensive traditional knowledge of medicinal plants, food plants, agriculture, and edible aquatic and terrestrial animals that was passed on orally through the indigenous language of their culture. Since refugees often migrate to a bordering country, many of the plants and animals are the same species or closely related species as those found in their home country. While refugee populations typically leave virtually all their material possessions behind, they mentally retain their traditional environmental knowledge and sometimes carry seeds of their favorite food plants and medicinal plants. When refugees are given access to the terrestrial and aquatic biosystems and agricultural land, they adapt ethnoecological and ethnobiological knowledge from their country of origin to the biological environments in their new host country. Traditional healers in the host country where the refugees have settled also often provide a significant amount of health care to these populations. In addition to knowledge on food plants and medicinal plants, ethnobiological knowledge that enhances the well-being of refugees includes plants used to build homes and thatched roofs, good firewood species to enable them to cook their foods and boil their water; and locally available aquatic and terrestrial animals that can be eaten.

PROVISION OF HEALTH CARE TO REFUGEES
BY LOCAL TRADITIONAL HEALERS

Traditional healers in the host country where the refugees have settled often provide a significant amount of health care to these populations. A salient example is from West Africa. In the early 1990s, over 750,000 people fled civil wars in Liberia and Sierra Leone and migrated as refugees to neighboring countries, for example, Guinea. Many of the refugees who came to Guinea settled into refugee camps, villages, and cities in the southern part of the country. While conducting ethnobotanical research with traditional healers in southern Guinea, I observed that the use of botanical resources for food, medicine, and housing materials was widespread throughout the villages and settlements of Liberian refugees. Some of the refugees who had been traditional healers in their home country of Liberia were providing medical care to the refugee populations. Since southern Guinea has many of the same medicinal plant species that neighboring Liberia has, the healers and other knowledgeable refugees (e.g., grandmothers and grandfathers) were able to collect and prepare many of the same botanical medicine species they used in their own countries.

Refugee traditional healers from Liberia and Sierra Leone, as well as Guinea traditional healers, play an important role in providing medical care for these refugees. These healers use plant medicines to treat a broad spectrum of physical ailments including infectious diseases such as diarrhea, malaria, and pneumonia. Some of the healers are also traditional psychiatrists who use botanical medicines to treat patients who are psychologically traumatized from atrocities they had observed or experienced.

While conducting research in April 1996, I collaborated with a traditional healer of the Kpelewo (also called Guerze) culture who lived in the village in the district of Nzerekore in Guinea, West Africa. Working closely with his brother, who was his apprentice, this healer had developed and ran a clinic that provided affordable or free health care to refugees and local people living in the area, using ethnomedical knowledge in his Kpelewo indigenous language. He cared for hundreds of patients with a variety of disorders, but a main focus for him was psychiatric illness. His psychiatric patients were able to stay in the healer's in-patient clinic until their illnesses were adequately treated. Some of his patients from Liberia were psychologically traumatized refugees. The modern medical diagnosis for these patients would be post-traumatic stress syndrome. Traditional botanical medicines were administered to treat the psychiatric illnesses. The clinic was also able to provide traditional psychiatric and spiritual care that was more culturally appropriate than modern medical psychiatric care. Patients at the clinic reported that they benefited from this healer's treatment.

On returning to the clinic in 1997, I was given the sad news that the healer had died two weeks earlier. His family and community were still in mourning. Fortunately, his brother had learned his Kpelewo ethnomedical skills well. He has succeeded his late brother as director of the clinic and is providing a similar quality of compassionate and culturally appropriate health care. The knowledge had been passed on from the late healer to his brother apprentice verbally in the Kpelewo language and had not been written down. The use of this language was essential in passing on this valuable ethnomedical information on the use of their botanical resources so that future generations can benefit from this ethnobiological knowledge.

The ability of refugees in Guinea to use their and the local healers' traditional knowledge to provide themselves with food, medicine, and shelter has taken some of the burden off the government of Guinea and international relief organizations (e.g., United Nations High Commissioner for Refugees and Médecins Sans Frontières) working to ensure that these people have appropriate food, medicine, and lodging. Whether or not a national or international relief program is established, the use of ethnobiological knowledge can significantly enhance the well-being of refugee populations. For instance, refugees from Laos and Cambodia that attended my medical clinic in California likewise reported that while they lived in refugee camps in Thailand, traditional healers from their cultures also played an important role in the treatment of many of the physical, psychological, and spiritual based illnesses experienced by the refugees.

ETHNOBIOLOGISTS' CONTRIBUTIONS TO PUBLIC HEALTH OF TROPICAL RURAL POPULATIONS

Local ethnobotanical knowledge allows for plants to be optimally used by humans for food and medicine. Traditional healers have detailed knowledge about the plants and diseases of their region and apply ethnopharmacological principles to prepare safe and effective botanical treatments for a variety of ailments. To understand and help conserve the knowledge of traditional healers, it is important to have a team of ethnobiologists with training in medicine, botany, pharmacology, anthropology, and field linguistics. These interdisciplinary teams can study the ethnobotanical knowledge of the local ethnolinguistic groups to understand the pharmacological activity of plants. Multiple studies cited in this chapter have demonstrated that when plants used ethnomedically to treat specific diseases are tested for biological activity, a significantly higher rate of pharmacological activity is seen compared with plants collected randomly. Ethnobiologists can collaborate with local populations to conserve their traditional food plants, medicinal plants, agricultural systems, and biological ecosystems. Scientific institutions or companies conducting ethnobiological research should contribute to capacity-building within the communities. This reciprocity should be based on the expressed needs of the community and contribute to the conservation of biocultural diversity and improvement of the public health of the people (King 1994; King and Carlson 1995; King, Carlson, and Moran 1996; Carlson et al. 1997b). If any marketable products are ever generated from ethnobiological research, the communities and countries should receive benefits as well as intellectual credit (King 1994; King and Carlson 1995; Iwu 1996; Carlson et al. in press).

It is essential that modern medical public health programs recognize the value of traditional ethnomedical knowledge, collaborate with traditional botanical healers, and harmoniously integrate appropriate modern medicine interventions into the existing traditional medicine infrastructure. For many diseases, traditional medicine is effective, much less expensive, more sustainable, and does not require foreign and governmental financial aid. It is important to understand the local infrastructure of traditional medicine before implementing a modern medical public health program. Enabling the traditional ethnomedical system to care for the diseases that it treats well frees governmental and international resources to focus on treating the diseases that traditional medicine is not well equipped to treat. Traditional healers can also assist in epidemiological surveys by monitoring the local populations and reporting the prevalence of specific diseases, including the emergence of new diseases (Groce and Reeve 1996; SEVA Foundation 1997). Ethnobiologists can play an important role in integrating modern medical and public health projects with the local infrastructure of traditional health care so they can work in a complementary way.

Ethnobiologists can also play a very important role in working with international relief organizations such as the United Nations High Commissioner for Refugees and Médecins Sans Frontières to help evaluate the needs of refugee populations and how these populations might best access biological resources for food, medicine, and shelter so that they can become more self-reliant. Ethnobiologists could also assist in the integration of traditional healers into the health care provided by the medical relief organizations. By integrating traditional healers into tropical public health programs, many of the common health problems can be treated inexpensively and sustainably with local botanical and cultural resources. Those diseases not treated well with traditional medicine can be reserved for the more expensive, less sustainable, and culturally invasive modern medicine.

REFERENCES

Abelson, P.H. 1990. Medicine from plants. *Science* 247:513.

Akerele, O. 1992. WHO guidelines for the assessment of herbal medicines. *Fitotherapia* 63(2):99–110.

Balick, M.J. 1990. Ethnobotany and the identification of therapeutic agents from the rainforest. In *Bioactive Compounds from Plants,* ed. D.J. Chadwick and J. Marsh. CIBA Foundation Symposium 154. Pp. 22–39. Chichester, England: John Wiley & Sons.

Balick, M.J. 1994. Ethnobotany, drug development, and biodiversity conservation: Exploring the linkages. In *Ethnobotany and the Search for New Drugs,* ed. D.J. Chadwick and J. Marsh. CIBA Foundation Symposium 185. Pp. 4–24. Chichester, England: John Wiley & Sons.

Bannerman, R.H. 1983. The role of traditional medicine in primary health care. In *Traditional Medicine and Health Care Coverage: A Reader for Health Administrators and Practitioners,* ed. R.H. Bannerman, J. Burton, and C. Wen-Chieh. Pp. 318–27. Geneva: World Health Organization.

Berkmans, P., et al. 1997. Inappropriate drug-donation practices in Bosnia and Herzegovina, 1992 to 1996. *New England Journal of Medicine* 337(25):1842–1845.

Berlin, B. 1992. *Ethnobiological Classification: Principles of Categorization of Plants and Animals in Traditional Societies.* Princeton, N.J.: Princeton University Press.

Berlin, B., D.E. Breedlove, and P.H. Raven. 1966. Folk taxonomies and biological classification. *Science* 154:273–275.

Bierer, D.E., et al. 1995. Isolation, structure elucidation, and synthesis of irlbacholine, 1,22-Bis[[[2-(trimethylammonium)ethoxy]-phosphinyl]oxy]docosane: A novel antifungal plant metabolite from *Irlbachia alata* and *Anthocleista djalonensis*. *Journal of Organic Chemistry* 60:7022–7026.

Bierer, D.E., et al. 1998. Ethnobotanical-directed discovery of cryptolepine from *Cryptolepis sanguinolenta*: Its isolation, synthesis, and antihyperglycemic activity. *Journal of Medicinal Chemistry* 41:894–901.

Caceres, A., et al. 1991. Plants used in Guatemala for the treatment of dermatophytic infections. 1. Screening for antimycotic activity of 44 plant extracts. *Journal of Ethnopharmacology* 31(3):263–276.

Caceres, A., et al. 1993. Plants used in Guatemala for the treatment of gastrointestinal disorders. 3. Confirmation of activity against enterobacteria of 16 plants. *Journal of Ethnopharmacology* 38(1):31–38.

Carlson, T.J., et al. 1997a. Modern science and traditional healing. Special publication, *Royal Society of Chemistry* 200 (Phytochemical Diversity): 84–95.

Carlson, T.J., et al. 1997b. Medicinal plant research in Nigeria: An approach to compliance with the Convention on Biological Diversity. *Diversity* 13(1):29–33.

Carlson, T.J., et al. In press. Case study on medicinal plant research in Guinea: Prior informed consent, focused benefit sharing, and compliance with the Convention on Biological Diversity. *Economic Botany.*

Carlson, T.J., and S.R. King. 1997. Ethnomedical field research methods to assess medicinal plants. In *Commercial Production of Indigenous Plants as Phytomedicines and Cosmetics*, ed. M.M. Iwu et al. Pp. 152–165. Lagos, Nigeria: BDCP Press.

Carvalho, L.H., and A.U. Krettli. 1991. Antimalarial chemotherapy with natural products and chemically defined molecules. *Memorias do Instituto Oswaldo Cruz* 86, Suppl. 11: 181–189.

Chen, J.L., et al. 1998. New iridiods with activity against respiratory syncytial virus from the medicinal plant *Barleria prionitis*. *Journal of Natural Products* 61:1295–1297.

Crockett, C.O., et al. 1992. *Cassia alata* and the preclinical search for therapeutic agents for the treatment of opportunistic infections in AIDS patients. *Cellular Molecular Biology* 38(5):505–511.

Durning, A.T. 1992. *Guardians of the Land: Indigenous Peoples and the Health of the Earth.* Worldwatch Paper 112. Washington, D.C.: Worldwatch Institute.

Elisabetsky, E., and R. Setzer. 1985. Caboclo concepts of disease, diagnosis, and therapy: Implications for ethnopharmacology and health systems in Amazonia. In *The Amazon Caboclo: Historical and Contemporary Perspectives*, ed. E. P. Parker. Pp. 243–378. Studies in Third World Societies Series, no. 32. Williamsburg, Va.: College of William and Mary, Department of Anthropology.

Elisabetsky, E., and L. Wannmacher. 1993. The status of ethnopharmacology in Brasil. *Journal of Ethnopharmacology* 38:137–143.

Etkin, N.L. 1996. Medicinal cuisines: Diet and ethnopharmacology. *International Journal of Pharmacognosy* 34(5):313–326.

Farnsworth, N.R., et al. 1985. Medicinal plants in therapy. *Bulletin of the World Health Organization* 63(6):965–981.

Gabriel, S.E., et al. 1999. A novel plant-derived inhibitor of cAMP-mediated fluid and chloride secretion. *American Journal of Physiology* 276(39):G58–G63.

Garrett, L. 1994. *The Coming Plague: Newly Emerging Diseases in a World Out of Balance.* New York: Penguin Books.

Groce, N.E., and M.E. Reeve. 1996. Traditional healers and global surveillance strategies for emerging diseases: Closing the gap. *Emerging Infectious Diseases* 2(4):351–353.

Grosvenor, P.W., et al. 1995. Medicinal plants from Riau Province, Sumatra, Indonesia. Part 1: Uses. *Journal of Ethnopharmacology* 45(2):75–95.

Hien, T.T., et al. 1996. A controlled trial of artemether or quinine in Vietnamese adults with severe Falciparum malaria. *New England Journal of Medicine* 335(2):76–83.

Holodniy, M., et al. 1999. A double-blind, randomized, placebo-controlled, Phase II study to assess the safety and efficacy of orally administered SP-303 for the symptomatic treatment of diarrhea in patients with AIDS. *American Journal of Gastroenterology* 94(11):3267–3273.

Ibrahim, D., and H. Osman. 1995. Antimicrobial activity of *Cassia alata* from Malaysia. *Journal of Ethnopharmacology* 45(3):151–156.

Iwu, M.M. 1993. *Handbook of African Medicinal Plants.* Boca Raton, Fl.: CRC Press.

Iwu, M.M. 1996. Biodiversity prospecting in Nigeria: Seeking equity and reciprocity in intellectual property rights through partnership arrangements and capacity building. *Journal of Ethnopharmacology* 51:209–219.

Johns, T. 1990. *With Bitter Herbs They Shall Eat It: Chemical Ecology and the Origins of Human Diet and Medicine.* Tucson: University of Arizona Press.

Kernan, M., et al. 1998. Antiviral phenylpropanoid glycosides from the medicinal plant *Markamea lutea*. *Journal of Natural Products* 61:564–570.

King, S.K. 1994. Establishing reciprocity: Biodiversity, conservation, and new models for cooperation between forest dwelling people and the pharmaceutical industry. In *Intellectual Property Rights for Indigenous Peoples: A Source Book,* ed. T. Greaves. Pp. 69–82. Oklahoma City: Society for Applied Anthropology.

King, S.K., and T.J. Carlson. 1995. Biomedicine, biotechnology, and biodiversity: The Western Hemisphere experience. *Interciencia* 20(3):134–139.

King, S.K, T.J. Carlson, and K. Moran. 1996. Biological diversity, indigenous knowledge, drug discovery, and intellectual property rights. In *Valuing Local Knowledge: Indigenous People and Intellectual Property Rights,* ed. S. Brush and S. Stabinsky. Pp. 167–185. Washington, D.C.: Island Press.

Krauss, M. 1992. The world's languages in crisis. *Language* 68(1):4–10.

Lewis, W.H., and M.P. Elvin-Lewis. 1995. Medicinal plants as sources of new therapeutics. *Annals of Missouri Botanical Garden* 82:16–24.

Lotlikar, M.M., and M. Rajarama Rao. 1966. Pharmacology of a hypoglycemic principle isolated from the fruits of *Momordica charantia*. *Indian Journal of Pharmacy* 28:129.

Lozoya, X. 1994. Two decades of Mexican ethnobotany and research in plant drugs. In *Ethnobotany and the Search for New Drugs,* ed. D.J. Chadwick and J. Marsh. CIBA Foundation Symposium 185. Pp. 130–152. Chichester, England: John Wiley & Sons.

Luo, J., et al. 1998a. Cryptolepine, a potentially useful new antihyperglycemic agent isolated from *Crytolepis sanguinolenta*: An example of the ethnobotanical approach to drug discovery. *Diabetic Medicine* 15:367–374.

Luo, J., et al. 1998b. Masoprocol: A new antihyperglycemic agent isolated from the creosote bush (*Larrea tridentata*). *European Journal of Pharmacology* 346:77–79.

Lutterodt, G.D. 1992. Inhibition of microlax-induced experimental diarrhoea with narcotic-like extracts of *Psidium guajava* leaf in rats. *Journal of Ethnopharmacology* 37(2):151–157.

Maffi, L. 1994. A Linguistic Analysis of Tzeltal Maya Ethnosymptomatology. Ph.D. diss., Ann Arbor: UMI.

Maikere-Faniyo, R., et al. 1989. Study of Rwandese medicinal plants used in the treatment of diarrhoea I. *Journal of Ethnopharmacology* 26(2):101–109.

Mossa, J.S. 1985. A study on the crude antidiabetic drugs used in Arabian folk medicine. *International Journal of Crude Drug Research* 23(3):137–145.

Obembe, A., et al. 1994. K20857 antipsychotic effect and tolerance of crude *Rauvolfia vomitoria* in Nigerian psychiatric inpatients. *Phytotherapy Research* 8(4):218–223.

SEVA Foundation. 1997. Saving the sight of children in Nepal. *Spirit of Service* Fall 1997:1–2.

Sousa Brito, A.R.M. 1996. How to study the pharmacology of medicinal plants in under-developed countries. *Journal of Ethnopharmacology* 54:131–138.

Spjut, R.W., and R.E. Perdue, Jr. 1976. Plant folklore: A tool of predicting sources of anti-tumor activity? *Cancer Treatment Report* 60:979–985.

Van Hensbroek, M.B., et al. 1996. A trial of artemether or quinine in children with cerebral malaria. *New England Journal of Medicine* 335(2):69–75.

Vlietnick, A.J., and D.A. Vanden Berghe. 1991. Can ethnopharmacology contribute to the development of antiviral agents? *Journal of Ethnopharmacology* 32(1–3): 141–154.

WHO. 1991. Guidelines for the Assessment of Herbal Medicines. Finalized at WHO consultation in Munich, Germany, June 19–21, 1991, and presented to Sixth ICDRA in Ottawa, Canada, 1991.

WHO. 1992. The use of essential drugs: Model list of essential drugs. Fifth Report of the WHO Expert Committee, 1992. *World Health Organization Technical Report Series* 825:1–75.

Zamora-Martinez, M.C., and C.N.P. Pola. 1992. Medicinal plants used in some rural populations of Oaxaca, Puebla, and Veracruz, Mexico. *Journal of Ethnopharmacology* 35(3):229–257.

Zent, S. 1996. Behavioral orientations toward ethnobotanical quantification. In *Selected Guidelines for Ethnobotanical Research: A Field Manual*, ed. M.N. Alexiades. Pp. 199–240. The Bronx, N.Y.: New York Botanical Garden Press.

31

INDIGENOUS PEOPLES AND THE USES
AND ABUSES OF ECOTOURISM

Ben G. Blount

According to a widely held view, ecotourism, by its very nature, should provide a means for endangered peoples to reverse the conditions that led to their problems in the first place. But a review of the literature on ecotourism reveals the many issues and problems that ecotourism practices raise in relation to indigenous peoples and biological diversity. Efforts are being made, however, to correct such problems and to reverse the trend of endangerment of cultures, languages, and biodiversity.

THE SCOPE OF TOURISM AND ECOTOURISM

By any account, tourism is big business. According to the annual reports of the World Tourism Organization (WTO) and the World Travel and Tourism Council (WTTC), tourism is among the world's largest industries. According to the WTTC 1994 Annual Report, tourism is the world's largest industry, is the world's largest employer and creator of jobs, and is responsible for 10 percent of world gross domestic product. The reports of the WTO are somewhat more tempered, indicating in 1992 that tourism was the third leading industry in the world, following petroleum and motor vehicles. By any standards, however, tourism has become a major economic and world activity. To illustrate by using international tourist arrivals, the number increased from 25 million in 1950 to 425 million in 1990, a 17-fold increase (WTO 1995). In developing countries, tourism income is especially important, sometimes accounting for more than 50 percent of the gross national product (Ziffer 1989; Lindberg 1991). In 1988, tourism expenditure for developing countries as a whole was estimated to be $55 billion (Ziffer 1989).

The proportion of tourism that includes the varieties of ecotourism is comparatively small, but the expenditures are large and growing. In one study, for example, the dollar value of "environmentally sensitive tourism" worldwide was placed at $10 billion in 1980, at $150 billion in 1995, and estimated to be $300 billion in 2000 (Jenner and Smith 1992). Those figures may, however, be far too low. Ceballos-Lascuráin (1996) noted that in 1988 the contribution of ecotourism to the national income of countries worldwide may have been as high as $233 billion. Whatever the correct amount may be, the growth rate of ecotourism exceeds that of tourism proper. Tourism appears to be increasing at approximately 4 percent annually worldwide, while the rate of ecotourism growth is estimated to be between 10 and 25 percent annually (Lindberg 1991). Ecotourism also accounts for much of the growth of tourism in developing countries. Of the $55 billion earned in 1988, up to $12 billion was considered to have come from ecotourism, and that ratio is expected to increase (Ziffer 1989), especially in countries with the most diverse flora, fauna, and ecosystems. Simply in terms of revenue generated, ecotourism would appear to hold promise as a source of funds for conservation of cultural, linguistic, and biological diversity.

HOW IS ECOTOURISM DEFINED?

The diversity of touristic practices in relation to the environment renders difficult any precise account of what ecotourism is. Since much of tourism, in general, has environmental aspects, the differentiation into ecotourism and nonecotourism hinges on how tourism actually relates to the environment. Only some of the types of relations qualify as ecotourism. One effort to characterize ecotourism is definitional. A frequently cited one derives from the Ecotourism Society, a U.S.-based organization founded in 1990: "purposeful travel to natural areas to understand the culture and natural history of the environment, taking care not to alter the integrity of the ecosystem, while producing economic opportunities that make the conservation of natural resources beneficial to local people" (Wood 1991). Much longer, prescriptive definitions are often employed. In those, the approach is to contrast ecotourism with other types of tourism, leading to an inventory of features that are contrastive and identificational. From a summary of those features, one can say that ecotourism typically differs from conventional or mass tourism in terms of group size, expectations for amenities, types of host sites, types of experiences, education about the environment, amount of interaction with local residents, and contributions toward conserving local resources (Nelson 1994). Moreover, ecotourists are expected to be more environmentally and socially aware than other types of tourists, and what they require, in return, is a much higher level of biodiversity, wildlife experience, and interaction with local populations in natural settings (Goodwin 1996; King and Stewart 1996).

Other, related attempts to clarify ecotourism have relied on features of classification. In those attempts, types of tourism are identified and classified in relation to each other. At the most generic level, two types of tourism are distinguished, conventional tourism and alternative tourism (Smith and Eadington 1992 [1977]). Alternative tourism includes any type of tourism that is not the typical mass tourism at recognized, major tourist sites. That would, of course, include ecotourism. Difficulties arise, however, in a precise designation of ecotourism because it shares characteristics with other types of alternative tourism. The tendency is to name a type by its most prototypical feature but, in a given instance, that feature may be shared by a number of other types of alternative tourism. Thus, through metonymy, ecotourism is sometimes referred to as "wilderness tourism," since ecotourism may involve a seeking of presumed intact wilderness. Not all "wilderness tourism," however, is ecotourism. A trip to the Rocky Mountains to engage in bow-hunting of mule deer may well be "wilderness tourism," but it is unlikely to be considered ecotourism.

Ecotourism is also sometimes called "nature tourism" or "nature-based tourism," again for obvious reasons, but also again, not all nature-based tourism is ecotourism. Watching elephants from a game lodge in Kenya may be nature tourism, but it is probably not ecotourism. Ecotourism is also sometimes called "sustainably managed tourism," "sustainable tourism," "environmentally responsible tourism," "conservation-supporting tourism," or simply "conservation tourism," all in reference to the conservation of the natural environment. To take yet another feature, ecotourism is also sometimes referred to as "environmentally educated tourism," "science tourism," and "ethical tourism," since educating oneself about and thus enhancing responsibility for the local biota, ecosystems, and local human populations are taken as goals of the travel. Still more terms are used synonymously with ecotourism, terms such as "green tourism," "soft tourism," "low-impact tourism," "natural-areas tourism," "ecotravel," and "adventure travel," among others.

What are we to make of the abundance of semisynonymous terms? And they are abundant. Butler (1993) identified 35 terms that have "links" to ecotourism, that is, some sharing of definitional features. Perhaps the best strategy is to recognize that ecotourism can be identified as a specific type of tourism, one that contains the fundamental units of the definition provided by the Ecotourism Society: (1) purposeful travel to a natural/wilderness area; (2) to appreciate and learn from the natural and human environments; (3) to promote conservation of the environment; and (4) toward sustainable management of local resources. Another way to consider the "content" of ecotourism is to say that it contains, uniquely, those four elements. Ecotourism, then, could be seen as the intersection of "nature-based tourism," "environmentally educated tourism," "conservation-supporting tourism," and "sustainably managed tourism" (Buckley 1994).

THE GREENING OF ECOTOURISM

A principled account of how ecotourism should be defined is one matter. The use of the term by the tourism industry, historically, is another matter altogether. Virtually as soon as the term ecotourism appeared in the early 1980s, it was appropriated by the travel industry. Products identified as "green" were seen as eco-friendly and to have a built-in commercial appeal. Ecotravel and ecotours began to appear, not only among the established travel agencies but among new, "cottage-industry" type travel firms that specialized in travel to areas where the expectations and requirements of ecotourism could be met. A host of ecotourism travel books also appeared.

It is clear that ecotourism quickly became exploited commercially. A new type of travel opportunity became available for the industry, and tours and travel venues already in place were called ecotourism, no matter how lacking in aspects of environmental tourism or even how destructive of the environment they were. Moreover, the cost of tours escalated, since one was expected to pay more for ecotourism than for conventional tourism. High prices came to be another feature of ecotourism, another form of "greening." Effects of the commercialization of ecotourism were exacerbated by the fact that the sites tended to be in developing countries. While ecotourism generated badly needed foreign currency in those countries, provided employment, and had other multiplier effects to stimulate economic growth (Harrison 1992; Smith and Eadington 1992 [1977]; Whelan 1991), the effects were certainly not all positive. Most of the money earned did not remain in the developing countries themselves, only one of the problems that have emerged.

NEGATIVE CONSEQUENCES OF ECOTOURISM

The "greening" of ecotourism has been symptomatic of a host of difficulties and problems, far beyond obscuring what a definition of the practice should be. For summary purposes, those can be cast here as economic, environmental, social, and cultural, although they are all closely interrelated. Environmental degradation can lead to serious resource loss, economic decline, and sociocultural disintegration, which in turn can lead to further resource depletion and environmental destruction, and so forth. Each of those categories can also be viewed in terms of more specific components, for example, for the environment, impact on soils, on water resources, on flora and fauna, and the like.

Economic Issues

On the economic side, a serious problem is that the revenues generated by eco-tourism tend to go to foreign-owned tourist firms. Boo (1990) estimated that 55 percent of the gross receipts from tourism are returned, one way or another, to foreign operations. Much of the remainder of the receipts is likely to go to national economies rather than to local organizations, and typically last on the list are the local people themselves. That practice constitutes what some have called the new wave of Western imperialism, where tourism and ecotourism have replaced colonialism but in a continually exploitative capitalist relationship (Nash 1989).

As serious as the lack of revenue sharing has been, there are many additional economic problems. The presence of large number of tourists, for example, can lead to increased cost of living at the local level, especially in relation to land speculation and associated taxes. Crandall (1987) estimated that in parts of Barbados land prices rose 50 percent annually in the early 1970s. Currently in the Bay Islands (Honduras) and on the Costa Maya of Quintana Roo, Mexico, based on my observations, land prices are increasing at an even faster rate, as parcels of land that sold for one or two thousand dollars a few years ago, are now selling for as much as $50,000. Local people tend to be displaced from the land (e.g., there are no Maya left on Costa Maya), being unable to afford the increased taxes. As prices increase from real estate speculation, other commodities also become more expensive. Increased cost of basic necessities—food and transportation—is also experienced by locals, accelerating poverty and associated problems from reduced standard of living.

Yet another problem is the diversion of funds from needed local projects, such as health care, to the creation of infrastructure for tourism, for example, the building of hotels, roads, airports, and the like. While infrastructure may provide some economic benefits to the local population, for example, roads for transport of economic goods and associated increase in jobs, that is not always the case. Moreover, labor may be diverted from needed local food production to tourist service positions, such as maids, cooks, custodians, and grounds-keepers (Hitchcock and Brandenburgh 1990).

Environmental Issues

Negative impacts and consequences of tourism and ecotourism on local environments may have more lasting effects than economic decline. Much of the problem resides in waste water disposal, especially, but not only, in coastal areas. Often there are no waste water disposal systems, other than dumping the waste directly into streams, rivers, or the coastal zone. Roatan (Bay Islands), for example, has no effective sewage disposal system, despite the recent rapid rise in nature tourism and scuba diving. Degradation of coral reef systems is especially problematic,

since the concentration of tourists in coastal zones for access to coral reefs intensifies waste water problems, which in turn increases the threats to the reef systems. Coral reef systems throughout the world, in fact, face similar problems, not
only from increased numbers of tourists but from other waste water runoff polluted with pesticides, fertilizers, and industrial products. Polluted runoff water
also, of course, affects coastal areas and marine life in general, not just coral reefs.

The sheer volume of tourists, including ecotourists, can negatively impact sensitive environmental areas. Mountain trekking in Nepal, for example, grew very
rapidly during the 1970s and 1980s. Whereas the number of trekking trips was
under 10,000 annually in the 1960s, the number increased to 110,000 in 1978 and to
240,000 in 1987 (Boo 1990). Problems of debris, trash, transport, accommodations,
and the like also increased substantially, as did trail erosion and disturbance of
local flora and fauna. The destruction of trees for the sale of firewood to tourists
has been especially damaging, resulting in the movement of the tree line in the
Annapurna Range by several hundred feet (Boo 1990). Environmental disturbance
also results from the large amount of trash left by hikers. An expedition to clean
up trash in the Chomolungma Nature Reserve (Mount Everest), Tibet, removed
2863 pounds of garbage over the course of 709 days (McConnell 1991).

Similar negative consequences have been reported for many sensitive natural
areas. Recent, extensive surveys of the literature on environmental issues and
tourism (Mieczkowski 1995; Ceballos-Lascuráin 1996) have identified a multitude
of negative impacts, classified as to whether the impact is on natural elements
such as air, soils, water, and wildlife or on ecosystems such as coastal, inland,
mountain, and polar. The problem is global, and the scale is immense.

Tourism has played a vital role in the establishment of national parks and reserves (Myers 1972), providing protection for local flora and fauna. Even in national
parks, however, considerable damage can occur. Nelson (1994), for example, reporting on the work of Dearden and Rollins (1993) noted that Banff National Park
(Canada) has been carefully managed for decades, following conservation laws and
policies. Despite those efforts, undesirable changes in the environment have occurred, especially in long-term perspective. Land-use patterns have changed significantly since 1930, as revealed in air photos, maps, and reports. Large areas of the
park that were free of roads, campsites, trails, and infrastructure in 1930, now have
those elements present, and the increased influx of tourists has had undesirable
consequences, including clearing of land and disturbance of wildlife habitat, particularly in the valleys where the grazing and wintering of elk occurs.

Social and Cultural Issues

Tourism and ecotourism can have a number of negative impacts on the social and
cultural well-being of people in host countries. Those range from socioeconomic

effects to a less tangible but crucially important sense of cultural identity and cultural integrity.

Socioeconomic effects can operate at various levels. In a worst case scenario, complete alienation of land and resources can occur, ranging from economic development to the setting aside of areas of lands for reserves, parks, and other protected areas. The relationships between tourism, protected areas, and local people have, in fact, been problematic and in some instances even destructive. Even if local populations, for example, are not excluded from protected areas, as has often occurred, access to resources can become restricted or even denied. If restrictions are enforced, local populations can lose a substantial portion, or even all, of the resources that were once available to them. But if closed areas are not monitored or patrolled, and often they are not, local inhabitants continue to use resources there, but illegally. They are, in effect, converted into criminals if they continue to use the resources. Seemingly irrational cases can occur, as in Tsavo National Park in Kenya, in which local populations were prohibited from hunting and thus were denied access to meat from wild game, but at the same time elephants were culled by park rangers to try to control overpopulation and environmental change, leading to a wastage of the meat.

Another common type of socioeconomic impact is accelerated differentiation of tourists, nonlocal nationals, and local people. The jobs available to locals, for example, are often the least desirable and lowest paid in the tourism enterprise, exacerbating the disparities of wealth and socioeconomic status between them, imported workers, and tourists. The megaresorts in Huatulco, Mexico, such as Club Med, are cases in point, where the managerial positions are occupied by individuals from Mexico City while the locals do the more menial work. A related problem is that the number of actual jobs generated by tourism/ecotourism may also be small, thus having only a relatively minor contribution to the local economy overall. In reserved areas, for example, the financial return to the local community is typically very small, if it exists at all. Goodwin (1996) notes, to take one instance, that at the Komodo National Park in Indonesia, foreign visitors pay $300 U.S. for a three-day trip, but relatively little of the funds are invested in the park, and the park is not managed for the benefit of the local population at all.

A related socioeconomic problem occurs when access to employment or other types of financial gain is not uniform in a community, which is often the case. Uneven introduction of cash or other economic goods typically leads to problematic socioeconomic differentiation within communities. Although the introduction of differences may not seem to be a serious problem, the communities that are affected by ecotourism are often small-scale and in marginal areas of developing countries. Typically in those contexts, social norms are heavily weighted toward prevention of emergence of individual differences, that is, leveling mechanisms operate to discourage salient differences in wealth and status. When an uneven flow of resources, especially of access to cash, occurs in such communities, divisiveness is the consequence. The emergence of differences in wealth can lead to conflict and social disintegration. In Cuyabeno Reserve in the Ecuadorian Amazon, for example, the local ecotouristism concession belongs to one extended family of Siona Indians, which alone among the local population contracts with a national tour company to manage the reserve. According to my observations, the economic gain from the concession does not appear to be large for the family, but in comparison with other local Siona Indians, the differences are significant, leading to resentment and open hostility.

Even if differential access to external income does not lead to serious socioeconomic differentiation, external income can become a mechanism for substantial cultural change. Goods purchased and introduced often have high status value, and if they are practical and make aspects of life easier, as for instance outboard motors clearly do, the value is further enhanced. The social relations within communities can thus be altered and realigned, as status increases for some members, leading to motivation for further changes as more and more individuals seek the means to obtain the new goods. The goods themselves can serve as foci for the restructuring of social relationships, leading away from the more traditional and local to the more modern and external. Changes can be dramatic, affecting much of the way of life of local communities. An example is in the Galápagos Islands, where some fishing communities have given up fishing completely and converted their boats from fishing craft to tourist boats (Sarmiento, personal communication 1997). Serving as guides for tourists is more profitable.

Social transformation of indigenous societies can occur through less economically direct and more subtle means. The process of contact itself can lead to altered senses of social and cultural identity. To illustrate, the destination of ecotourists tends to be natural, "wilderness" areas, where plants, animals, and even landscape can be seen as "wild." Indigenous people, by their very presence, can also be seen as part of the nature and wilderness, from the perspective of ecotourists. In fact, for indigenous people to be present in a "wilderness," they have to be "wild," or else the concept of wilderness itself as a place without human occupation or impact is violated. The cultural conceptualization of a native environment as "wilderness" is, in fact, at the core of ecotourism, and there are profound consequences.

At the most basic level, "wilderness" becomes a commodity. Ecotourists pay for experiencing "wilderness," and through that process, an area takes on values that otherwise would not be present (King and Stewart 1996). Commodification leads to a redefinition of an area, converting it from its current state to "nature," "wilderness," "virgin forest," and the like. Areas so designated tend to be seen as objectified and, in an extreme version, are thought of as existing independent of people and culture (Hayles 1995). The local populations are then seen as intruders, as for example, tourists in Ngorongoro Crater in Tanzania see the resident Maasai pastoralists. In less extreme versions, local residents are objectified and, as noted, seen as "wild" themselves and possessors of "primitive" culture. The place of local culture in the ecotouristism enterprise then, ideally, is to be seen, appreciated, and maintained, a perspective at odds with the notion that participatory relationships should exist.

The commodification of the "local" has the potential, perhaps inevitability, of changing the way the indigenous people perceive the area, the environment, and themselves. The "lived-in environment" becomes a market commodity, altering the way in which the local people relate to, evaluate, and utilize the area. A shift in the relationship with the environment can bring about profound changes in both the culture and the environment.

Reserved areas are also often managed to make them appear natural, pristine, and thus more authentic (Allen and Hoekstra 1992). Parks, especially, tend to be contrived in that regard (Soulé 1995). MacCannell (1973), building on the work of Erving Goffman (1959), presented the argument that tourist settings often represent a staged authenticity in which "front-stage" and "back-stage" behavior is modified. Front-stage is the setting or meeting place for tourists and locals, whereas back-stage is where locals interact among themselves away from the tourists. MacCannell claimed that most tourist settings are "false backs," that is, front-stages in the guise of back-stages. In that regard, as King and Stewart note (1996), the dioramas of tourist productions are like museum displays, that is, socially defined constructs rather than "real" or "true" experiences of the natural and pure. The illusion of authenticity, however, is maintained in "natural" settings, and it may be essential to the whole process. However contrived the arrangement may be, the staging of the authentic reinforces the value of a "wilderness" area and makes genuine the necessity to protect it. The travel itself is validated. The culturally constructed goals of an ecotourist can be reinforced, including a contribution to the conservation of a "natural area." Noteworthy in this regard was the need to reduce the number of tours in the Galápagos as a response to the criticism of ecotourists. Individuals in ecotour groups sometimes became annoyed when other groups were encountered, complaining that the sense of being alone in a wilderness was violated.

Commodification of the local culture and environment also includes traditional material goods. Even the sale of mundane cultural artifacts can have impact if they become produced for sale to tourists. Differential access to the tourist market, again, can introduce socioeconomic difficulties. Overuse of native plants and other resources can also occur, to the detriment to the environment. For example, Maasai gourds used for storing milk and drinking milk have become a popular item for sale to tourists, placing new demands on resources and on local social relationships. Craft production, however, can have positive benefits, aside from the income generated.

Craft traditions that have diminished can be revived by the potential for sale to tourists, as Hitchcock and Brandenburgh (1990) report for the Baswara of Botswana.

A number of other sociocultural problems may be produced by tourist and eco-tourist incursions into indigenous areas. A common problem is the introduction of behavioral traits or cultural artifacts that are offensive to some segments of local communities but that are adopted by other segments of the community. Those can involve matters of personal clothing, adornment, food preferences and habits, inter-generational disrespect, and exacerbation of gender relations. In regard to the latter, Mueller (personal communication 1997) noted that in Costa Rica a major complaint of local communities was the mutual attraction of local men and tourist women, with a consequence of social dislocation in local gender relations. Similar problems have been reported elsewhere, for example, in the Indian Ocean resort of Malindi (Kenya). They are likely to be widespread, wherever tourism / ecotourism occurs.

PROBLEM-SOLVING: DIRECTIONS

Given the litany of problems, difficulties, and issues identified above, what can be proposed as steps toward solutions? What has been done, what is underway, and what could be done in regard to ecotourism that would diminish if not eliminate the endangerment of indigenous peoples, their cultures and languages, and the environments in which they live? At the broadest level, an answer would be that efforts have been underway on many fronts for several years to address the whole gamut of problems (Mieczkowski 1995; Ceballos-Lascuráin 1996; Miller 1996). The Ecotourism Society, in particular, has been in the forefront of efforts to find solutions, providing published guidelines and information especially for managers and planners (Hawkins, Wood, and Bittman 1995; Eagles and Nilsen 1997). To focus the present discussion, however, two arenas can be identified in which problems that are especially pressing and consequential can be addressed. Those are the practices of ecotour commercial enterprises and the inclusion of local people themselves in the decision-making and management process.

Environmental organizations and travel organizations have taken the lead in addressing problems, although many others, from governmental organizations to private individuals, have contributed. The "players" in ecotourism include not only environmental organizations and travel and tour organizations, but researchers, governmental agencies, NGOs, the indigenous populations, and, of course, the ecotourists. Management and policy issues need to be addressed to and among each of the groups, toward better integration. A variety of steps are underway in those regards. For example, management of protected areas is currently a central issue, and a number of different management plans have been proposed. In 1993, a set of guidelines for the management of such areas emerged from an international conference, Sustainable Tourism in Islands and Small States, held in Malta

(Jafari and Wall 1994). The guidelines included proposals to provide information and advice on policy from research results to political authorities, who in turn should devise and enforce standards based on those results. They also included proposals for the long-term monitoring of the effects of tourism/ecotourism on the economy, environment, and cultures of the host country.

Other guidelines have come not from tourism researchers but from the travel industry, which is highly appropriate, since the commercial exploitation of ecotourism by travel firms has been at the core of problems. The Pacific Asia Travel Association, for example, developed a code in 1992 for environmentally responsible tourism (PATA 1992). The code calls for priority in travel to be placed on sustainable use of renewable resources and conservation of nonrenewable resources. Adoption of a code of environmental ethics was also promoted. A number of other guidelines with comparably desirable goals have been adopted. The National Audubon Society has its "Traveler's Code of Ethics" (Ingram and Durst 1989), and the American Society of Travel Agents (ASTA) has its "Ten Commandments on Ecotourism" (WTO 1993). Hotels are also responding to some of the demands of ecologically minded tourists. The Intercontinental Hotels Group, for example, has a 220-page operating manual for "ecocorrect" hotels (Sitnik 1996).

Perhaps the most significant recent effort at guidelines is the Charter on Sustainable Tourism adopted jointly by the WTO, UNEP, and UNESCO Man and the Biosphere Program in 1995. The charter applies the recommendation of the 1992 Rio Declaration on Environment and Development and establishes a set of imperatives for major changes in the tourism industry as a whole, proposing that tourism development "must be ecologically sound in the long term, economically viable, as well as ethically and socially equitable for the local communities" (WTO 1995). Although these may be only statements of principle, they are important steps toward a more environmentally responsible ecotourism. A reconceptualization of ecotourism in those terms would aid in all facets of the people-environment equation.

A host of issues surrounds the question of the place of indigenous people in the entire ecotourism enterprise. For example, are there sites that are environmentally fragile, or damaged, to the extent that tourism should not be allowed there? Who should make decisions of that type? Serious questions of political economy emerge. Should local populations, for instance, have the right to tell their national government that their environmental resources and way of life is being degraded to the extent that they close down the sites, even if the sites are major revenue generators? Should national governments have the right to close environmentally sensitive areas against the wishes of the local population, who stand to lose revenue and even access to a cash economy as a consequence? In cases where sharing of power and decision making should take place in ongoing ecotours, how are indigenous populations to be integrated into the development of policies for the sites and in the execution of policies in the management and protection of them? If local societies do, in fact, participate in ecotouristic activities, what kind of steps could or should be taken to minimize negative impacts such as culture loss and to maximize their well-being?

One conclusion that emerges repeatedly from the extensive literature on economic planning and development, including tourism, is that "top-down," governmental approaches do not work. No matter how well-intentioned, how systematic, or how organized plans are, they fail to work if they are directed from a source exterior to a local community without involvement from the community. Although it would seem to be apparent that if a local community is not involved in plans for development, members are not likely to have any commitment to them, that lesson has had to be learned again and again. Planners for ecotourism could learn from the lesson; indigenous populations that are directly associated with ecotourism have to be involved in planning and development from the beginning and at each step of the way. In fact, that principle can be generalized; all of the interested and affected parties, the stakeholders, have to be involved in decision making and implementation throughout the process.

Multiple levels of control throughout the ecotourism enterprise are also necessary. At one level is it necessary to ensure that indigenous peoples' claims for equal participation in the control of resources be recognized and honored by governments, conservation organizations, and travel firms. In return, the local population should assume primary responsibility for technical management of protected areas, since they, theoretically, are best placed to accomplish that. All of the members of a local community, however, are unlikely to have equal access to and control of resources in protected areas and therefore to share in a reasonable way the income generated from ecotourism. Multiple levels of control and integration are commonly needed to facilitate local involvement in the management of resources, which is necessary both to enhance stability and to enable positive change. One type of positive change, for example, results from giving more power and voice to indigenous people as to the amount, location, timing, and nature of tourist visitation. This has the potential to empower them (King and Stewart 1996), as well as the potential for greater local control over the cultural degradation that can occur from commodification. Although cultural contact in guest and host relationships may result, inevitably, in culture modification or loss, the empowerment of local populations for greater control over their own destiny is both desirable and necessary.

SUMMARY

The tying of ecotourism to reduced endangerment of cultures, environments, and biodiversity was a natural development. Ecotourism generates money on a large scale, it occurs in environments that are relatively remote and pristine, those environments tend to be high biodiversity areas, and the indigenous people tend to be poor and marginalized. It seems clear that arrangements to apply some proportion of ecotourism funds to ecosystem preservation and to incorporate local populations into the process of management would be a natural step to take. A framework would be created to manage the pressures that endanger local environments and the indigenous people who live in them. At the same time, the survival of ecotourism would also be enhanced. The very existence of ecotourism is dependent on the preservation of the "wilderness" in which it is played out, and if the wilderness goes, the ecotourism goes. Given this critical mutuality of stakeholder interests, cooperation toward consensual goals would be beneficial on all fronts and make headway in the conservation of indigenous cultures, indigenous languages, and biodiversity.

REFERENCES

Allen, T., and T. Hoekstra. 1992. *Toward a Unified Ecology.* New York: Columbia University Press.

Boo, E. 1990. *Ecotourism: The Potentials and Pitfalls.* Vols. 1 and 2. Washington, D.C.: World Wildlife Fund.

Buckley, R. 1994. A framework for ecotourism. *Annals of Tourism Research* 21(3):661–669.

Butler, R.W. 1993. Tourism—an evolutionary perspective. In *Tourism and Sustainable Development: Monitoring, Planning, and Managing,* ed. J.G. Nelson, R.W. Butler, and G. Wall. Pp. 27–43. Waterloo, Ontario: Department of Geography Publication Series, no. 37.

Ceballos-Lascuráin, H. 1996. *Tourism, Ecotourism, and Protected Areas.* Cambridge: International Union for Conservation of Nature and Natural Resources.

Crandall, L. 1987. The social impact of tourism on developing regions and its management. In *Travel, Tourism, and Hospitality Research,* ed. J.R.B. Ritchie and C.R. Goeldner. Pp. 373–383. New York: John Wiley.

Dearden, P., and R. Rollins. 1993. *Parks and Protected Areas in Canada: Planning and Management.* New York: Oxford University Press.

Eagles, P.F.J., and P. Nilsen, eds. 1997. *An Annotated Bibliography for Planners and Managers.* 4th ed. North Bennington, Vt.: Ecotourism Society.

Goffman, E. 1959. *The Presentation of Self in Everyday Life.* Garden City, N.Y.: Doubleday.

Goodwin, H. 1996. In pursuit of ecotourism. *Biodiversity and Conservation* 5:277–291.

Harrison, D., ed. 1992. *Tourism and the Less Developed Countries.* London: Belhaven.

Hawkins, D.E., M.E. Wood, and S. Bittman. 1995. *The Ecolodge Sourcebook for Planners and Developers.* North Bennington, Vt.: Ecotourism Society.

Hayles, N. 1995. Searching for common ground. In *Reinventing Nature: Responses to Postmodern Deconstruction,* ed. M. Soulé and G. Lease. Pp. 47–64. Washington, D.C.: Island Press.

Hitchcock, R.K., and R.L. Branderburgh. 1990. Tourism, conservation, and culture in the Kalahari Desert, Botswana. *Cultural Survival Quarterly* 14(2):20–24.

Ingram, C.D., and P.D. Durst. 1989. Nature-oriented tour operators: Travel to developing countries. *Journal of Travel Research* 28(2):11–15.

Jafari, J., and G. Wall. 1994. Sustainable tourism. *Annals of Tourism Research* 21(3):667–669.

Jenner, P., and C. Smith. 1992. *The Tourism Industry and the Environment.* London: Economist Intelligence Unit.

King, D.A., and W.P. Stewart. 1996. Ecotourism and commodification: Protecting people and places. *Biodiversity and Conservation* 5:293–305.

Lindberg, K. 1991. *Policies for Maximizing Nature Tourism's Ecological and Economic Benefits.* Washington, D.C.: World Resources Institute.

MacCannell, D. 1973. Staged authenticity: Arrangements of social space in tourist settings. *American Journal of Sociology* 79:589–603.

McConnell, R.M. 1991. Solving environmental problems caused by adventure travel in developing countries: The Everest Environmental Expedition. *Mountain Research and Development* 11(4):359–366.

Mieczkowski, Z. 1995. *Environmental Issues of Tourism and Recreation.* Lanham, Md.: University Press of America.

Miller, J.A., ed. 1996. *The Ecotourism Equation: Measuring the Impacts.* New Haven: Yale School of Forestry and Environmental Studies Bulletin Series, no. 99.

Myers, N. 1972. National parks in savannah Africa: Ecological requirements of parks must be balanced against socio-economic constraints in their environment. *Science* 178:1255–1263.

Nash, D. 1989. Tourism as a form of imperialism. In *Hosts and Guests: The Anthropology of Tourism,* ed. V. Smith. 2d ed. Pp. 33–48. Philadelphia: University of Pennsylvania Press.

Nelson, J.G. 1994. The spread of ecotourism: Some planning implications. *Environmental Conservation* 21(3):248–255.

PATA. 1992. *Endemic Tourism: A Profitable Industry in a Sustainable Environment.* San Francisco: Pacific-Asia Travel Association.

Sitnik, M. 1996. Sustainable ecotourism: The Galápagos balance. In *The Ecotourism Equation: Measuring the Impacts,* ed. J.A. Miller. Pp. 89–94. New Haven: Yale School of Forestry and Environmental Studies Bulletin Series, no. 99.

Smith, V.L., and W.R. Eadington. 1992 (1977). *Tourism Alternatives: Potentials and Problems in the Development of Tourism.* 2d ed. Philadelphia: University of Pennsylvania Press.

Soulé, M. 1995. The social siege of nature. In *Reinventing Nature: Responses to Postmodern Deconstruction,* ed. M. Soulé and G. Lease. Pp. 137–170. Washington, D.C.: Island Press.

Whelan, T., ed. 1991. *Nature Tourism: Managing for the Environment.* Washington, D.C.: Island Press.

Wood, M.E. 1991. Formulating the Ecotourism Society's regional action plan. In *Ecotourism and Resource Conservation,* vol. 1, ed. J.A. Kusler. Pp. 8–89. Madison, Wis.: Omnipress.

WTO. 1993. *Tourism in 1992: Essential Data.* Madrid: World Tourism Organization.

WTO. 1995. *WTO's 1994 International Tourism Overview.* Madrid: World Tourism Organization.

WTTC. 1994. *Travel and Tourism.* Brussels: World Travel and Tourism Council.

Ziffer, K.A. 1989. *Ecotourism: The Uneasy Alliance.* Washington, D.C.: Conservation International.

32

PROTECTORS, PROSPECTORS, AND PIRATES
OF BIOLOGICAL RESOURCES

Stephen B. Brush

Just as the "information age" has commenced, two of the world's great stores of information, the diversity of biological organisms and of human languages, are imperiled. The loss of species and human languages is an emblem of a subtler and pervasive impoverishment of biological and cultural information that follows the widening embrace of the global economy. The precipitous declines in species and language diversity represent a crisis both to local cultures and to the larger public, which benefits from genetic resources and local knowledge. One proposal directed at conserving both biological diversity and indigenous or local knowledge has been to encourage "biological prospecting"—the search for commercially viable genetic sequences and natural compounds to be obtained under contracts (Reid et al. 1993). The rationale is that, by compensating indigenous peoples for sharing knowledge and biological resources, bioprospecting increases the individual or private value of these resources and promotes conservation. Moreover, it is argued that this policy addresses a perceived inequity between indigenous peoples and industrial states.

Bioprospecting contracts raise two fundamental questions: Who owns indigenous knowledge? And should such knowledge be commodified? The answer to the first question is straightforward—indigenous peoples own their knowledge in that it cannot be taken from them without their consent. The second question is more difficult to answer, because giving the right to one person or group to commodify knowledge may effectively deny the same right to others with similar knowledge. Intellectual property law and jurisprudence are established to sort out contested claims of ownership and exclusion, a process that emphasizes the role of the state (Gollin and Laird 1996). Bioprospecting thus relies on an increased integration into political, legal, and economic spheres beyond the locality and a larger role of the state in determining how indigenous knowledge and biological resources are controlled. At the local level, bioprospecting calls into question the nature of authorship and the circulation of knowledge, plants, and seeds.

The issues of commodification of and compensation for biological resources and indigenous knowledge can be profitably analyzed by examining the case of crop genetic resources—including local varieties, or landraces, folk nomenclature and traditional knowledge—from traditional agricultural systems in regions of crop origins (Vavilov regions). This analysis questions the possessive individualism of bio-prospecting and suggests that the focus should be shifted toward bio-cooperation among the many users and producers of genetic resources.

BIOPROSPECTING CONTRACTS

Biological resources and indigenous knowledge have historically been collected under the principle of common heritage as part of the public domain. Genetic resources and indigenous knowledge are noncompetitive—one person's use of elements within the public domain does not deprive others of use. Problems with some public goods are that the costs of maintaining them are hard to calculate and individuals who are directly involved in producing or maintaining public goods may not be fairly compensated. Theoretically, the deterioration of public goods, such as water pollution or soil erosion, is attributable to the imbalance between private and social values of public goods. The genetic resources of crop landraces (locally selected and maintained crop populations) in Vavilov centers of diversity satisfy the definition of public goods and illustrate the problems of estimating and equilibrating public and private values. Landraces have public value as resources to improve crop production. Thus, maize farmers in Iowa or Africa receive part of the public value of maize landraces cultivated by Zapotec farmers in Mexico. Yet Zapotec farmers bear an uncompensated private cost for keeping maize landraces.

Bioprospecting seeks to solve this imbalance by establishing a mechanism to transfer some of the public value of maize landraces to Zapotec farmers. Contracts can involve direct benefits (money, training, community development assistance) and long-term benefits (a share in royalties). These agreements satisfy the spirit of the Convention on Biological Diversity (Gollin and Laird 1996). Contracts also acknowledge control by indigenous peoples over genetic resources in that they require consent in how the resources are obtained and used. Such contracts are already established for pharmacological and biochemical prospecting (Reid et al. 1993; Mays et al. 1996), and they are actively being developed for crop genetic resources as "material collection agreements" or "material transfer agreements." To date, virtually all bioprospecting contracts have relied on private companies, but there is no reason why public institutions (e.g., U.S. Department of Agriculture's National Germplasm System) cannot enter into such agreements.

While the contract is usually written between the indigenous community and the private and/or public institution, it carries implications far beyond the immediate context. Gollin and Laird (1996) observe that the global-local link is only viable with direct involvement and support of national legislation, to provide for enforcement of the contract and intellectual property. These authors argue that companies investing in bioprospecting contracts expect an intellectual property framework to protect their investment. Following the 1994 conclusion of worldwide trade negotiations (Uruguay Round-GATT and TRIPS), intellectual property is protected in virtually every country. The reliance on intellectual property to back up bioprospecting agreements presents a major ethical dilemma, because it suggests that if knowledge and genetic resources are collected under contract and lead to a patentable product, other communities, which are not part of the contract but which have the same resources, can be deprived of the opportunity to commercialize their knowledge. Hypothetically, an exclusively Zapotec contract that results in a patented discovery in hybrid maize from Bolita landraces, can deprive Mixtec farmers with the same landraces of the same commercial possibility. While Zapotecs may not directly claim intellectual property in their contract, the exclusionary effect of the contract can be the same because the seed company will claim it.

GLOBAL RESTRUCTURING AND THE LOSS OF GENETIC RESOURCES

The integrative force of market relations has largely replaced direct military and political expansion by the state, amplified by large populations and communication technology. The plight of the world's languages and biological diversity suggests that contemporary globalization is having effects equally disastrous as the integration of distinct cultures into a world system following the European expansion after 1400. Ironically, the problems of modern integration cannot be addressed without reliance on outside forces. Local autarchy would deepen the crisis of endangered species and languages because it would do nothing to increase the private value of biological resources and cultural knowledge, nor link local value to wider public value, nor stem local forces weighed against biological and cultural diversity.

Crop species and germplasm, the biological basis of agriculture, were globalized long before modern times, as shown by the movement of crops in prehistoric times within Old and New Worlds. The post-Colombian exchange of crops (e.g., maize, potatoes, chile peppers, beans, tobacco, coffee, bananas, coconuts, yams, cacao) between the Old and the New World and beyond shows the local impact of global communication before the modern era. Exotic species enriched diets, displaced local species, reorganized local and regional economies, and provided new mediums for surplus extraction and exploitation. Modern globalization has accomplished a thorough and virtually simultaneous linkage of spatially dispersed communities into complex networks of markets, communication, and international flows of technology and capital. Replacement of local species has occurred more rapidly and more extensively. Genetic erosion is the agricultural equivalent of the clearing of tropical rain forests for pastures, reducing biological diversity within and among populations. The spread of modern cultivars and other agricultural inputs is a response to population growth and economic change, but it also reflects the global diffusion of technology linked to input pricing and credit policy, national seed systems and commodity boards, and international subsidies for basic science research.

Biological resources from local crop populations are an essential form of information capital to meet the demands of population growth and the needs of public and private agricultural science. Both the loss of crop genetic resources from farming systems and the need for crop germplasm have been met by creating gene banks, herbaria, and botanical gardens. Nevertheless, off-site conservation cannot save all biological resources or, more important, the knowledge of local farming cultures or the ecological processes that generate genetic diversity. A recurrent theme in crop conservation is the need to increase the value of local genetic resources to the farmers who maintain them, in order to promote on-farm (in situ) conservation.

Like other forms of capital, crop genetic resources are fluid and often more valuable on the global level than on the local level. Moreover, like other forms of information capital in the global economy, biological resources can be appropriated as intellectual property in some economies and legal systems. As seeds in farming communities, landraces and the diversity in their chromosomes are treated as common property or as public goods. But once landraces enter the global system as genetic resources, they can be disassembled and used as components in new plants that are privately owned as protected varieties and other forms of intellectual property. This contrast between common and intellectual property has spawned a long debate over the creation of "farmers' rights" that might be the equivalent of breeders' rights embodied in plant variety protection in the U.S. and Europe (Brush 1992).

CONTESTED AGENDAS

Since 1980, four parties have joined in an international debate concerning biological diversity and indigenous rights: (1) indigenous peoples' organizations, (2) private users of biological resources, (3) public users of biological resources, and

(4) states. The 1992 Convention on Biological Diversity (CBD) was a milestone in the debate about conservation and compensation of biological resources, but the CBD has not been fully implemented, and an international debate continues about its implementation in relation to other international agreements, such as the General Agreement on Tariffs and Trade (GATT). The continuing debate articulates four fundamental issues: control, access, compensation, and conservation.

Indigenous Peoples

Farmers who maintain rich stores of crop genetic resources are often members of minority cultures embedded into states dominated by other ethnic groups. Consequently, the interests of these farmers are easily merged analytically with those of tribal or indigenous peoples. Three claims are raised in various iterations of indigenous peoples' rights over knowledge and biological resources—control, compensation, and conservation. Posey and Dutfield (1996) document the fact that the ownership and control of biological resources are the central concerns of indigenous peoples. This concern is summarized in Article 29 of the Draft UN Declaration on the Rights of Indigenous Peoples: "Indigenous peoples are entitled to the recognition of the full ownership, control and protection of their cultural and intellectual property" (in Posey and Dutfield 1996:186).

Under public domain practices, indigenous peoples lose control of their knowledge and resources that have been shared with others. Once shared, indigenous knowledge may be used in offensive ways or may enrich nonindigenous people without benefit to its originators (Greaves 1996). Article 29 of the Draft UN Declaration (above) proposes to remove indigenous knowledge from the public domain. This right might be construed as the equivalent of the basic right of privacy, but privacy is usually not applied to socially held knowledge. Privacy can be enforced by indigenous groups themselves, by refusing to allow outsiders to obtain information or to collect biological specimens on group land, except under informed consent. Nevertheless, knowledge is volatile and protean, and few groups have the means to control access or use of knowledge without the support of the state, through such mechanisms as trade secrets. Dependence on the state to protect privacy is implied in the Draft UN Declaration, but such dependence means that the desire for privacy will be weighed against other social interests. In Australia, for instance, the Supreme Court of the Northern Territory enjoined publication of a book which revealed tribal, cultural, and religious secret ceremonies shown in confidence to C. P. Mountford (Foster and Others v. Mountford and Rigby, Ltd. 1976). In this case, the group's request was upheld because of its contention that serious social harm would result from publication, not because of the right to privacy per se.

Another central item on indigenous peoples' agendas for managing their knowledge and biological resources is the right to intellectual property, suggesting an interest in commercializing their knowledge. The ease of reproducing biological resources and ideas requires extraordinary means to make them the exclusive commercial property of a single person or group. A variety of intellectual property mechanisms are available for plant breeders: trade secrets, plant variety protection, and patents. These mechanisms may not be suitable, however, for the plant resources and knowledge of indigenous peoples, raising the possibility that novel forms of intellectual property should be created to protect the interests of indigenous peoples (Posey and Dutfield 1996).

Private Users of Biological Resources

Private users of biological resources include seed and pharmaceutical companies. Access to biological diversity is the single overriding objective of private users. Access to crop germplasm has been secured through the creation of national and international crop gene banks and the policy of free access granted to bona fide genetic researchers. Since the drafting of the CBD, however, open access to uncollected germplasm is uncertain, because of states' inexperience and lack of policy for exercising sovereign control over biological resources. Dependability of supply is arguably more important than cost. There appears to be little resistance by private users to the idea that the suppliers of biological resources be compensated for these resources (Mays et al. 1996). If compensation to a supplier leads to dependable access to the germplasm and information, with clearly articulated rules and conditions, it is a small cost indeed. Moreover, compensation is not likely to affect profits seriously. The supply of genetic resources remains large enough to keep costs relatively low, and raw germplasm from farms and forests represents a small fraction of the overall research effort. Commercial seed varieties and drugs are the end product of research chains that often require up to 10 years and millions of dollars to complete. The cost of obtaining biological resources could increase without appreciably affecting the total cost of producing new crops and natural products. Because the number of suppliers is large and genes are usually not highly endemic to specific communities, users have a wide choice of sources of biological resources.

Public Users of Biological Resources

Public users are not-for-profit entities and include national and international research programs and universities, applied researchers, and basic researchers. Public research remains an important source of agricultural innovation and improved crops, and the beneficiaries of public research are the billions of people whose food supply is less expensive and more secure because of this research. The distinction between private and public users can be blurred at times, for instance in patenting, but the dichotomy is true and important. Although the discussion of the equity of the current system of using genetic resources has focused on the contrast between indigenous peoples and private users, the largest groups of users of crop genetic resources are public.

Like private users, public users of biological resources are interested in access to these resources. Since no state is self-sufficient in genetic resources, national programs everywhere depend on open access to each other states' germplasm, a need satisfied by the creation of public gene banks. Besides access, public users of genetic resources have also made conservation an important agenda item. This is in keeping with the broad basic research agenda of many public users. While access and conservation were conflated in the creation of the international system of gene banks, conservation has become a more autonomous objective, strengthened by the CBD and the development of conservation biology.

States

The final party to the international discussion about genetic resources are states, representing the interests of private and public users of genetic resources, consumers who benefit from research, and, potentially, the interests of indigenous peoples. The role of the state in the management and flow of genetic resources is affirmed in several ways. The CBD recognizes states' sovereignty over the biological resources within their territory. The international system of financing and monitoring conservation is based on the state, and national research programs are among the leading users of genetic resources. Intellectual property is the purview of states.

States' agendas for biological resources commonly share two items—control and access. Because public users are often part of the state apparatus, conservation may also be included on a state's agenda. The sovereignty clause of the CBD testifies to the importance afforded to control of biological resources. The universal reliance on exotic germplasm means that all states share a common interest in access to resources, and dealing with other states has been the logical way to obtain access. Few states have attempted to block access to some genetic resources, and open access rather than competitive bargaining is the general rule for the movement of germplasm among states. Consequently, national legislation regulating genetic resources is likely to resemble norms created by industrial countries, which emphasize open access and a narrow interpretation of eligibility to intellectual property protection, exemplified by breeders' rights. Conversely, protection of indigenous knowledge is less likely to be in accord with national agendas.

Compensation to states for biological resources is problematic. Sovereignty provides a means to seek compensation at the state level, but several obstacles block this (Brush 1996). No international market functions for genetic resources per se, and the existence of large public collections, which freely distribute genetic resources, makes a market unlikely. Finally, seeking compensation may affect access to genetic resources from other places. To date, the objective of access appears to outweigh the interest in compensation.

KNOWLEDGE, RESOURCES, AND COMMUNITY: SOCIAL ISSUES

Discussants in the dialog on biological resources and indigenous rights share certain agenda items (e.g., control) but not others (e.g., conservation). Bioprospecting contracts establish a mechanism for producers and users of genetic resources to align agendas and balance their separate interests, but this balancing act may be jarred by other parties with different agendas. The terrain of balancing equities among indigenous peoples, private and public users of biological resources, and states is poorly charted and rough. Gollin and Laird (1996) observe that "the devil is in the details" in designing national legislation for bioprospecting. Lack of attention to details may obscure unintended, negative consequences of bioprospecting agreements. Successful balancing of equities, through bioprospecting agreements or other means, necessitates social research and analysis in two areas: ownership of indigenous knowledge and management of biological resources.

Appropriation of Indigenous Knowledge

Long-term reliance on bioprospecting to finance conservation will depend on the capacity of bioprospectors to protect their investment through intellectual property (Gollin and Laird 1996). Whether exclusive rights are claimed at the farm gate or in the industrial laboratory, the effect of bioprospecting will be to appropriate an element of culture as private, marketable property. Robust markets in archaeological artifacts, antiquities, and "ethnic" art and the repatriation of human remains and grave objects to native North Americans have made cultural property a familiar category (Merryman 1989; Coombe 1993; Greaves 1996). Cultural property is defined by international treaties as objects of artistic, archaeological, ethnological, or historic value (Merryman 1989)—Mayan stelae, Parthenon marble reliefs (the "Elgin Marbles"), burial objects from the tombs of pharaohs or Mochica nobility, sacred textiles of Bolivian communities. Environmental items—rare fauna, flora, or fossils—are also included in the UNESCO Convention for the Protection of Cultural Property (Merryman 1989). Nonetheless, the usefulness of cultural property concepts to the control of genetic resources is limited. It is ambiguous whether cultural property rights obtain for states or cultural groups (Merryman 1989; Coombe 1993). Existing cultural property is intended for unique objects whose possession is competitive, unlike knowledge that is noncompetitive. The provenience of cultural property that is now protected is generally well known and easily established, unlike indigenous knowledge that is often hybrid. Cultural property is not intended for commercial purposes (Coombe 1993), and the concept of cultural property may conflict with embedded notions of intellectual property, especially authorship.

Limited discussion of the appropriation of cultural property points to a complexity of issues that arise when cultural property is construed beyond its original boundedness in objects to include ideas and other nonmaterial goods (Coombe 1993; Woodmansee and Jaszi 1994; Strathern 1996). The essentialization of culture, the nature of authorship, and arrogation of the public domain emerge as elementary and embedded issues. While rudimentary, these issues have not received thorough attention by anthropologists, indigenous spokespersons or the proponents of bioprospecting. While some anthropologists (e.g., Greaves 1996) have championed the right of indigenous peoples to own and commercialize their knowledge, the field in general has moved away from the idea of culture as a bounded object that might be possessed, let alone commercialized. Characteristics that are usually associated with property (either material or intellectual)—boundedness, continuity, authorship, homogeneity—are conspicuously problematic in our current definition of culture (Handler 1991). Coombe (1993) notes the rich ironies of "possessive individualism" in the arena of appropriating cultural symbols for political and commercial purposes, where non-Western people are obliged to embrace Western ideological constructions of possessive individualism if they are to limit the use of their cultural knowledge. The logic allowing companies to "own" cultural symbols and names as trademarks (e.g., Cherokee™, Winnebago™) extends also to the right of cultures to claim objective identity (Coombe 1993). Both trademarks and ownership of cultural symbols are utilitarian devices that constrict the public domain in order to meet socially negotiated goals. A similar logic would seem also to extend to biological resources and cultural knowledge about plants. Landraces and ethnobotanical knowledge are, after all, cultural artifacts, and ownership meets a social goal—stewardship of environmental goods. The fact that natural compounds—genetic sequences and whole organisms—have been classified as intellectual property in industrial countries lends credence to the idea that local, ethnobotanical knowledge, medicinal plants, and crop species might also fit this category (Greaves 1996).

The proposal for ownership for cultural objects raises the issue of partibility. Can ownership be given without also granting the right to commercialize culture? Other values than the market undoubtedly drive certain groups to claim ownership of cultural objects (Handler 1991; Coombe 1993), but market values are central to claims of control over biological resources. Partibility can, of course, be limited, as in the case of national monuments, but which classes of cultural property should be treated as nonpartible? Partibility implies a form of essentialism that may be unproblematic for objects whose authorship is unambiguous, whose use is competitive, and that demonstrate the qualities of continuity, stability, boundedness. Nonmaterial cultural goods, however, do not share these essential qualities. Ethnobotanical knowledge, design motifs in material arts, musical styles, and other nonmaterial elements pass through cultural boundaries with ease, making it all but impossible to attribute authorship. Indeed, a fair assumption in regarding any element of ethnobiological knowledge is that it is not exclusive to one culture. The permeability of cultural boundaries suggests that authorship regarding cultural knowledge has an unusual, if not impossible, burden of proof.

Extending the characteristics of alienable property to cultural goods, such as ethnobotanical knowledge, however, rests on an essentialist's concept of cultural authorship—one that has been largely rejected in anthropology. The essentialist's concept of author springs from the doctrine of "possessive individualism" (Handler 1991) and embraces the "hero inventor" quality of authorship, a quality fixed in law but challenged by writers, artists, and ethnographers (e.g., Woodmansee and Jaszi 1994; Mays et al. 1996). Coombe (1993) notes that copyright laws are intended for individuals and their specific creations, not for "ideas, cultural themes, practices and historical experiences," and she rejects the idea that intellectual property can be extended to "collective rights, collective authors, and claims of intergenerational creation." In other words, indigenous ownership of cultural symbols might, in fact, be subject to the same limits as experienced by authors or inventors seeking copyright or patents under Western laws.

Limits to collective rights evolve from concepts of authorship of cultural knowledge and from the interest to protect the public domain. While collective individualism is commonplace, the juristic persons of modern capitalism (e.g., corporations, states) are very different entities compared with cultural collectivities (tribes, communities, ethnic groups). Authorship by juristic persons is intentional, dedicated, and specifically constituted. Authorship by tribes, communities, and ethnic groups may have these qualities, but it is also collective, incremental, ambiguous, tacit, and socially distributed (Strathern 1996). Juristic persons make it their business to designate authorship; tribes, villages, and ethnic groups have no such business. Strathern observes that the contrast between authorship in possessive individualism and in collective, cultural entities is analogous to distinct modes of knowledge in industrial countries. The qualities of monopoly, exclusion, and partibility flow easily from the form and function of juristic persons but not from the other cultural forms.

Strathern points out that social communities, and not cultures, claim cultural identity and rights in corporate images and that societies and not cultures determine boundaries. Since cultures are fluid, permeable, and protean, the boundaries that societies may draw around cultural identities are artifices. A social claim of exclusive cultural identity may bemuse persons in other societies who share the culture or within the same society who construct separate identities. Cultural identity is not dependent on exercising monopoly over the objects of cultural property. Thus, Diaspora communities are not discouraged from retaining a sense of ownership in the cultural identity of their original societies. The tolerance of blurred ownership is, however, untenable when the aspects of partibility and commerce are thrown into the equation.

Management of Biological Resources

The rapidity with which potatoes spread to Europe and maize to Africa and Asia after the Colombian contact with the New World bears witness to the ubiquitous circulation of useful plants. Indeed, mixing rather than isolation appears to be the rule for crop varieties, from field and farm village levels to the global one. Human ingenuity in altering agricultural landscapes and selecting varieties, the need to renew seed stocks, and constant reconfiguration of the factors of production and consumption make cosmopolitanism rather than endemism the rule for crops and varieties. The worldwide system of plant exploration and introduction of exotics is based on the fact that useful traits and whole organisms can be easily moved and fit into new environments. This same quest operates fully at small biological and spatial scales, in the exchange of crops between farmers and the migration of traits through crop populations of villages and microregions (Brush et al. 1995).

The circulation of seed is a result of several factors. Both seeds and crop phenotypes have finite viability. Seed stocks degenerate as they become infected with pests and pathogens; the competitive advantages of crop varieties dull against the host of diseases that plague them; the physical and human conditions that favor a particular variety inevitably change to its disadvantage. Tastes go in and out of fashion. Seeds from different strains are mixed on common threshing floors and by wages in kind, gifts, sale, and bartering between households. The circulation of seeds through villages, market networks, and regions is both inevitable because of the practices of everyday production and necessary for the viability of agriculture. Moreover, periodic and small exchanges of seed between farms and villages rapidly compounds into significant genetic changes in crop populations. The biological result at the regional level is that most genetic diversity is found within farms rather than between them (Brush et al. 1995), a pattern that confounds drawing cultural boundaries around crop populations. The result of the needs and practices of farmers to shuffle their seed stock, combined with the biological forces of outcrossing and the economic forces of crop breeding and technological diffusion, is that crop populations look very much like cultures—composites of traits and elements from many sources, some local but many exotic. Just as cultures cannot be described as bounded, essentially derived entities, neither can crops. Like cultures, crops survive by transformation in natural and cultural contexts with permeable boundaries, without definite continuity or authorship, and as heterogeneous rather than homogeneous entities.

CONCLUSION

Bioprospecting is predicated on a cultural construction of possessive individualism, in its view of both how genetic resources are created and how they can best be preserved. This construction is buttressed by the successes of late capitalism in framing our imagination of how individuals and societies relate to one another and to the natural world. This ideological construction is so pervasive that the public domain has become all but invisible, and the noncontractual collection and use of biological resources is stigmatized as "biopiracy," a form of theft (Odek 1994). Bioprospecting rests on the appropriation of culture and biological resources whose origin is ambiguous. The urgency of preserving biological resources may prompt us to overlook this trespass onto the commons, but neither the nature of culture nor that of biological resources have been fully weighed in designing this strategy to save them. The intended impact of contracts is to encourage stewardship and accessibility, but unintended impacts will surely follow, especially since contracts disregard the nature of cultures and the biological resources they are intended to save. For instance, it is possible that the normal give and take of cultural ideas and genetic resources between farmers and villages will be obstructed or otherwise distorted by property relations and the expectation of profit that are introduced by bioprospecting.

The two information systems discussed in this paper, culture and crop genetic resources, share the attribute of transparency. To remain vital, both depend on the constant acquisition of new information—on the exchange of information across ephemeral boundaries. The notion of public dominion for cultural knowledge and genetic resources is logical and widely held, deduced from the facts that all cultures and farming systems must absorb and give off information and that authorship springs from the stream of ideas that flow through cultures and villages.

Defending the public domain seems quaint in the glare of late capitalism, but in the villages and farms where crop genetic resources thrive, "biocooperation" is the common ethic. Biocooperation beyond the village level also exists, reflected in the work of national and international agricultural research programs aimed at developing knowledge and technology. Some of this knowledge and technology is privatized as intellectual property, but the greater amount exists as public goods that are widely shared and immensely beneficial. Examples are improved crop varieties that have alleviated the crisis of food production in many areas. Biocooperation can also guide our efforts to conserve genetic resources and local knowledge in the agroecosystems that continue to generate genetic resources. Increasing the individual benefit or private value of traditional crops can be achieved in many ways other than selling rights to use their germplasm. Cooperative research between farmers and scientists can help to increase the productivity of traditional crops, thus increasing their value without privatization. Developing local and regional markets where demand for traditional crops exists but is unmet, is another form of biocooperation that can benefit both farmers and conservation. The recovery of "lost" seeds, from gene banks or other regions, can be accomplished through cooperation between agricultural scientists and farmers. Finally, information technology can greatly improve the exchange of crop varieties and information among networks of farmers who are often isolated from one another.

The dilemma is whether biocooperation at the village level and between farmers and agricultural scientists can survive in a context where ownership and contracting dominate the exchange of seeds and farmer knowledge beyond the village. There is overwhelming evidence that crop genetic resources come from culturally and biologically open systems. Bioprospecting proposes introducing a commercial ethic, which contradicts the historic pattern of biocooperation; it is but one alternative to address the issue of increasing the private value of genetic resources to their current and future stewards. Compensation for genetic resources does not have to be cast into the mold of possessive individualism. Enhancing other public goods—education, health, and agricultural technology—which affect traditional farmers can also give recognition to their contributions to world agriculture. The logical framework of bioprospecting implies that the non-contractual collection and use of genetic resources constitutes "biopiracy." This formulation is mistaken because it misconstrues the nature of creating genetic resources. Other social groups besides indigenous producers and private users have interests and positions in the conservation of biological resources. The challenge of declining species and languages can be met with alternative frameworks than the possessive individualism implied in bioprospecting. Biocooperation between the stewards of genetic resources in peasant villages and users of those resources in industrial countries rests on identifying a broad spectrum of interests, values, and methods of recognition and compensation. Monetary compensation is but one of many alternatives in this spectrum.

REFERENCES

Brush, S.B. 1992. Farmers' rights and genetic conservation in traditional farming systems. *World Development* 20:1617–1630.

Brush, S.B. 1996. Valuing crop genetic resources. *Journal of Environment and Development* 5:416–433.

Brush S.B., et al. 1995. Potato diversity in the Andean center of crop domestication. *Conservation Biology* 9:1189–1198.

Coombe, R.J. 1993. The properties of culture and the politics of possessing identity: Native claims in the cultural appropriation controversy. *Canadian Journal of Law and Jurisprudence* 6:249–285.

Foster and Others v. Mountford and Rigby, Ltd. 1976. *Australian Law Reports* 14 ALR:71–76.

Gollin, M.A., and S.A. Laird. 1996. Global politics, local actions: The role of national legislation in sustainable biodiversity prospecting. *Boston University Journal of Science and Technology Law* (electronic document available from: Lexus: 2 B.U.J. Sci. and Tech. L. 16).

Greaves, T.A. 1996. Tribal rights. In *Valuing Local Knowledge: Indigenous People and Intellectual Property Rights,* ed. S.B. Brush and D. Stabinsky. Pp. 25–40. Washington, D.C.: Island Press.

Handler, R. 1991. Who owns the past? History, cultural property, and the logic of possessive individualism. In *The Politics of Culture,* ed. B. Williams. Pp. 63–74. Washington, D.C.: Smithsonian Institution Press.

Mays, T.D., et al. 1996. Quid pro quo: Alternatives for equity and conservation. In *Valuing Local Knowledge: Indigenous People and Intellectual Property Rights,* ed. S.B. Brush and D. Stabinsky. Pp. 259–280. Washington, D.C.: Island Press.

Merryman, J.H. 1989. The public interest in cultural property. *California Law Review* 77: 339–364.

Odek, J.O. 1994. Bio-piracy: Creating proprietary rights in plant genetic resources. *Journal of Intellectual Property Law* 2:141–181.

Posey, D.A., and G. Dutfield. 1996. *Beyond Intellectual Property: Toward Traditional Resource Rights for Indigenous Peoples and Local Communities.* Ottawa: International Development Research Centre.

Reid W.V., et al. 1993. *Biodiversity Prospecting: Using Resources for Sustainable Development.* Washington, D.C.: World Resources Institute.

Strathern, M. 1996. Potential property: Intellectual property and property in persons. *Social Anthropology* 4:17–32.

Woodmansee, M., and P. Jaszi, eds. 1994. *The Construction of Authorship: Textual Appropriation in Law and Literature.* Durham, N.C.: Duke University Press.

Part 4

A VISION FOR THE FUTURE, AND A PLEA

33

POSSIBILITIES AFTER PROGRESS

Richard B. Norgaard

It is early December and I am in Uppsala for a few days rather than somewhere else. The sun is low on the horizon at midday as I head for the oldest cathedral in Scandinavia. Walking past buildings stuccoed in traditional deep reds, greens and yellows with their south walls basking in the clean sunlight, the castle on the hill against a crisp blue sky, the bridge is coated in ice condensed from the mist of the stream. Upon entering the cathedral, I find once again that Catholicism has its local particulars. The vastness is light, the pews birch blond, the walls and marble caskets as clean-lined as Scandinavian furniture today. Out into the cold again, I stroll past the university of Celsius and Linnaeus and then head into the city center to mingle with early Christmas shoppers. Main street, reclaimed for foot traffic, is lined with tasteful shops selling domestically produced clothes for the rich. The department stores sell shirts made in Portugal and shoes made in China, just like everywhere else in Europe. I step determinedly past a MacDonald's overflowing with happy shoppers, reassuring myself that the local bakeries, cafes, and delis are doing a brisk business today too. Emerging from a gallery of intermingled smaller shops, I hear the familiar music from the Andes and seek out the four men from Peru.

They are well positioned near an intersection, strumming, drumming, wailing on flutes and singing their hearts out. Like me, they are in Uppsala rather than Barcelona, Berkeley, Chicago, Frankfurt, New York, São Paulo, Sydney, Tokyo, Toronto, or wherever else I have heard them and their compañeros in the last few years. Sharing a thousand years of culture, crassly making a buck, joining a multicultural global discourse, freezing their butts off, sending money home to Peru to sustain the indigenous resistance to the same, all of the above. And now I am more aware that the bundled people strolling past me include a surprising number of colorful faces from around the globe, political and economic refugees from failed efforts to prompt progress, evidence of Sweden's cultural confidence and willingness to provide a band-aid to a badly bungled world.

The future was not supposed to turn out like this. It was supposed to simply get better, smoothly for all, as we merged into some higher sociocultural life form. We were supposed to merge into one people, not reidentify as Basques, Catalans, Padanians, and Silesians, not frenetically unfold in all of these locally fragmented, globally interconnected, mixed histories. Western people's belief in progress and their ability to infect others with the same belief has defined the last two centuries. Western science, technology, social organization, and even religion were expected to provide all with a future increasingly free of material want and to foster international harmony. Belief in progress rationalized the drive and softened the acceptance of the imperialism of Western science, technology, governance, economics, and religion. And these forces drove the loss of biological, cultural, and linguistic diversity.

The idea of progress is now being actively contested by new and global environmental, multicultural, and epistemological discourses. Evidence is accumulating that we are moving into another era. Just how this age after the idea of progress will be defined is still open, but it will surely carry new possibilities as well as new risks for diversity.

Among the historical roots of the idea of progress was a Christian belief that with greater material well-being there would be both less temptation to sin and more opportunity to study the Bible and prepare oneself for heaven. While personal and collective moral advance was initially central to the idea of progress, material well-being became an end in itself around the beginning of the nineteenth century. Western beliefs about science were critical to this transition. Science promised control over nature and material well-being through new technologies. With ever increasing technical control through scientific progress, people did not have to worry about living with nature or saving it for future generations. Quite to the contrary, transforming nature was evidence of technological prowess and scientific progress.

With the demonstrable advance of knowledge, technology, and social organization, there was less need for moral progress. Moral concern and personal sacrifice for the poor and for the future became less important. The march of progress would surely take care of all. With Western culture being clearly superior, respect, or even simple tolerance, for the cultures of others became obstacles in their path to a brighter future. Nation-states arose in Europe from the feuding, feudal remnants anchored in Catholicism and Protestantism around new beliefs in the possibilities for rational governance. The formation of nation-states continued through the twentieth century on the promises of rational order, evidenced, for example, in Brazil's motto, "Ordem e Progresso."

With Western science clearly being superior, other cultures' ways of knowing were rationally dismissed. The documentation by anthropologists of traditional knowledge systems was supported not because such systems were thought to be of value, but as raw entertainment and evidence of Western advance. We were convinced that cultural differences would diminish to the French cooking with more butter and the Thai with more spices, but the important stuff—calories, protein, and vitamins—would be made available for all through the application of modern scientific knowledge. This unification of cultures, furthermore, would bring peace as cultural conflicts faded and a shared cosmology brought opportunities for all to work together toward even further material improvement.

The commitment to materialism has been much of the problem. Belief in the importance of material progress was so great over the twentieth century that it spanned capitalist and socialist rhetoric and agendas while it silently pervaded all within these supposedly opposed ways of socially organizing. Production has been primary and gross domestic product our aggregate symbol of well-being. But there have been no benchmarks. Happiness has been presumed to stem from an ever increasing GDP per capita and living in a nation whose GDP per capita ranks above those of others. Coached by incessant advertising, we have defined ourselves by our ability to purchase the latest commercial products. Our identity has been linked to our rung on the ladder of material ascension. Even our own academic community of scholars elevates natural scientists' knowledge of the material world and denigrates those in the humanities as dabbling in feelings. By portraying history as materially driven, Marx intended to give history real bite. That which is unimportant is "immaterial." Materialism, by being so central to our sense of being modern, left biological, cultural, and linguistic subtleties on the margins to collapse. Other cultural constructions of life with less emphasis on raw materialism withered. The drive to consume hastened the transformation of nature, destruction of biodiversity, and loss of cultures and languages. People who resisted material destinies have been slaughtered, infected with diseases, pushed off their land, and forced into the few nooks and crannies of minimally transformed world that remain. Production and consumption is but a part of life, but the utilitarian language of materialism is poorly suited for the larger part. The public dominance of utilitarian language has been a unifying and *disenriching* form of imperialism.

Obviously, materialism has had a strong economic component to it. We constructed economies to produce more and more and have only recently emerged from a global confrontation between particular forms of capitalism and socialism competing to see which could increase material production and consumption the fastest. And each tried to draw materially poorer nations into their spheres. Thus we set out to develop nations as separate political projects. But at least it was a process of building whole nations with complete economies and governance systems. Lately, we have moved into a very different phase of history. Now we are talking about development as a global market in which nations are simply the places in which particular industries happen to locate for the time being. Nations no longer develop; they simply host and accommodate corporate production facilities and consume name-brand distributed goods. It is the globe that will develop; the economists are sure that somehow we will all be better in the end, but the early statistics look grim. In this new phase, there is little sense of moral obligation to help the poor even in the midst of now dampened expectation that progress will eventually see to their needs. The reification of markets and their dominance in every discourse suggests the modern era is closing in a paucity of ideas. This is not a triumph for social organization through markets. Thinking only in terms of market relationships is the collapse of economic, let alone social, analysis, discourse, and development.

In retrospect it is easy to wonder at our historical belief in progress. Material gains have been offset by new dependencies and increasingly complex social and environmental problems. The homogenization of cultures proved neither so easy nor so desirable as expected. And peace only became plausible because the prospect of full-scale war could no longer be imagined. Yet biologically, culturally, and linguistically destructive regional wars abounded and continue still. The new economic globalization threatens what is left. Most people, however, seem to accept the last throes of modernity as the way life will forever be. A few strive to set us on a better path. Others realize we are individually and collectively on a treadmill. Some advocate throwing ourselves into the gears of the progress machine. But I do not think we can fight it. I awake each day with the hope, and even some anticipation, that there is not much more that can be globalized, that another regime is emerging, that there will be new possibilities after progress.

In spite of globalization, it is clear that we are also in a period of reculturalization, not 7000 languages and cultures coming back to life, but a period in which at least some are coming back and the remaining ones are proving somewhat more resistant. Cultural pluralism, religious diversity, and languages from the "East" have become the norm in the big cities of the most Western of nations. Several nations have broken apart along cultural fault lines, others appear on the brink, and smaller states are emerging in their dust. Simultaneously, we are seeing an explosion of nongovernmental organizations (NGOs) arising to fill new needs, to solve the problems that modern forms of organization have created. We see a global "NGOization" with numerous new alliances being formed between environmentalists, laborers, and indigenous peoples. This spontaneous social organization is proving surprisingly powerful in the face of global capital. The fact that the World Bank can no longer work effectively without participating with NGOs and bringing in other speakers to their gatherings is a positive development. We still, of course, have the World Bank. Academics, the bearers of the banners of modern progress and guardians of Western ideals, are rather hesitant to talk about the transition. At the same time, those of us who choose to live in this multicultural, globally interlinked, NGOized realm work up to half of the time along lines different from the proper academic model.

Transition periods are risky. This cannot be overstressed. National breakups have resulted in the reemergence of cultural warfare with disastrous consequences for many. After centuries of fighting over which economic banner to unify under, we are ill-equipped to handle diversity as a state of being and goal for the future. Science has also played a central role, an authoritative voice, in resolving disputes. But the voice of science is no longer as monolithic nor a voice of the state. In part this is because individual scientists and whole research institutions have accepted the task, increasingly consciously, of learning and speaking for special interests. With science being less and less accepted as a resolving, progressive, public voice, teaching and research institutions have been losing stature, direction, and broad support, making them also ever more susceptible to and dependent on special interests. Meanwhile, there is a mass popular environmental understanding that science has not led to the control of nature and that nature's response to new technologies will be our demise. As we look to science for guidance, we largely see scientocratic paralysis, the enlightenment hope all divided up into little disciplines that can't speak to each other and help us put the great picture together. And when scientists do establish a transdisciplinary discourse—as exemplified by the exceptional interaction of astrophysicists, ecologists, oceanographers, economists, and many others under the auspices of the Intergovernmental Panel on Climate Change—the scientists who are not a part of the shared discourse get equal publicity for their shriveled understanding.

There are other risks as well. Both World Bank leaders and environmentalists talk about sustainable development as if all countries were as socially and politically stable as Costa Rica. Few modernists see the significant cultural, economic, and environmental consequences of social conflict within and between nations. We have many nations breaking apart and more that are undoubtedly going to dissolve. We are going to experience more war. These disruptions will have tremendous biological, cultural, and linguistic consequences. It is especially risky because we can no longer pretend to share a common language of modern rationality or common hopes to bring us together.

I think it is also risky because there is some chance that this impoverished economic ideology present today, this very narrow way of thinking about social organization, will succeed long enough to have serious impact. I do not think it will succeed over a very long period, but nomadic capital comes with tremendous risks in our overcrowded, environmentally stressed world. I am very concerned that the vast majority of people are still thinking as if we were in the world we tried to establish after World War II and that our responses to current social and environmental problems will only exacerbate our situation. The lag in our interpretation of where we seem to be may be the greatest risk to diversity at this time.

Scientists learning together is an important step in recognizing that we are in a phase of change. Gatherings such as the Endangered Languages, Endangered Knowledge, Endangered Environments conference, bringing anthropologists, biologists, and linguists together to share common concerns, are a part of the transition, and these are going on all over. Again, the elaboration of the process and consequences of climate change that is taking place among a group of scientists from around the world and from almost all of the natural and social sciences is an incredible model of how scientists can learn together, indeed *must* learn together. It demonstrates that we are capable of coming to a consensus in the face of uncertainties. We have also seen over the past two decades thousands of biologists switch from supposedly "being objective" and letting change fall where it may to "having an objective," the conservation of biological diversity. They have made an overt decision to come together and "be subjective" to work toward an objective. The conservation biologists are not only doing the science of diversity but have produced new narratives about our relation to the natural world that represent some of the finest scientific writing of the twentieth century. The rise of ethnobiology is exciting and is nicely complemented by the development of agroecology. New participatory research techniques entail learning with local peoples and sharing in mutually designed and implemented research and demonstration projects. Economists have joined with ecologists to develop an ecological economics. Diverse groups are also uniting to share values and understandings. Conflicts are being resolved through new mediation techniques and through the use of science juries as well as value juries. There is a lot of change going on that is moving in the right direction.

Drastic change has its risks, but there are also new opportunities. We need to start thinking about what the possibilities after progress will be. We need to be able to identify where positive changes are occurring as well as negative developments so that we can help guide the times. We must find spaces where we can work effectively, perhaps simply to hide from destructive change, but eventually to show a new direction, a new vision, so that as the collapse of progress occurs, we do not simply end up with total disaster. We can help construct interesting spaces in the transition to a new world. It is unclear what this will be like, but I think in 25 to 50 years people will have a rallying cry other than "Progress." I hope there will not be too much collapse in the process, but I am sure there are going to be lots of problems. In an era of considerable change, however, there also will be numerous niches opening up where good things can happen.[1]

NOTE

1. *Editor's note:* Many of the ideas alluded to in this short epilogue are developed further in R.B. Norgaard, *Development Betrayed: The End of Progress and a Coevolutionary Revisioning of the Future* (London and New York: Routledge, 1994).

34

SILENT NO MORE

California Indians Reclaim Their Culture—
And They Invite You to Listen

L. Frank Manriquez

I come to you, but some people may think I am a little late for this party.[1] I come to you with an extinct language and no homeland. Periodically, there are people who say things that make me feel not so desperate. But for the most part, it doesn't sound too promising. I think that before we can fix anything bigger out there we should really understand the indigenous person–scientist relationship.

More than understand it, actually, we should accept things we don't understand from the indigenous persons themselves. I hear a lot of things. I work with a lot of natives in California but I have traveled with natives all across the United States. I know a little bit about how they are feeling about scientists. I also have a lot of friends who are scientists, but I want to show you a native perspective of why we fear you. [Draws on the blackboard.]

These lines represent people, lineages. Biologically, we get some DNA from this old Indian lady. And it is going to prove whether the people over here are really the people that they say they are. Government study, we find out this, this, and that. Equals dead Indians. What we get is dead Indians. All of your work somehow ends up making us dead. You come visit our villages. Who asked you? Yes, sometimes when you are there we find out we need you a great deal. I myself have a great need for linguists, biologists, ethnobotanists. I can use all the -ologists I can find. We become very good friends because we have a common goal. But most of the time to natives this is the outcome: science is proving that we are dead. You only have five of these people, and the cut-off limit for extinction is four. The five of us left standing think, wow, we don't feel extinct!

I know it is a matter of economics to governments—we can save this species, but we can't save this other species, and this species is much more rare, far more abundant—but what do the natives feel about this species that we are not looking at? I think that a lot of time we go into native communities and come out thinking we know what is good for them, but we never listen to the native community. Don't get me wrong, as I said I have a great need for scientists because I am trying to bring back my culture. When you are "extinct" you are deep down in the water, you know. And you are reaching up, and sometimes what reaches down is this white hand with a lot of knowledge that holds the memories of your grandparents—mostly, your grandfathers', because it was always men out in the field, and you don't care what women say, even though women are responsible for most of the language and most of the culture that you see today, because women hung on. So I implore you that when you go out in the field you look at what the women have to say. They may not know about the trees, but they know about the insects, and it takes 500 insects to feed us. . . .

The reason I'm saying this flat out is that I want you all to be useful. There is no more time to waste, I can tell you this from an "extinct" point of view, you know. Listen to us. Hear what we have to say when you are there with us. A lot of times all my -ologist friends don't really want a contemporary native with them. A lot of times they want someone who can build a thatched hut. Well, I can build a thatched hut but I don't live in one at this point. We are not allowed to be living now. Our cultures, our words, our concepts—people always want to find what was, but you have to integrate and find out what is.

You are all looking for answers, so I'm going to give you some of ours. When you talk about making your work accessible, it doesn't mean in a library, or in a store for $10.95, because sometimes I can't come up with $10.95. Accessible means that if you take a photo of this one person you give them a copy. If you make a tape of these people you give them a tape immediately. Your first priority is to give immediately back. This way we will trust you, because for years and years people have come to us and just taken. I was walking through the California State Indian museum with a friend, and there was this beautiful Karuk canoe, and he said: "That was in our backyard. These anthropologists pulled up with a pickup truck while we were having dinner and put it in, and now here it is." This goes on every single day, it is not something that only happened in the past with anthropologists, archeologists, linguists.

On the good side, we have somebody like Leanne Hinton.[2] We got hold of her. She is *our* anthropologist. Most of the time, anthropologists, linguists have their Indians—well, she is our anthropologist. If New Mexico or Arizona tried to get her we would have a fit. She has taken the step to throw herself in that river and roll with those boulders—get small with us down to the ocean. It is not always easy. I have come home with her from meetings, we rode in the car, and she would comment: "One more time I am the dreaded anthropologist linguist scientist who took something for nothing." So you guys have a long history to get over with us. It is just the way it is. Nobody gave it to you like nobody gave me extinction—well, actually somebody did give it to you like somebody gave me extinction. It is ours to deal with. So really, listen to what the natives need.

We are doing documentation. I am on the Board of the Advocates for Indigenous California Language Survival—maybe you have read about it in the magazine *News from Native California,* or maybe you have read *Flutes of Fire.* A group of people got together and said, nothing is happening with our languages. Well, now something *is* happening: a program with a lot of work, a lot of people involved. We did have a good sum of money. I hear you talk about funding. Well, the natives are going to do this with or without your help, with or without your input. They are doing it, but it's coming out in little bits and pieces. With your help it could come out in full force, if you help us do what we want.

There is a sentence here that I thought was just wonderful [reading]: "Often the first step is to point out that each language is a store of intellectual capital."[3] Well, it is not so much each language, I would like to say that to scientists who are always measuring things. If I know *one word* in my language my creator will let me go to where I have to go when I pass away. I don't have a whole language. It is silly to think I will bring an extinct language back to fluency with only 300 people in an extinct tribe. I talk to my computer. I feed the language in, and when I make mistakes the computer talks back.

But if I have *one word,* it is the power of one word, and whoever is at the garden gate—the pearly gates, the happy hunting grounds—will recognize me and it will be enough for me to go in. There is so much power in just one word. Somebody asked: "Why save the language or save your dance, why bother? Why don't you just give it up, become one of us?" Well, you can't give up the color of your eyes. You can't give up what has been running through your blood for ages.

I am an urban Native. I come from the city of Santa Monica, California. I am only 44, so I don't remember the olden days. But I learn one word and I have native knowledge, as you call it. We don't need all your books and your dictionaries. We need to go faster. With one word . . . even if it is "rabbit" or "house," I can see how that word is constructed and it is so different from this world. It teaches me so much. I always try to explain to my scientist friends that native knowledge can come from within, from dreams. I am always presented as nonrational, but the nonrational is very rational to us. When you look at native knowledge, a lot of people don't accept that we dreamed it, that it's why we are dancing this way. Some of my friends brought back a dance, and they decided to dance this way. An anthropologist said: "I read in an eighty-year-old book that you danced this other way." Then they go off and write papers. They are not letting us make our own lives again. You know, it took us three hundred years to get to this point. It is going to take us a while to get back. We are going to make some anthropological mistakes for a while, but we will get home eventually. We need you all to help us go there. Because the world is too small. I am not imploring for a little native, please help us save the planet. That is not what it is all about. It is for your kids too.

We need your help. I don't really have a bad relation with scientists. [Shows an illustration in *News from Native California*.] It is called "Coyote tests the prevailing winds of theory."[4] I sit in anthropology courses just to see what is going on. (I mean all the -ologies when I say "anthro.") But we natives in California are not waiting for anybody to help us anymore. We went out and we found Leanne. I dragged her out. Others dragged her out. So if some native comes knocking on your door for some real help think about giving it to them. It can really turn into personally rewarding work, I think.

I am also with the California Indian Basketweavers Association. We had a first gathering four or five years ago, and it was incredible, because traditionally you stay in your own village, you don't go out very far. You stay in your own home. So this was the first time we had ever gotten together as basketweavers. It turns out there is one in this section, one in this section, one in this section of California, and we are all isolated. We are all picking our native plants. We don't quite know what we are doing, but we think it is the right way. Some people know exactly what they are doing, but the majority of us really wonder, why are we doing this? Why are we risking our health to weave these baskets? So there was a gathering and 150 women got in the room together and cried. They cried for ages.

The next year there was no more crying. There were teachers and there were students who were weaving. But we wondered if we should not give up weaving. That was a big question we had. I feel and a lot of us feel that weaving is the physical manifestation of us. That is our native lives woven in that basket. It is the whole world, our cosmology, the plants, it is everything. So weavers' lips are falling off, weavers' teeth are falling off, weavers' jaws are falling off. We take the babies with us so they will learn and we sit them with us in the patches where we gather. And they are so full of pesticides that we are killing ourselves and killing our children. We ask ourselves, should we continue weaving? Is a culture worth it? Our answer is yes. So there go the pitiful Indians with their pieces of body parts falling off, and they are dying of cancer left and right, so what do we do about it? We find land that is relatively clean and we knock on the door and say: "Hey, you have land here. Can we put materials on there? Can we plant materials to make our baskets and keep our culture alive?" And so it has turned into a humongous project. People are coming out of the woodwork. People do want to help. So we do deal with a lot of pesticide issues. A lot of people are still weaving in areas that are heavily polluted, but we are trying to plant gardens all over.

This is a Native Californian Indian gathering policy. For the past few years we have been knocking on doors of Washington, Sacramento, each others' homes, everyone's homes, saying that we want to go and gather in the forest. But they don't want us to gather in the forest because they say we will wreck it—as if it weren't stupidly managed anyway. Now they have figured out they don't know how to manage a forest. We say, you can't manage a forest, a forest is a forest. They always try to manage forests and rivers and people's lives. So they won't let us go in the forests. Ninety-year-old women in this very state have been thrown in jail for collecting medicinal herbs, for picking plants. So we say that we have to develop policies and we meet with people. They shake hands with us, they take photos with us, they say they will affect policy. Then we go home, and the administration changes, or they get a better job, and we have another group of people we have to shake hands with. It is just like the old time photos, you know. Everybody gets a chance to be in a picture with an Indian.

But natives are not lying down anymore, being passive. We have learned how to use this system. Slowly, we will be all right, and with us being all right, we will all be all right. I would like to say something about singing and dancing and healing the earth—another reason why the language is so important. This ground here[5] is Ohlone land. They are extinct too, or so the story goes. Actually, they were just up here last weekend. This land does not understand the words I am using in my language. So if I say, "Heal, land, heal," it really does not know what to do. It has a different language; it comes from creation. So every land has its own language. You know, natives always talk about a roundness, a holistic thing. Well, that is part of it. You take away a language, take away the very last word, and this land is going to shudder and shake you off. We are getting close to that in a lot of places. We need the language because it is part of our mandate from our creators to take care of the Earth. That is something that people gloss over and get all Pocahontasy-eyed about.[6]

As you can see, we are not all painted-up Indians anymore, but we are still here.

Original art by L. Frank Manriquez

NOTES

1. This chapter is a slightly edited transcript of the talk given by L. Frank Manriquez at the Endangered Languages, Endangered Knowledge, Endangered Environments conference. Transcription from tape was done by Steve Bartz, whose help is gratefully acknowledged. Editing was done collaboratively by L. Frank Manriquez and Luisa Maffi. Explanatory footnotes were added by Maffi.

2. Leanne Hinton is professor of linguistics at the University of California, Berkeley, and director of the Survey of California and Other Indian Languages, a research unit that focuses on documentation of Native American languages. She consults with Native Americans in California and the Southwest on language maintenance and restoration programs, including the very successful Master-Apprentice program. She is the author of the book *Flutes of Fire: Essays on California Indian Languages* (Berkeley, Calif.: Heyday Books, 1994), mentioned later on in Manriquez's talk.

3. Manriquez was reading from Andrew Pawley's conference paper. See Pawley (this volume), where virtually the same sentence appears.

4. The illustration was one piece of Manriquez's own artwork, which is regularly featured in the *News from Native California* column "Acorn Soup."

5. In the San Francisco Bay Area.

6. Manriquez is referring to mainstream revisionist, romanticized versions of Native American history, such as can be found in school textbook and Disney cartoon accounts of the story of the seventeenth-century Native American woman Pocahontas.

CODE OF ETHICS OF THE INTERNATIONAL SOCIETY OF ETHNOBIOLOGY

As approved at the 6th International Congress of Ethnobiology

Whakatane, Aotearoa/New Zealand, 23–28 November 1998

PREAMBLE

This Code of Ethics has its origins in the Declaration of Belém agreed upon in 1988 at the Founding of the International Society of Ethnobiology (in Belém, Brazil).

It is acknowledged that much research has been undertaken in the past without the sanction or prior consent of indigenous and traditional peoples and that such research has resulted in wrongful expropriation of cultural and intellectual heritage rights of the affected peoples causing harm and violation of rights.

The ISE is committed to working in genuine partnership and collaboration with indigenous peoples, traditional societies and local communities to avoid these past injustices and build towards developing positive, beneficial and harmonious relationships in the field of ethnobiology.

The ISE recognises that culture and language are intrinsically connected to land and territory, and cultural and linguistic diversity are inextricably linked to biological diversity. Therefore, the right of Indigenous Peoples to the preservation and continued development of their cultures and languages and to the control of their lands, territories and traditional resources are key to the perpetuation of all forms of diversity on Earth.

PURPOSE

The Purpose of this Code of Ethics is:

to optimise the outcomes and reduce as much as possible the adverse effects of research (in all its forms, including applied research and development work) and related activities of ethnobiologists that can disrupt or disenfranchise indigenous peoples, traditional societies and local communities from their customary and chosen lifestyles; and

to provide a set of principles to govern the conduct of Ethnobiologists and all Members of the International Society of Ethnobiology (ISE) engaged in or proposing to be engaged in research in all its forms, especially collation and use of traditional knowledge or collections of flora, fauna, or any other element found on community lands or territories.

The ISE recognises, supports and prioritises the efforts of indigenous peoples, traditional societies and local communities to undertake and own their research, collections, databases and publications. This Code is intended to enfranchise indigenous peoples, traditional societies and local communities conducting research within their own society, for their own use.

It is hoped that this Code of Ethics will also serve to guide ethnobiologists and other researchers, business leaders, policy makers, and others seeking meaningful partnerships with indigenous peoples, traditional societies and local communities and thus to avoid the perpetuation of past injustices to these peoples. The ISE recognises that, for such partnerships to succeed, all relevant research activities must be collaborative. Consideration must be given to the needs of all humanity, and to the maintenance of robust and vigorous scientific standards.

It is desirable that scientists, international citizens and organisations, and indigenous peoples and local communities collaborate to achieve the purpose of this Code of Ethics and the objectives of the ISE.

PRINCIPLES

The Principles of this Code are to embrace, support, and embody the many established principles and practices of international law and customary practice as expressed in various international instruments and declarations including, but not limited to, those documents referred to in Appendix 1 of this Code of Ethics. The following Principles are the fundamental assumptions that form this Code of Ethics.

Principle of Prior Rights

This principle recognises that indigenous peoples, traditional societies, and local communities have prior, proprietary rights and interests over all air, land, and waterways, and the natural resources within them that these peoples have traditionally inhabited or used, together with all knowledge and intellectual property and traditional resource rights associated with such resources and their use.

Principle of Self-Determination

This principle recognises that indigenous peoples, traditional societies and local communities have a right to self-determination (or local determination for traditional and local communities) and that researchers and associated organisations will acknowledge and respect such rights in their dealings with these peoples and their communities.

Principle of Inalienability

This principle recognises the inalienable rights of indigenous peoples, traditional societies and local communities in relation to their traditional territories and the natural resources within them and associated traditional knowledge. These rights are collective by nature but can include individual rights. It shall be for indigenous peoples, traditional societies and local communities to determine for themselves the nature and scope of their respective resource rights regimes.

Principle of Traditional Guardianship

This principle recognises the holistic interconnectedness of humanity with the ecosystems of our Sacred Earth and the obligation and responsibility of indigenous peoples, traditional societies and local communities to preserve and maintain their role as traditional guardians of these ecosystems through the maintenance of their cultures, mythologies, spiritual beliefs and customary practices.

Principle of Active Participation

This principle recognises the crucial importance of indigenous peoples, traditional societies and local communities to actively participate in all phases of the project from inception to completion, as well as in application of research results.

Principle of Full Disclosure

This principle recognises that indigenous peoples, traditional societies and local communities are entitled to be fully informed about the nature, scope and ultimate purpose of the proposed research (including methodology, data collection, and the dissemination and application of results). This information is to be given in a manner that takes into consideration and actively engages with the body of knowledge and cultural preferences of these peoples and communities.

Principle of Prior Informed Consent and Veto

This principle recognises that the prior informed consent of all peoples and their communities must be obtained before any research is undertaken. Indigenous peoples, traditional societies and local communities have the right to veto any programme, project, or study that affects them. Providing prior informed consent presumes that all potentially affected communities will be provided complete information regarding the purpose and nature of the research activities and the probable results, including all reasonably foreseeable benefits and risks of harm (be they tangible or intangible) to the affected communities.

Principle of Confidentiality

This principle recognises that indigenous peoples, traditional societies and local communities, at their sole discretion, have the right to exclude from publication and/or to have kept confidential any information concerning their culture, traditions, mythologies or spiritual beliefs. Furthermore, such confidentiality shall be guaranteed by researchers and other potential users. Indigenous and traditional peoples also have the right to privacy and anonymity.

Principle of Respect

This principle recognises the necessity for researchers to respect the integrity, morality and spirituality of the culture, traditions and relationships of indigenous peoples, traditional societies, and local communities with their worlds, and to avoid the imposition of external conceptions and standards.

Principle of Active Protection

This principle recognises the importance of researchers taking active measures to protect and to enhance the relationships of indigenous peoples, traditional societies and local communities with their environment and thereby promote the maintenance of cultural and biological diversity.

Principle of Precaution

This principle acknowledges the complexity of interactions between cultural and biological communities, and thus the inherent uncertainty of effects due to ethnobiological and other research. The Precautionary Principle advocates taking proactive, anticipatory action to identify and to prevent biological or cultural harms resulting from research activities or outcomes, even if cause-and-effect relationships have not yet been scientifically proven. The prediction and assessment of such biological and cultural harms must include local criteria and indicators, thus must fully involve indigenous peoples, traditional societies, and local communities.

Principle of Compensation and Equitable Sharing

This principle recognises that indigenous peoples, traditional societies, and local communities must be fairly and adequately compensated for their contribution to ethnobiological research activities and outcomes involving their knowledge.

Principle of Supporting Indigenous Research

This principle recognises, supports and prioritises the efforts of indigenous peoples, traditional societies, and local communities in undertaking their own research and publications and in utilising their own collections and data bases.

Principle of the Dynamic Interactive Cycle

This principle holds that research activities should not be initiated unless there is reasonable assurance that all stages of the project can be completed from (a) preparation and evaluation, to (b) full implementation, to (c) evaluation, dissemination and return of results to the communities, to (d) training and education as an integral part of the project, including practical application of results. Thus, all projects must be seen as cycles of continuous and on-going dialogue.

Principle of Restitution

This principle recognises that every effort will be made to avoid any adverse consequences to indigenous peoples, traditional societies, and local communities from research activities and outcomes and that, should any such adverse consequence occur, appropriate restitution shall be made.

ADOPTION OF PRINCIPLES BY ISE

The above Preamble, Purpose and Principles ("the Principles") of the ISE Code of Ethics were adopted by resolution of the Annual General Meeting of the ISE held at Whakatane, Aotearoa/New Zealand on Saturday, 28 November 1998.

The resolution was in these terms:

Resolved that the ISE adopts the Preamble, Purpose and Principles of the Code of Ethics as amended (at the ISE AGM at Whakatane, War Memorial Hall on 28 November 1998) with the understanding that the Ethics Committee receive and review any further proposed changes or amendments to the Principles of the Code which will be collated and presented to the AGM of the next ICE Meeting to be held in the Year 2000 in Athens, Georgia, United States of America.

Noted by the AGM that the Principles form the first part of the ISE Code of Ethics and that the second part comprising the more detailed Standards of Practice are to be developed by the Ethics Committee for discussion and presentation at the next ICE.

STATEMENT OF MISSION

Terralingua: Partnerships for Linguistic and Biological Diversity

A. TERRALINGUA RECOGNIZES:

1. That the diversity of languages and their variant forms is a vital part of the world's cultural diversity;

2. That cultural diversity and biological diversity are not only related, but often inseparable; and

3. That, like biological species, many languages and their variant forms around the world are now faced with an extinction crisis whose magnitude may well prove very large.

B. TERRALINGUA DECLARES:

4. That every language, along with its variant forms, is inherently valuable and therefore worthy of being preserved and perpetuated, regardless of its political, demographic, or linguistic status;

5. That deciding which language to use, and for what purposes, is a basic human right inhering to members of the community of speakers now using the language or whose ancestors traditionally used it; and

6. That such usage decisions should be freely made in an atmosphere of tolerance and reciprocal respect for cultural distinctiveness—a condition that is a prerequisite for increased mutual understanding among the world's peoples and a recognition of our common humanity.

C. THEREFORE, TERRALINGUA SETS FORTH THE FOLLOWING GOALS:

7. To help preserve and perpetuate the world's linguistic diversity in all its variant forms (languages, dialects, pidgins, creoles, sign languages, languages used in rituals, etc.) through research, programs of public education, advocacy, and community support.

8. To learn about languages and the knowledge they embody from the communities of speakers themselves, to encourage partnerships between community-based language/cultural groups and scientific/professional organizations who are interested in preserving cultural and biological diversity, and to support the right of communities of speakers to language self-determination.

9. To illuminate the connections between cultural and biological diversity by establishing working relationships with scientific/professional organizations and individuals who are interested in preserving cultural diversity (such as linguists, educators, anthropologists, ethnologists, cultural workers, native advocates, cultural geographers, sociologists, and so on) and those who are interested in preserving biological diversity (such as biologists, botanists, ecologists, zoologists, physical geographers, ethnobiologists, ethnoecologists, conservationists, environmental advocates, natural resource managers, and so on), thus promoting the joint preservation and perpetuation of cultural and biological diversity.

10. To work with all appropriate entities in both the public and private sectors, and at all levels from the local to the international, to accomplish the foregoing.

Terralingua: Partnerships for Linguistic and Biological Diversity
P. O. Box 122
Hancock, MI 49930-0122, USA
http://www.terralingua.org

Dr. Scott Atran is Researcher at the National Center for Scientific Research, CREA-École Polytechnique, Paris, France, and Research Scientist at the Institute for Social Research, as well as Adjunct Professor of Psychology and Anthropology at the University of Michigan, Ann Arbor. Address: Institute for Social Research, University of Michigan, Ann Arbor, MI 48106-1248, USA. E-mail: satran@umich.edu.

Dr. William L. Balée is Professor of Anthropology at Tulane University. Address: Dept. of Anthropology, Tulane University, New Orleans, LA 70118, USA. E-mail: wbalee@mailhost.tcs.tulane.edu.

Dr. Herman Batibo is Professor of African Linguistics and Head of the Department of African Languages and Literature at the University of Botswana, Gaborone, Southern Africa. Address: Dept. of African Languages and Literature, University of Botswana, Private Bag 0022, Gaborone, Botswana. E-mail: batibohm@mopipi.ub.bw.

Dr. Ben G. Blount is Professor of Anthropology at the University of Georgia, Athens. Address: Dept. of Anthropology, University of Georgia, Athens, GA 30602-1619, USA. E-mail: bblount@arches.uga.edu.

Dr. Stephen B. Brush is Professor in the Department of Human and Community Development at the University of California, Davis. Address: Dept. of Human and Community Development, University of California, Davis, CA 95616, USA. E-mail: sbbrush@ucdavis.edu.

Dr. Thomas J. Carlson, M.D., M.S., is Associate Adjunct Professor, Department of Integrative Biology, University of California, Berkeley, and director of the Health, Ecology, Biodiversity, and Ethnobiology (HEBE) center at the Berkeley Natural History Museums, University of California. Address: 1001 Valley Life Sciences Building # 2465, University of California, Berkeley, CA, 94720-2465, USA. E-mail: tcarlson@socrates.berkeley.edu.

Dr. Greville G. Corbett is Professor of Linguistics and of Russian Language at the University of Surrey. Address: Surrey Morphology Group, SLLIS, University of Surrey, Guilford, Surrey GU2 5XH, UK. E-mail: g.corbett@surrey.ac.uk.

Dr. Margaret Florey is Lecturer in Linguistics in the Department of Linguistics at Monash University, Victoria, Australia. Address: Department of Linguistics, School of Philosophy, Linguistics, and Bioethics, Monash University, P.O. Box 92, Victoria 3800, Australia. E-mail: margaret.florey@arts.monash.edu.au.

David Harmon is Executive Director of the George Wright Society, a nonprofit association of park and protected area professionals. Address: The George Wright Society, P.O. Box 65, Hancock, MI 49930-0065, USA. E-mail: dharmon@georgewright.org.

Dr. Jane Hill is Regents' Professor of Anthropology and Linguistics at the University of Arizona, Tucson. Address: Dept. of Anthropology, University of Arizona, Tucson, AZ 85721, USA. E-mail: jhill@u.arizona.edu.

Dr. Eugene S. Hunn is Professor of Anthropology at the University of Washington, Seattle. Address: Dept. of Anthropology, Box 353100, University of Washington, Seattle, WA 98195-3100, USA. E-mail: hunn@u.washington.edu.

Dr. Manuel Lizarralde is Assistant Professor of Anthropology and Botany at Connecticut College. Address: Dept. of Anthropology, Connecticut College, 270 Mohegan Ave., Mail Box 5407, New London, CT 06320, USA. E-mail: mliz@conncoll.edu.

Dr. Luisa Maffi is President of the international NGO Terralingua: Partnerships for Linguistic and Biological Diversity and a Research Collaborator at the Smithsonian Institution's National Museum of Natural History, Dept. of Anthropology, in Washington, D.C. She is currently Secretary of the International Society of Ethnobiology. Address: Terralingua, 1766 Lanier Place NW, Washington, DC 20009, USA. E-mail: maffi@terralingua.org.

Dr. Ian Saem Majnep is a Kalam from Papua New Guinea who lives in the Kaironk Valley, Madang Province, PNG. He holds an honorary doctorate in science from the University of Papua New Guinea. Address: c/o Kaironk Community School, Simbai, Madang Province, Papua New Guinea.

L. Frank Manriquez is a Tongva/Ajachmem Native Californian artist and tribal activist. She is Board Member of the California Indian Basketweavers Association, the Advocates for Indigenous California Language Survival, and the Native California Network. Address: Native California Network, 1670 Bloomfield Rd., Sebastopol, CA 95472, USA.

Dr. Douglas Medin is Professor of Psychology at Northwestern University. Address: Northwestern University, Dept. of Psychology, 102 Swift Hall, 2029 Sheridan Road, Evanston, IL 60208, USA. E-mail: medin@northwestern.edu.

Dr. Katharine Milton is Professor of Physical Anthropology in the Department of Environmental Science, Policy and Management at the University of California, Berkeley. Address: Dept. of Environmental Science, Policy and Management, Division of Insect Biology, 201 Wellman Hall, University of California, Berkeley, CA 94720-3112, USA. E-mail: kmilton@socrates.berkeley.edu.

Dr. Brent D. Mishler is Director of the University and Jepson Herbaria and Professor, Department of Integrative Biology, at the University of California, Berkeley. Address: Dept. of Integrative Biology, 1001 Valley Life Sciences Bldg. # 2465, University of California, Berkeley, CA 94720-2465, USA. E-mail: bmishler@socrates.berkeley.edu.

Felipe Molina is a Yoeme (Yaqui) from Arizona. He is Diabetes Project Coordinator for Native American Communities at Native Seeds/SEARCH, a nonprofit devoted to the preservation of native plant resources. Address: Native Seeds/SEARCH, 526 N 4th Ave., Tucson, AZ 85705, USA. E-mail: nss@azstarnet.com.

Dr. Denny Moore is a Researcher with the Brazilian National Council for Scientific and Technological Development and Coordinator of the Linguistics Division in the Department of Human Sciences at the Museu Emílio Goeldi in Belém, Brazil. Address: Museu Emílio Goeldi-DCH, Avenida Magalhaes Barata 376, C.P. 399, 66.040 Belém, Pará, Brazil. E-mail: moore@amazon.com.br.

Dr. Peter Mühlhäusler is Foundation Professor of Linguistics at the University of Adelaide, Australia. Address: Dept.of Linguistics, University of Adelaide, Adelaide, South Australia 5005, Australia.

Dr. Gary P. Nabhan is Director, Center for Sustainable Environments, Northern Arizona University. Address: Center for Sustainable Environments, P. O. Box 5765, Flagstaff, AZ 86011-5765, USA. E-mail: gary.nabhan@nau.edu.

Dr. James Nations is Conservation International's Vice President for Mexico and Central America. Address: Conservation International, 1919 M St. NW, Washington, DC 20036, USA. E-mail: j.nations@conservation.org.

Dr. Richard B. Norgaard is Professor of Energy Resources and of Agricultural and Resource Economics at the University of California, Berkeley. Address: Energy and Resources Group, 310 Barrows Hall, University of California, Berkeley, CA 94720-3050, USA. E-mail: norgaard@socrates.berkeley.edu, norgaard@igc.org.

Dr. Christine Padoch is Matthew Calbraith Perry Curator at the Institute of Economic Botany of the New York Botanical Garden. Address: Institute of Economic Botany, New York Botanical Garden, 200 St. and Southern Boulevard, The Bronx, NY 10458-1526, USA. E-mail: cpadoch@nybg.org.

Dr. Andrew Pawley is Professor of Linguistics in the Research School of Pacific and Asian Studies, Australian National University, Canberra. Address: Dept. of Linguistics, RSPAS, Australian National University, Canberra, ACT 0200, Australia. E-mail: apawley@coombs.anu.edu.au.

Dr. Miguel Pinedo-Vasquez is Associate Research Scientist in the Center for Environmental Research and Conservation at Columbia University. Address: CERC, Columbia University, 1200 Amsterdam Ave., MC 5557, New York, NY 10027, USA. E-mail: map57@columbia.edu.

Dr. Darrell A. Posey is Director of the Programme for Traditional Resource Rights of the Oxford Centre for the Environment, Ethics, and Society, University of Oxford, as well as Visiting Researcher with the Brazilian National Council for Science and Research Professor at the Federal University of Maranhão, São Luis, Maranhão. Address: Oxford Centre for the Environment, Ethics, and Society, Mansfield College, University of Oxford, Oxford OX1 3TF, UK. E-mail: darrell.posey@mansf.ox.ac.uk.

Dr. Tove Skutnabb-Kangas is a Researcher at Roskilde University, Denmark. Address: Roskilde University, Dept. of Languages and Culture, 3.2.4., P.O.Box 260, 4000 Roskilde, Denmark. E-mail: skutnabb-kangas@vip.cybercity.dk, tovesk@babel.ruc.dk.

Dr. Eric A. Smith is Professor of Anthropology at the University of Washington, Seattle. Address: Dept. of Anthropology, Box 353100, University of Washington, Seattle, WA 98195-3100, USA. E-mail: easmith@u.washington.edu.

Dr. Victor M. Toledo is a Researcher at the Center of Ecology of the National Autonomous University of Mexico, Morelia Campus. Address: Instituto de Ecología, UNAM, Apdo. Post. 41-h, Sta. María Guido, Morelia, Michoacán 58090, Mexico. E-mail: vtoledo@oikos.unam.mx.

Dr. Michael Warren was, until his death in December 1997, Professor of Anthropology and Director, Center for Indigenous Knowledge for Agriculture and Rural Development at Iowa State University, Ames, IA, USA.

Dr. Phillip Wolff is Assistant Professor of Psychology at the University of Memphis. Address: Dept. of Psychology, University of Memphis, 3699 Southern Ave., Bldg. 300, Memphis, TN 38152, USA. E-mail: pwolff@memphis.edu.

Dr. Jeffrey Wollock is Research Director of the Solidarity Foundation, an information service affiliated with the Native American Council of New York City. Address: Solidarity Foundation, 221 West 57th Street, New York, NY 10019-6201, USA. E-mail: jeff@nativecouncil.com.

Dr. Stanford Zent is a Researcher in the Department of Anthropology, Venezuelan Institute for Scientific Research. Address: Depto. de Antropología, Instituto Venezolano de Investigaciones Científicas, Apdo. Post. 21827, Caracas 1020-A, Venezuela. E-mail: szent@ivic.ve.

INDEX

Page numbers in **boldface** indicate figures or tables

environmental degradation, 26; from ecotourism, 507–508

environmental management: imperialism and, 252–253; indigenous peoples' participation in, 278; language loss and, 251–252; prehistoric, 24

ethics and social justice, 38

ethnobiological knowledge, 18, 20; cultural support for, 213, 214–215, 219, 222; of indigenous peoples, 146, 147–148; linguistic encoding of, 148, 154; local vs. Western scientific, 152–153; loss, formal education and, 273, 278; loss, subsistence practices' change and, 273, 274–275, **275;** refugees' adaptation to new environments and, 495–496; in urban environments, 213. *See also* traditional ecological knowledge

ethnobiologists, 8, 105

ethnobotanical knowledge, 23, 190–191; bilingual ability and, 275–277, **276;** formal education and, 276–277, **277,** 278; formal schooling and, 209; lifestyle change and, 207–209; pharmacological salvage missions, 147; residency length in territorial group and, 274, **274;** subsistence practices and, 274, **275;** of U.S. college students, 212–213. *See also* traditional ecological knowledge

ethnobotanical plot surveys, 197–198

ethnocentrism: defined, 54

ethnoecological knowledge, 19, 26; Alune, 329–331; impact upon biodiversity, 17–18

ethnoecology: defined, 149

ethnographic dictionaries, 237

ethnolinguistic groups: Austronesian, 326; Native North America, 102, **103;** South American Indian, 23

ethnophysical nomenclature, 255–256

European Charter for Regional or Minority Languages, 404

European Union, 323

evolution, 74–75; biodiversity and, 72; biological, 14, 15, 39, 79; cultural, 38, 78, 79, 96, 122; descent with modification, 129; devolution of folk knowledge and,

212–227; of knowledge of English tree terms, 220–222; lineage splitting, 72; linguistic, 78; natural selection, 75, 76, 78; phylogenetic, 72–75; random drift, 75

evolutionary theory, 37

extinction: of biological and linguistic lineages, 15; of endemic species, 119; of ethnolinguistic groups, 16; of experience, 7, 18, 23; rate in biodiversity, 71

farming. *See* agriculture

field methods, 20, 159–162; among Kalam, 352–354; continental-scale case study, 95. *See also* tape documentation

floodplain management: Amazon Basin, 365–366, 372

folk knowledge. *See* indigenous knowledge; traditional ecological knowledge

folklore performers: protections for, 425–426

Food and Agricultural Organization of the United Nations (FAO), 447

forest management, 28; by Amazonian smallholders, 368–370; California Indians' views on, 543–544; Mesoamerican milpa system, 478–479; in Mexico, 483

forests: loss in Amazonia, 266. *See also* rainforests; tropical forests

Foundation on Inter-Ethnic Relations, 407

Four Directions Council of Canada, 381, 385

freezing of complex formations, 182–183, 186

French language: singular-plural number, 86, 89

frog, tree (*Phylomedusa bicolor*): secretion, hunting ritual uses, 292–293; secretion, physiological effects, 292

functional domain attrition, 179

Galapagos Islands: ecotourism impacts, 510, 511

Gavião language, 434

Gciriku language, 313

gender systems, 88

General Agreement on Tariffs and Trade (GATT), 519, 521